Climate Change and Food Security in South Asia

Rattan Lal · Mannava V.K. Sivakumar
S.M.A. Faiz · A.H.M. Mustafizur Rahman
Khandakar R. Islam
Editors

Climate Change and Food Security in South Asia

Springer

Editors
Rattan Lal
Coffey Rd 2021
43210-1043 COLUMBUS Ohio
USA
lal.s1@osu.edu

Mannava V.K. Sivakumar
World Meteorological Organization
Climate Prediction and Adaptation Branch
7bis, Avenue de la Paix
1211 Geneva 2
Switzerland
msivakumar@wmo.int

S.M.A. Faiz
University of Dhaka
Dept. of Soil, Water and Environment
1000 Dhaka
Bangladesh
duregstr@bangla.net

A.H.M. Mustafizur Rahman
University of Dhaka
Dept. of Soil, Water and Environment
1000 Dhaka
Bangladesh
dmrahman@agni.com

Khandakar R. Islam
OARDC Piketon Station
Shyville Road 1864
45661 Piketon
islam27@osu.edu

ISBN 978-90-481-9515-2 e-ISBN 978-90-481-9516-9
DOI 10.1007/978-90-481-9516-9
Springer Dordrecht Heidelberg New York London

Library of Congress Control Number: 2010935664

Printed on acid-free paper

Springer is part of Springer Science+Business Media (www.springer.com)

Foreword

Across South Asia large populations depend on semi-subsistence agriculture for their livelihoods. Rainfall in the semi-arid and sub-humid regions of South Asia is highly variable and undependable and influences agricultural productivity. Farming practices in these regions have developed as a response to such climatic risks.

According to the Fourth Assessment Report of the WMO/UNEP Intergovernmental Panel on Climate Change (IPCC) released in 2007, future projections of climate change indicate that South Asia is very likely to warm significantly during this century. Also, freshwater availability is projected to decrease particularly in large river basins due to climate change and coastal areas will be at greatest risk due to sea level rise and increased flooding from the sea and rivers. In some South Asian countries, a substantial reduction in crop yields from rainfed agriculture could occur. Additionally, dramatic changes in the land use patterns in South Asia compound the problem of climate change.

To cope with climate variability and change more effectively in South Asia, it is necessary to identify integrated adaptation and mitigation options for a range of agro-ecosystems so as to enable a favourable policy environment for the implementation of the framework. It is important to identify/understand impacts, vulnerability and adaptation; select and implement adaptation actions; enhance cooperation among South Asian countries to better manage climate change risks; and enhance integration of climate change adaptation with sustainable agricultural development in South Asia.

It is with this background that WMO, FAO and the Ohio State University organized the International Symposium on Climate Change and Food Security in South Asia from 25 to 29 August 2008 in Dhaka, Bangladesh. The Symposium was co-sponsored by the UN Economic and Social Commission for Asia and Pacific (ESCAP), the University of Dhaka and the Government of Bangladesh. The Symposium was attended by around 250 participants from 17 countries. It identified several key recommendations, knowledge gaps and opportunities for the design of programmes aimed at minimizing short- and long-term vulnerability of the region to climate change.

We also wish convey our sincere thanks to Prof. S. M. A. Faiz, former Vice-chancellor, University of Dhaka and President, National Organizing Committee and all his colleagues for their untiring efforts in ensuring the full success of this Symposium.

Several papers presented at the Symposium are brought together in this book and we hope that this volume will serve as a significant source of information to all agencies and organizations interested in the subject of climate change and food security and in promoting effective adaptation strategies to cope with climate change in South Asia.

Michel Jarraud
Secretary General
World Meterological Organization
7 bis, Avenue de la Paix
1211 Geneva 2, Switzerland

Jacques Diouf
Director-General
Food and Agriculture Organization
viaie delle Terme di Caracalla
0153 Rome, Italy

Preface

South Asia (SA) includes the region surrounded by Western Asia, Central Asia, Eastern Asia, Southeastern Asia and the Indian Ocean. It consists of Iran, Afghanistan, Pakistan, India, Nepal, Bangladesh and Sri Lanka. The region is also referred to as the Indian Sub-Continent or the Sub-Himalayan region, because of its unique and distinct geographical and physiographic setting. It was once a small continent which collided with mainland Asia about 50–55 million years ago, giving rise to the Himalayan mountains and the Tibetan Plateau. The SA region has diverse climate and terrains ranging from glaciers and tropical rainforest, the highest rainfall in the world to desert, and flat alluvial plains to hills and undulating terrains. The region has a population of 1.62 billion (24% of the world's population) living in 642 million hectares or 5% of the world's geographical area. The region is developing rapidly, both industrially and economically. Yet, there is a widespread problem of poverty, hunger and malnutrition. As much as 75% of the poor and malnourished people live in rural areas and are dependant on subsistence and small scale agriculture. Yet, globalization has accelerated economic growth. By 2007, the regions GDP growth had reached 9%/year which extended to all countries of SA. The growth rate in 2009 was 5.6%, which was the smallest decline compared with all other regions of the world.

Climate change is a major concern in SA because of alterations in temperature and precipitation, rise of sea level, melting of the Himalayan glaciers, and degradation of natural resources and the environment. According to the Fourth Assessment Report of the IPCC (2007), future projections of climate indicate that SA is very likely to warm during the twenty-first century. Also, the fresh water availability is projected to decrease and coastal areas will be at greatest risk due to increased flooding at the sea and rivers. In some SA countries, a substantial decrease in crop yields from rainfed agriculture could occur. Additionally, dramatic changes in land use patterns in SA compound the problem of climate change. To cope with climate change more effectively in SA, it is necessary to identify integrated adaptation and mitigation options for a range of agroecosystems so as to enable a favorable policy environment for the implementation of Regional Climate Change Adaptation Network. Majority of the poor people in SA are at risk because of the increase in frequency of extreme events, and especially the drought, floods and variability in climate. Glacier melting is a cause of concern because of its impact on the availability of water, and thus agronomic productivity. Climate change may also impact the on-set, distribution and amount of monsoons, as was the case in 2009 when monsoons failed in India. It is also feared that crop yields and agronomic production will be adversely affected, thereby exacerbating the food insecurity. The U.N. Millennium Development Goals, especially of cutting hunger and poverty by half, may not be met by 2015.

Soil degradation is a serious problem in SA. A total of 83 Mha is affected by water erosion and 59 Mha by wind erosion. In addition, loss of nutrients and organic matter content is a widespread problem throughout the regions, and affect about 26 Mha of cropland area. Nutrient depletion is exacerbated by erosion. The loss of cereal production in India is estimated at four million tonnes (Mt) on moderately degraded land, and 11 Mt on strongly regarded land, which together amount to 8% of India's total cereal production. This loss is equivalent to $2.3 billion/year. Salinization is another problem, especially in the irrigated areas of the Indo-Gangetic Plains.

Therefore, an international symposium entitled "Climate Change and Food Security in South Asia" was organized. The symposium was held from 25 to 30 August 2008 at the University of Dhaka, Bangladesh, and was jointly sponsored by the Ohio State University, World Meteorological Organization, University of Dhaka, Economic and Social Commission for Asia and Pacific, and Food and Agriculture Organization of the U.N. The objectives of the symposium were:

- To provide a central forum to develop an improved understanding and assessment of the climate change impacts on agriculture and the associated vulnerability in South Asia
- To identify and discuss integrated mitigation and adaptation win-win options for the agricultural sector in different agroecosystems of South Asia
- To discuss and propose a regional Agricultural Mitigation and Adaptation Framework for Climate Change in South Asia
- To discuss and recommend policy and financial innovations to enable smooth implementation of the regional framework and it's integration into the sustainable development planning of SA countries
- To discuss appropriate options for strengthening information exchange on climate change impacts and cooperation on agriculture mitigation and adaptation actions among SA countries

The Symposium was attended by more than 250 participants from 17 countries. The symposium was opened by His Excellency, Dr. Iajuddin Ahmend, President of Bangladesh, and closed by His Excellency Dr. Ólafur Ragnar Grímsson, President of Iceland. All presentations were organized into nine technical sessions. Papers submitted for publications were reviewed and revised for publication in this volume.

The editors thank all of the authors for their outstanding contributions to this volume. Their efforts will allow others to gain from their work, and will, we hope, lead to development of new policies to help mitigate the greenhouse effect while providing many other benefits to agriculture and society. These efforts have led to a merging of science and policy.

Thanks are due to the staff of Springer Verlag for their timely efforts in publishing this information and making it available to the scientific and the policy communities. In addition, numerous colleagues, graduate students and staff at the Ohio State University made valuable contributions. We especially thank Ms. Theresa L. Colson for her efforts in handling all of the papers included here from the first draft through the peer review process to provide information to the publisher. Ms. Alli Curry and Ms. Sara Klips helped in typing and proofing the manuscripts. The efforts of many other also were important in publishing this relevant and important work.

Rattan Lal
Mannava V.K. Sivakumar
S.M.A. Faiz
A.H.M. Mustafizur Rahman
Khandakar R. Islam

Dhaka Symposium Declaration

The International Symposium on Climate Change and Food Security in South Asia was held at the University of Dhaka from 25 to 30 August, 2008. It was jointly sponsored by the Ohio State University, the World Meteorological Organization (WMO), the Food and Agriculture Organization (FAO) of the United Nations, the UN Economic and Social Commission for Asia and Pacific (UNESCAP), and University of Dhaka and the Government of Bangladesh. The symposium was attended by around 250 participants from 17 countries.

Climate change has multi-dimensional impacts on agro-ecosystems in South Asia, including increases in temperature, declines in fresh water availability, sea level rise, glacial melting in the Himalayas, increased frequency and intensity of extreme events, and shifting of cropping zones. They all impact agriculture and the related food sector as well as the general economies, societies and the environment in South Asia.

Agriculture is a bearer, a contributor, as well as a mitigator of climate change. Small landholders (<2 ha) and resources poor, subsistence farmers predominate in the region and contribute to CO_2 emissions. The per capita land area is <0.1 ha in many countries in South Asia and is rapidly decreasing because of conversion of land to non-agricultural uses, soil degradation and continued population growth. The serious problems of soil degradation and desertification are likely to be exacerbated by climate change through accelerated erosion, fertility depletion, salinization and acidification. Subsistence agriculture, characterized by low productivity and extractive farming, is extremely vulnerable to climate change. The latter may constrain attainment of food production targets in the South Asian countries.

The symposium identified several key recommendations, knowledge gaps, and opportunities for policy makers. Researchers and extension systems, international organizations, and NGOs to implement programs designed to minimize short- and long-term vulnerability of the South Asian region to climate change. Principal recommendations are to:

- Create a Climate Change and Food Security in South Asia Network (CCFSSANet) and establish a South Asia Climate Outlook Forum (SACOF).
- Stimulate multi-disciplinary research on climate change and food security in South Asia and identify effective mitigation and adaptation options, including carbon sequestration in different ecosystems.

- Initiate and strengthen cooperation among academic and research institutions, international organizations, and NGOs to provide opportunities for strengthening institutions, human resource development and capacity building.
- Develop innovative financial mechanisms to scale up technical and financial support for the adaptation efforts of the South Asian countries.
- Promote adoption of mitigation and adaptation options through payments for ecosystem services such as carbon trading.
- Strengthen regional institutional and policy mechanisms to promote and facilitate implementation of location-specific adaptation and mitigation practices.

The symposium urges the development partners and the private sector to fund the implementation of programs that reflect the recommendations outlined about that deal with the mitigation and adaptation to climate change while advancing food security in South Asia. The participants thank the University of Dhaka and the Bangladesh Government for hosting the symposium and providing all the necessary facilities.

30 August 2008, Dhaka, Bangladesh

One World – One ChallengeA Foreword by the President of Iceland Ólafur Ragnar Grímsson

Climate change has altered the way we view the world. The challenge it presents brings countries and regions together in ways we have never known before. The fate of our nations, the future of all, rich and poor, in Europe, in Asia and in every corner of the world, have become closely intertwined.

If everything goes as the scientists now predict, the melting of the glaciers in Iceland and Greenland and the disappearance of the Arctic and Antarctic ice sheets will transform the oceans, leading to rising sea levels in distant countries, threatening the livelihood of farmers and fishermen, city dwellers and villagers in Asia and Africa and resulting in serious consequences in both the Americas. Never before has the whole world been so interconnected in the face of a major adversity. Our common fate is the core of the defining challenge of the twenty-first century.

Our efforts must be unified and successful in formulating a comprehensive global strategy which will enable us to prevent this impending world-wide disaster.

Many small island states are already giving high priority to these concerns. For them, the prospect of rising sea levels and destructive hurricanes poses a greater threat than any military scenarios have done up to now.

Bangladesh needs no call to action. More seriously than any other country it faces disastrous consequences of rising sea levels; could possibly loose in the coming decades up to one third of the land mass.

This is indeed a grave scenario. My participation in the symposium is an acknowledgement of the fact that Bangladesh has truly become a frontline state in the fight against climate change.

In recent years we have gained increasing awareness of how our eco-world is in fact a single system, how developments in a particular area of the grand mechanism of our existence may have hitherto undreamt-of consequences in another. The most dramatic contemporary manifestation of this interdependence is the relationship we have come to understand between climate change and the destruction of the soil, and how this constitutes a vicious circle.

We must acknowledge that all nations, wherever they are in the world, will be affected by climate change. It is therefore necessary that every state become a constructive partner in a global dialogue on the security implications of climate change.

An ever-changing natural environment confronts us with enormously complex and difficult challenges, demonstrating clearly the imperative need for fresh approaches, new ways in which the international community must address urgent policy decisions, translating scientific knowledge into improved and more effective ways of solving practical problems.

We must seek guidance from the heritage which has grown out of earlier global crises and model our actions with respect to the frameworks already in existence, on the treaties and institutions, both regional and global, which provide the pillars of the existing international community.

Dialogue on how this should be done, how to proceed from analysis to preventive action, how to extend and develop our international security framework, is now a clear priority.

The international symposium in Dhaka is an important contribution to such a global cooperation.

Closing Address

By

Ad Spijkers
FAO Representative in Bangladesh
Food and Agriculture Organization of the United Nations (FAO)
Delivered at the
International Symposium on
Climate Change and Food Security in South Asia
Dhaka, Bangladesh, 30 August 2008

Your Excellency Dr. Fakhruddin Ahmed, Hon'ble Chief Advisor, Government of the People's Republic of Bangladesh,
Your Excellency Olafur Ragnar Grimsson, Hon'ble President of Iceland,
Honourable Ministers,
Distinguished participants and guests,
Ladies and Gentlemen

1. It is a great honour and special privilege for me, on behalf of Mr. He Changchui, Assistant Director-General and Regional Representative for Asia and the Pacific, Food and Agriculture Organization of the United Nations, to welcome you all to the closing session of the International Symposium on Climate Change and Food Security in South Asia.

   The inauguration of the symposium by the President of Bangladesh and today the presence of Chief Advisor, Government of Bangladesh and the President of Iceland signify the strong support and the deep commitment to issues relating to climate change, food security and sustainable agriculture development.

A. It is estimated that the current food production in South Asia has *increased three times from 117 million tonnes in 1961 to 348 million tonnes in 2006*, but the dietary energy consumption has improved not enough emphasising the fact that the growth rate is not sufficient to tackle the emerging challenges in addition to population pressure. Ensuring food security in the future requires a great deal of *additional efforts in yield improvement*, with limited scope for expanding area under cultivation.

B. FAO has revealed recently that the Boro *rice output in Bangladesh* is estimated at record *17.54 million tonnes,* increased by some *17.2%* from the previous year and *29.3%* above the 5-year average. This increase of production was mainly due to favourable weather conditions and *extra efforts made by farmers and Government* in response to the high food prices and production loss of *1.4 million tonnes in 2007 Aman season* following severe flood and Cyclone Sidr.

Ladies and Gentlemen:

2. The majority of the South Asian countries *face multiple challenges* including frequent occurrence of *natural hazards, excessive dependence on agriculture,* widespread *poverty,* malnourishment, low income, soaring food prices and vulnerability to climate change.

   Climate change poses a major concern to food security in South Asia. Several regional and national initiatives have been undertaken in the recent past on climate change adaptation and mitigation.

C. We in FAO strongly feel that it is not enough; *adaptation and mitigation requires* socio-institutional learning process and participatory community based actions for technology refinement and transfer. Location-specific technologies and *good practices* need to be built upon an improved understanding of the links between climate change and food provision, while *promoting socio-economic development* and limiting further environmental degradation.

   FAO assists all its member countries in the region in identifying potential adaptation and mitigation options most applicable to their particular circumstances and in integrating climate change responses in food and agricultural policies, *National Programmes for Food Security (NPFSs), Special Programmes for Food Security (SPFSs) and National Forest Programmes (NFPs).*

Mr. Chairman,

3. In the aftermath of the emerging challenges and taking stock of the existing initiatives, we WMO, ESCAP, the Ohio State University, the University of Dhaka and FAO have organized the regional symposium on climate change and food security focusing on long-term implications of climate change. Indeed, the regional symposium has provided a *wealth of insights* on the linkages between climate change impacts and multiple dimensions of food security.

FAO is very pleased to note that the symposium is concluded with a series of recommendations and we are committed to implement them. *Let's take the main thrust* of the symposium to mobilise the national, regional and international opinion to counter the climate change impacts and join hands and move forward together for ensuring food security.

4. The Government of Bangladesh recognised the need for regional collaboration and recommended the development partners to assist in establishing an *International Think Tank for Adaptation* which will provide a forum to study the vulnerability of countries to climate change, scope and constraints to adaptation, develop relevant data bases, and provide a *network among countries* and professionals.

Ladies and Gentlemen:

5. Finally, we would like to thank all our partners and the Government of Bangladesh for organizing a successful symposium. Given that a large portion of the population in South Asia is undernourished, climate change and global warming,

highlighted by the Secretary General of the UN as top priorities, should be particularly high in our agenda. This makes it a challenge for all of us to get the work done in this region with nearly a quarter of mankind. *Shall we fail them?* Thank you.

Contents

Part I
Climate Change in South Asia

Chapter 1
Climate of South Asia and the Human Wellbeing

Rattan Lal

Whether the weather be cold or whether the weather be hot, we'll weather the weather whatever the weather, whether we like it or not.

Anonymous

Contents

Abstract South Asia is a large landmass of 642 million hectares (Mha) and comprising of diverse climates. The population of about 1.62 billion distributed in eight countries is strongly impacted by the climate. The Indus Valley Civilization of Harappa and Mohenjodaro vanished because the rainfall declined from 400 to 800 mm/year between 2500 and 2000 BC to less than 200 mm/year in 1700 BC. The fall of Indus civilization is also attributed to degradation of soil and water resources and to flooding of Indus caused by deforestation of the gallery forest. The rapid decline in population of South Asia between AD 1000 and 1700 was also attributed to meteorological drought. The present population of 1.62 billion is drastically influencing the regional and global climate by deforestation, land use conversion and biomass burning, and fossil fuel combustion. There has been a significant anthropogenic warming in Asia since 1950s. The problem is confounded by the acceleration of economic development jeopardizing natural resources which

R. Lal (✉)
The Ohio State University, 2021 Coffey Road, 210 Kottman Hall, Columbus, OH 43210
e-mail: lal.1@osu.edu

R. Lal et al. (eds.), *Climate Change and Food Security in South Asia*,
DOI 10.1007/978-90-481-9516-9_1, © Springer Science+Business Media B.V. 2011

are already under stress by the Asian Brown Cloud, caused by soot and aerosol can strongly impact monsoons. The biomass, especially animal dung, must be used as a soil amendment rather than a source of energy and household fuel. The fate of 10,000 or so of the Himalayan glaciers also depends on the human-induced climate change. Thus, adaptation to climate change is essential, in which conversion to a judicious land use and widespread adoption of recommended soil/water/crop/vegetation management practices are important strategies. Building soil/ecosystem/social resilience is important to weathering the weather.

Keywords Indus valley civilization • Mohenjodaro • Harappa • Ghaggar • Famines • Monsoons • Himalayan glaciers

Abbreviations

AI	Aridity Index
D	Dryness ratio
ET	Evapotranspiration
Gg	Giga gram (billion gram)
Gt	Giga tonnes (billion tonnes)
L	Latent heat of water
Mha	Million hectares
Myr	Million years
P	Precipitation
R	Radiation energy
SA	South Asia
SOM	Soil organic matter

1.1 Climate and Physiography of South Asia

South Asia or Southern Asia is the region comprising sub-Himalayan countries. It consists of the Indian sub-continent south of the Himalayas and The Hindu Kush, and includes Afganistan, Bangledesh, India, Iran, Nepal, Pakistan and Sri Lanka. It has a population of 1.62 billion and land area of 642 million hectares (Mha). The SA region, a world within the world is characterized by diverse climates and physiographic features. Climate of a region is defined by the precipitation and evaporation balance, in which the latter is influenced by the temperature. Two commonly used indices of climate are the Aridity Index (AI = P:ET) where P is precipitation and ET is evapotranspiration, and the Dryness Ratio (D = R/LP) where R is annual net radiation, L is the latent heat of evaporation, and P is average annual precipitation. Using the AI, the ratio is <0.05 for hyper-arid, 0.05–0.20 for arid, 0.20–0.50 for semi-arid, 0.5–0.65 for dry sub-humid, and >0.65 for humid climates. Typically, hyper-arid areas receive <200 mm of precipitation annually, arid areas receive <200 mm of winter rainfall or <400 mm of summer rainfall, semi-arid areas 200–500 mm of winter rainfall or 400–600 mm of summer rainfall, and dry sub-humid regions from 500 to 700 mm of winter rainfall or 600–800 mm of summer rainfall (FAO 1993). In addition to total amount of rains, the length of the growing season is defined by its distribution or the period of favorable water balance and temperature regime to support plant growth. The growing season duration is <75 days for arid climates, 75–120 days for semi-arid regions, 120–150 days for sub-humid climates, >150 days for humid regions.

The SA region, spanning from Iran in the west to Sri Lanka in the east, is characterized by diverse climates (Fig. 1.1). The arid climate, where evaporation exceeds the precipitation for most of the year, predominates in the western region. In contrast, the humid climates prevail in the eastern region. The rainfall is monsoonal, characterized by rainstorms of high intensity, high energy load and high erosivity. Monsoonal rains are highly variable and erratic. Most of the region has high temperature during the summer, with soil temperature often reaching 50°C at 1 cm depth.

Landscape of the region comprises of three distinct physiographic features. The Himalayas and the Hindu Kush mountains comprise the mountainous terrain. Two predominant alluvial plains include the interior basin of Iran, and the Indo-Gangetic Plains. The uplands include the Deccan Plateau of India and the central hill massif of Sri Lanka. Both Mountainous and Upland regions are prone to water erosion and gullying.

There are two types of river systems. Rivers flowing from the Himalayas and the Hindu Kush mountains originate from glacial melt and feed the alluvial plains. Predominant among these rivers are the Indus, Ganges, Brahmaputra, and Meghna Thus, a rapid retreat of the Himalayan glaciers is considered a threat to the flow of these rivers. The flow of rivers originating from the uplands depends on the rainfall.

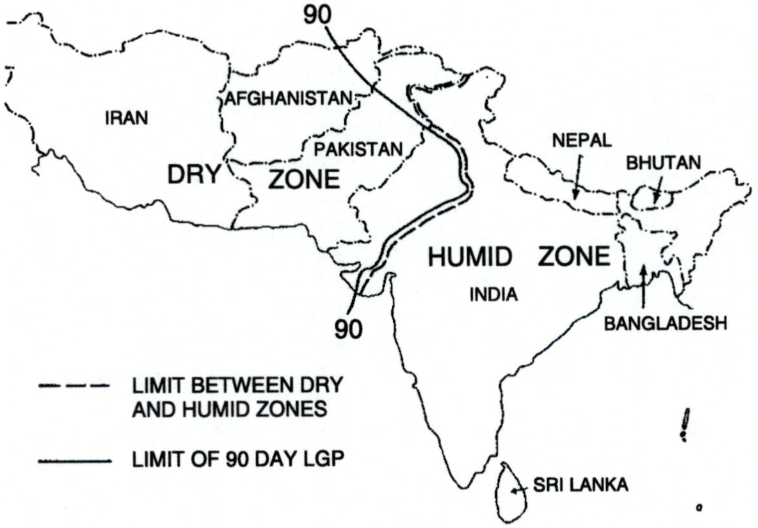

Fig. 1.1 South Asia region. 90-day (LGP = 90-day length of growing period (FAO 1994)

Important among these rivers of central and southern India are Krishe, Kaveri, Narmada and Trapti. Thus, land use and protection of head waters in the watersheds is critical to their flow.

This chapter outlines the historic and present day climate change with reference to drought, monsoon failure, and human wellbeing.

1.2 Indus Valley Civilization and Climate

Climate has played a major role throughout the human history in South Asia. The rise and fall of the Indus Valley Civilization is strongly linked to climate. The ancient cities of Harappa and Mohenjodaro in the northwestern region of ancient India flourished between 2800 and 2600 BC. While the present climate of the region is the Thar Desert covering Rajasthan (and partly the adjacent regions of Punjab, Haryana, Sindh, and Gujrat), the Indus Valley Civilization flourished on cultivation of wheat (*Triticum aestivum*), barley (*Hordeum vulgare*), Melon (*Citrullus* spp.), dates (*Phoenix dactylifera*) and cotton (*Gossypium hirsutum*). The earliest villages appeared along the Ravi in Punjab around 3300 BC (Kenoyer 2003), and grew wheat, barley, legumes and sesame (*Sesamum indicum*) (Posshl 2004). The city of Harappa flourished between 2800 and 2600 BC on agriculture and grazing. There were large granaries in Harappa and Mohenjodaro, an evidence of strong agriculture. However, cities were abandoned in 1900 BC. Among many hypothesis, climate drying is considered a plausible explanation (Singh 1971). The rainfall at that time ranged between 400 and 800 mm per annum (Lamb 1982), compared with the present day rainfall of <200 mm per annum (Table 1.1).

Table 1.1 Rainfall patterns during pre-historic era of the Indus Valley civilization in the ancient northwestern India (Recalculated from Lamb 1974)

Period (thousand years before present)	Rainfall (mm)		
	Annual	Summer	
10	720	630	
9	520	430	
8	430	370	
7	960	560	
6–5	570	420	
6	1,200	630	
5	800	600	} Harappan Era
4	730	460	
3.5	200	100	
2	200	100	
1	400	300	
0.5	500	400	
0.1	250	220	

The Indus Valley Civilization covered the region larger than the Nile Valley and Mesopotamian civilizations combined (Lamb 1982). Between 7,500 and 5,500 years ago, the rainfall of the region then ranged between 1,000 and 1,200 mm per annum. The region had dense vegetation which was a habitat for elephants, rhinoceros, and water buffalo. The region then was characterized by a stronger northward development of the summer monsoon rainfall which was adequate to support a flourishing agricultural civilization. It is hypothesized that rise of the Harappan civilization between 2800 and 2600 BC corresponded with good monsoons, and its decline was due to onset of the drought about 1900 BC (3,900 years ago), as is evidenced by the disappearance of many rivers. Climate had gotten worse and one of the major rivers of the Indus valley (Ghaggar-Hakra) had shifted and dried up. The extreme dry period lasted between 1,800 and 3,500 years ago (Lamb 1982). By the time, the Aryan people migrated into the Indus Valley about 1500 BC, the region was already too dry to support cultivation of winter cereals. The post-Harappan society eventually expanded to the eastern regions (Punjab, Haryana, northern Rajasthan and western U.P.) which were wetter and had favorable moisture regimes. Singh et al. (1990), Wasson et al. (1984) and Enzel et al. (1955) studied the climate of the region by identifying the pollens preserved in the salt lakes in the Thar desert. Monsoonal records, changes in precipitation over a multi-million year time scale, have been studied by a range of technologies. The past record can also be assessed through study of the rate of sediment transport into the ocean. The rates of erosion and transport of sediments deposited in the Bay of Bengal and the Arabian Sea have been widely studied for the Himalayan region. The deposition of sediments began ~40 Myr ago, increased after 20 Myr ago and sharply increased 8–10 Myr ago (Rea 1992; Ruddiman 2001). The rate of sediment transport in relation of climate change in recent years, the end of Harappan civilization about 1900 BC, has also been studied to infer the changes in monsoonal rains.

In addition to the possible impact of climate change, some researchers have argued that the climate during the period of the greatest vitality of the Harappan civilization was not very different from that of today (Wasson 2006). Thus, demise of the Harappan Civilization is also attributed to degradation of soil, water, and other natural resources. Fairservis (1967) concluded that too many demands were being made on the soil and water resources of the region. Deforestation of the galley forest (riparian forest) may have exacerbated flooding of the cities developed along the river and the floodplains (Fairservis 1967). Damming of the Indus by tectonic activity may also have been a cause (Raikes 1965).

1.3 Drought and Famines in South Asia

Famines have been experienced throughout the history of human settlement in SA. The emperor Janak was advised to plow a field to break the famine and drought. The event is connected with the birth of Sita, the heroine of the epic Ramayana. Occurrence of famines in SA is another example of the adverse effects of drought on decline in agricultural production and well being of the people (Dyson 1991; Mahahatna 1996). The intensity and frequency of famines is closely related to the extent and severity of drought. The term drought refers to the lack of adequate supply of water due to failure of rains, low storage of water in surface and sub-surface reservoirs, and particularly low storage in the root zone. It is an important factor affecting agronomic production, and, thus, food security. There are three types of droughts (Maybank et al. 1995). The meteorological drought is caused by long-term decline in rainfall (Street and Findlay 1981). The hydrological drought is caused by a prolonged period of decline in surface runoff and depletion of the ground water reserve. The agronomic drought refers to reduction in soil moisture storage because of losses of water by runoff and evaporation. The agronomic drought is exacerbated by decline in soil quality caused by depletion of soil organic matter and clay contents and the attendant decline in soil structure and aggregation, and the truncation of the topsoil by accelerated erosion. The agronomic drought specifically refers to reduction in soil moisture availability during the growing season (Dracup et al. 1980).

During the Middle Ages, the population of South Asia increased to a maximum of 200–300 million about AD 1000 (Lal 1973). Thereafter, it decreased to about 190–200 million by AD 1200, to 170 million by 1388, followed by a sharper decline to merely 120 million between AD 1525 and 1550. By the time when Taj Mahal was built in 1631, the population of South Asia was merely 130–140 million. This decline in population is primarily attributed to the occurrence of famines (Lal 1973; Lamb 1982), caused by droughts and failure of monsoons. It was the combination of meteorological, hydrological and agronomic droughts that caused this sharp decline in population between AD 1000 and 1650. Indeed, the Mogul empire experienced numerous famine during the sixteenth and seventeenth centuries.

During the seventeenth century, marked by the Mogul dynasties under the emperor Akbar and his predecessors, the meteorological droughts were common with frequent failure of Monsoons. The historical city of Fatehpur Sikri was built by emperor Akbar in 1570, and was the capital from 1571 to 1585. The city, located about 40 km southwest of Agra, was abandoned in 1588. The abandonment of the newly constructed capital city of Fatehpur Sikri is a prime example of the fallout from meteorological drought caused by a long term decline in rainfall. Lamb (1982) observed that this era of meteorological drought in South Asia coincided with the expansion of the polar cap and greater warmth in Siberia. There were strong failures in the summer monsoons in Bengal in 1770 that killed 10 million of the Bengai's 30 million people. Famines were also experienced in 1816 and 1819, and later between 1876 and 1879. The crisis of 1816 also caused the pandemic of cholera in Bengal. The deaths due to famine in the late 1870s in India and China have been estimated at 14–18 million. There were also serious famines during the twentieth century, especially in 1943–1944 in Bengal which killed 2.1 million people out of a total population of 60 million. There was a severe food shortage in Bihar in 1965–1966 in Maharashtra in 1970, and in west Bengal in 1974–1975.

1.4 Monsoons in South Asia

The process of seasonal transfers of heat between the tropical ocean and land is called Monsoons (Ruddiman 2001). The process is driven by the fact that water in the ocean responds more slowly than land to seasonal changes in solar heating because water has a heat capacity of 1 cal/g/C and land (soil) has a heat capacity of 0.2 cal/g/C. Therefore oceans have a high thermal inertia. Summer monsoon circulation, a predominant source of rainfall in South Asia, involves in-and-up flow of moist (water-laden) air over the Indian sub-continent that produces heavy rainfall between June and September. The differential heating between the ocean and the South Asian landmass sets in motion a large-scale summer monsoon over the Indian sub-continent by a two stage process (Ruddiman 2001): (i) initially a dry process due to the rising of sensible heat, and (ii) later a wet process linked to ocean moisture and release of latent heat. The winter monsoon, reversal of the summer monsoon, comprises down-and-out motion of cold and dry air from the South Asian landmass to the ocean. The summer precipitation is generally high in regions of low pressure and upward motion.

The failure of monsoons in SA is attributed to numerous natural and anthropogenic factors. An important among numerous anthropogenic factors is the emission of soot through widespread use of traditional biofuels in the subcontinent. The use of traditional biofuels produces a continental scale plume (brown cloud comprising of soot and aerosols) that migrates over the Indian Ocean during the winter months. This pollution plume may affect both regional and global climate. Habib et al. (2004) reported that black C emissions from biofuel combustion are estimated at 172–340 Gg C/year with the corresponding organic C emissions of 582–1683 Gt C/year (Venkataraman et al. 2005). Emissions of black C and CO_2 from biomass burning in SA have increased sixfold since 1930 (Ramanathan et al. 2005). These emissions have been linked to large reduction in surface solar radiation, surface evaporation and summer monsoon rainfall. The atmospheric brown cloud envelopes vast regions of the sea reducing solar radiation, suppressing evaporation, cooling the northern ocean surface, weakening the latitudinal sea-surface temperature gradients, and stabilizing the lower atmosphere (Ramanathan et al. 2005). These effects interfere with the regional hydrologic cycle. It has been estimated that the India-averaged summer rainfall has decreased by 5% since 1960s, with a large reduction of 1 mm/day in July. The observed frequency of droughts has increased steadily from the 1930–1950 values of 1/decade to 4/decade in 1980s, and 3/decade in 1990s. The average decadal rainfall has decreased over the entire period. This anthropogenic effect on monsoon variability and uncertainty in SA may intensify in the future to drought frequency of 6/decade if the particulate and soot emissions continue to increase. Black C (soot) also absorbs solar radiation and has a direct warming effect, similar to that of CO_2 which absorbs infrared terrestrial radiation. Therefore, limiting emission of black C would mitigate/moderate global warming.

1.5 The Himalayan Glaciers and the Climate Change

The global climate change may also impact the Himalayan glaciers, likely leading to a rapid retreat in the face of global warming. It is widely believed that India's 10,000 or so Himalayan glaciers are shrinking rapidly. Mayweski and Jascheke (1979) reported that the Himalayan glaciers have been retreating since 1850, but the rate of retreat has increased since the 1960s (Jangpang and Vohra 1962; Kurien and Munshi 1972; Srikanta and Panhi 1972; Vohra 1981). Since 1960, almost a fifth of the Indian Himalayan ice coverage has disappeared. The 30-km long Gangotri glacier retreated an average of 22 m/year and shed a total of 5% of its length between 1934 and 2003. However, the retreat was 12 m/year during 2004 and 2005 and standstill since 2007. Some studies have shown that the Gangotri glacier has retreated at the rate of 12–34 m/year since 1971 (Kumar et al. 2008; Naithni et al. 2001). The downwasting of glaciers may adversely impact the supply of water to the densely populated regions of the Indus and Ganges–Baghmeputra basins. In addition to supply of water for irrigation, glacier retreat may adversely impact the supply of drinking water and that needed for industrial development and urbanization. In contrast to the general belief; however, a recent report indicates no sign yet of the Himalayan meltdown (Bagla 2009). Reportedly, both Gangotri and Saichan glaciers are at standstill. Many glaciers in the Karakoram mountains have also stabilized or undergone an aggressive advance (Bagla 2005). Some argue that rates of retreat have been less in the past 30 years than the previous 60 years. Even if the glaciers are melting, the report says, it would not entail a drastic water shortage because Ganges results primarily from monsoon rainfall. The flow in Ganges will be affected more strongly by anomalies of monsoons than by glacier meltdown (Bagla 2009). The complex process of glacier dynamics is influenced by a range of factors including the altitude, snowfall, and changes in summer weather such as cloudiness. Thus, there is a need for additional studies.

1.6 Recent Climate Change in South Asia

The Asian continent has experienced a significant warming since 1950s (Ramaswamy 2009). The SA region is extremely vulnerable to drought, variability in monsoons, floods and other extreme events. Thus, energy, agriculture, water, soil and other natural resources are extremely vulnerable to climatic shifts, especially in densely populated regions of SA (e.g., The Indo-Gangetic Plains). The problem is exacerbated by soot and black carbon (Ramanathan et al. 2005; Menon et al. 2002), along with sulfate and other aerosols (Lau et al. 2008; Randles and Ramaswamy 2008; Shiu et al. 2009). While the temperature anomalies are easy to model, it is difficult to predict impact of climate disruption on precipitation changes and alteration in the hydrological cycle. Thus, development and validation of models (ground truthing) is essential.

1.7 Conclusions

Being mostly an agrarian population, dependant on rainfed agriculture and that with supplemental irrigation involving rain-harvested water, climate change had in the prehistoric past, in the present era and will continue to strongly impact wellbeing of the vast majority of people living in South Asia. Climate change and the overall aridization of the region was probably an important cause of the fall of the Indus civilization about 1900 BC. The problem was then exacerbated by degradation of soil and water resources and flooding due to deforestation. Similarly, the recurrence of drought-induced famines was the cause of decrease in population of northern India between 1000 and 1700 AD, and the great famine of Bengal in 1940s. With high population of 1.7 billion and increasing, the region is extremely vulnerable to any climate change that adversely impacts water supply and exacerbates frequency and intensity of droughts. Adaptation, based on choice of judicious land use and adoption of recommended soil/crop/nutrient/water management practices, is an important strategy. Restoring degraded soils and ecosystems and enhancing soil/ecosystem/social resilience is the goal. Sustainable management of soil and water resources can make a difference in successful adaptation to vagaries of changing climate.

References

Bagla P (2009) No sign yet of the Himalayan meltdown, Indian report finds. Science 326:924–925

Bailis R, Ezzati M, Kammen DM (2005) Mortality and greenhouse gas impacts of biomass and petroleum energy futures in Africa. Science 308:98–103

Dracup JA, Lee KS, Paulson EG (1980) On the definition of droughts. Water Resour Res 16:297–302

Dyson T (1991) The demography of south Asian famines. Popul Stud 45(5–25):279–297

Enzel Y, Ely LL, Mishra S, Ramesh R, Amit R, Lazar B, Rajaguru SN, Baker VR, Sandler A (1955) High resolution Holocene environmental changes in the Thar Desert, northwestern India. Science 284:125–128

Fairservis WA (1967) The origin, character and decline of an early civilization. Novistates 2302:1–48

FAO (1993) Key aspects of strategies for sustainable development of Drylands. FAO, Rome, Italy

FAO (1994) Land degradation in South Asia: its severity/causes and effects on the people. World Soil Resources Report 78. FAO, Rome, Italy

Habib G, Venkataraman C, Shrivastava M, Banergee R, Stehr JW, Dickerson RR (2004) New methods for estimating biofuel consumption for cooking: atmospheric emissions of black carbon and SO_2 from India. Global Biogeochem Cycles 18:1–11

Haritahya UK, Singh P, Kumar N, Gupta RP (2006) Suspended sediments from the Gangotri glacier: quanitification, variability and associations with discharge and air temperature. J Hydrol 321:116–130

Jangpang BS, Vohra CP (1962) The retreat of the Skunkulpa (Ralam) Glacier in the Cenertal Himalaya, Pithoragarh District, UP, I India, Inst. Assoc. Sci. Hydrol, Pupl. No. 58 pp 234–238

Kenoyer JM (2003) Uncovering keys to the lost Indus cities. Sci Am 289:58–67

Kumar K, Dumka RK, Miral MS, Satyal GS, Pant M (2008) Estimation of the retreat rate of Gangotri glacier using rapid static kinematics GPS survey. Curr Sci 94:258–262

Kurien MN, Munshi MN (1972) A survey of Sonapani Glacier, Lahaul District, Punjab, Geol. Surv. India, Misc., Publ. No. 15, pp 83–88

Lal KS (1973) Growth of the Muslim population in the Medieval India. Research Publications, Delhi, India

Lal R (2009) Soil degradation as a reason for inadequate human nutrition. Food Sec 1:45–57

Lamb HH (1982) Climate, history and the modern world. Methuen, London, p 387

Lau KM et al (2008) The joint aerosol monsoon experiment: a new challenge for monsoon climate research. Bull Am Meteorol Soc 89:369–383

Mahahatna A (1996) The demography of famines – an Indian historical perspective. Oxford University Press, New Delhi

Maybank J, Bonsal B, Jones K, Lawford R, O'Brien EG, Ripley EA, Wheaton E (1995) Drought as a natural disaster. Atmos-Ocean 33:195–222

Mayweski P, Jascheke PA (1979) Himilaya and trans Himilayan glacier flucations since AD 1812. Arctic Alpine Res 11:267–287

Menon S, Hansen J, Nazarenko L, Luo Y (2002) Climate effects of black carbon aerosols in China and India. Science 297:2250–2253

Naithani AK, Nainwal HV, Sati KK, Parsad C (2001) Geomorphological evidences of retreat of the Gangotri glacier and its characteristics. Current Sci 80:87–94

Posshl GL (2004) India, ancient-Indus Valley. In: Krech S, McNeill JR, Merchant C (eds) Encyclopedia of world environmental history. New York, Routledge, pp 672–675

Raikes RL (1965) The Mohenjodaro floods. Antiquity 39:196–203

Ramanathan V, Crutzen PJ, Kiehl JJ, Rosenfeld D (2001) Aerosols, climate and the hydrologic cycle. Science 294:2119–2124

Ramanathan V et al (2005) Atmospheric brown clouds: impact on South Asian climate and hydrologica cycle. Proc Natl Acad Sci USA 102:5326–5333

Ramaswamy V (2009) Anthropogenic climate change in Asia: key challenges. EOS 90:469–471

Randles CA, Ramaswamy V (2008) Absorbing aerosols over Asia: a geophysical fluid dynamics laboratory general circulation model sensitivity study of model response to aerosol optical depth and aerosol absorption. J Geophys Res 113:D21203

Rashid H, Flower BP, Paore RZ, Quinn TM (2007) A 25ka monsoon variability record from the Andaman Sea. Quat Sci Rev 26:2586–2597

Rea DK (1992) Delivery of the Himalayan Seidments to the northern Indian ocean and its relation to global climate, sea level, uplift and sea water strontium. In: Duncan RA et al (eds) Synthesis of results from scientific drilling of the Indian Ocean. American Geophysical Union, Washington, DC

Ruddiman WF (2001) Earth's climate: past and future. W.H. Freeman, New York, p 465

Shiu C-J, Liu SC, Chen J-P (2009) Diurnally asymmetric trends of temperature, humidity, and precipitation in Taiwan. J Clim 22:5635–5649

Singh G (1971) The Indus Valley culture seen in the context of post-glaciel climatic and ecological studies in northwest India. Rev Paleobot Paleontol 64:351–388

Srikanta SV, Panhi RN (1972) Recession of the Bada Shigri Glacier. Geological Survey of India, Misc., Publ. No. 15, pp 97–100

Street RB, Findlay BF (1981) An objective climatological study of prolonged dry spells (meteorological drought) in the Canadian Praries. Canadian Climate Center, AES, Downsview, ON, Canada, Report #81–10, pp 29

Venkataraman C, Habib G, Eiguren-Fernandez A, Miguel AH, Friedlander SK (2005) Residential biofuels in South Asia: carbonaceous aerosol emissions and climate impacts. Science 307:1454–1456

Vohra CP (1981) In: Lall JS (ed) Himalayan glaciers in the Himalaya: aspects of change. Oxford University Press, New Delhi, pp 138–151

Wasson RJ (2006) Exploitation and conservation of soil in the 300 year agricultural and forestry history of South Asia. In: McNeill JR, Winiwarter V (eds) Soils and Societies: perspectives from environmental history. The White Horse Press, UK, pp 13–50

Wasson RJ, Smith GI, Agarwal DP (1984) Late quaternary sediments, minerals, and infrared geochemical history of Didwana Lake, Thar Desert, India. Paleogeogr Paleoclimatol Paleoecol 46:345–372

Chapter 2
Climate Change in South Asia

Mannava V.K. Sivakumar and Robert Stefanski

Contents

Abstract South Asia, is home to over one fifth of the world's population and is known to be the most disaster prone region in the world. The high rates of population growth, and natural resource degradation, with continuing high rates of poverty and food insecurity make South Asia one of the most vulnerable regions to the impacts of climate change. In general, past and present climate trends and variability in South Asia can be characterized by increasing air temperatures and there is an increasing trend in the intensity and frequency of extreme events in South Asia over the last century. Temperature projections for South Asia for the twenty-first century suggest a significant acceleration of warming over that observed in the twentieth century. Recent modelling experiments indicate that the warming would be significant in Himalayan Highlands including the Tibetan Plateau and arid regions of Asia. An increase in occurrence of extreme weather events including heat wave and intense precipitation events is projected in South

M.V.K. Sivakumar (✉) and R. Stefanski
Climate Prediction and Adaptation Branch, World Meteorological Organization,
7bis Avenue de la Paix, 1211, Geneva 2, Switzerland
e-mail: msivakumar@wmo.int

R. Lal et al. (eds.), *Climate Change and Food Security in South Asia*,
DOI 10.1007/978-90-481-9516-9_2, © Springer Science+Business Media B.V. 2011

Asia, along with an increase in the interannual variability of daily precipitation in the Asian summer monsoon. The projected impacts of climate change in South Asia will vary across sectors, locations and populations. Temperature rise will negatively impact crop yields in tropical parts of South Asia where these crops are already being grown close to their temperature tolerance threshold. While direct impacts are associated with rise in temperatures, indirect impacts due to water availability and changing soil moisture status and pest and disease incidence are likely to be felt. The most significant impacts are likely to be borne by small-holder rainfed farmers who constitute the majority of farmers in this region and possess low financial and technical capacity to adapt to climate variability and change. The projected impacts of climate change in different parts of South Asia are described. The coping capacity of the rural poor, especially in the marginal areas, is low and there is a need to mainstream the good practices for adaptation to climate change into sustainable development planning in the region. Improved understanding of the climate change impacts, vulnerability and the adaptation practices to cope with climate change could help this process.

Keywords Indian sub-continent • Mangroves • El-Niño • ENSO • Rainfall variability • Productivity • Fisheries • Sea level rise

Abbreviations

ENSO	El Niño Southern Oscillation
FAR	Fourth Assessment Report
GDP	Gross Domestic Product
GHG	Greenhouse gas
GLOF	Glacial Lake Outburst Flood
HDR	Human Development Report
NAO	North Atlantic Oscillation
UNEP	United Nations Environment Program
WMO	World Meteorological Organization

2.1 Introduction

South Asia, comprising of eight countries i.e., Afghanistan, Bangladesh, Bhutan, India, the Maldives, Nepal, Pakistan and Sri Lanka, is home to over one fifth of the world's population (Table 2.1) and is the most densely populated geographical region in the world. South Asia is known to be the most disaster prone region in the world (UNEP, United Nations Environment Programme 2003). Although agricultural sector in South Asia is continuing to grow, it is declining in relative importance, both in terms of its contribution to GDP (Table 2.1) and its share of the labour force. Urbanization is increasing, and farm households are diversifying their sources of income beyond agriculture. This relative decline of agriculture is inevitable in countries that experience

Table 2.1 Statistics of various South Asian countries for 2008

Country	Area (km^2)	Population (millions)	Arable land (%)	GDP growth rate (%)	Agric. contribution to GDP (%)
Afghanistan	652,230	28.40	12.13	3.4	31.0
Bangladesh	143,998	156.05	55.39	4.9	19.1
Bhutan	38,394	0.69	2.3	21.4	22.3
India	3,287,263	1,166.08	48.83	7.4	17.6
The Maldives	298	0.40	13.33	5.7	7.0
Nepal	147,181	28.57	16.07	4.7	32.5
Pakistan	796,095	176.24	24.44	2.7	20.4
Sri Lanka	65,610	21.32	13.96	6.0	13.4

economic growth, which has been widespread in the region. Nevertheless, a significant percentage of the economically active population is still involved in agriculture in South Asia, and agricultural employment is especially important for the livelihoods of the poor. South Asia is also home to a majority of the world's poor. According to FAO (2009), 1.02 billion people are undernourished worldwide in 2009. About 456 million people in South Asia are estimated to be undernourished.

In the recent past, climate change emerged as the single most pressing issue facing society on a global basis, with serious implications for the food security of billions of people in the developing countries. The inter-annual, monthly and daily distribution of climate variables (e.g., temperature, radiation, precipitation, water vapor pressure in the air and wind speed) affects a number of physical, chemical and biological processes that drive the productivity of agricultural, forestry and fisheries systems (Easterling et al. 2007). The high rates of population growth, and natural resource degradation, with continuing high rates of poverty and food insecurity make South Asia one of the most vulnerable regions to the impact of climate change. Land use, land degradation, urbanization and pollution, affect the ecosystems in this region directly and indirectly through their effects on climate. These drivers can operate either independently or in association with one another (Lepers et al. 2005). Complex feedbacks and interactions occur on all scales from local to global. Cassman et al. (2003) emphasize that climate change will add to the dual challenge of meeting food (cereal) demand while at the same time protecting natural resources and improving environmental quality in these regions. In the long run, climate change impacts, such as changes in temperature, shifts in growing seasons, storms, floods, droughts, and changed rainfall patterns, risk the livelihood of drylands populations. Therefore, adaptation to the adverse effects of climate change through sustainable land management is a crucial, though simultaneously challenging, task.

2.2 Climate Change

The climate system is a complex, interactive system consisting of the atmosphere, land surface, snow and ice, oceans and other bodies of water, and living things. The Intergovernmental Panel on Climate Change (IPCC) was established in 1988, by the World Meteorological Organization (WMO) and the United Nations Environment Program (UNEP), to assess scientific information on climate change, as well as its environmental and socioeconomic impacts, and to formulate response strategies. Climate change is defined by the IPCC as any change in climate over time, whether due to natural variability or as a result of human activity (IPCC 2007). Evidence from observations of the climate system has led to the conclusion that human activities are contributing to a warming of the earth's atmosphere. Human activities – primarily burning of fossil fuels and changes in land cover – are modifying the concentration of atmospheric constituents or properties of the Earth's surface that absorb or scatter radiant energy. In particular, increases in the concentrations of greenhouse gases (GHGs) and aerosols are strongly implicated as contributors to climatic changes observed during the twentieth century and are expected to contribute to further changes in climate in the twenty-first century and beyond. In the 1950s, the greenhouse gases of concern remained CO_2 and H_2O, the same two identified by Tyndall (1861) a century earlier. It was not until the 1970s that other greenhouse gases – CH_4, N_2O and CFCs – were widely recognized as important anthropogenic greenhouse gases (Ramanathan 1975; Wang et al. 1976; IPCC 2007). By the 1970s, the importance of aerosol-cloud effects in reflecting sunlight was known (Twomey 1977), and atmospheric aerosols (suspended small particles) were being proposed as climate-forcing constituents. The amount of carbon dioxide in the atmosphere has increased by about 35% in the industrial era, and this increase is known to be due to human activities, primarily the combustion of fossil fuels and removal of forests. These changes in atmospheric composition are likely to alter temperatures, precipitation patterns, sea level, extreme events, and other aspects of climate on which the natural environment and human systems depend.

As climate science and the Earth's climate have continued to evolve over recent decades, increasing evidence of anthropogenic influences on climate change has been found. Correspondingly, the IPCC has made increasingly more definitive statements about human impacts on climate. IPCC released its Fourth Assessment Report (FAR) in 2007 that focused on observed climate change and the potential impacts of future climate change. The sections below on observed and future climate change borrow heavily from the information provided in FAR.

2.2.1 *Observed Climate Change*

At the time of the Third Assessment Report of IPCC, scientists could say that the abundances of all the well-mixed greenhouse gases during the 1990s were greater than at any time during the last half-million years (Petit et al. 1999), and this record now extends back nearly 1 million years (IPCC 2007). In 2005, the concentration of carbon dioxide exceeded the natural range that has existed over 650,000 years. Evidence from observations of the climate system show an increase of 0.74 ± 0.18°C in global average surface temperature during the 100 year period from 1906 to 2005 and an even greater warming trend over the 50 year period from 1956 to 2005 than over the entire 100 year period i.e., 0.13°C ± 0.03°C vs. 0.07°C ± 0.02°C per decade (IPCC 2007). Eleven of the 12 year period between 1995 and 2006 are among the 12 warmest years since the instrumental record of global surface temperature was started in 1850 (IPCC 2007). Land regions have warmed at a faster rate than the oceans. Warming has occurred in both land and ocean domains, and in both sea surface temperature (SST) and nighttime marine air temperature over the oceans. However, for the globe as a whole, surface air temperatures over land have risen at about double the ocean rate after 1979 (more than 0.27°C per decade vs 0.13°C per decade).

The following information on the observed climate change in South Asia was summarized from the report of the Working Group II chapter on Asia (Chapter 10 – Cruz et al. 2007) of IPCC. In general, past and present climate trends and variability in Asia can be characterized by increasing air temperatures which are more pronounced during winter than in summer. During recent decades, the observed increases in some parts of Asia have ranged between less than 1–3°C per century. Across all of Asia, interseasonal, interannual and spatial variability in rainfall trend has been observed during the past few decades. Decreasing trends in annual mean rainfall have been observed in the coastal belts and arid plains of Pakistan and parts of North-East India with increasing trends in Bangladesh. Table 2.2 provides a detailed list on observed characteristics in surface air temperature and rainfall for some of the countries in South Asia.

As for extreme climatic events, there has been new evidence on recent trends, particularly on the increasing tendency in the intensity and frequency of these events in South Asia over the last century. There have been significantly longer

Table 2.2 Summary of key observed past and present climate trends and variability in some countries in South Asia (Cruz et al. 2007)

Country	Change in temperature	Change in precipitation
Bangladesh	Increasing trend of about 1°C in May and 0.5°C in November from 1985 to 1998	Decadal rain anomalies above long term averages since 1960s
India	0.68°C increase per century with increasing trends in annual mean temperature and warming more pronounced during post monsoon and winter	Increase in extreme rains in north-west during summer monsoon in recent decades and lower number of rainy days along east coast
Nepal	0.09°C increase per year in Himalayas and 0.04°C in Terai region with more in winter	No distinct long-term trends in precipitation records for 1948–1994
Pakistan	0.6–1.0°C increase in mean temperature in coastal areas since early 1900s	10–15% decrease in coastal belt and hyper arid plains and increase in summer and winter precipitation over the last 40 years in northern Pakistan
Sri Lanka	0.016°C increase per year between 1961 to 90 over entire country and 2°C increase per year in central highlands	An increase trend in February and decrease trend in June

heat waves in many countries of South Asia with several cases of severe heat waves. The AR4 report noted that, in general, the frequency of more intense rainfall events in many parts of Asia has increased, causing severe floods, landslides, and debris/mud flows. It is interesting that at the same time, the number of rainy days and total annual amount of precipitation has decreased. Therefore, the total amount rainfall has decreased but the rain has been concentrated in few days. Analysis of rainfall data for India highlights the increase in the frequency of severe rainstorms over the last 50 years. The number of storms with more than 100 mm rainfall in a day is reported to have increased by 10% per decade (UNEP, United Nations Environment Programme 2007).

An example of this can be demonstrated by the extreme rainfall event which occurred in Mumbai, India on 26 and 27 July 2005. In a matter of 18 h, 944 mm of rain was recorded which was the highest rainfall ever recorded in the last 100 years in India. Mumbai and adjacent areas of Maharashtra experienced one of their worst floods in history (Government of Maharashtra 2005).

In many parts of South Asia, there have been an increasing frequency and intensity of droughts. The linear trends of rainfall decreases for 1900–2005 were 7.5% in South Asia (significant at <1%). Droughts have become more common, especially in the tropics and subtropics, since the 1970s (IPCC 2007). Observed marked increases in drought in the past three decades arise from more intense and longer droughts over wider areas, as a critical threshold for delineating drought is exceeded over increasingly widespread areas. Decreased land precipitation and increased temperatures that enhance evapotranspiration and drying are important factors that have contributed to more regions experiencing droughts, as measured by the Palmer Drought Severity Index. Also, there has been a noted decrease in the number of cyclones originating from the Bay of Bengal and Arabian Sea since 1970 but the intensity of these storms has increased and the damage caused by intense cyclones has risen significantly in India and the Tibetan Plateau. Table 2.3 details a summary of observed changes in extreme events and severe climate anomalies in South Asia.

During the twentieth century, the changes in temperature and precipitation described above caused important changes in hydrology over large regions. One change was a decline in spring snow cover. Less snow generally translates to lower reservoir levels. The earlier onset of spring snowmelt exacerbates this problem. Snowmelt started 2–3 weeks earlier in 2000 than it did in 1948 (Stewart et al. 2004). Particularly worrisome is the reduction in the mass balance of the glaciers and this has serious implications for the availability of water for over 500 million people in South Asia.

Another manifestation of changes in the climate system is a warming in the world's oceans. The global ocean temperature rose by 0.10°C from the surface to 700 m depth from 1961 to 2003 (IPCC 2007). Warming causes seawater to expand and thus contributes to sea level rise. This factor, referred to as thermal expansion, has contributed 1.6 ± 0.5 mm per year to global average sea level over the last decade (1993–2003). Other factors contributing to sea level rise over the last decade include a decline in mountain glaciers and ice caps (0.77 ± 0.22 mm per year) (IPCC 2007). In the coastal areas of Asia, the current rate of sea-level rise is reported to be between 1 and 3 mm/year which is slightly greater than the global

Table 2.3 Summary of observed changes in extreme events and severe climate anomalies in South Asia (Cruz et al. 2007)

Climatic event	Observed change
Heat waves	Frequency of hot days and multiple-day heat wave has increased in past century in India with an increase in deaths due to heat stress in recent years.
Intense rains and floods	Serious and recurrent floods in Bangladesh, Nepal and north-east states of India during 2002, 2003 and 2004; floods in Surat, Barmer and in Srinagar of India during summer monsoon season of 2006; 17 May 2003 floods in southern province of Sri Lanka were triggered by 730 mm rain.
Droughts	50% of droughts associated with El Niño; consecutive droughts in 1999 and 2000 in Pakistan and Northwest India led to sharp decline in water tables; consecutive droughts between 2000 and 2002 caused crop failures, mass starvation and affected ~11 million people in Orissa, India; droughts in Northeast India during summer monsoon of 2006.
Cyclones/typhoons	Frequency of monsoon depressions and cyclones formation in Bay of Bengal and Arabian Sea on the decline since 1970 but intensity is increasing causing severe floods in terms of damages to life and property.

average. The rate of sea-level rise of 3.1 mm/year has been observed over the past decade compared to 1.7–2.4 mm/year over the twentieth century which suggests that the rate of sea level rise has accelerated relative to the long-term average.

Chapter 10 of the IPCC Working Group II report also summarized the impacts of observed changes in climate. For agriculture, the report noted that the production of rice, maize and wheat over recent decades has declined in many parts of Asia due to increasing water stress partly due to increasing temperature, increasing frequency of El Niño's and a lower number of rainy days. The chapter summarizes a study by the International Rice Research Institute which observed a 10% decrease in rice yield for every 1°C increase in growing-season minimum temperature (Peng et al. 2004). For water resources, there is concern about melting glaciers, since they account for over 10% of freshwater supplies the drier parts of Asia. There have been observations of Asian glaciers melting faster than in the past, especially in Central Asia, Western Mongolia and North-West China. However, studies in northern Pakistan indicate that glaciers in the Indus Valley region may be expanding, due to increases in winter precipitation over western Himalayas during the past 40 years. In India, Pakistan, Nepal and Bangladesh, water shortages have been attributed to rapid urban growth, industrialization, population growth and inefficient water use, which are exacerbated by a changing climate and its negative impacts on water demand, supply and quality.

The coastlines of South Asia are highly prone to cyclones and the combination of extreme climatic and non climatic events have caused coastal flooding, resulting in substantial economic losses and fatalities. In the major river deltas of the region, wetlands have been significantly altered in recent years due to large scale sedimentation, land-use conversion, logging and human settlement. Also, salt water is reported to have penetrated 100 km or more inland along tributary channels of the Bay of Bengal during the dry season. Along the South Asian coastlines, a large portion of the mangroves, which help prevent salt-water intrusion have been reportedly lost during the last 50 years largely due to human activities.

In the coastal areas of Asia, the current rate of sea-level rise is reported to be between 1 and 3 mm/year which is slightly greater than the global average. The rate of sea-level rise of 3.1 mm/year has been observed over the past decade compared to 1.7–2.4 mm/year over the twentieth century which suggests that the rate of sea level rise has accelerated relative to the long-term average.

2.2.2 Future Climate Change

Looking ahead, IPCC (2007) projects increases in global mean surface air temperature (SAT) continuing over the twenty-first century, driven mainly by increases in anthropogenic greenhouse gas concentrations, with the warming proportional to the associated radiative forcing. An expert assessment based on the combination of available constraints from observations and the strength of known feedbacks simulated in the models used to produce the climate change projections indicates that the equilibrium global mean SAT warming for a doubling of atmospheric CO_2, or 'equilibrium climate sensitivity', is likely to lie in the range 2–4.5°C, with a most likely value of about 3°C (IPCC 2007). Warming in the twenty-first century is expected to be greatest over land and at the highest northern latitudes. It is very likely that heat waves will be more intense, more frequent and longer lasting in a future warmer climate.

Increasingly reliable regional climate change projections are now available due to advances in modeling and understanding of the physical processes of the climate system. IPCC (2007) projections show that drier subtropical regions are warming more than the moister tropics. Warming is likely to be above the global mean in South Asia. The temperature projections for South Asia for the twenty-first century suggest a significant acceleration of warming over that observed in the twentieth century. The increase in temperature is least rapid, similar to the global mean warming, in South-East Asia, larger over South Asia and East Asia and greatest in the continental Asia (Central, West and North Asia). Also, in general, the projected warming over Asia is higher during northern hemisphere winter than during summer for all time periods. Recent modelling experiments indicate that the warming would be significant in Himalayan Highlands including the Tibetan Plateau and arid regions of Asia.

Mean winter precipitation will very likely increase in northern Asia and the Tibetan Plateau and likely increase in West, Central, South-East and East Asia. Summer precipitation will likely increase in North, South, South-East and East Asia but decrease in West and Central Asia. Droughts associated with summer drying could result in regional vegetation die-offs (Breshears et al. 2005) and contribute to an increase in the percentage of land area experiencing drought at any one time, for example, extreme drought increasing from 1% of present-day land area to 30% by the end of the century (Burke et al. 2006).

Climate extremes encompass both extreme weather, with durations of minutes to days (the synoptic timescale), and extreme climate events with durations of months, in the case of periods of wet/stormy weather, or years, in the case of drought (McGregor et al. 2005). In all cases, the frequency of extreme events may be affected by seasonal to inter-annual fluctuations of large scale climate variations such as El Niño/Southern Oscillation (ENSO) and the North Atlantic Oscillation (NAO) (Schwierz et al. 2006).

The frequency of occurrence of climate extremes is expected to change during the next century, with increases in the frequency of heat waves and heavy precipitation events, and decreases in the frequency of frost days, as a consequence of anthropogenically-forced climate change (Easterling et al. 2000). An increase in occurrence of extreme weather events including heat wave and intense precipitation events is projected in South Asia (Lal 2003) along with an increase in the interannual variability of daily precipitation in the Asian summer monsoon (Lal et al. 2000; May 2004; Giorgi and Bi 2005). For tropical cyclones, there is a projected increase of 10–20% in the intensity of storms with an increase in sea-surface temperature of 2–4°C relative to the current temperatures in East Asia, South-East Asia and South Asia. An increase in heights of storm-surges could result from the occurrence of stronger winds, increases in sea-surface temperatures and low pressures associated with tropical cyclones resulting in an enhanced risk of coastal disasters along the coastal regions of the countries of East, South and South-East Asia.

Annual average river runoff and water availability are projected to decrease by 10–30% over some dry regions at mid-latitudes and in the dry tropics. The areas suitable for rainfed agriculture are expected to significantly decrease affecting adversely land productivity potential of the continent (Fischer et al. 2002).

2.3 Climate Change Impacts in South Asia

Projections indicate that climate variations in South Asia will be varied and heterogeneous, with some regions experiencing more intense precipitation and increased flood risks, while others encounter sparser rainfall and prolonged droughts. The impacts will vary across sectors, locations and populations. Temperature rise will negatively impact rice and wheat yields in tropical parts of South Asia where these crops are already being grown close to their temperature tolerance threshold. While direct impacts are associated with rise in temperatures, indirect impacts due to water availability and changing soil moisture status and pest and disease incidence are likely to be felt. The most significant impacts are likely to be borne by small-holder rainfed farmers who constitute the majority of farmers in this region and possess low financial and technical capacity to adapt to climate variability and change.

Following is a short summary of the expected impacts across this diverse region.

2.3.1 Impacts of Enhanced Temperatures

While plant response to elevated CO_2 is positive, recent studies confirm that the effects of elevated CO_2 on plant growth and yield will depend on photosynthetic pathway, species, growth stage and management regime, such as water and nitrogen (N) applications (Jablonski et al. 2002; Ainsworth and Long 2005). Increased temperatures may also reduce CO_2 effects indirectly, by increasing water demand. Rain-fed wheat grown at 450 ppm CO_2 demonstrated yield increases with temperature increases of up to 0.8°C, but declines with temperature increases beyond 1.5°C; additional irrigation was needed to counterbalance these negative effects (Xiao et al. 2005).

Temperature rise will negatively impact rice and wheat yields in tropical parts of South Asia where these crops are already being grown close to their temperature tolerance threshold (Kelkar and Bhadwal 2007). Kumar and Parikh (2001) show that even after accounting for farm level adaptation, a 2°C rise in mean temperature and a 7% increase in mean precipitation will reduce net revenues by 8.4% in India. Wheat yields are predicted to decline by 6–9% in sub-humid, semiarid, and arid areas with 1°C increase in temperature (Sultana and Ali 2006), while even a 0.3°C decadal rise could have a severe impact on important cash crops like cotton, mango, and sugarcane (MoE 2003). In Srilanka, half a degree temperature rise is predicted to reduce rice output by 6%, and increased dryness will adversely affect yields of key products like tea, rubber, and coconut (MENR, Ministry of Environment and Natural Resources 2000). In the hot climate of Pakistan, cereal crops are already at the margin of stress. An increase of 2.5°C in average temperature would translate into much higher ambient temperatures in the wheat planting and growing stages. Higher temperatures are likely to result in decline in yields, mainly due to the shortening of the crop life cycle especially the grain filling period. The National Communication (MoE 2003) highlighted that crops like wheat, cotton, mango, and sugarcane would be more sensitive to increase in temperatures compared to rice.

Drylands and mountain regions are likely to be more vulnerable than others (Gitay et al. 2001) and ecosystem degradation is largest in these regions (Hassan et al. 2005). Climate change is likely to cause additional inequities, as its impacts are unevenly distributed over space and time and disproportionately affect the poor (Tol 2001; Stern 2007).

2.3.2 Impacts of Precipitation Variability and Water Resources

Current vulnerabilities to climate are strongly correlated with climate variability, in particular precipitation variability. These vulnerabilities are largest in the semi-arid and arid low-income countries with large tracts of drylands, where precipitation and stream flow are concentrated over a few months, and where year-to-year variations are high (Lenton 2004). In such regions a lack of deep groundwater wells or reservoirs (i.e., storage) leads to a high level of vulnerability to climate variability, and to the climate changes that are likely to further increase climate variability in future.

Water resources are inextricably linked with climate, so the prospect of global climate change has serious implications for water resources and regional development (Riebsame et al. 1995). Tendencies of increase in intense rainfall with the potential for heavy rainfall events spread over few days are likely to impact water recharge rates and soil moisture conditions. A warmer climate, with its increased climate variability, will increase the risk of both floods and droughts (Wetherald and Manabe 2002).

South Asia is endowed with great rivers, which are the lifelines of the regional economy. The ice mass covering the Himalayan-Hindu Kush mountain range is the source of the nine largest rivers of Asia, including the Ganges, Brahmaputra, and Indus. These rivers provide water to more than half of the world's population. Many people in Asia are dependent on glacial melt water during dry season. Accelerated glacial melt questions the very perennial nature of many of the Himalayan flowing rivers. This is likely to have huge implications on those dependent on the resource affecting water availability for agricultural purposes (Kelkar and Bhadwal 2007). In Nepal and Bhutan, melting glaciers are filling glacial lakes beyond their capacities contributing to Galcial Lake Outburst Floods (GLOFs) (UNEP, United Nations Environment Programme 2007). Of 2,323 glacial lakes in Nepal, 20 have been found to be potentially dangerous with respect to GLOFs. The most significant such event occurred in 1985, when a glacial lake outburst flood caused a 10–15 m high surge of water and debris to flood down the Bhote Koshi and Dudh Koshi rivers for 90 km, destroying the Namche Small Hydro Project (Raut 2006).

Semi-arid and arid areas are particularly exposed to the impacts of climate change on freshwater. In semi-arid areas, climate change may extend the dry season of no or very low flows, which particularly affects water users unable to rely on reservoirs or deep groundwater wells (Giertz et al. 2006). Kundzewicz et al. (2007) explain that many of these areas will suffer a decrease in water resources due to climate change. The population of Maldives mainly depends on groundwater and rainwater as a source of freshwater. Both of these sources of water are vulnerable to changes in the climate and sea level rise. With the islands of the Maldives being low-lying, the rise in sea levels is likely to force saltwater into the freshwater lens. The groundwater is recharged through rainfall. Although the amount of rainfall is predicted to increase under an enhanced climatic regime, the spatial and temporal distribution in rainfall pattern is not clear (Ministry of Environment and Construction 2005).

Agricultural irrigation demand in arid and semi-arid regions of Asia is estimated to increase by at least 10% for an increase in temperature of 1°C (Fischer et al. 2002; Liu 2002). Efforts to offset declining surface water availability due to increasing precipitation variability will be hampered by the fact that groundwater recharge will decrease considerably in some already water-stressed regions, where vulnerability is often exacerbated by the rapid increase in population and water demand.

The greatest impact will continue to be felt by the poor, who have the most limited access to water resources. Rapid depletion of water resource is already a cause for concern in many countries in South Asia. About 2.5 billion people will be affected with water stress and scarcity by the year 2050 in South Asia (HDR 2006). In the drylands, farmers and pastoralists also have to contend with other extreme natural resource challenges and constraints such as poor soil fertility, pests, crop diseases, and a lack of access to inputs and improved seeds. These challenges are usually aggravated by periods of prolonged droughts and/or floods and are often particularly severe during El Nino events (Vogel 2005; Stige et al. 2006). The impact of changes in precipitation and enhanced evaporation could have profound effects in some lakes and reservoirs.

2.3.3 *Impacts of Increased Frequency of Extreme Events and Natural Disasters*

South Asia suffers an exceptionally high number of natural disasters. Between 1990 and 2008, more than 750 million people – 50% of the region's population – were affected by a natural disaster, leaving almost 60,000 dead and resulting in about $45 billion in damages.

Several studies showed that generally, the frequency of occurrence of more intense rainfall events in many parts of South Asia has increased, causing severe floods, landslides, and debris and mud flows, while the number of rainy days and total annual amount of precipitation has decreased (Mirza 2002; Lal 2003). Increasing frequency and intensity of droughts in many parts of South Asia are attributed largely to a rise in temperature, particularly during the summer and normally drier months, and during ENSO events (Lal 2003). An increase in the frequency of droughts and extreme rainfall events could result in a decline in tea yield, which would be the greatest in regions below 600 m (Wijeratne 1996). With the tea industry in Sri Lanka being a major source of foreign exchange and a significant source of income for laborers the impacts are likely to be grave. On an average during the period 1962–1988, Bangladesh lost about 0.5 million tonnes of rice annually as a result of floods that accounts for nearly 30% of the country's average annual food grain imports (Paul and Rashid 1993).

Short-term natural extremes, such as storms and floods, interannual and decadal climate variations, as well as large-scale circulation changes, such as the El Niño Southern Oscillation (ENSO), all have important effects on crop, pasture and forest production (Tubiello 2005). Increased climate extremes may promote plant disease and pest outbreaks (Gan 2004).

There is growing evidence that the frequency and extent of drought has increased as a result of global warming. A global analysis has shown that abrupt changes in rainfall are more likely to occur in the arid and semi-arid regions, and that this susceptibility is possibly linked to strong positive feedbacks between vegetation and climate interactions (Narisma et al. 2007). The socio-economic impacts of droughts may arise from the interaction between natural conditions and human factors, such as changes in land use and land cover, water demand and use. Excessive water withdrawals can exacerbate the impact of drought. Changes in the frequencies of extreme events will have an impact on land degradation processes such as floods and mass movements, soil erosion by both water and wind, and on soil salinization.

2.3.4 Impacts on Crop, Pasture and Forest Productivity

Agriculture is the mainstay of several economies in South Asia. It is also the largest source of employment. The sector continues to be the single largest contributor to the GDP in the region. As three-fifth of the cropped area is rainfed, the economy of South Asia hinges critically on the annual success of the monsoons (Kelkar and Bhadwal 2007). In the event of a failure, the worst affected are the landless and the poor whose sole source of income is from agriculture and its allied activities. Cruz et al. (2007) concluded that the crop yield in many countries of Asia has declined, partly due to rising temperatures and extreme weather events and that future climate change is likely to affect agriculture, risk of hunger and water resource scarcity with enhanced climate variability and more rapid melting of glaciers.

For Asia, the results of recent studies suggest that substantial decreases in cereal production potential could be likely by the end of this century as a consequence of climate change. Cruz et al. (2007) stressed, however, that regional differences in the response of wheat, maize and rice yields to projected climate change could likely be significant. Results of crop yield projections, using the HadCM2 climate model, indicate that crop yields could likely increase up to 20% in East and South-East Asia while likely decrease up to 30% in Central and South Asia even if the direct positive physiological effects of CO_2 are taken into account. In South Asia, there could be a significant decrease in non-irrigated wheat and rice yields for a temperature increase of greater than 2.5°C which could incur a loss in farm-level net revenue of between 9% and 25%. One study points out that in Bangladesh, production of rice and wheat might drop by 8% and 32%, respectively, by the year 2050. Studies show that a 0.5°C rise in winter temperature could reduce wheat yield by 0.45 tons per hectare in India. Other studies suggest that 2–5% decrease in Indian wheat and maize yield potentials for temperature increases of 0.5–1.5°C could occur.

For countries in South Asia, the net cereal production is projected to decline at least between 4% and 10% by the end of this century under the most conservative climate change scenario. The changes in cereal crop production potential suggest increasing stress on resources induced by climate change in many of the developing countries of Asia. Climate change could affect not only the crop production per unit area but also the area of production. More than 28 Mha in South and East Asia will require a substantial increase in irrigation for sustained productivity and the demand for agricultural irrigation in arid and semi-arid regions of Asia is estimated to increase by at least 10% for an increase in temperature of 1°C.

Grasslands in the drylands of South Asia consisting of fast-growing, often short lived species, are sensitive to CO_2 and climate change, with the impacts related to the stability and resilience of plant communities (Mitchell and Csillag 2001). Experiments support the concept of rapid changes in species composition and diversity under climate change. In dry regions, there are risks that severe vegetation degeneration leads to positive feedbacks between soil degradation and reduced vegetation and rainfall, with corresponding loss of pastoral areas and farmlands (Zheng et al. 2002). The natural grassland coverage and the grass yield in Asia, in general, are projected to decline with a rise in temperature and higher evaporation (Lu and Lu 2003). Thermal stress reduces productivity, conception rates and is potentially life-threatening to livestock (Easterling et al. 2007).

Although climate change will impact the availability of forest resources, the anthropogenic impact, particularly land-use change and deforestation in tropical zones, is likely to be extremely important (Zhao et al. 2005). The mangrove forests along the Indus Delta in Pakistan are an especially diverse ecosystem. They provide fuelwood and fodder to local inhabitants and are breeding grounds for an estimated 90% of shrimps that are exported. Pakistan's national communication report states that detrimental impacts of climate change on rural livelihoods would result in more people being forced to seek employment in urban areas (MoE 2003). In Sri Lanka, Somaratne and Dhanapala (1996) estimate a decrease in tropical rain forest of 2–11% and an increase in tropical dry forest of 7–8%. This study also indicates that increased temperature and rainfall would result in a northward shift of tropical wet forest into areas currently occupied by tropical dry forest. Droughts combined with deforestation increase fire danger (Laurance and Williamson 2001).

2.3.5 Impacts on Crop Pests and Diseases

According to Cruz et al. (2007), some studies have shown that higher temperatures and longer growing seasons could result in increased pest populations in temperate regions of Asia. Warmer winter temperatures would reduce winter kill and increase insect populations. Overall temperature increases may influence crop pest and disease interactions by increasing pest and disease growth rates which would then increase the number of reproductive generations per season and by decreasing pest and disease mortality due to warmer winter temperatures, would make the crop more vulnerable. The report stated that climate change along with changing pest and disease patterns will affect how crop production systems react in the future.

2.3.6 Impacts on Fisheries

For fisheries, an increased frequency of El Niño events could likely lead to measurable declines in fish larvae abundance in the coastal waters of South and South-East

Asia (Cruz et al. 2007). This and other factors are expected to contribute to a general decline in fishery production in the coastal waters of East, South and South-East Asia. There is a potential to substantially alter fish breeding habitats and fish food supply and therefore the abundance of fish populations in Asian waters due to the response to future climate change to the following factors: ocean currents; sea level; sea-water temperature; salinity; wind speed and direction; strength of upwelling; the mixing layer thickness; and predator response.

2.3.7 Impacts of Sea Level Rise

Low-lying coastal cities will be at the forefront of impacts; vulnerable to the risks of sea level rise and storms. These cities include Karachi, Mumbai, and Dhaka – all of which have also witnessed significant environmental stresses in recent years. Higher seawater levels would also increase the risk of flooding due to rainstorms, by reducing coastal drainage. A rise in sea level would raise the water table, further reducing drainage in coastal areas. All these effects could have possibly devastating socioeconomic implications, particularly for infrastructure in low lying deltaic areas.

Noronha et al. (2003) provided a coastal district level ranking of vulnerability to one metre sea level rise in India by constructing a weighted index as an average of the share of land area affected in the total area of district; the share of population affected in the total population of the district; and the index of relative infrastructure development. The most vulnerable districts were found to be the metropolises of Chennai and Mumbai.

The population of Maldives mainly depends on groundwater and rainwater as a source of freshwater. Both of these sources of water are vulnerable to changes in the climate and sea level rise. With the islands of the Maldives being low-lying, the rise in sea levels is likely to force saltwater into the freshwater lens (Ministry of Environment and Construction 2005).

2.4 Conclusions

South Asia is one of the most vulnerable regions in the world to climate change in view of the huge population, the large number of poor facing food insecurity, inappropriate soil and management practices on marginal lands in the semi-arid regions leading to increasing rates of land degradation and the projected impacts of climate change on the agricultural, forestry and fisheries sectors. The coping capacity of the rural poor, especially in the marginal areas, is poor and there is a need to mainstream the good practices for adaptation to climate change into sustainable development planning in the region. Improved understanding of the climate change impacts, vulnerability and the adaptation practices to cope with climate change could help this process.

References

Ainsworth EA, Long SP (2005) What have we learned from 15 years of free-air CO_2 enrichment (FACE)? A meta-analysis of the responses of photosynthesis, canopy properties and plant production to rising CO_2. New Phytol 165:351–372

Breshears DD, Cobb NS, Rich PM et al (2005) Regional vegetation die-off in response to global-change-type drought. Proc Natl Acad Sci USA 102:15144–15148

Burke EJ, Brown SJ, Christidis N (2006) Modelling the recent evolution of global drought and projections for the twenty-first century with the Hadley Centre climate model. J Hydrometeorol 7:1113–1125

Cassman KG, Dobermann A, Walters DT, Yang H (2003) Meeting cereal demand while protecting natural resources and improving environmental quality. Annu Rev Environ Resour 28:315–358

Cruz RV, Harasawa H, Lal M, Wu S, Anokhin Y, Punsalmaa B, Honda Y, Jafari M, Li C, Huu N (2007) Asia. Climate change, 2007, impacts, adaptation and vulnerability. In: Parry ML, Canziani OF, Palutikof JP et al (eds) Contribution of Working Group II to the Fourth Assessment Report of the Intergovernmental Panel on Climate Change. Cambridge University Press, Cambridge, pp 469–506

Easterling W, Aggarwal PK, Batima P, Brander KM, Erda L, Howden SM, Kirilenko A, Morton J, Soussana JF, Schmidhuber J, Tubiello FN (2007) Food, fibre and forest products. Climate Change 2007: impacts, adaptation and vulnerability. In: Parry ML, Canziani OF, Palutikof JP (eds) Contribution of Working Group II to the Fourth Assessment Report of the Intergovernmental Panel on Climate Change. Cambridge University Press, Cambridge, pp 273–313

Easterling DR, Meehl GA, Parmesan C, Changnon SA, Karl TR, Mearns LO (2000) Climate extremes: observations, modelling, and impacts. Science 289:2068–2074

FAO (2009) The State of Food Insecurity in the World 2009. Food and Agriculture Organization of the United Nations, Rome, Italy

Fischer G, Shah M, van Velthuizen H (2002) Climate change and agricultural vulnerability. International Institute for Applied Systems Analysis, Vienna, Austria. Available online at: www.iiasa.ac.at/Research/LUC/JB-Report.pdf

Gan J (2004) Risk and damage of southern pine beetle outbreaks under global climate change. Forest Ecol Manag 191:61–71

Giorgi F, Bi X (2005) Regional changes in surface climate interannual variability for the 21st century from ensembles of global model simulations. Geophys Res Lett 32:L13701. doi:10.1029/2005GL023002

Giertz S, Diekkruger B, Jaeger A, Schopp M (2006) An interdisciplinary scenario analysis to assess the water availability and water consumption in the Upper Oum catchment in Benin. Adv Geosci 9:1–11

Gitay H, Brown S, Easterling W, Jallow B (2001) Ecosystems and their goods and services. Climate Change,2001, Impacts, Adaptation, and Vulnerability. In: McCarthy JJ, Canziani OF, Leary NA, Dokken DJ, White KS (eds) Contribution of Working Group II to the Third Assessment Report of the Intergovernmental Panel on Climate Change. Cambridge University Press, Cambridge, pp 237–342

Government of Maharashtra (2005) Maharashtra floods 2005. http://mdmu.maharashtra.gov.in/pdf/Flood/statusreport.pdf (Accessed 3 Nov 2009)

Hassan R, Scholes R, Ash N (eds) (2005) Ecosystems and human wellbeing: volume 1: current state and trends. Island Press, Washington, DC, p 917

HDR (2006) Human Development Report 2006. Beyond scarcity: power, poverty and the global water crisis. United Nations Development Programme, New York

IPCC (2007) Climate Change 2007: the physical science basis. In: Solomon S, Quin D, Manning M, Chen X, Marquis M, Averyt KB, Tignor HL, Miller M (eds) Contribution of Working Group I to the Fourth Assessment Report of the Intergovernmental Panel on Climate Change. Cambridge University Press, Cambridge, pp 1–996

Jablonski LM, Wang X, Curtis PS (2002) Plant reproduction under elevated CO_2 conditions: a meta-analysis of reports on 79 crop and wild species. New Phytol 156:9–26

Kelkar U, Bhadwal S (2007) South Asian regional study on climate change impacts and adaptation: implications for human development. Human Development Report Office, Occasional Paper. United Nations Development Programme

Kundzewicz ZW, Mata LJ, Arnell NW, Döll P, Kabat P, Jiménez B, Miller KA, Oki T, Sen Z, Shiklomanov IA (2007) Freshwater resources and their management. In: Parry ML, Canziani OF, Palutikof JP, van der Linden PJ, Hanson CE (eds) Climate Change 2007: impacts, adaptation and vulnerability. Contribution of Working Group II to the Fourth Assessment Report of the Intergovernmental Panel on Climate Change. Cambridge University Press, Cambridge, pp 173–210

Kumar KSK, Parikh J (2001) Indian agriculture and climate sensitivity. Global Environ Change 11(2):147–154

Lal M (2003) Global climate change: India's monsoon and its variability. J Environ Stud Policy 6:1–34

Lal M, Meehl GA, Arblaster JM (2000) Simulation of Indian summer monsoon rainfall and its intraseasonal variability. Reg Environ Change 1:163–179

Laurance WF, Williamson GB (2001) Positive feedbacks among forest fragmentation, drought, and climate change in the Amazon. Conserv Biol 15:1529–1535

Lenton R (2004) Water and climate variability: development impacts and coping strategies. Water Sci Technol 49:17–24

Lepers E, Lambin EF, Janetos AC, Defries R, Achard F, Ramankutty N, Scholes RJ (2005) A synthesis of information on rapid land-cover change for the period 1981–2000. BioScience 55:115–124

Liu CZ (2002) Suggestion on water resources in China corresponding with global climate change. China Water Resour 2:36–37

Lu XR, Lu XY (2003) Climate tendency analysis of warming and drying in grassland of Northeast Qingzang Plateau of China. Grassland of China 24:8–13

May W (2004) Simulation of the variability and extremes of daily rainfall during the Indian summer monsoon for present and future times in a global time slice experiment. Clim Dynam 22:183–204

McGregor GR, Ferro CAT, Stephenson DB (2005) Projected changes in extreme weather and climate events in Europe. In: Kirch W, Menne B, Bertollini R (eds) Extreme weather and climate events and public health responses. Springer, Dresden, pp 13–23

MENR (Ministry of Environment and Natural Resources) (2000) Initial national communication under the United Nations Framework Convention on Climate Change: Sri Lanka. MENR, Government of Sri Lanka, Colombo

Ministry of Environment and Construction (2005) State of the Environment – Maldives 2005

Mirza MQ (2002) Global warming and changes in the probability of occurrence of floods in Bangladesh and implications. Global Environ Chang 12:127–138

Mitchell SW, Csillag F (2001) Assessing the stability and uncertainty of predicted vegetation growth under climatic variability: northern mixed grass prairie. Ecol Model 139:101–121

MoE (Ministry of Environment (2003) Pakistan's initial national communication on climate change. Islamabad: MoE, Government of Islamic Republic of Pakistan. http://unfccc.int/resource/docs/natc/paknc1.pdf

Narisma GT, Foley JA, Licker R, Ramankutty N (2007) Abrupt changes in rainfall during the twentieth century. Geophys Res Lett 34:L06710. doi:doi:10.1029/2006GL028628

Noronha L, Choudhri BS, Nairy KS (2003) Relative vulnerability of districts to a potential sea-level rise along the coastline of India. In TERI. 2003. Environmental threats, vulnerability and adaptation: case studies from India. TERI, New Delhi, pp 121–142

Paul B, Rashid H (1993) Flood damage to rice crop in Bangladesh. Geogr Rev 83(2):151–159

Peng S, Huang J, Sheehy JE, Laza RE, Visperas RM, Zhong X, Centeno GS, Khush GS, Cassman KG (2004) Rice yields decline with higher night temperature from global warming. P Natl Acad Sci USA 101:9971–9975

Petit JR, Jouzel J, Raynaud D, Barkov NI, Barnola JM, Basile I, Bender M, Chappellaz J, Davis M, Delaygue G, Delmotte M, Kotlyakov VM, Legrand M, Lipenkov VY, Lorius C, Pépin L, Ritz C, Saltzman E, Stievenard M (1999) Climate and atmospheric history of the past 420,000 years from the Vostok ice core, Antarctica. Nature 399:429–436

Ramanathan V (1975) Greenhouse effect due to chlorofluorocarbons: climatic implications. Science 190:50–52

Raut A (2006) Climate impacts on Nepal. Tiempo 60:3–5

Riebsame WE, Strzepek KM, Wescoat L Jr, Perrit R, Graile GL, Jacobs J, Leichenko R, Magadza C, Phien H, Urbiztondo BJ, Restrepo P, Rose WR, Saleh M, Ti LH, Tucci C, Yates D (1995) Complex river basins. In: Strzepek KM, Smith JB (eds) As climate changes, international impacts and implications. Cambridge University Press, Cambridge, pp 57–91

Schwierz C, Appenzeller C, Davies HC, Liniger MA, Müller W, Stocker T, Yoshimori M (2006) Challenges posed by and approaches to the study of seasonal-to-decadal climate variability. Clim Chang 79:31–63

Somaratne S, Dhanapala AH (1996) Potential impact of global climate change on forest in Sri Lanka. In: Erda L, Bolhofer W, Huq S, Lenhart S, Mukherjee SK, Smith JB, Wisniewski J (eds) Climate change variability and adaptation in Asia and the Pacific. Kluwer, Dordrecht, The Netherlands, pp 129–135

Stern N (2007) The economics of climate change: the stern review. Cambridge University Press, Cambridge, p 692

Stewart IT, Cayan DR, Dettinger MD (2004) Changes in snowmelt runoff timing in Western North America under a 'business as usual' climate change scenario. Clim Change 62:217–232

Stige LC, Stave J, Chan KS, Ciannelli L, Pretorelli N, Glantz M, Herren HR, Stenseth NC (2006) The effect of climate variation on agro-pastoral production in Africa. P Natl Acad Sci USA 103:3049–3053

Sultana H, Ali N (2006) Vulnerability of wheat production in different climatic zones of Pakistan under climate change scenarios using CSM-CERES-Wheat Model. Paper presented in the Second International Young Scientists' Global Change Conference, Beijing, 7–9 November 2006, organised by START (the global change System for Analysis, Research and Training) and the China Meteorological Administration

Tol RSJ (2001) Equitable cost–benefit analysis of climate change policies. Ecol Econ 36:71–85

Tubiello FN (2005) Climate variability and agriculture: perspectives on current and future challenges. In: Knight B (ed) Impact of climate change, variability and weather fluctuations on crops and their produce markets. Impact Reports, Cambridge, UK, pp 45–63

Twomey S (1977) Influence of pollution on shortwave albedo of clouds. J Atmos Sci 34: 1149–1152

Tyndall J (1861) On the absorption and radiation of heat by gases and vapours, and on the physical connection. Philos Ma 22:277–302

UNEP (United Nations Environment Programme) (2003) GEO year book 2003. UNEP, Nairobi

UNEP (United Nations Environment Programme) (2007) GEO yearbook 2007: an overview of our changing environment. UNEP, Nairobi, p 86

Vogel C (2005) Seven fat years and seven lean years? Climate change and agriculture in Africa. IDS Bull 36:30–35

Wang WC, Yung YL, Lacis AA, Mo T, Hansen JE (1976) Greenhouse effects due to man-made perturbations of trace gases. Science 194:685–690

Wetherald RT, Manabe S (2002) Simulation of hydrologic changes associated with global warming. J Geophys Res 107(D19):4379. doi:doi:10.1029/2001JD001195

Wijeratne MA (1996) Tea: plucking strategies. Planters chronicle: September 1996, p 443

Xiao GW, Liu Q, Xu Z, Sun, Wang J (2005) Effects of temperature increase and elevated CO_2 concentration, with supplemental irrigation, on the yield of rain-fed spring wheat in a semiarid region of China. Agr Water Manage 74:243–255

Zhao Y, Wang C, Wang S, Tibig L (2005) Impacts of present and future climate variability on agriculture and forestry in the humid and sub-humid tropics. Climatic Change 70:73–116

Zheng YQ, Yu G, Qian YF, Miao M, Zeng X, Liu H (2002) Simulations of regional climatic effects of vegetation change in China. Q J Roy Meteor Soc 128:2089–2114, Part B

Chapter 3
Solar Irradiance of the Earth's Atmosphere

Sultana N. Nahar

Contents

Abstract The sun is the primary driver of terrestrial atmospheric phenomena and energy source for the earth. It emits radiation over a large energy band and ejects highly energetic plasma fluxes of charged particles into space. The sun is an active star that (i) goes through a 12-year maximum–minimum emission cycle, (ii) has huge, non-periodic eruptions in solar flares and coronal mass ejections, and (iii) has nearly equipartition of energy between particle and radiation fluxes. Variations in these emissions interact with all atmospheric layers down to the earth surface. The precise nature of these interactions can be examined through microscopic physics at atomic and molecular levels for illustration of physical and chemical processes, and their impact on macroscopic problems such as global climate change, and more localized manifestations such as the atmospheric brown cloud (ABC) prominent in Asia. I will describe some of these calculations for atomic and molecular species such as carbon, nitrogen, oxygen, sulfur and their compounds. While the visible and near-infrared solar radiation penetrates through, more energetic components in the ultraviolet (UV) and the x-ray are absorbed by the upper layers of the atmosphere and thus provide protecting shields for life on earth. The atmosphere has been maintaining a fine energy balance of solar radiation entering the earth by radiating the same amount into space. Certain atmospheric gases trap radiation energy and reflect back near earth' surface to heat it up in an energy cycle. This phenomenon

S.N. Nahar (✉)
Department of Astronomy, The Ohio State University, Columbus, OH 43210, USA
e-mail: nahar@astronomy.ohio-state.edu

R. Lal et al. (eds.), *Climate Change and Food Security in South Asia*,
DOI 10.1007/978-90-481-9516-9_3, © Springer Science+Business Media B.V. 2011

is known as the Greenhouse effect and has maintained the average earth surface temperature at 14°C. However, for over a 100 years the balance is being perturbed due to changes in atmospheric compositions of greenhouse gases. More energy is being trapped than released, leading to global warming and climate change. Depletion of atmospheric ozone molecules has created holes for harmful radiation to reach earth's surface. The basic scientific data in current atmospheric models lack accurate parameters for fundamental atomic and molecular processes, and hence provide predictions which have large uncertainties. We aim to explore the sensitivity of numerical simulations to accurately predict the response of the earth's atmosphere to changes in elemental and molecular composition. To wit, what is the effect of including new high-accuracy photoionization and radiative transition rates in C, N, O, H_2O, etc. in climate models? How do temporal-spatial and temperature-density dependencies of fundamental physical and chemical parameters and rates affect the absorption of solar radiation by the ABC? Such studies should lead to an improved understanding of global warming and climate change processes, and help in the future steps.

Keywords Radiation • Ultraviolet radiation • Asian brown cloud • Greenhouse effects • CFCs • HCFCs

Abbreviations

ABC	Atmospheric brown cloud
CFC	Chlorofluorocarbon
CSIRO	Council of Scientific & Ind. Research
DES	Dielectric satellite
HFC	Hydrofluorocarbon
INODEX	Indian Ocean Experiment
IPCC	International Government Panel on Climate Change
OCO	Orbiting Carbon Observatory
UNEP	UN Environment Program
UV	Ultraviolet

3.1 Chemical Compositions and Global Warming

Global warming is directly related to solar irradiance. The earth is our home planet and the sun is an integral part of our lives. The sun is responsible for virtually all energy that reaches the earth's surface. The interaction of solar radiation with atmospheric gases establishes the atmospheric temperature. The atmosphere contains number of component gases: 78.08% nitrogen (N_2), 20.95% oxygen (O_2), 0.93% argon, 0.038% carbon dioxide (CO_2) and traces of some others. About 1–4% could be water vapor depending on the climate. The atmospheric gases which absorb and trap energy causing the earth to heat up in a cyclic process are the greenhouse gases.

The natural constituent greenhouse gases are water vapor, which causes about 36–70% of the heating, carbon dioxide (CO_2) about 9–26%, methane (CH_4) about 4–9%, and ozone (O_3) about 3–7%. These gases have maintained a constant average earth temperature by a natural balance as they radiate about the same amount of energy that they absorb. Even a slight change in the concentration of particular gases in the atmosphere can prevent heat from being radiated out into space and upset this fine balance. The current global warming is a result of such imbalance. Over the last 100 years, the earth is seen going through a climate change, that is, changes in its weather systems, rainfall and temperatures. The average temperature of the earth's near-surface air and oceans has increased by 0.74 ± 0.18 C (1.33 ± 0.32°F) during the last 150 years ending in 2005. These changes can be caused naturally either, as a result of changes in the way oceans and the atmosphere interact, or from changes in the amount of energy received from the sun. Recent measurements indicate that the earth is presently absorbing solar energy of 0.85 ± 0.15 W/m^2 more than it is emitting into space.

Atmospheric concentrations of various greenhouse gases have been increasing slowly over the years and are believed to causing global warming. These concentrations are considerably higher than at any time during the last 650,000 years, the period for which reliable data has been extracted from ice cores. Concentration of greenhouse gas CO_2 has seen a steady rise. Currently about 6.5 billion tonnes of carbon dioxide are emitted globally each year, mostly through burning fossil fuel such as, coal, oil and gas by industries and domestic transportations, and through land use by deforestation and agricultural processes. The Intergovernmental Panel on Climate Change (IPCC) Special Report on Emissions Scenarios predicts that this increment in atmosphere may increase to from 541 to 970 ppm by the year 2100. Since the industrial revolution the concentrations of greenhouse gases have increased by human activities. The increases are +35% for carbon dioxide since 1,700, +150% for methane and +20% for nitrous oxide. Other greenhouse gases are also on the rise, such as, aerosol gases like hydrofluorocarbons (HFC) and chlorofluorocarbon (CFC). The conclusion of IPCC research indicated that most of the temperature increase since the mid-twentieth century is due to the increase in these anthropogenic gases. However, increase of CO_2 is believed to be the main gas accountable for global warming. Present technology cannot prevent the release of carbon dioxide, although there is some that can reduce the amount of gas being released.

The effect of global warming through heating up the earth's surface, oceans and atmosphere is causing devastating damage for the inhabitants of our planet. A rise in temperature of just one or two degrees will result in changing weather patterns and lead to increased flooding, desertification, crop failures, freshwater shortages, increase of salinity in inland water, and storms. The polar snow caps are retreating and sea levels are rising. United Nations Environment Programme (UNEP) has shown that glaciers melting has more than doubled in places. Some of the biggest losses have occurred in the Alps and Pyrenees mountain ranges in Europe. Estimates for 2006 indicate shrinkage of 1.4 m of 'water equivalent', compared to half a meter in 2005. A huge Antarctic ice sheet, 160 square mile piece which is more than seven times the size of Manhattan, of the Wilkins ice shelf collapsed into the ocean near the end of February 2008. The piece had been attached to Antarctica for hundreds, or may be even 1,500 years. Retreat of glaciers have created a number of glacier lakes in the Himalayas mountains. Bhutan has about 24 such lakes. The swollen lakes pose the risk of outbursts and flooding to the populated areas. Sea level is rising by 0.8 mm/year. IPCC estimates from its research on around 100 glaciers that the rise may be over 20 cm by 2100, where the increase of 1–23 cm will be due to melting of glaciers alone. Its data shows significant shrinkage taking place in Asia, European countries including Austria, Norway, Sweden, Italy, Spain and Switzerland. The overall headline figures from the IPCC expect global sea levels to rise by between 11 and 88 cm this century, and to rise further after that. However, these figures are based on computer modeling. With more accurate data on how solar radiation is being trapped, the models are expected to provide more precise predictions.

Another cause for concern in the climate change crisis is the depletion of ozone, O_3, in the atmosphere. O_3 is a molecule with three oxygen atoms which interacts efficiently with solar UV radiation. In amount ozone is rare in our atmosphere, averaging about three molecules for every 10 million air molecules. Despite of this small amount, O_3 plays a vital role in the upper atmosphere (stratosphere) and the lower atmosphere (troposphere). Stratospheric ozone (often referred to as "good ozone") plays a beneficial role by absorbing most of the biologically damaging ultraviolet, called UV-B, sunlight. Without the filtering action of the ozone layer, more UV-B would penetrate the atmosphere and would reach the earth's surface.

The absorption of ultraviolet radiation by ozone creates a source of heat, and is the reason for temperature rise as one goes to higher altitudes. Ozone thus plays a key role in the temperature structure of the earth's atmosphere. On the other hand, surface-level tropospheric ozone (often called "bad ozone") is destructive to living systems when comes in direct contact. Studies show harmful effects of ozone on crop production, forest growth, and human health. Concern has been raised as ground-based and satellite-borne instruments measurement showed decreasing amount of stratospheric ozone, while increase in tropospheric ozone. Over some parts of Antarctica, up to 60% of the total overhead of ozone depleted during Antarctic spring (September–November). This phenomenon is known as the Antarctic ozone hole. In the Arctic polar regions, similar processes occur that have also led to significant chemical depletion of the column ozone during late winter and spring in seven out of the last 11 years. Smaller, but still significant, stratospheric decreases have been seen over other more populated regions of the earth. Increases in surface UV-B radiation have been observed in association with local decreases in stratospheric ozone. In addition, near-surface ozone is causing problem as a key component of photochemical "smog", especially in many cities. The scientific evidence from studies by the international research community has shown that human-produced chemicals are responsible for the observed depletions of the ozone layer. The ozone-depleting compounds, such as carbon tetrachloride, methyl chloroform which are commonly known as halocarbons, contain various combinations of the chemical elements chlorine, fluorine, bromine, carbon, and hydrogen. One example is the halons which contain carbon, bromine, fluorine, and (in some cases) chlorine, used as fire extinguishants. The other compound, known as chlorofluorocarbon contain only chlorine, fluorine, and carbon. These human-produced gases are used in many applications, such as, refrigeration, air conditioning, foam blowing, cleaning of electronics components, and as solvents. Upon usage, these float in the atmosphere up to the stratosphere and break down ozone into oxygen which then ionizes and escapes into space.

3.2 Atmospheric Brown Cloud (ABC)

Recently, another problem which is worsening the climate has been of much concern. It is the atmospheric brown cloud (ABC) which blocks solar radiation from reaching the earth surface by partly reflecting out and partly absorbing the sunlight. ABC is more pronounced in Asia – especially China, Arabian Peninsula, India, Bangladesh, Korea, Japan etc. where it is enhancing the global warming. ABC is the thick haze formed in humid condition, usually in winter when the monsoon is with no rainfall to wash the pollution. The ingredients are airborne particles and pollution due to coal powered industrial soot, biomass burning, vehicle emissions, burning of woods, dung, and crops. This pollution layer was detected during intensive field observation under the Indian Ocean Experiment (INODEX) in 1999. Later the United Nations Environment Program (UNEP) supported a project called ABC to study the pollution. The brown cloud that hangs over southern Asia could precipitate an environmental disaster that could affect billions of people. A recent CSIRO study found that the Asian brown cloud is also affecting rainfall in Australia. A new study shows that getting rid of it could actually help increase the rice harvests in the subcontinent. ABC is believed to have reduced sunlight, by about 20% since 1970s, by reflecting part of it back into space. This cools the surface and thus reduces the evaporation and causes less monsoon rainfall. It also absorbs the sunlight and thus raises solar atmospheric heating. Model suggests 50% temperature raised in the area is due to ABC (Ramanathan et al. 2007) and is believed to cause melting the Himalayan glaciers. However, the model included the data for solar heating with uncertainty of about a factor of four. The study of solar heating of the earth's atmosphere is crucial in precise modeling of its impact on the environment, and future plans. The following sections describe and discuss briefly our present day understanding, and how further study can lead to our projected aim.

3.3 Solar Irradiation of the Earth

Study of interaction of sunlight with atmospheric gases is an integral part of understanding global warming and future predictions. The sun brightens up and warm up the day as the earth spins every day, and causes seasons as the earth moves along its orbit. The earth is much smaller than the sun, its radius being 110 times smaller than that of the sun and only a small fraction of solar emission irradiates the earth. The sun, is however, an "unQuiet" star. During active period of its cycle of 11 years of maximum and minimum activity, it erupts with explosions that eject large amount of particles and radiation into space as solar flares which can affect the earth considerably. Features in solar activity are being elucidated through multi-wavelength spectroscopy using observations by space based satellites, such as, SOHO, Chanda x-ray Observatory, aided by detailed theoretical atomic and molecular calculations. The "Halloween" solar storm that occurred on October 28, 2003, was well documented by the space observatories Chandra and SOHO. Sun's active spots were detected especially by SOHO. Its mass spectrometer, LASCO, detected large coronal mass ejections around the sun, which is covered at the center by the detector. Eight hours after ejection from the sun LASCO was swarmed by the ejected particles in the form of a proton shower. Emission of x-rays in KeV energy range peaked in the radiation spectrum of solar flares. The emission bumps were found to be produced by He-like Ca, Fe, and Ni. Solar x-ray emission spectra may be analyzed by observing dielectronic satellite (DES) lines that form when an electron colliding with a two-electron He-like ion forms an excited 3-electron system. An atomic state with two electrons in excited levels is known as a doubly excited autoionizing state. An autoionizing state decays quickly by emitting a photon, which forms a DES line and a stable bound state. Study of DES lines (e.g. Nahar and Pradhan 2006) provides various diagnostics, such as temperature, density, and chemical abundances of the plasma surrounding the sun.

Materials, such as electrons, protons, and heavy ions ejected out into space by the solar explosions can be dangerous as these can damage tissue, break strands of DNA, and lead to diseases like cancer. The powerful electromagnetic pulses during a solar storm can also affect satellites and communications, and can even disrupt electrical service over long distances. However, earth holds protective shields around it. Its magnetic field channels these particles around the earth, funneling some of it to the poles to produce the most commonly noticed effect, the glowing aurorae. The upper layers of earth's atmosphere deflect and block part of radiation. For example ozone in stratosphere blocks most of ultraviolet, x-rays, and Gamma rays, and the lower atmosphere scatters and burns most of the incoming particles.

While the earth's atmosphere blocks most of the harmful radiation, it lets visible light, most radio waves, and some infrared light through, making it possible to study the universe with ground-based telescopes at these wavelengths. Atmospheric opacity gives the measure of radiation transport such that less opacity means more radiation can pass through. Most of the infrared light coming to earth is absorbed by water vapor and carbon dioxide in the earth's atmosphere. Observations of astronomical objects with radiation heavily attenuated by the atmosphere are carried out by space-based telescopes.

3.4 Solar Energy Distribution and Greenhouse Effect

The most important scientific input for climate change models is related to the greenhouse effect. Of the total solar radiation reaching the earth, 30% is reflected back to space (6% by air, 20% by clouds, and 4% by the surface of the earth),

19% is absorbed by atmosphere (16% by atmospheric gases, 3% by clouds) and 51% by the earth surface. Fifty-one percent of energy absorbed by the earth's surface is reradiated into the atmosphere (21% as infrared emission, 7% in sensible heat flux, 23% in latent heat flux). The atmosphere absorbs this 51% of long wavelength infrared radiation, plus the original 19% of short wavelength radiation absorbed directly from the sun, and then reradiates the total 70% back to space (6% radiated directly and 64% radiated from clouds and atmosphere). Hence, a stable relation is maintained between the atmospheric thermal structure and solar radiation.

In the energy cycle of the greenhouse effect, part of the energy is trapped to heat the earth surface. Direct overhead sunlight at the top of the atmosphere provides 1,366 W/m^2. Due to geometric effects and reflective surfaces only 235 W/m^2 is absorbed, 67 W/m^2 by air and 168 W/m^2 by land and water. Earth's surface temperature is raised only to −18 C by 168 W/m^2. However, the greenhouse gases absorb the outgoing and reflected radiation, and thus trapping more radiation in the atmosphere and subsequently reflecting it back to the lower layers of the atmosphere closer to the surface which raises the earth's temperature. The gases absorb 452 W/m^2 thermal infrared radiation emitted by the earth's surface. Of the total energy 519 W/m^2 (= 67 + 452) it delivers 324 W/m^2 (62%) back to earth and transmits the remainder 195 W/m^2 (38%) to space. The total energy 492 W/m^2 (168 from sunlight +324 from atmosphere) raises earth's surface temperature to +14°C. This energy recycling process, that is the greenhouse effect, is an essential piece of the earth's climate. Greenhouse effect has maintained abundant liquid water for the earth's habitants. A more detailed study of the effect requires one to know how atmospheric atoms and molecules interact with radiation, discussed next.

3.5 Atomic and Molecular Processes of Photo-Absorption, Photon-Emission, and Atmospheric Opacity

The macroscopic phenomena due to solar irradiance of the earth are related to a number of atomic and molecular processes involving light and atmospheric gases which absorb and emit radiation at different wavelengths, that is, photons of different energy. Followings are the processes of absorption and emission of a photon. A photon is described by hν where ν is the photon frequency and h is the Planck's constant. Its interaction with an atomic or molecular species, X^{+Z} of charge Z, can increase the charge with loss of an electron.

1. Photoexcitation – an electron in an atomic or molecular system, X^{+Z}, absorbs the photon and jumps to a higher excited state while remaining in the atomic or molecular system:

$$X^{+Z} + h\nu \rightarrow X^{+Z*}$$

   The asterisk (*) denotes an excited state.

2. De-excitation (inverse of excitation) – an electron in an excited atomic or molecular state emits energy in the form of a photon and drops down to the ground level:

$$X^{+Z^*} \rightarrow X^{+Z} + h\nu$$

3. Photoionization/photo-dissociation/photo-electric effect – an electron absorbs a photon and exits the atom/molecule:

$$X^{+Z} + h\nu \rightarrow X^{+Z+1} + e$$

4. Electron-ion recombination (inverse of photoionization) – a free electron loses its energy as a photon and combines to an ion:

$$X^{+Z} + e \rightarrow X^{+Z+1} + h\nu$$

5. Collisional excitation – an atom or a molecules goes to an excited state by the impact of a free electron, and decays by emitting a photon.

$$X^{+Z} + e \rightarrow X^{+Z^*} + e'$$

The first two processes are confined to the atomic system and hence emit or absorb photons of particular energy related to quantized atomic levels. The parameters of interest relevant to the processes are called oscillator strengths (f-values) and radiative decay rates (A-values) respectively. They produce only lines in the spectrum. However, the other three processes show characteristic features over a range of energies in their cross sections. Study of these features in cross sections are crucial for any precise modeling. For example, sun's ultraviolet radiation breaks down O_2 and N_2 molecules into atoms, and then photo-ionizes them in the ionosphere. Nahar (1998) and Nahar and Pradhan (1997) carried out calculations for photoionization cross section (σ_{PI}) with photon energy which showed that the ionization does not vary smoothly with energy, but goes through resonances at different energies. The resonant peaks indicate huge enhancement of ionization at particular energies (Fig. 3.1).

Nahar (1995, 1996) studied the characteristic patterns for electron combining with a doubly ionized (two electrons stripped off) sulfur forming a singly ionized sulfur. Her work showed that at low temperature the recombination rate is very high, and goes down with higher temperatures since high velocity electrons have less time to combine. However, when the temperature is high enough to form resonant states, the recombination increases, and forms a peak which is followed by a smooth decay (Fig. 3.2).

Due to various processes between the gases and sunlight, earth's atmospheric opacity spectrum shows various windows of radiation transmission reaching the earth. For example, (i) oxygen molecules, O_3 and O_2, and atomic oxygen (O) absorb ultraviolet radiation, (ii) water and oxygen absorb some infrared, microwave and radio frequencies, and (iii) carbon dioxide absorbs infrared frequency photons, Determination of atmospheric opacity, which provides the measures of radiation transfer, depends on the above processes. For example, monochromatic opacity (for a particular photon frequency), k_ν, depends on the photo-excitation parameter, f_{ij}, *oscillator strength* for transition from level i to level j,

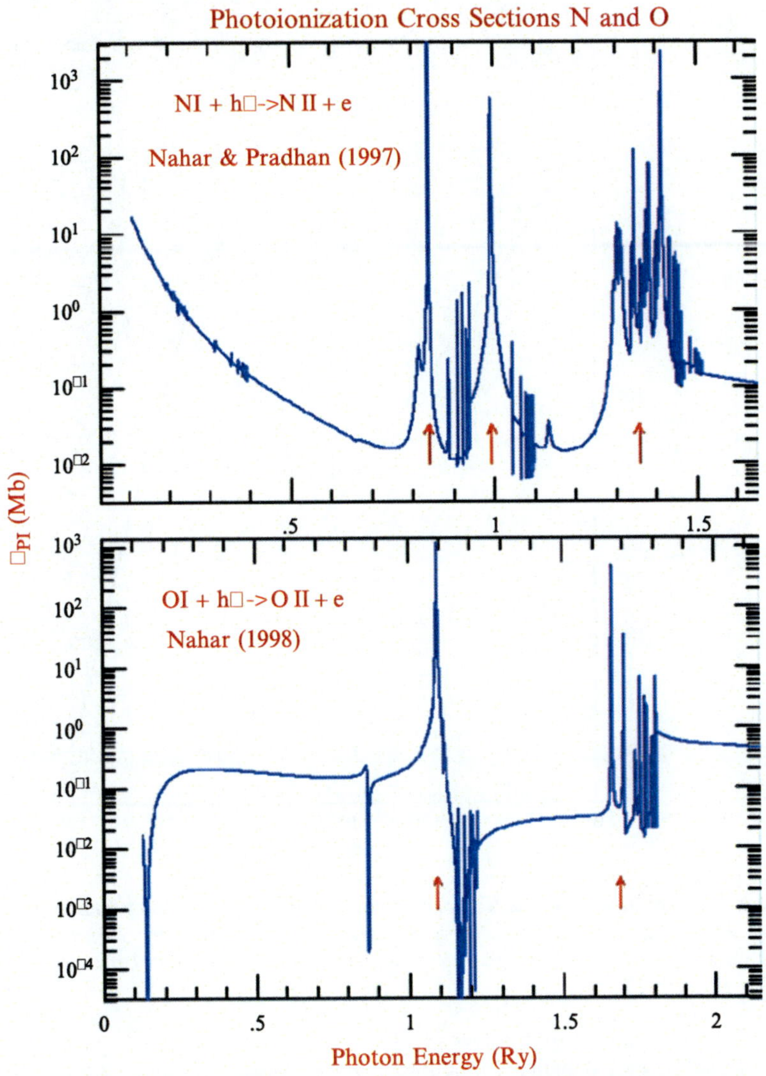

Fig. 3.1 Photoionization cross sections (sPI) of oxygen and nitrogen (Nahar 1998; Nahar and Pradhan 1997). At energies (pointed by *arrows*) the resonance peaks indicate enhancement of ionization of the atom by orders of magnitude

$$k_v(i \rightarrow j) = [\pi e^2/mc] N_i f_{ij} \varphi$$

where N_i = ion density in state i, φ is a profile factor, and the rest are constants, and on photoionization cross sections σ_{PI}

$$K_v = N_i \sigma_{PI}$$

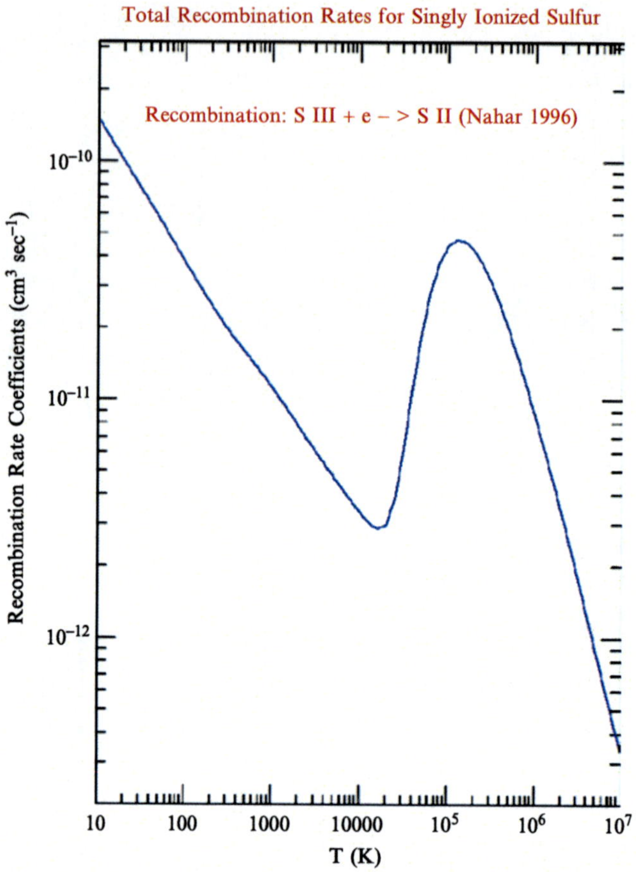

Fig. 3.2 Total recombination rate for singly ionized sulfur, S II, from doubly ionized sulfur S III (Nahar and Pradhan 2006). The recombination rates shows that starting high at very low temperature it decreases until at a higher temperature where it peaks due to resonances in the process

The total opacity depends on the interaction of radiation with all atoms and molecules in the atmosphere. Complete atmospheric modeling requires opacities and parameters of all other processes. Two international projects have undertaken large-scale computations for precise understanding of astronomical objects, the Opacity Project (1995, 1996), and the Iron Project (Hummer et al. 1993), involving collaborators from the US (Ohio State U, NASA-Goddard, Rollins), U.K., France, Germany, Belgium, Venezuela, and Canada. These two projects entail all radiative and collisional processes described above using an ab initio R-matrix method. A huge amount of accurate atomic data for many atoms and ions from hydrogen to nickel are available through databases: TOPbase, TIPbase at CDS (France) at http://vizier.u-strasbg.fr/topbase/topbase.html, Ohio Supercomouter Center at http://opacities.osc.edu, and NORAD-Atomic-Data at: www.astronomy.ohio-state.edu/~nahar/nahar_radiativeatomicdata/index.html. These resources can also be utilized for atmospheric models.

A detailed solar spectrum at the earth's surface after attenuation by the chemical constituents of the atmosphere and without consideration of change of change in greenhouse gases (e.g. R. Kurucz, CfA, Harvard University, kukurucz.harvard.edu/sun/fluxatlas) shows that the lines corresponding to various photon absorptions are closely lying for ultra-violet photons indicating high absorption of the radiation and less dense for optical radiation indicating less absorption. Yellow is absorbed minimum (reason for the Sun to look yellow). Absorption of lines increases in the infrared range due to absorption mostly by water. The best calculations for H_2O opacity in atmosphere were done using over 800M transitions.

With changes in the atmospheric composition the spectrum will obviously change. It may be noted that some bands that are saturated (i.e. 100% of radiation in that band is absorbed) can be affected by further increases in greenhouse gas concentrations. The greenhouse gases will cause the radiation to be captured closer to the earth's surface and increase the temperature. Detailed and accurate study of atmospheric opacities will therefore provide more accurate radiation transfer with varying components, solar heating of the earth, and more precise predictions of global warming and climate change. Such calculations will require high-performance large-scale atomic and molecular calculations, as, for example, carried out at the Ohio Supercomputer Center in Columbus, Ohio.

Finally, we mention a few space observatories engaged in atmospheric research. NASA, along with space agencies from other countries, has a fleet of satellites known as the A-Train to study the atmosphere affecting the global temperature. Satellite Aqua collects information on earth's water cycle, Cloudsat studies clouds for their role in regulating the climate, Calipso studies the way aerosols interact with clouds, Parasol can distinguish natural from human-produced aerosols and measures polarized light, Glory measures the energy budget of the earth and determines the global temperature, and Aura maps global pollution. NASA's Orbiting Carbon Observatory (OCO) was launched to pinpoint key locations on the earth's surface where CO_2 is being emitted and absorbed, but was lost in space while setting in its orbit. However, Japan's Gosat is carrying out similar observations on global greenhouse effects through spectroscopy of the sunlight reflected off the earth's surface, in which the constituent colors determine how much carbon dioxide and molecular oxygen are present. Two carbon observatories, A-SCOPE (Advanced Space Carbon and Climate Observation of Planet Earth) and BIOMASS, are planned to be launched from Europe in 2016.

3.6 Conclusion

- The sun is the main source of energy for living conditions that depend on the thermal structure in the atmosphere.
- The relation between solar irradiation and earth's atmosphere needs to be understood; atomic and molecular processes are inherent to atmospheric models. Numerical simulation requires complex quantum-mechanical calculations for atomic and molecular processes using high-performance computing platforms.

- Large number of atomic parameters for radiative processes in the atmosphere are needed. A concerted MULTI-DISCIPLINARY effort is crucial in solving the problem of global warming and protect our home planet.
- PLAN: Calculation of accurate atmospheric opacities and models.

Acknowledgements Partially supported by a NASA grant.

References

Hummer DG, Berrington KA, Eissner W, Pradhan Anil K, Saraph HE, Tully JA (1993) Atomic data from the IRON project. 1: goals and methods. Astron Astrophys 279:298–309

Nahar SN (1995, 1996) Electron-ion recombination rate coefficients for Si I, Si II, S II, S III, C II and C-like ions C I, N II, O III, F IV, Ne V, Na VI, Mg VII, Al VIII, Si IX, and S XI. Astrophys Suppl J 101:423–434, 106:213–214

Nahar SN (1998) Photoionization cross sections and oscillator strengths for oxygen ions: O i-O vii. Phys Rev A 58:3766–3782

Nahar SN, Pradhan AK (1997) Electron-ion recombination rate coefficients, photoionization cross sections, and ionization fractions for astrophysically abundant elements I. carbon and nitrogen. Astrophys J Suppl 111:339–355

Nahar SN, Pradhan AK (2006) Dielectronic satellite spectra of helium like iron and nickel from the unified recombination method. Phys Rev A. 73:062718-1-8

Ramanathan V, Ramana MV et al (2007) Warming trends in Asia amplified by brown cloud solar absorption. Nature 448:575–578

The Opacity Project Team (1995, 1996) The Opacity Project, vol 1 & 2. Institute of Physics, London

Chapter 4
Mass Balance Approach to Study Glaciers Dynamics in the Himalayas

Dagfinnur Sveinbjörnsson and Helgi Björnsson

Contents

Abstract There is a strong evidence that the glaciers of the Himalayas are in the process of a rapid meltdown. The implications of this development will be broad and sweeping for the river systems and more generally for the water resources of South Asia. The consequences will be grave for food production and livelihood of hundreds of millions of people. Despite this alarming forecast about the glaciers in the Himalayas, in particular on the southern site of the mountain range, these glaciers have not been researched to a serious degree and are in need of in-depth research initiative. Among numerous aspects of the Himalayan glaciers which have not been studied to the extent they should is the mass balance. This chapter briefly surveys the literature on the state of the glaciers on the Indian Himalayas, outlines in general terms, the method of mass balance modeling as it has been applied to glaciers in Iceland, and finally, suggests ways in which the modeling procedure can be applied in the Himalayas.

Keywords Modeling glacier dynamics • Himalayas • South Asian river systems • Glacier retreat • Downwasting

D. Sveinbjörnsson (✉)
e-mail: dagfinnur@gmail.com

H. Björnsso
Institute of Earth Sciences, University of Iceland, Iceland

R. Lal et al. (eds.), *Climate Change and Food Security in South Asia*,
DOI 10.1007/978-90-481-9516-9_4, © Springer Science+Business Media B.V. 2011

Abbreviations

ISRO	Indian Space Research Organization
ICSI	International Commission for Snow and Ice's Working Group
GLOFs	Glacial lake outburst floods
GSI	Geological Survey of India

4.1 The Himalayan Glaciers

The Himalayan glaciers are the least studied and least known glaciers in the world. Yet, in a comparative perspective, these glaciers are extremely vital for the livelihood of a large number of people as a unique reservoir that feeds more than a dozen major rivers. Important among these are the mighty perennial rivers of Ganges, Indus and Brahmaputra which sustain hundreds of millions of people downstream. Therefore, it is a matter for great concern, which is widely recognized, that the Himalayan glaciers are prone to a process glaciologists refer to as downwasting, i.e., the melting, receding and ultimate withering of glaciers (Figs. 4.1–4.3).

Several studies indicate that downwasting is taking place with a vengeance in the Himalayas. The Indian Space Research Organization (ISRO) has recently, based on

Fig. 4.1 Example of a lowland glacier, Iceland

Fig. 4.2 Example of a lowland glacier, Iceland

Fig. 4.3 Example of a lowland glacier, Iceland

the data of satellite imaging, discerned the changes in 466 glaciers. The ISRO observed more than 20% reduction in size between 1962 and 2001. As bigger glaciers disintegrate into smaller pieces, each piece retreats faster than its parent. This trend has been evident for quite some time.

Direct observations of a select few snout positions among the thousands of Himalayan glaciers indicate that they have been in a general state of decline and retreat since 1850 (Mayweski and Jaschke 1979). However, the cause for alarm is the present very rapidly accelerating rate of retreat. Studies by Jangpang and Vohra (1962), Kurien and Munshi (1972), Srikanta and Pandhi (1972), Vohra (1981), and many others about the snout fluctuation of the Himalayan glaciers since the early 1960s indicate dramatic increase in the rate of retreat. By studying the longitudinal profiles of the river Ganges, Owen and Sharma (1998) reported that between 1971 and 1996 (Fig. 4.4), the Gangotri Glacier retreated by about 850 m. This study indicates a post 1971 retreat rate of 34 m/year. For the 61 years (1935–1996), data of the Geological Survey of India (GSI) also shows that the retreat rate is about 28 m/year. These studies clearly indicate an increase in the rate since 1971 (Kumar 2005). Similar trends were reported by Naithani et al. (2001).

Similar to Gangotri, the Chorabari glacier has also retreated at the rate of 9 m/year for the 3 year period since 2006. Although, 3 years may be too short a period to draw firm scientific conclusions about the health of the glacier, the study also confirms a disquieting pattern of glacial retreat across the Himalayas. In a separate

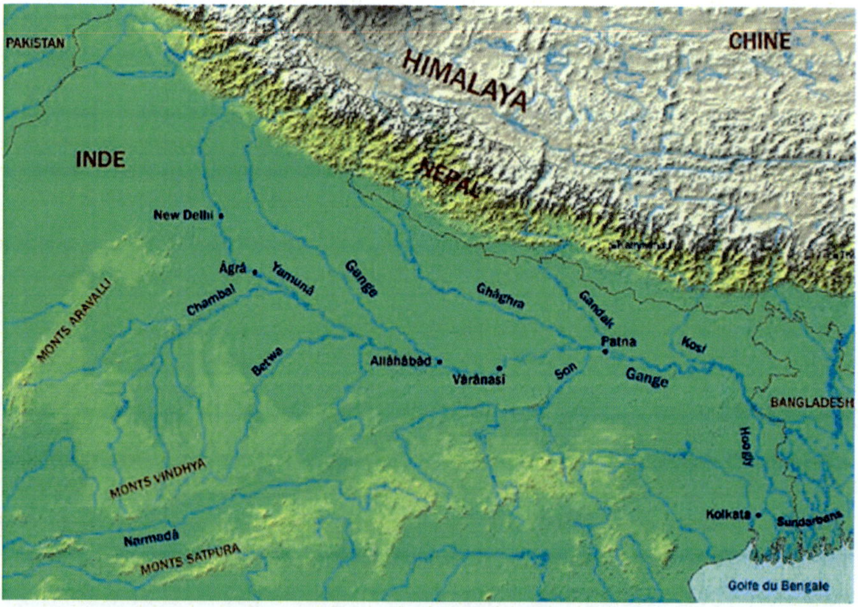

Fig. 4.4 Map showing the course of the Ganges and selected tributaries

study on the Parbati glacier, one of the largest in the area, it was found to be retreating by 52 m/year during the 1990s. Another glacier, named Dokriani, lost 20% of its size in three decades and between 1991 and 1995 its snout retreated back almost 17 m/year. Thus, there is an ample evidence that the glaciers in the Himalayas are in the process of a rapid meltdown (Fig. 4.4).

4.2 Consequences

The Intergovernmental Panel on Climate Change (IPPC 2007) reported that the likely consequences of these trends are, in the most general terms, an increased likelihood of flooding over the next three decades and then, as glaciers further recede, extensive drying up of the rivers they feed (Parry 2007, p. 13). This will in turn bear heavily on food security in South Asia, as it will have drastic consequences for the regions drinking water supply and irrigation for agricultural production. Changes in water supply will also adversely affect a variety of ecological systems including those especially consequential for the spreading of diseases. A range of industries will also be adversely affected. In the Ganges alone, the loss of glacier melt water would reduce July–September flows by two-thirds, causing water shortages for over 500 million people and 37% of India's irrigated land (Singh et al. 1994). This could readily translate into malnutrition, food insecurity and risks of famine.

Another reason for immediate concern is the danger of so-called glacial lake outburst floods (GLOFs): catastrophic discharges of water resulting primarily from melting glaciers. Observations indicate that the frequency of GLOF's in the Himalayas increased during the second half of the twentieth century and have cost lives, property and infrastructure in India, Nepal and China. For example, the catastrophic flooding of Bihar in August of 2008 was in all likelihood a consequence of a glacial outburst flood in the mountains. The data gathering efforts required for a scientific study of the glaciers in the Himalayas therefore, needs to be supplemented with a systematic evaluation of the livelihood of the population. What areas will be most affected by the change in river flow, floods and droughts? Where are ice-dammed lakes most likely to form and cause threats of abrupt outburst floods, with dire consequences for populations living downstream?

The consequences of melting glaciers are indeed multi-dimensional and serious. A response on behalf of governments, as well as regional and local institutions, is deemed necessary. For such a response to become informed and effective, it is necessary to increase and improve research on the glaciers dynamics. Among the many fields of scientific engagement which needs to be improved is the so-called mass balance modeling, relating glacier mass balance to meteorological parameters and thereby to climate. This approach or method can be easily used in the Himalayas.

4.3 Mass Balance Modeling: An Illustration from Iceland

Glaciers cover 11% of the surface area of Iceland, which amounts to 20 years of precipitation. These glaciers are also rapidly shrinking. Glaciers in Iceland are accessible and lend themselves particularly well to efforts to gather data, which in turn makes it possible to engage in modeling of their development over time. The basic data required for an analysis of glacial evolution are: (1) glacier geometry: surface and bedrock, (2) glacier mass balance, (3) ice dynamics, (4) glacio-meteorology, and (5) drainage of melt-water. With data on these parameters in hand it becomes possible to model the mass balance, the ice dynamics and, thus the response of the glacier to climate scenarios.

The modeling process is explained by a series of figures presented herein. Figure 4.5 shows the topography of the glacial distribution in Iceland. Figure 4.6 illustrates in broad and general terms the fundamental elements of the analysis of glacial response to climate scenarios, namely the path from field observations and modeling to predictions given specific climate scenarios. Figure 4.7 illustrates the temporal variations in the mass balance of Vatnajökull and Langjökull while Fig. 4.8 demonstrates the accelerating volume loss of those two glaciers. Figure 4.9 shows the predicted climate scenario for the Icelandic highlands. Figure 4.10 illustrates the model response for three glaciers, namely Vatnajökull, Hofsjökull and Langjökull. Finally, Fig. 4.11 is a computer generated picture of the likely evolution of Hofsjökull until the end of the twenty-second century.

The analysis presented above indicates that the glaciers in Iceland may disappear within 100–200 years, and that the discharge from the glaciers will increase over the next 50 years and then decrease. It is also possible to outline how the seasonal rhythm in discharge will change and some rivers may disappear. Eventually, river discharge may consist primarily of precipitation. The sediment load of rivers will also drastically change.

Fig. 4.5 Topography displaying the glacier distribution of Iceland. The glaciers cover 11% of Iceland and the volume of ice equals 20 years of precipitation on Iceland (Björnsson and Pálsson 2008)

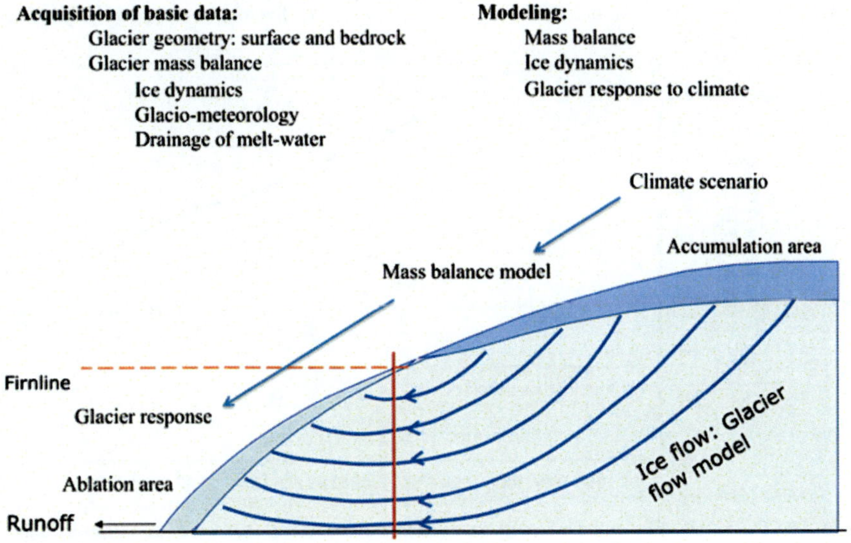

Fig. 4.6 Analysis of glacial response to climate scenarios

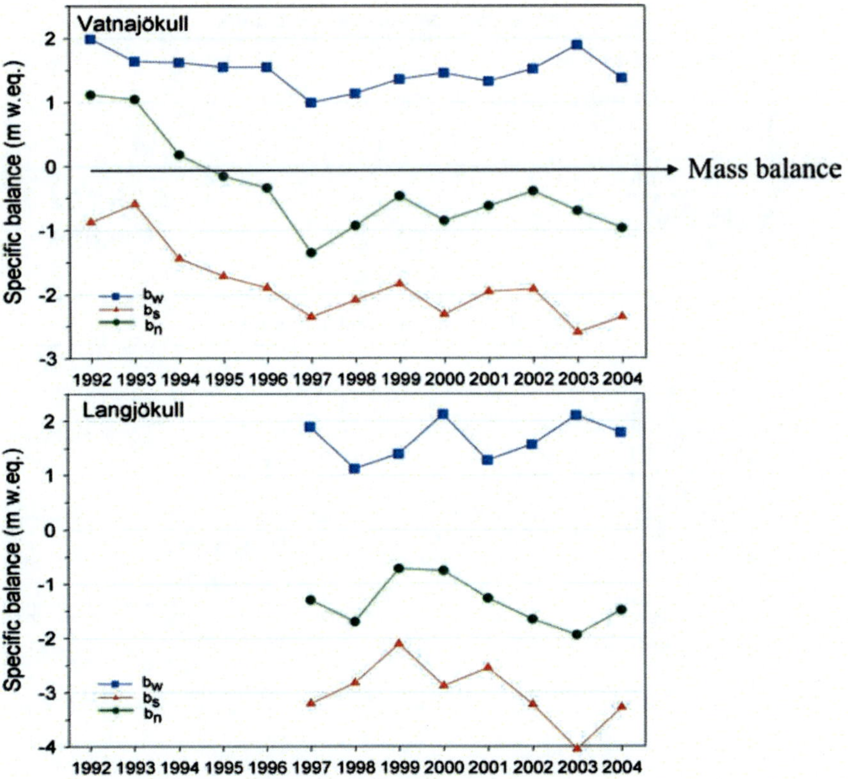

Fig. 4.7 Temporal variations in Vatnajökull and Langjökull mass balance (for glaciological years 1991/92 to 2005/06). Specific mass balance is given in m water equivalent, where b_w stands for winter balance, b_s summer balance and b_n annual net balance (Björnsson and Pálsson 2008)

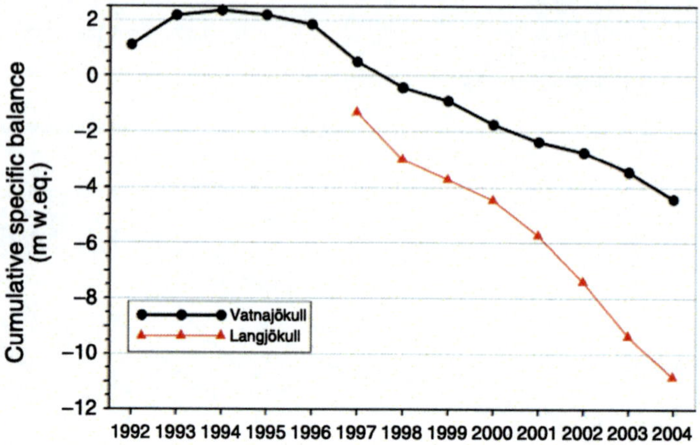

Fig. 4.8 The accelerating volume loss of Vatnajökull andLangjökull (Björnsson and Pálsson 2008)

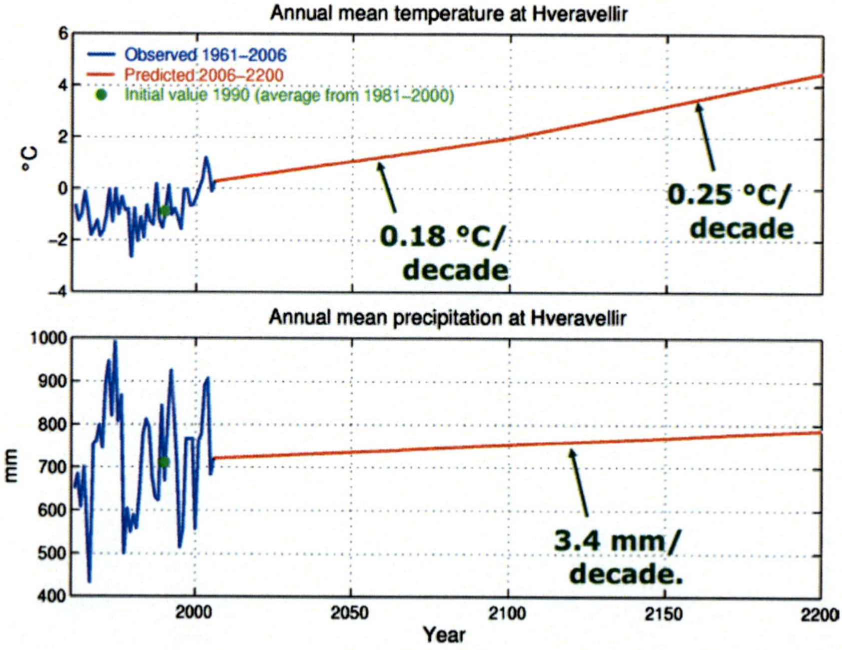

Fig. 4.9 Climate change scenario for the Icelandic highland (Jóhannesson et al. 2007)

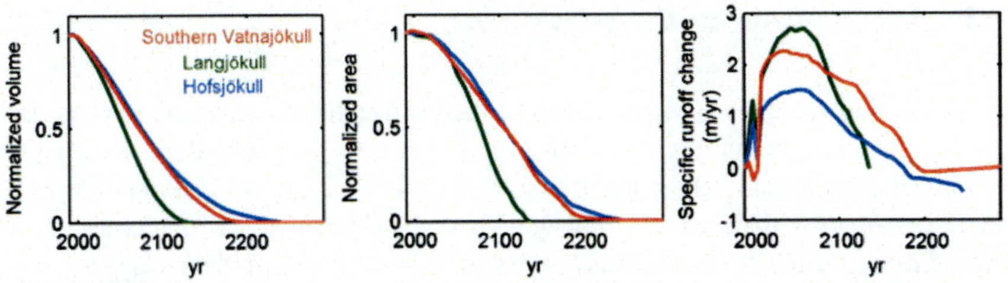

Fig. 4.10 Model responses shown for three Icelandic glaciers: Vatnajökull, Langjökull and Hofsjökull. Volumes and areas are normalized to present day values and specific runoff is from the present day glacier covered area (Björnsson and Pálsson 2008)

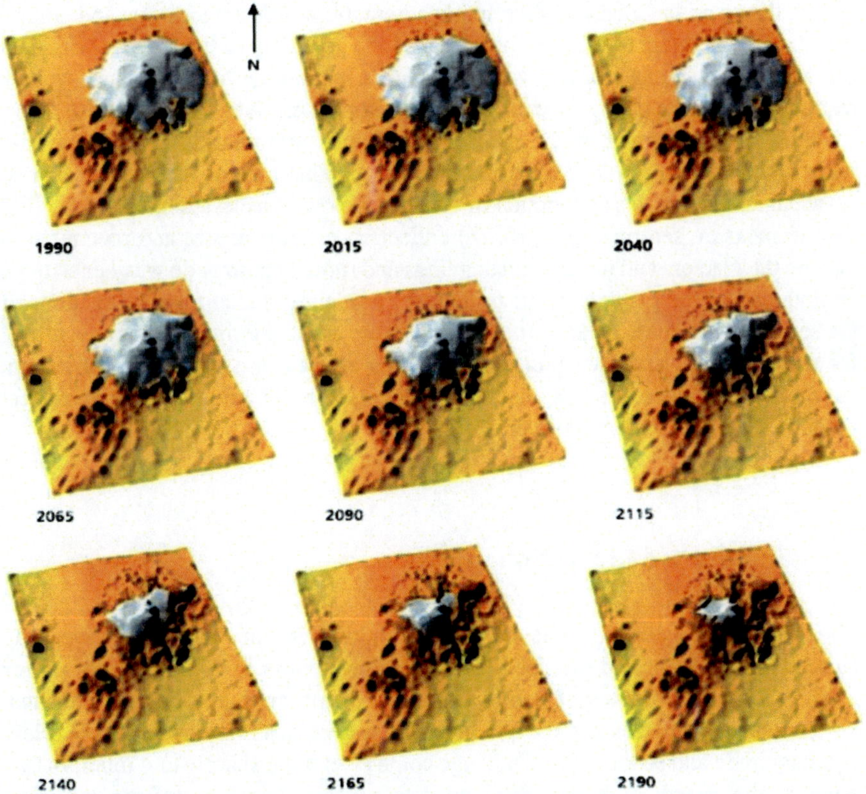

Fig. 4.11 The likely evolution of Hofsjökull until the end of the twenty-second century given the predicted climate scenario (Björnsson 2009)

4.4 Applications in the Himalayas

For an application of the above modeling method in the Himalayas, it will be necessary to map the territory; to obtain an up-to-date overview of the present distribution of glaciers and classify them according to elevation, range and size, and describe the larger trends in their rate of shrinkage. These data lead to an informed selection of particular target ice fields that may serve as representative of their category. This entails chiefly: (a) collection of information on geometry, in particular elevation range, and (b) systematic collection of satellite images of Himalayan glaciers.

When the territory has been mapped it is then possible to embark upon the collection of mass balance data. Information is required about the present mass balance of the target glaciers as a function of elevation and its relation to climate variables. This requires improving upon the existing data gathering of the annual mass balance on the glaciers and running automatic weather stations on and around target glaciers. Integrating the information on the geometry of the glaciers with data on mass balance and climatic variables makes it possible to model the glacier's response to climate change scenarios and the impact on river run-off.

Work on this aspect of the project entails the combination of existing data with future annual mass balance measurements. This part of the project entails: (a) accumulation rates as a function of elevation, snow line elevation, use of satellite data to measure seasonal changes, (b) collection of temperature and precipitation data on the glaciers and in their surroundings and running automatic weather stations on and around the glaciers, (c) mass balance modeling including investigations of the sensitivity of mass balance to air temperature and the tuning of a degree model for mass balance, and factoring in the shading effect of a steep mountain terrain, (d) modeling of glacier's response to climate change scenarios through numerical modeling, and (e) impact on river runoff.

4.5 Imperative for Action

The far-reaching implications of glacial melting in the Himalaya have only slowly simmered in and the need for greater awareness in the making of policy as well as the importance of public action have only gradually been established. As awareness increases and the need for action becomes imperative the complexity, magnitude and scope of the challenge come even more sharply into relief. In that context, the dearth of knowledge and data on this theme is nothing less than alarming. For example, river flow data is so scantly and recent that scientists cannot predict how the way in which the glaciers currently retreat will affect the river volume and flow. There is, therefore, an urgent need for research to establish scientific knowledge. In the absence of a serious effort by the scientific community an informed and constructive reaction by the governments and people of South Asia will be impossible.

4.6 Conclusions

The above discussion has two main elements. On the one hand it entails a general survey of the state of knowledge on the Himalayan glaciers. On the other hand it presents an outline of the way in which the mass balance of the glaciers in Iceland has been analyzed. Rigorous methods of geophysics have made it possible to conclude with reasonable certainty that the glaciers of Iceland are melting at an alarming rate and will all but disappear given the predicted scenario of climate warming within 200 years. Evidence on the glaciers in the Himalayas is quite scarce and at some level anecdotal. However, the studies surveyed above indicate in no uncertain terms that the glaciers in the Himalaya are receding at quite an alarming rate.

It is clear that planning and preparation for the consequences of imminent melting of the glaciers in the Himalaya will be impossible in the absence of a more serious and intense scientific analysis of the evolution of the glaciers. Therefore, it stands to reason to make efforts to apply in the Himalaya the method of mass balance analysis as applied to the glaciers of Iceland. This would offer an interesting opportunity for an international scientific collaboration.

References

Björnsson H, Pálsson F (2008) Icelandic glaciers. Jökull 58:365–386

Björnsson H (2009) Jöklar á Íslandi. Opna Reykjavík

IPPC (2007) Climate change 2007: physical science basis. Cambridge University Press, Cambridge

Jangpang BS, Vohra CP (1962) The retreat of the Skunkulpa (Ralam) glacier in the Central Himalaya. Pithoragarh District, UP, India. Inst Assoc Sci Hydrol Pupl 58:234–238

Jóhannesson TG, Aðalgeirsdóttir H, Björnsson P, Crochet et al (2007) Effect of climate change on hydrology and hydro-resources in Iceland. Final report of the VO-project OS-2007/011

Kumar R (2005) Status review of possible impacts of climate change on Himalaya glaciers, glacier retreat and its subsequent impacts on fresh water regime. HIGHICE-India glacier research group, SES, JNU, New Delhi

Kurien MN, Munshi MN (1972) A survey of Sonapani Glacier, Lahaul District, Punjab. Geol Surv India Misc Publ 15:83–88

Mayweski P, Jaschke PA (1979) Himalaya and Trans Himalayan glacier fluctuations since A.D. 1812. Arctic Alpine Res 11(3):267–287

Naithani AK, Nainwal HV, Sati KK, Parsad C (2001) Geomorphological evidences of retreat of the Gangotri glacier and its characteristics. Curr Sci 80(1):87–94

Owen LA, Sharma MC (1998) Rates and magnitudes of paraglacial fan formation in the Garhwal Himalaya: implications for landscape evolution. Geomorphology 26:171–184

Parry ML (2007) Summary for policy makers. In: Parry ML, Canziani OF, Palutikof JP, van der Linden PJ, Hanson CE (eds) Climate Change 2007: impacts, adaptation and vulnerability: contribution of Working Group II to the fourth assessment report of the Intergovernmental Panel on Climate Change. Cambridge University Press, Cambridge

Singh P, Jain SK, Kumar N, Singh UK (1994a) Snow and glacier contribution in the chenab river at Akhnoor, CS (AR)-131. National Institute of Hydrology, Roorkee, India

Singh P, Jain SK, Kumar N, Singh UK (1994b) Snow and glacier contribution in the Ganga River at Devprayag, CS (AR)-132. National Institute of Hydrology, Roorkee, India

Srikanta SV, Pandhi RN (1972) Recession of the Bada Shigri glacier. Geolog Surv India 15:97–100

Vohra CP (1981) In: Lall JS, Moddie AD (eds) Himalayan glaciers in the Himalaya: aspects of change. Oxford University Press, New Delhi, pp 138–151

Part II
Climate Change and Water Resources

Chapter 5
Understanding Surface Water Flow and Storage Changes Using Satellites: Emerging Opportunities for Bangladesh

Faisal Hossain and Douglas Alsdorf

Contents

Abstract This paper overviews the monitoring of surface water flow and storage using satellites. The overview is cast in the context of surface water-related problems of Bangladesh and South Asia. The paper then provides a basic introduction of a planned space-borne mission for surface water called SWOT (Surface Water and Ocean Topography) mission suggested for launch in 2015. The opportunities offered by SWOT for enhancing the capacity for flood hazards monitoring and adaptation to climate change for Bangladesh are also overviewed.

Keywords Bangladesh • Space-borne discharge • SWOT • Surface water • Floods • Climate change

Abbreviations

GBM Ganges, Brahmaputra, Meghna
FFWC Flood Forecasting and Warming Center

F. Hossain (✉)
Department of Civil and Environmental Engineering, Tennessee Technological University, Cookeville, TN 38505-0001
e-mail: fhossain@tntech.edu

D. Alsdorf
School of Earth Sciences, The Ohio State University, Columbus, OH 43210
e-mail: alsdorf.1@osu.edu

R. Lal et al. (eds.), *Climate Change and Food Security in South Asia*,
DOI 10.1007/978-90-481-9516-9_5, © Springer Science+Business Media B.V. 2011

IWM	Institute of Water Modeling
GIS	Geographic Information System
SWOT	Surface Water and Ocean Topography
WSOA	Wide Swath Ocean Altimeter
KaRin	Ka band Radar Interferometer

5.1 Introduction

Bangladesh is largely a riverine delta situated at the most downstream end of three large river basins – the Ganges, the Brahmaputra and the Meghna, i.e. the GBM basin (Fig. 5.1; Paudyal 2002). The flood plains of these three international rivers, together with smaller rivers and streams, account for about 80% of the area of Bangladesh (Hofer 1998). Yet only around 7–8% of the total drainage area of the GBM basins is situated inside the boundaries of Bangladesh.

Problems related to surface water availability are very widespread in Bangladesh. In general, the geographic location and average land elevation of Bangladesh are conducive to the four major water related problems: (1) Flood; (2) Erosion; (3) Drought and (4) Storm Surges. For about 7 months of the year (the non-Monsoon period spanning October–May), drought, exacerbated by the impoundment of the Ganges river in upstream India, creates acute water shortage in the Western regions of Bangladesh (Fig. 5.2). Estimates also indicate that about 3,000 km of the river banks will have been

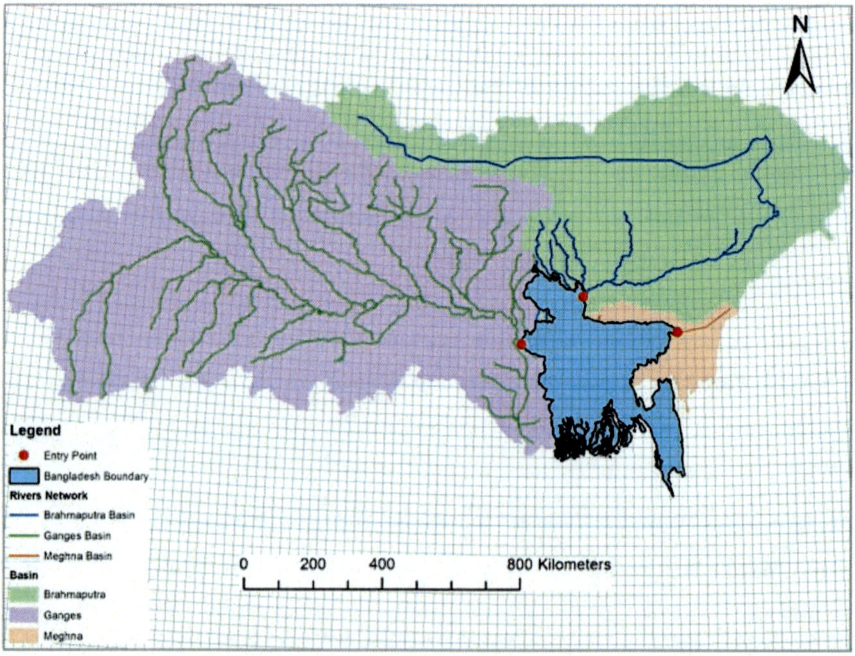

Fig. 5.1 Map produced by Institute of Water Modeling (IWM), Bangladesh

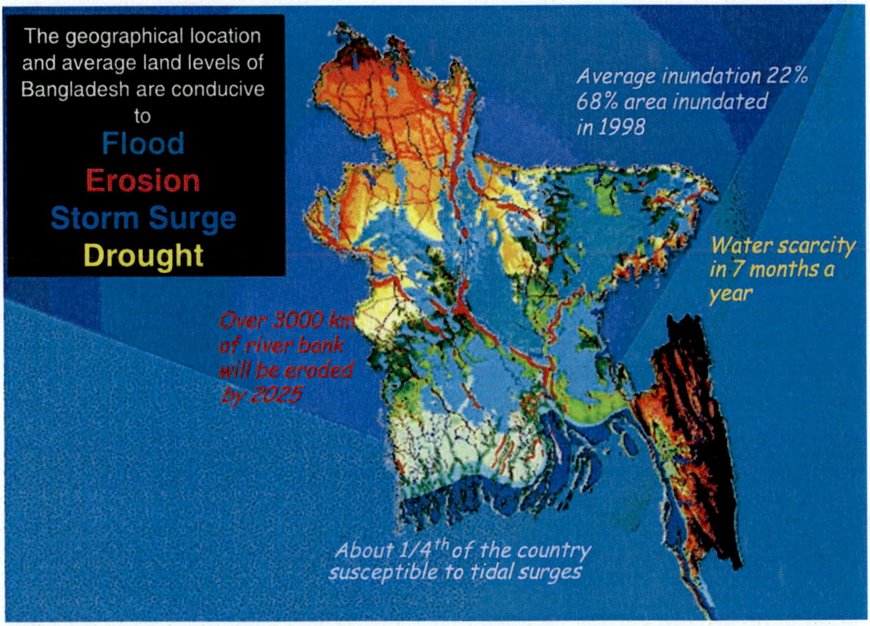

Fig. 5.2 Overview of surface water related environmental hazards of Bangladesh. (Map courtesy of Institute of Water Modeling, Bangladesh)

eroded by 2025, while in any given year about 22% of the total land area is usually inundated by Monsoon-driven flooding by the major rivers. For extreme wet years, this inundation can swell to 68% of the total land area as has been witnessed during the recent catastrophic floods of 1998 (Chowdhury 2000). Finally, about one-fourth of the country is susceptible to tidal surges from cyclones (Fig. 5.2).

Because many of the water-related problems are 'routine', Bangladesh government has invested in a monitoring network and a forecasting system. The most notable system in this regard is the one designed for floods, called the 'FLOOD WATCH', developed and operated by the Flood Forecasting and Warning Center (FFWC) of the Bangladesh Government Ministry of Water Resources. FLOOD WATCH is currently operated in real-time by the FFWC and believed to be the world's largest flood forecasting model (4,200 grid points) presently in operation (Kjelds and Jorgensen 1997). Technical support for system modifications and upgrades are continuously provided by the Institute of Water Modeling (IWM) of Bangladesh. Every day during most of the monsoon season the model simulates the water level conditions during the previous 7 days (hind-cast simulations) and during the coming 3 days (forecast simulation). These forecasted levels are then converted for decision support to either one of the five outputs for public dissemination (e.g., internet: www.ffwc.gov.bd; radio, TV and news dailies) and for disaster management (e.g., Prime Minister's Office and Disaster Management Bureau). These five outputs are: (1) normal (river stage expected to remain below the danger level); (2) warning (river stage projected to threaten the danger level);

(3) danger (river stage projected to exceed the danger level); (4) severe (over bank flooding in progress) (5) no data. The danger level at a river location is the level above which it is likely that the flood may cause damages to nearby crops and homesteads. Currently, FLOOD WATCH issues flood warnings for 30 river stations in the country using in-situ rainfall information from 84 stations (see Fig. 5.3).

Despite the significant reduction of the average number of deaths associated with catastrophic floods after the implementation of FLOOD WATCH in 1995 (flood related deaths in 1988 were 2379, compared to 918 in 1998 floods; Ninno et al. 2001; Chowdhury 2000), a significant amount of annual damage of lives and property by floods is nevertheless endemic in Bangladesh. Two challenging factors identified in this regard suggest that an even better decision support or monitoring with satellite data might be possible. These are: (1) increasing the lead time of river flow forecast beyond 3 days by early estimation of surface flow conditions further upstream of Bangladesh within India, Nepal and Bhutan (Hossain and Katiyar 2006); and (2) reducing high operational costs of daily in-situ rainfall measurements to maintain long-term sustainability of FLOOD WATCH. An increase in lead time has potential significance for reducing the country's agricultural vulnerability to flooding hazards. For example, 7- to 10-day forecasts are much more useful than daily forecasts in agricultural decision support as they inform farmers of the potential benefits of delayed sowing or early reaping of crops, while a 21-day forecast is considered most ideal for South and Southeast Asian nations (Asian Disaster Preparedness Center – ADPC 2002).

However, basin level hydrological modeling of large international river catchments is gradually becoming a more challenging task due to the complexity in collecting and handling information and data such as rainfall, river discharge, topographic parameters, land use, cropping pattern etc. For example, the general trend on global data collection for discharge and rainfall measurements is reported to be on the decline (see Stokstad 1999; Shiklomanov et al. 2002). Similarly, topography, accessibility and to a large extent economic considerations restrict the routine data collection which can consequently limit the accuracy of the data. In international rivers, such as the GBM, further restriction in the availability of data and information beyond borders is a major hurdle to any large scale water resources modeling work (Hossain et al. 2007).

5.2 The Potential Role of Satellites for Bangladesh: The SWOT Mission

Recently, satellite based remotely sensed data have found increasing use in support of a wide variety of applications in water resource management, disaster emergency preparedness, weather and flood forecasting. Remote sensing can provide data at various scales (from the meso-scale level up to the global earth coverage) and at regular temporal resolutions. Most of the remotely sensed data is also regularly updated and freely accessible on the internet. The data is mostly available in grid

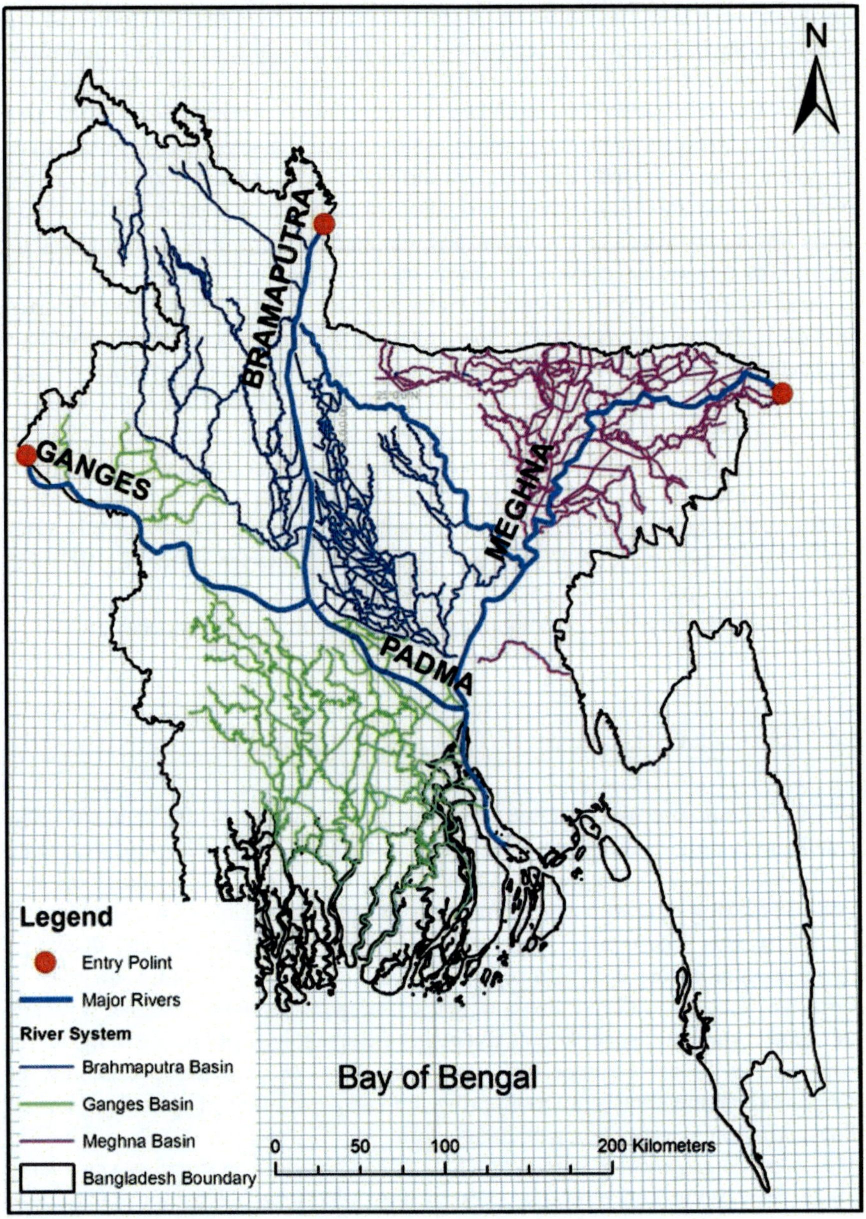

Fig. 5.3 River network inside Bangladesh where water level is forecasted by FLOOD WATCH and released to public daily as qualitative warning levels

format and easy to extract with Geographic Information System (GIS) tools and can be used to develop more realistic models, and identify changes in water resources or assess 'what-if' scenarios more accurately. In particular, given the unique vantage of space that is possessed by satellites (which allows them to cover a large area overcoming the hurdles on the ground), water-measuring satellites, therefore, have the potential to: (1) extend the accuracy and range of forecasted flood levels in the lowermost riparian nation through early assessment of the surface runoff evolution in the upstream nations and (2) minimize the negative impact of unavailable data and/or high operational costs of in situ networks.

In particular, the proposed Surface Water and Ocean Topography (SWOT) space mission to estimate surface water flow and storage changes will have tremendous implications for Bangladesh (Alsdorf et al. 2007). The SWOT satellite mission concept is an international effort by the scientific community founded on many years of research heritage on surface water monitoring from space (Alsdorf et al. 2007). SWOT recognizes that the ideal instrument for measuring surface water hydraulics (i.e., movement as well storage) is a device capable of providing image based measurements of water levels and its temporal and spatial derivatives of this water level. The technology for SWOT is a Ka-band Radar Interferometer (KaRIN) that has been developed from the efforts of the Wide Swath Ocean Altimeter (WSOA). More details on the SWOT mission may be found at http://swot.jpl.nasa.gov/.

5.3 How Useful Can Be SWOT for Bangladesh?

SWOT will measure water surface elevations using near-nadir radar interferometry. It will aim to provide discharge of large rivers at least 10 days or less (Alsdorf et al. 2007). With SWOT's suggested launch in 2015, Bangladesh may anticipate two ground breaking developments in its effort to monitor and forecast surface water related hazards. These are: (1) the availability of water elevations and discharge information from the upstream regions of the Ganges and Brahmaputra river in India and Nepal; and (2) a more complete coverage of the large seasonal wetlands (or 'haors') and other water bodies/rivers that are sparsely gaged inside Bangladesh. In particular, #1 can be expected to improve the flood forecasting capability of Bangladesh tremendously by monitoring the early evolution of river flow several thousand kilometers upstream of Bangladesh.

In order to gauge the potential of a SWOT-like mission, an exercise was carried out to understand how useful a mission like SWOT would be for Bangladesh discharge measurement. Readers can refer to the work of Jung et al. (2009) for details of this assessment. Herein, only a summary is provided below.

The Brahmaputra River was chosen for an investigation of the utility of the SWOT-like mission (Fig. 5.4). The Manning's equation (Equation below) was used for estimation of river discharge. Manning's equation yields water flow velocity as a function of water surface slope, hydraulic radius and an empirical value, Manning's roughness coefficient, n. (Albertson and Simons 1964). It is an empirical

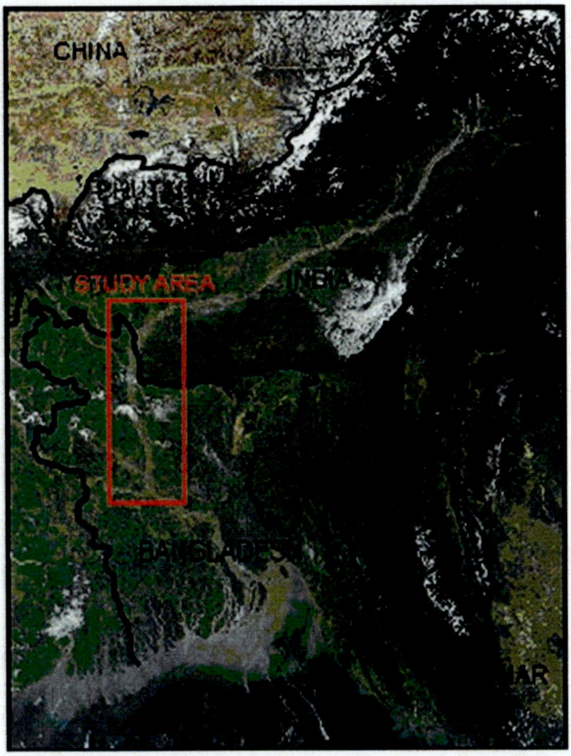

Fig. 5.4 Study area (shown in *red box*) of the Brahmaputra river for assessing the usefulness of the SWOT mission for Bangladesh

equation, but nevertheless well used in the hydrological sciences (i.e. LeFavour and Alsdorf 2005; Bjerklie et al. 2005). Multiplying flow velocity by channel cross-sectional area yields discharge (Fig. 5.5).

$$Q = \frac{AR^{2/3}\left(\frac{\partial h}{\partial X}\right)^{1/2}}{n}$$

A = river flow cross-section (width × depth); R = hydraulic radius (A × (width + 2 × depth)$^{-1}$); $\partial h/\partial x$ = water surface slope; n = Manning's roughness coefficient; Q = discharge.

The Brahmaputra River is sand bedded, without vegetation, therefore n is estimated to range from 0.018 to 0.035 (Ashworth et al. 2000; Albertson and Simons 1964; Coleman 1969). The value n = 0.025 was used for discharge estimations in this study because it is suggested for natural streams in fair condition by Albertson and Simons (1964).

The Shuttle Radar Topography Mission (SRTM), which flew during the month of February 2002, provided elevation at 90 m resolution for the study region. This was

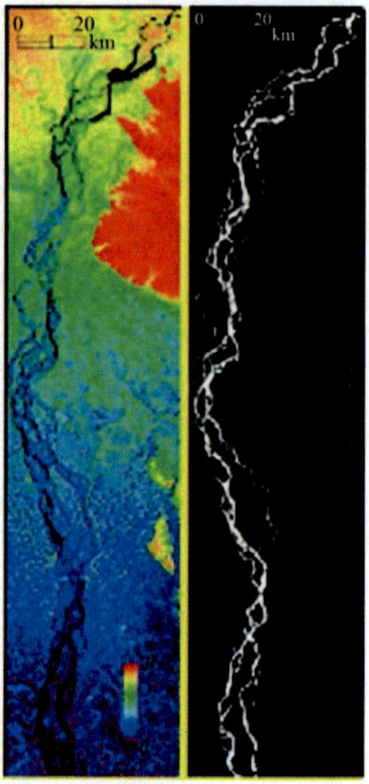

Fig. 5.5 *Left panel* – SRTM derived elevation of the Brahmaputra river reach. *Right panel* – land-water classification using LandSat data

overlaid with Landsat data at 30 m resolution to identify the pixels that were land or water (Fig. 5.5). IWM provided detailed bathymetry and measured discharge for the river reach as part of a 5-year technical arrangement with the host institution of the first author. Using knowledge of bathymetry and the space-derived river width, the hydraulic radius was computed and then used for calculation of the Manning's discharge. Figure 5.6 shows the estimated discharge as a function of flow distance using space-borne data. Comparison with observed measurement of flow shows that a SWOT-like mission is capable of estimating the low-flow (non Monsoon) river discharge of a braided river like Brahmaputra within reasonable confidence.

5.4 Implications of SWOT for Developing Adaptation Strategies for Climate Change

It is difficult to foresee how exactly the routine measurements from SWOT will facilitate developing strategies for climate change for Bangladesh and the greater South Asian region. However, one aspect that is clear is that the spatially distributed nature on surface water levels from SWOT, which is often missing but valuable in the current state-of-the art, can identify the low-lying coastal regions most vulnerable

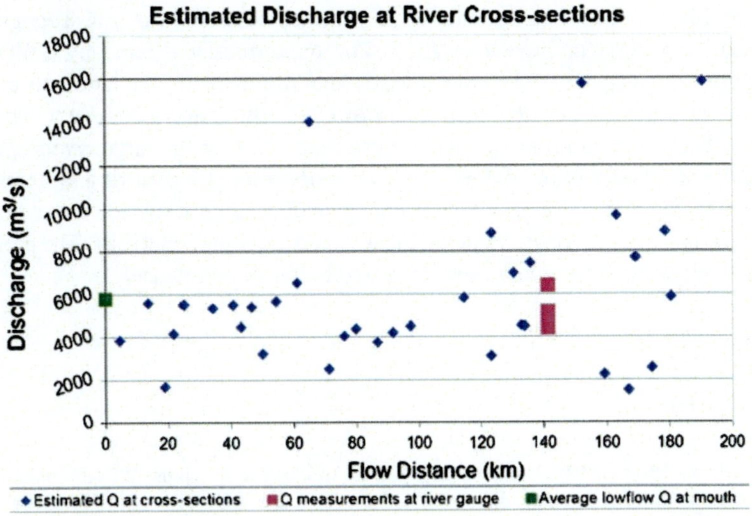

Fig. 5.6 Discharge estimation from space for Brahmaputra river. *Blue diamonds* indicate the estimated discharge using satellite-derived elevations. *Pink rectangles* are the measured flow on the Brahmaputra river reach. Flow distance is measured from upstream region of the river reach. (Figure taken from: Jung et al. 2009 and Hamski et al. 2008)

to sea level rise or where the reclamation by sea is in progress. Consequently, this can help the Bangladesh government prioritize areas in greater need of adaptation strategies. For example, for farmers in the low lying regions, floating garden beds for growing vegetables have lately become increasingly common. With SWOT's detection of water levels, such adaptive farming strategies can be proactively made available to the inhabitants of the region as the first instances of sea level rise is detected by SWOT.

Also, SWOT measurements of inland water bodies can help identify storage and storage change of freshwater resources in lakes, ponds, 'haors' or wetlands. With a more accurate assessment of storage change, drought management can be made more effective. For example, most often times, drought is forecasted on the basis of future rainfall patterns (such as an on-going or future anticipated failure of the Monsoons). Usually, the amount of water availability from atmospheric sources is not assessed in conjunction with the dynamic nature of freshwater storage changes. SWOT has potential to provide the governments of South Asian region with an additional tool to make more accurate assessment of the hydrology question – 'how much water will be available?'

5.5 Conclusion

This paper overviewed the monitoring of surface water flow and storage using satellites. The overview was cast in the context of surface water-related problems of Bangladesh. The paper provided a brief introduction of a planned space-borne mission for surface water called SWOT (Surface Water Ocean and Topography) mission suggested for launch in 2015. The opportunities offered by SWOT for enhancing the capacity for flood hazards monitoring and adaptation to climate change for Bangladesh were also overviewed. Preliminary assessment indicated that a SWOT-like mission can indeed estimate the flow of large rivers like the Brahmaputra river with confidence for use in operational forecasting systems.

Around the world, we have a poor understanding of both surface water flows in rivers and the changes in waters stored in lakes, wetlands, and reservoirs. The problems are not unique to Bangladesh, but are certainly felt more intensely.

References

Albertson ML, Simons BB (1964) Section 7: fluid mechanics. In: Chow VT (ed) Handbook of applied hydrology: a compendium of water-resources technology. McGraw-Hill, New York

Alsdorf DE, Rodriguez E, Lettenmaier DP (2007) Measuring surface water from space. Rev Geophys 45:RG2002 (doi: 10.1029/2006RG000197)

Ashworth PJ et al (2000) Morphological evolution and dynamics of a large, sand braid-bar, Jamuna River, Bangladesh. Sedimentology 47:533–555

Asian Disaster Preparedness Center – ADPC (2002) Application of climate forecasts in the agriculture sector. Climate Forecasting Applications in Bangladesh Project, Report 3, Asian Disaster Preparedness Center (ADPC) Bangkok

Bjerklie DM, Moller D, Smith LC, Dingman SL (2005) Estimating discharge in rivers using remotely sensed hydraulic information. J Hydrol 309:191–209. doi:10.1016/l.lhydrol.2004.11.022

Chowdhury MR (2000) An assessment of flood forecasting in Bangladesh: the experience of the 1998 flood. Natural Hazards 22:139–163

Coleman JM (1969) The Brahmaputra river: channel processes and sedimentation. Sediment Geol 3:129–139

Hamski J et al (2008) American society of limnology and oceanography-ASLO conference, March 2–7, Orlando, Florida

Hofer T (1998) Floods in Bangladesh: a highland-lowland interaction? University of Berne, Switzerland

Hossain F, Katiyar N, Wolf A, Hong Y (2007) The emerging role of satellite rainfall data in improving the hydro-political situation of flood monitoring in the under-developed regions of the world. Natural Hazards (Special Issue) INVITED PAPER 43, pp 199–210 (doi 10.1007/s11069-006-9094-x)

Hossain F, Katiyar N (2006) Improving flood forecasting in international river basins. EOS Trans AGU 87(5):49–54

Jung HC, Hamski J, Durand M, Alsdorf D, Hossain F, Lee H, Hossain AKMA, Hasan K, Khan AS, Hoque AKMZ (2009) Characterization of complex fluvial systems via remote sensing of spatial and temporal water level variations, Earth Surface Processes and Landforms, SPECIAL ISSUE-Remote Sensing of Rivers, (doi: 10.1002/esp)

Kjelds KT, Jorgensen GH (1997) Flood watch: GIS based management system for flood forecast applications, 17th ESRI user's conference paper 486

LeFavour G, Alsdorf DE (2005) Water slope and discharge in the Amazon River estimated using the shuttle radar topography mission digital elevation model. Geophys Res Lett 32:L17404. doi:10.1029/2005GL023836

Ninno del C, Dorosh PA, Smith PC, Roy DK (2001) The 1998 floods in Bangladesh: disaster impacts, household coping strategies, and response, International Food Policy and Research Institute, Research Report 122 (Washington DC). ISBN 0-89629-127-8
Paudyal GN (2002) Forecasting and warning of water-related disaster in a complex hydraulic setting – the case of Bangladesh. Hydrol Sci J 47(S):S5–S18
Stokstad E (1999) Scarcity of rain, stream gages threatens forecasts. Science 285:1199
Shiklomanov AI, Lammers RB, Vörösmarty CJ (2002) Widespread decline in hydrological monitoring threatens pan-arctic research. EOS Transactions-AGU 83(2):16–17

Chapter 6
Inter-annual Water Storage Changes in Asia from GRACE Data

C.K. Shum, Jun-Yi Guo, Faisal Hossain, Jianbin Duan, Douglas E. Alsdorf, Xiao-Jun Duan, Chung-Yen Kuo, Hyongki Lee, Michael Schmidt, and Lei Wang

Contents

Abstract Time-varying gravity field solutions from the Gravity Recovery and Climate Experiment (GRACE) satellite mission have been used to investigate the inter-annual changes of hydrologic water storage (ΔS) within Asia, focusing on the

C.K. Shum (✉), J.-Y. Guo, J. Duan, D.E. Alsdorf, H. Lee, and L. Wang
School of Earth Sciences, Ohio State University, 275 Mendenhall Lab,
125 S. Oval Mall, Columbus, OH 43210, USA
e-mail: ckshum@osu.edu; guo.81@osu.edu; duan.29@osu.edu; alsdorf.1@osu.edu; lee.2444@osu.edu; wang.1115@buckeyemail.osu.edu

F. Hossain
Department of Civil & Environmental Engineering, Tennessee Technological University, Prescott Hall 332, 1020 Stadium Drive. Box 5015, Cookerville, TN 38505, USA
e-mail: fhossain@tntech.edu

X.-J. Duan
Department of Mathematics & Systems Science, National University of Defense Technology, Changsha, Hunan 410073, China
e-mail: xiduan@nudt.edu.cn

C.-Y. Kuo
Department of Geomatics, National Cheng Kung University, No. 1, Ta-Hsueh Road, Tainan 701, Taiwan
e-mail: kuo70@mail.ncku.edu.tw

M. Schmidt
Deutsches Geodätisches Forschungsinstitut, Alfons-Goppel-Strasse 11, 80539, Muenchen, Germany
e-mail: schmidt@dgfi.badw.de

R. Lal et al. (eds.), *Climate Change and Food Security in South Asia*,
DOI 10.1007/978-90-481-9516-9_6, © Springer Science+Business Media B.V. 2011

India–China–South Asia region. Instead of computing GRACE monthly ΔS from geopotential coefficients, we choose to compute the annually averaged ΔS before data smoothing, which improved the accuracy for the resulting inter-annual water storage changes. We then applied a novel method of decorrelation, filtering and land signal leakage reduction to the data, which yielded more accurate and higher spatial resolution (200 km or longer, half-wavelength) GRACE storage change observables over the study region. The technique provides a tool for future more in-depth studies of terrestrial hydrology in this region or globally. GRACE inter-annual variations in water storage change (2002–2007) exhibit large extremes over the region: droughts in Eastern China in 2004 and Indo-China and Bangladesh in 2005; and flooding in Indo-China and India in 2006. In general, GRACE inter-annual ΔS variations have significantly larger amplitudes (increase or decrease of water storage change) than the values predicted by the Global Land Data Assimilation System (GLDAS) hydrologic model, and than the total precipitation observed by the Topical Rainfall Measurement Mission (TRMM).

Keywords Terrestrial hydrology • GRACE • Temporal gravity field • TRMM • GLDAS

Abbreviations

NASA	National Aeronautics and Space Administration
DLR	Deutsches Zentrum für Luft- und Raumfahrt joint satellite mission
GRACE	Gravity Recovery and Climate Experiment
TRMM	Topical Rainfall Measurement Mission
KBR	K/Ka-band low-low inter-satellite ranging
ENSO	El Niño-Southern Oscillation
GLDAS	Global Land Data Assimilation System
GIA	glacial isostatic adjustment
EOF	empirical orthogonal functions
SCs	Stokes coefficients

6.1 Introduction

Monitoring, understanding, and quantifying the global water cycle budget in light of global climate change represents an important scientific research topic with significant societal impacts. The National Aeronautics and Space Administration (NASA) and Deutsches Zentrum für Luft- und Raumfahrt (DLR) joint satellite mission, the Gravity Recovery and Climate Experiment (GRACE) mission launched in March 2002, is designed to measure small mass changes within the Earth over a large spatial scale. The GRACE twin-satellite mission uses primarily a microwave K/Ka-band low-low inter-satellite ranging (KBR) system at μm precision and have been demonstrated to be able to measure climate-change signals, which manifest in the form of temporal gravity variations resulting from Earth's mass redistributions (Tapley et al. 2004a, b). The GRACE instrument represents one of newest observational system to improve the estimate of hydrologic, glacier, ice-sheet and oceanic mass changes with unprecedented accuracy, ~a few *cm* in the form of water thickness change (Wahr et al. 2006). GRACE is currently measuring the Earth's mass redistributions with a spatial resolution longer than 300–600 km (half-wavelength) or finer and at monthly temporal resolution. GRACE is capable of observing the total (both surface and subsurface) water thickness change over an entire watershed or basin (Wahr et al. 1998), and although at relatively coarse spatial and temporal resolutions, GRACE represents a revolutionary tool to address contemporary research problems in terrestrial hydrology.

GRACE primarily measures storage change, ΔS, over a large basin. $\Delta S = P - E - R$, where P is the precipitation, E is the evapotranspiration, and R is the runoff. From the terrestrial hydrologic perspective, it is difficult to even study basins as a whole (even at monthly scale), for example, in the Ganges/Brahmaputra/Meghna region because of the need of inputs from in situ sources for calibration to run conventional hydrologic modeling. GRACE provides a more comprehensive and basin-wide time series of ΔS, as opposed to the range of approximations and well-known difficulties in large-scale hydrologic modeling in data-sparse/ungauged regions. Constraining the water storage change over an entire basin using GRACE and other data (e.g., model data) allows one to estimate basin-scale evapotranspiration (Rodell et al. 2004), or to estimate basin-scale discharge for example for the Amazon and Mississippi River basins (Syed et al. 2005). Other GRACE hydrologic studies primarily follow the method established by Rodell et al. (2004), including the Amazon basin (Han et al. 2005b, Schmidt et al. 2006, 2008), extreme weather conditions in Europe (Seitz et al. 2008), Congo basin (Crowley et al. 2006), and annual water storage changes in Bangladesh (Andersen et al. 2008).

In this work we focus on the India–China–South Asia region. This is an important region for global hydrology because of the South and South East Asian Monsoon that controls the regional water cycle for three large river basins – Ganges, Brahmaputra and Irrawady. The South Asian monsoon is also known to have an inverse relationship with El Niño-Southern Oscillation (ENSO) (Kumar et al. 1999) and snow cover in Eurasia (Bamzai and Shukla 1999). From the societal point of view, the South Asian region is projected to be most water stressed in a changing climate (Vorosmarty et al. 2000), large scale contamination of shallow groundwater resources (Nahar et al. 2008) and proposed man-made diversion of surface water from wet basins of the northern region to dry basins in the South (Misra et al. 2007). In particular, the detection of significant anthropogenic depletion of ground water in India using GRACE data was recently reported (Rodell et al. 2009).

Here, we study the inter-annual hydrological mass changes using GRACE in the India–China–South Asia region in terms of year-to-year change based on yearly averages. An empirical decorrelation-Gaussian filtering-leakages reduction approach is adapted (Duan et al. 2009; Guo et al. 2010). The primary motivation for the study include: (1) The mass changes from year to year reveal more hydrologic information such as extremes (drought or flooding) than using the "trend" method which is obtained via least squares fit of GRACE data simultaneously removing annual and semi-annual signals, (2) the yearly averaged GRACE geopotential (or Stokes) coefficients have less error than each of the monthly coefficients. Hence, a less strong decorrelation filter and a shorter spatial smoothing radius may be adopted to produce less distorted and yielding water storage change with higher spatial resolution, (3) to examine the effectiveness of the adopted decorrelation/filtering and land signal leakage repair methods to the post-processing of the GRACE data, and (4) to compare inter-annual variations which are arguably less known than the seasonal variations in this study region. Here we compare the GRACE observed Asian inter-annual net water storage change, ΔS, with the values predicted by the Global Land Data Assimilation System (GLDAS) hydrologic model, which is a dynamic state-space prediction system based on hydrologic modeling that can be sensitive to input data errors from the assimilation platforms (Rodell et al. 2004). Finally, we compare the GRACE observed and GLDAS predicted inter-annual ΔS qualitatively with the precipitation data from NASA's Tropical Rainfall Measurement Mission (TRMM).

6.2 Methods

GRACE observations are routinely processed at several data centers, including the University of Texas at Austin Center for Space Research (CSR), GeoForschungs Zentrum Potsdam (GFZ), NASA/Jet Propulsion Laboratory (JPL), and Centre National d'Etudes Spatiales/Groupe de Recherches de Géodésie Spatiale (CNES/GRGS). The final results provided by these data centers are monthly geopotential models, referred to as the level 2 (L2) data products. The CSR release 4 (RL04) L2 data product is used in this study.

The L2 products are the most commonly used GRACE data in the study of mass changes at the Earth's surface. It is well known that these data have systematic errors, including the spurious north-south stripes in the monthly maps of Earth's mass changes. Hence, the GRACE results are usually smoothed in the space domain using a Gaussian filter (e.g., Jekeli 1981; Wahr et al. 1998). Swenson and Wahr (2006) found that the Stokes coefficients (SCs) of the spherical harmonics of the GRACE geopotential changes (e.g., the difference between two monthly models) are correlated. More specifically, the SCs of the geopotential changes of the same order and the same parity in degrees are correlated with each other. Swenson and Wahr (2006) designed an empirical moving window-polynomial fit filter to compute the correlated errors and remove them from the SCs. Chamber (2006) and Chen et al. (2007) followed this idea, but chose to fit all the correlated coefficients using one polynomial. Here we refer this correlation removal procedure as decorrelation. Duan et al. (2009) designed a more sophisticated scheme, extending the Swenson and Wahr (2006) decorrelation filter, with more flexible choice of decorrelation parameters, based on the standard deviations (SDs) of the geopotential coefficients of the monthly GRACE models. However, the SCs after decorrelation still contain non-negligible errors. To produce meaningful results of mass changes, smoothing in the spatial domain is still necessary. But the smoothing radius required is shorter, and hence, the results have relatively higher spatial resolution than the case in which no decorrelation is performed.

Recently, more decorrelation methods have been developed, including the Tikhonov-type regularization (Kusche 2007), the use of empirical orthogonal functions (EOF) (Schrama et al. 2007; Wouters and Schrama 2007), the use of statistics for trend and annual variations in spherical harmonic coefficients (Davis et al. 2008), and optimal filtering by minimizing an objective function (Klees et al. 2008).

As for spatial smoothing, apart from the classical isotropic filter (Jekeli 1981; Wahr et al. 1998), non-isotropic Gaussian filters were also proposed (Han et al. 2005a; Guo et al. 2010). Guo et al. (2010) defined explicitly the non-isotropic smoothing kernel in the spatial domain, and devised an algorithm to reduce the land-ocean signal leakage caused by smoothing if the signal over land/ocean could be assumed far larger than that over ocean/land. Chen et al. (2006) discussed the optimal choice of smoothing radius for the isotropic Gaussian filter. But the idea could be generalized to non-isotropic Gaussian filters.

In our study, we adopt the methods developed by Duan et al. (2009) for decorrelation and by Guo et al. (2010) for spatial Gaussian smoothing and leakage reduction. Other decorrelation methods are less suitable for the problem. For the EOF method (Schrama et al. 2007; Wouters and Schrama 2007), there is too few temporal sampling in our case. The statistical method of Davis et al. (2008) handles only the trend and annual components. The methods of Kusche (2007) and Klees et al. (2008) require an a priori signal covariance matrix. The signal covariance matrix should be accurate enough so that it does not bias the results (Klees et al. 2008). However, for our case, it is not evident that we are capable to compute an accurate enough a priori signal covariance matrix. As it will be shown later, the inter-annual water storage change predicted by the GDLAS model significantly disagrees with the GRACE observation.

We now briefly describe the decorrelation and spatial Gaussian smoothing and leakage reduction algorithms by Duan et al. (2009) and Guo et al. (2010), respectively used in this study to post-process GRACE Level 2 data products.

6.2.1 Decorrelation

The decorrelation process is to remove the correlated errors in the SCs of geopotential changes. The correlation lies among the SCs of the same order and the same parity in degrees. In the approach of Duan et al. (2009), a portion of lower degree and order SCs is left unchanged, and the rest are filtered using a moving window-polynomial fit high pass filter. They fix the polynomial to be quadratic, and change the window width to adjust the strength of the filter. The SCs kept unchanged and the window width depend on the pattern of SDs of the geopotential coefficients provided in the L2 product.

The portion of SCs kept unchanged is characterized by the curve of the empirical formula (Duan et al. 2009):

$$l = l_0 + \beta m^r \tag{6.1}$$

where l and m are the degree and order, respectively, of the SCs. The parameter r is chosen empirically so that the curve follows approximately an isoline of the SD map, and l_0 and β are determined by the end-point pair of the curve. For the CSR RL04 data, we have $r = 3.5$. In Fig. 6.1 (Left panel), we show a map of the SDs of the CSR RL04 data, and two choices of the curve of the unchanged portion of SCs defined by the end point pairs [(35, 0), (10, 10)] and [(45,0), (15, 15)]. As the curve is to approximately follow an isoline, and it has $m = 0$ at one end, and $l = m$ at the other end, it is in fact chosen by only one parameter (e.g., the value of l at the end point with $m = 0$).

The filter window width formula of Duan et al. (2009) is

$$w = \max\left\{30e^{-\frac{[(1-\gamma)m^p+\gamma l^p]^{1/p}}{K}} + 1, 5\right\} \tag{6.2}$$

which was modified from the one used by Swenson and Wahr (2006). The value of w is rounded to an odd integer. The parameters γ and p are chosen empirically, so that the pattern of w follows that of the SD, i.e., their isolines approximately. For the CSR RL04 data, we have $\gamma = 0.1$ and $p = 2$. In Fig. 6.1 (Right panel), we show w computed for $K = 15$. Actually, K is the only parameter to be altered for choosing the window width.

If we write $w = 2\alpha + 1$, the correlated error to be removed form C_l^m (similarly for S_l^m) is obtained by fitting $C_{l-2\alpha}^m, \ldots, C_{l-2}^m, C_l^m, C_{l+2}^m, \ldots, C_{l+2\alpha}^m$ with a quadratic polynomial. The result of the fit is to be subtracted from the original value of C_l^m to obtain the filtered value.

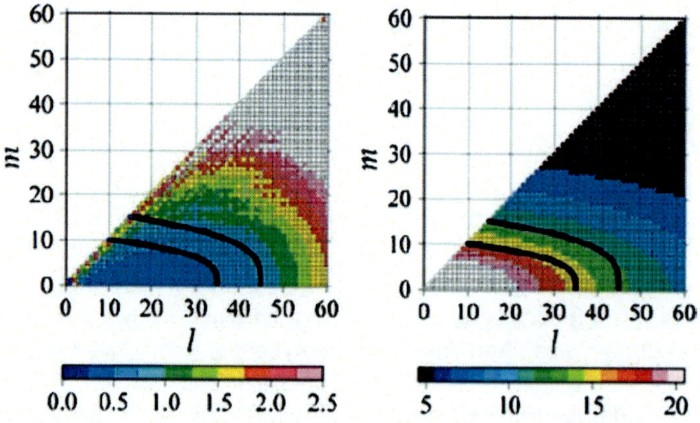

Fig. 6.1 *Left*: error (scaled by × 10^{12}) pattern of C_l^m of a GRACE monthly gravity model (July 2005, the pattern from any month is similar). The two dark curves are used to define the portion kept unchanged in the decorrelation algorithm. *Right*: window width w as function of l and m used in our decorrelation algorithm computed with $K = 15$. The dark curves are the same as those in the left panel

6.2.2 Gaussian Smoothing and Leakage Reduction

Due to the appearance of north-south stripes, it can be inferred that GRACE L2 data have better resolution in south-north direction than in east-west direction, especially at lower latitude regions. Hence, non-isotropic spatial smoothing may potentially keep more signal information by choosing a larger smoothing radius along the longitude than that along the latitude direction. The non-isotropic filter (Guo et al. 2010) is defined as:

$$f_s(\bar{\theta},\bar{\lambda}) = Y^{-1}(\bar{\theta},\bar{\lambda})\int_0^{\pi} d\theta \int_{\bar{\lambda}-\pi}^{\bar{\lambda}+\pi} f(\theta,\lambda)G(\theta-\bar{\theta})G(\lambda-\bar{\lambda})d\lambda, \tag{6.3}$$

$$Y(\bar{\theta},\bar{\lambda}) = \int_0^{\pi} d\theta \int_{\bar{\lambda}-\pi}^{\bar{\lambda}+\pi} G(\theta-\bar{\theta})G(\lambda-\bar{\lambda})d\lambda \tag{6.4}$$

where the Gaussian function is defined by

$$G(x) = e^{-\frac{x^2}{2\sigma^2}}, \tag{6.5}$$

and the parameter σ is related to the smoothing radius in the following form:

$$\sigma = \frac{r}{\sqrt{2\ln(2)}}. \tag{6.6}$$

The discrete formulae for the computation are given in Guo et al. (2010). Different smoothing radii are choosing for $G(\theta-\bar{\theta})$ and $G(\lambda-\bar{\lambda})$. Hence, as compared to the isotropic filter, a reasonably smooth map of signal may be obtained without excessive filtering along the latitude.

The spatial Gaussian smoothing is a method of weighted averaging. Signals over land and ocean are averaged together along coasts to compute the smoothed values at both the land and ocean side. For this reason, the signal is leaked from the larger value side to the smaller value side. However, if the signal over one inner domain (assumed land here) is much larger than the signal over the outer domain (assumed ocean here), the signal leakage may be reduced by rescaling the signal over the inner domain, which has the larger signal. The result of this kind of leakage reduction is as if the Gaussian smoothing were done for land and ocean separately, but without the appearance of side lobes (e.g., Wahr et al. 1998). The leakage reduction method is best explained using the schematic for the simplified one-dimensional case shown in Fig. 6.2. Detailed description of the computation formalism can be found in Guo et al. (2010).

6.3 Data and Results

In this work, we processed the CSR RL04 data product from January 2003 through December 2007. The geopotential coefficients are first averaged yearly. As the GRACE geopotential solution is missing the month of June 2003, we linearly interpolated using

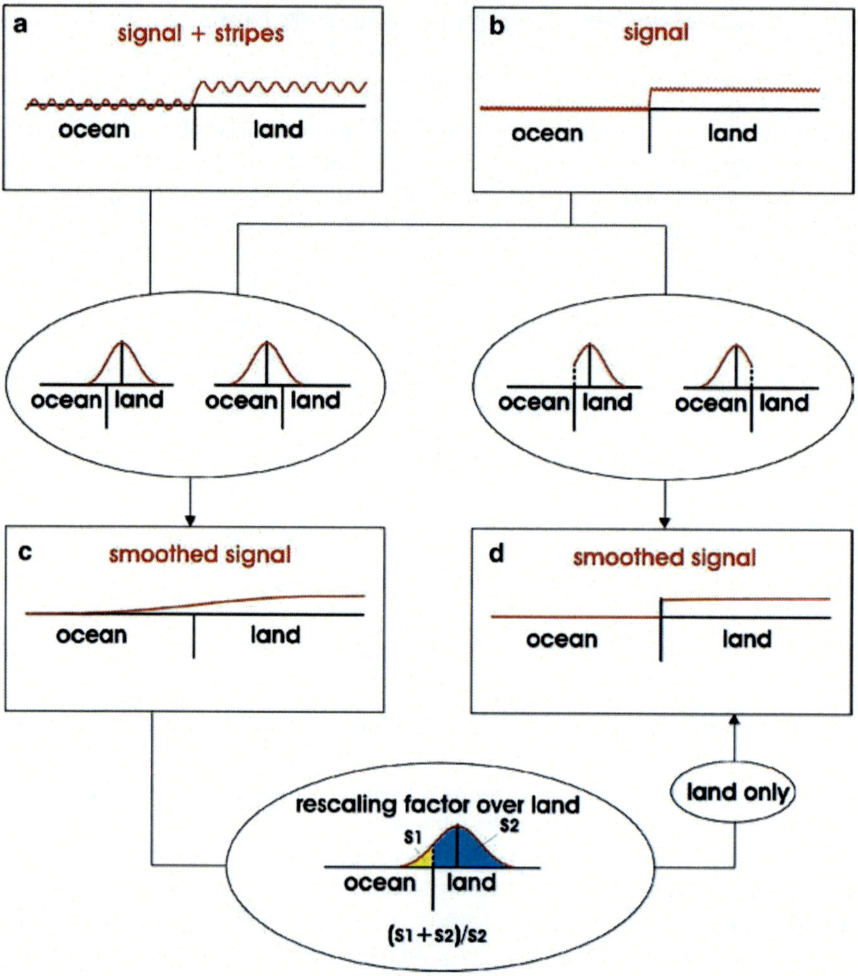

Fig. 6.2 Schematic of our algorithm for reducing leakage caused by Gaussian smoothing is shown here. Assuming that (1) the signal over land is far larger than that over ocean, and that (2) the usual Gaussian smoothing eliminates the effect of stripes [(**a**) and (**b**) leading to approximately the same result shown in (**c**) after smoothing]. Our idea is to rescale land signal [shown in (**c**)] to obtain results as if Gaussian smoothing is applied to land and ocean separately for the signal part alone [shown in (**d**)]. To obtain leakage reduced signal over ocean, we smooth the rescaled signal over land [shown in (**d**)] with ocean signal set to zero using the same Gaussian filter in computing the smoothed signal (**c**), and then subtract the result from the smoothed signal (**c**)

the May and July models to fill in the data gap. The year-to-year differences, i.e., the differences between successive yearly averages, of the geopotential coefficients are first decorrelated, and then corrected for geopotential changes due to glacial isostatic adjustment (GIA) processes using the ICE5G (VM4) GIA model (Peltier 2004). Then, yearly differenced mass changes at the Earth's surface are computed (Wahr et al. 1998). The Love numbers used for the loading computation are adopted from Guo et al. (2004).

First, we show in Fig. 6.3 the effect of decorrelation and spatial smoothing using the annual average mass (water storage over land, sea level over ocean) changes from 2005 to 2006 without performing signal leakage reduction. As expected, the stronger the filters are used for decorrelation and spatial smoothing, more stripes are removed, and resulting in smoother mass change maps. However, as stronger filtering for decorrelation distorts more of the signal, and a stronger filtering for spatial smoothing reduces spatial resolution, a compromise between error reduction (stripe removing or dccorrclation), signal distortion and resolution should be considered depending on the nature of the problem to be studied. In this work, we choose for decorrelation the unchanged portion defined by degree-order end point pair [(45,0), (15,15)], and window width computed using $K = 20$, as a stronger filter does not bring in significant reduction of stripes. Our experience indicates that this choice leads to minimal signal distortion (Duan et al. 2009).

In Fig. 6.4 we show the results after signal leakage reduction computed with decorrelation parameters specified in the last paragraph. Here a trend is computed from the year-to-year variations. In the leakage reduction method, the signal over land is rescaled by the factor $(S1+S2)/S2$. The parameters used in the decorrelation algorithm are: the portion of low degree-order portion of SCs kept unchanged are defined by the end point pair [(45,0), (15,15)], and the window width is computed with K = 20, spatial smoothing radii is 2d × 3d (or 2° latitude by 3° longitude, or 220 km by 330 km). This case is illustrated in Fig. 6.3, top row, middle panel. To illustrate the effect of the signal leakage reduction, we computed a trend using the yearly differenced maps (Fig. 6.4, right, top panel), then conduct signal leakage repair (Fig. 6.4, left, top panel). It is evident that from the geographical maps, in Regions A & B, that the leakage reduction case (Fig. 6.4, left, top panel) shows more abundant signals than the case without leakage correction. Figure 6.4 (bottom panel) shows the averaged GRACE ΔS time series exhibit larger seasonal amplitudes in the leakage repaired cases than the cases with no repair for Regions A and B.

Figure 6.5 shows the year-to-year changes of annually averaged water storage change from 2003 to 2007 computed using GDLAS model (1d × 1d, Fig. 5, left panels), and using GRACE data (Fig. 5, middle panels). Year-to-year changes of TRMM 3B43 V6 monthly precipitation data (right panels) are computed to qualitatively compare with GRACE and GLDAS. The decorrelation and filtering parameters used are: the portion of low degree-order portion of SCs kept unchanged are defined by the end point pair [(45,0), (15,15)], and the window width is K = 20, and the filter with radii 2d × 3d.

Unlike the GLDAS data that are directly comparable with GRACE results, the TRMM data are used primarily for qualitative comparisons, without considering the evaporation and flow/runoff (Fig. 6.5). In the study region of India-South Asia, the seasonal Monsoons dominate the annual surface water change as compared to evaporation and runoff. The difference in water storage between two successive years is expected to be somewhat proportional to the total rainfall of the second year. Hence, the rainfall pattern in 2004 shown in Fig. 6.5 (right panels) can be expected to follow the 2004–2003 water storage change observed by GRACE or predicted by GLDAS, and similarly for other interannual variations. However, the

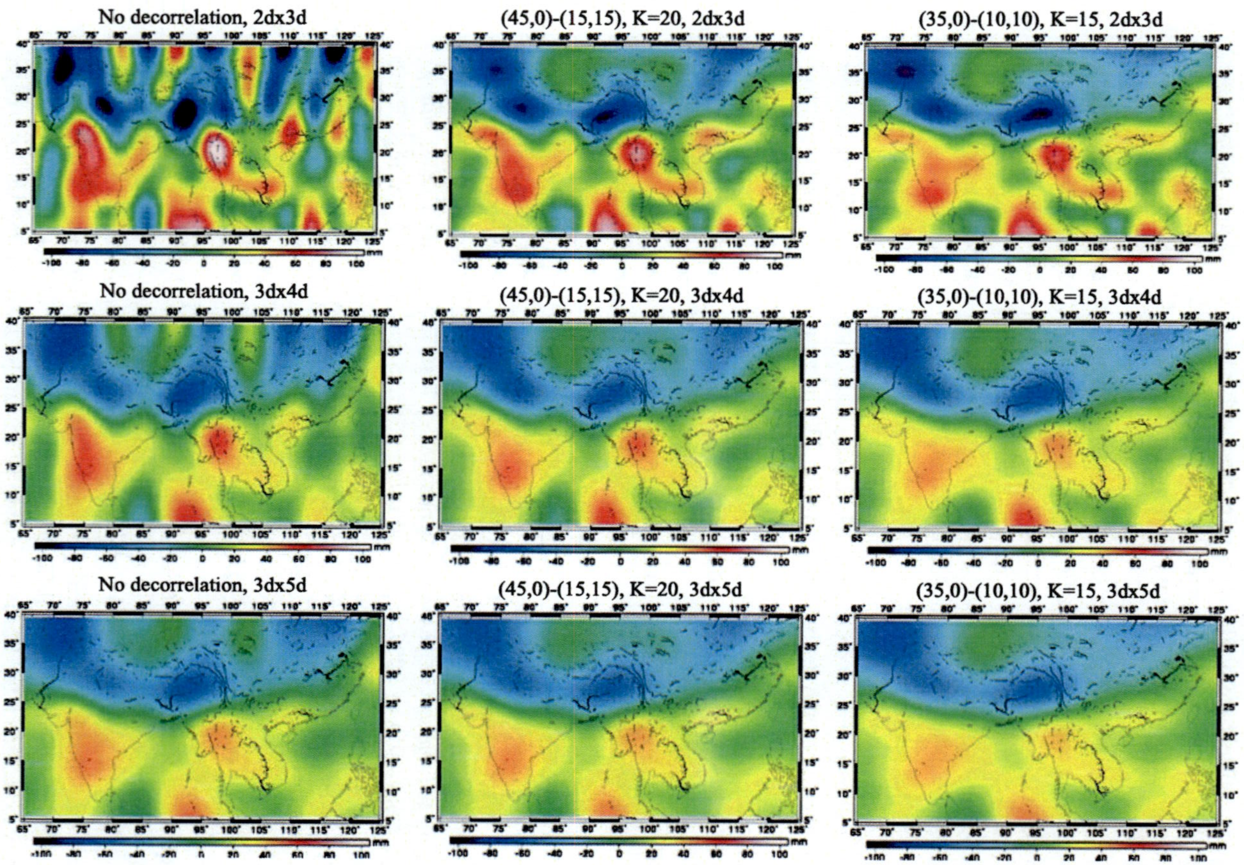

Fig. 6.3 Changes of annually averaged water storage over land/sea level over ocean from 2005 to 2006 obtained from GRACE data using different decorrelation parameters and spatial smoothing radii. We use the figure in plain middle to explain the labels: (45,0)–(15,15) is the degree-order end point pair defining the portion kept unchanged, K = 20 is used to compute window width, and 3d × 4d means a non-isotropic Gaussian filter of 3° along latitude and 4° along longitude is used for spatial smoothing. As there is no leakage reduction is performed, leakage caused by spatial smoothing is present

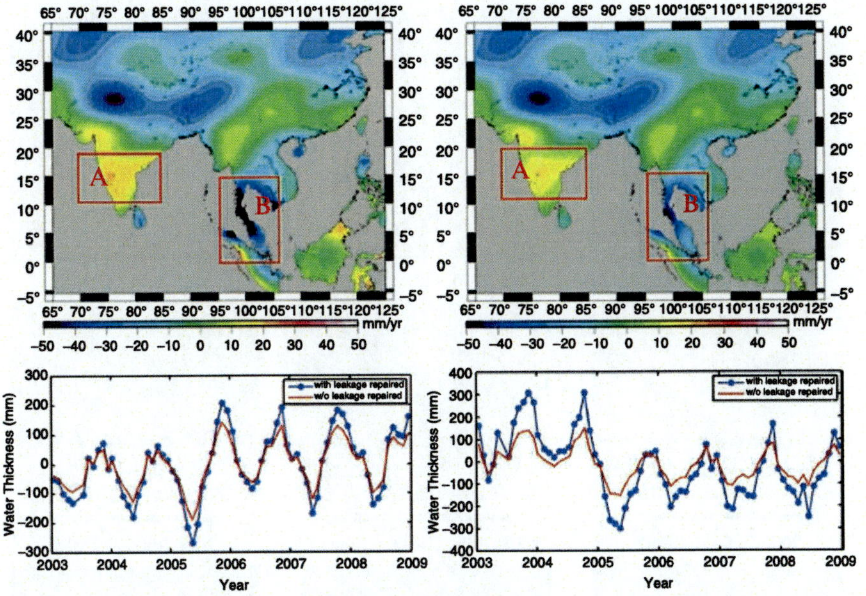

Fig. 6.4 GRACE water thickness change trend (2002–2008) in mm/year: with signal leakage correction (*top left*), and without signal leakage correction (*top right*). GRACE monthly water thickness change time series indicating the significant differences after signal leakage correction for India (Region A, *left*), and for Indo-China (*right*)

response of water storage to rainfall is expected to be delayed. Hence, we also computed annual averages by shifting 1 and 2 months to the past (as in Fig. 6.5), respectively. Since the differences between the data of different shifts are not significant, the detailed results in this experiment are not shown here.

In terms of hydrology, the GRACE inter-annual ΔS qualitatively agree over several regions with the GLDAS predicted values for North East China, Southern India, and Borneo Island (Fig. 6.5). The mostly year-to-year oscillating negative (2004–2003, 2006–2005) and positive anomalies (2005–2004) in the Northwestern India is the subject of the Rodell et al. (2009) study which indicated significant anthropogenic pumping of water. Here is seen that significant inter-annual variations exist in this region. Overall the GRACE signal exhibits much stronger amplitudes than the GLDAS model values, which is similar to conclusions from other hydrologic studies comparing GRACE data and model predictions, e.g., Han et al. (2005b), and that GLDAS model is under-predicting hydrologic signals. The agreement appears better synoptically for the latter years, where a distinct wetting trend (increase in water storage) over India and Indo-China is evident in the GRACE observations. This might be explained by a possible wetter than normal period from 2005–2007 in TRMM rainfall (Fig. 6.5) observed over Southern India indicating the recession of a prolonged drought. The increase in reservoir level of the Narmada Dam (constructed in Western India) by more than 30 m over the last decade (1999–2009)

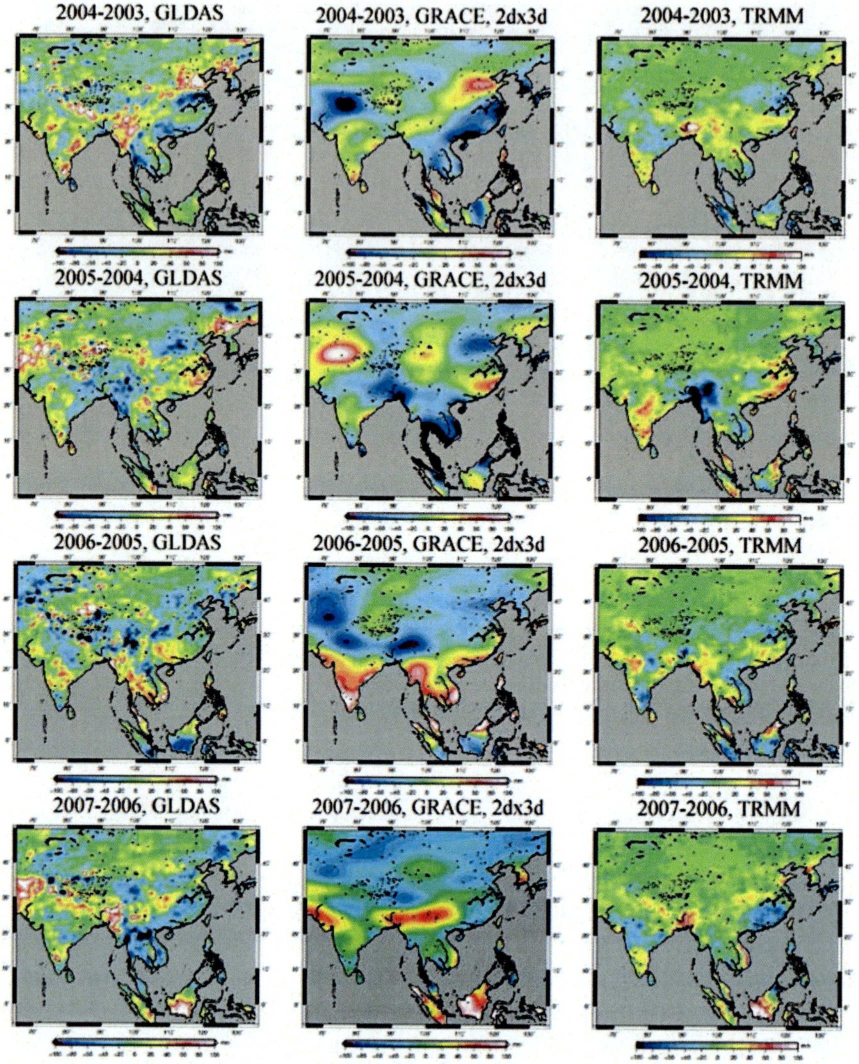

Fig. 6.5 Year-to-year changes of annually averaged water storage change from 2003 to 2007 computed using GLDAS model (1d × 1d, *left panels*), and using GRACE data (*middle panels*). Year-to-year changes of TRMM precipitation data (*right panels*) are computed to qualitatively compare with GRACE and GLDAS. The decorrelation parameters are: the portion of low degree-order portion of SCs kept unchanged are defined by the end point pair [(45,0), (15,15)], and the window width is K = 20, and the filter with radii 2d × 3d

could be a potential reason for the increasing trend in storage depicted by GRACE maps. Overall, GRACE observations appear to be agreeing in general patterns with the water storage change simulated by GLDAS. However, significant differences remain. Notably, GRACE observed large inter-annual changes in Eastern China (2004–2003, a drought), in Indo-China and Bangladesh (2005–2004, droughts), and in Indo-China and India (2006–2005, excess water or flooding), significantly disagreeing with GLDAS model predictions in terms of amplitude, or GLDAS significantly under-estimates the increased or decreased inter-annual water storage changes. TRMM inter-annual rainfall and GRACE inter-annual storage change have some agreements, e.g., 2005–2004, Bangladesh drought, but for the most part, there are disagreements. Slightly better similarity between the GLDAS and TRMM data is observed than between GRACE and GLDAS, though the GRACE and GLDAS data represent the same quantity, while TRMM not. This may be due to the fact that TRMM data are used in deriving the GLDAS model.

6.4 Specific Recommendations for Policy Makers

It is difficult to foresee how exactly the routine measurements from GRACE will facilitate developing strategies for policy makers in the South Asian region. However, recent work by Rodell et al. (2009) using GRACE has indicated that the Northwestern Indian region is being systematically depleted of ground water through pumping for anthropogenic uses. The situation is deemed very analogous to the case of the Ogalalla aquifer of the mid-western US. Thus, in essence, the GRACE measurements can allow policy makers a more accurate assessment of available water and consequently improve long-term drought management. For example, most often times, drought is forecasted on the basis of future rainfall patterns (such as an on-going or future anticipated failure of the Monsoons). Usually, the amount of water availability from atmospheric sources is not assessed in conjunction with the trends observed in underground water storage. GRACE has the potential to provide the governments of South Asian region with an additional tool to make more accurate assessment of the hydrology question – *how much water will be available in the long run?* – by integrating gravity-based storage data with other hydrologic variables of precipitation and stream flow.

In this paper, GRACE demonstrates its potential to observe extreme weather conditions including droughts and flooding or excess water in the entire water basin with continental scale through examining its interannual or year-to-year variations. It is conceivable that policy makers can use this information to possibly identify regions with frequency extreme weather conditions, and design policy to mitigate or to adapt to these extreme events.

6.5 Conclusions

We have compared the year-to-year changes of water storage computed from GRACE geopotential solutions with the GLDAS water storage model and the TRMM rainfall model. We have used techniques to obtain more higher spatial resolution GRACE results of yearly water storage changes. The method of decorrelation, filtering and in particular, land signal leakage repair yields more accurate GRACE storage change observables over the India–China–South Asia study region. The technique provides a tool for further in depth studies of terrestrial hydrology in this region or globally. GRACE exhibits much stronger extreme signals (increase or decrease in water storage) than the values predicted by GDLAS. There are some general agreements however, significant discrepancies are observed between inter-annual variations of water storage change from GRACE and from GLDAS hydrologic model, and rainfall observations from TRMM. For example, the increase of water storage in India and Indo-China from 2005 to 2006 seen in GRACE results is not seen in GLDAS or in TRMM data.

Acknowledgments We used the monthly GLDAS data provided by the IERS Global Geophysical Fluids Center's Special Bureau for Hydrology hosted by University of Texas at Austin Center for Space Research (UT/CSR). The GRACE data products are from the NASA/DLR GRACE project and we used the data products from UT/CSR. The TRMM data product is provided by NASA/GSFC. We thank Lóránt Földváry for his help in generating a figure in the paper and for valuable discussions. This research is supported by grants from NSF (EAR-0440007), from NASA (NG04GN19G), from National Geospatial-Intelligence Agency's University Research Initiative (NURI) Program, and from the Ohio State University's Climate, Water and Carbon Program.

References

Andersen O, Berry P, Freeman J, Lemoine FG, Lutsckhe S, Jakobsen F, Butts M (2008) Satellite altimetry and GRACE gravimetry for studies of annual water storage variations in Bangladesh. Terr Atmos Ocean Sci 19(1–2):47–52

Bamzai AS, Shukla J (1999) Relation between Eurasian snow cover, snow depth, and the Indian summer monsoon: an observational study. J Climate 12:3117–3132

Chambers DP (2006) Evaluation of New GRACE time-variable gravity data over the ocean, Geophys Res Lett 33:L17603, doi:10.1029/2006GL027296

Chen JL, Wilson CR, Seo KW (2006) Optimized smoothing of Gravity Recovery and Climate Experiment (GRACE) time-variable gravity observations. J Geophys Res 111:B06408. doi:10.1029/2005JB004064

Chen JL, Wilson CR, Tapley BD, Grand S (2007) GRACE detects coseismic and postseismic deformation from the Sumatra-Andaman earthquake. Geophys Res Lett 34:L13302. doi:10.1029/2007GL030356

Crowley JW, Mitrovica JX J, Bailey RC, Tamisiea ME, Davis JL (2006) Land water storage within the Congo Basin inferred from GRACE satellite gravity data. Geophys Res Lett 33:L19402. doi:10.1029/2006GL027070

Davis JL, Tamisiea ME, El'osegui P, Mitrovica JX, Hill EM (2008) A statistical filtering approach for Gravity Recovery and Climate Experiment (GRACE) gravity data. J Geophys Res 113:B04410. doi:10.1029/2007JB005043

Duan XJ, Guo JY, Shum CK, ven der Wal W (2009) On the postprocessing removal of correlated errors in GRACE temporal gravity field solutions. J Geodesy. doi:10.1007/s00190-009-0327-0

Guo JY, Li YB, Huang Y, Deng HT, Xu SQ, Ning JS (2004) Green's function of the deformation of the Earth as a result of atmospheric loading. Geophys J Int 159:53–68. doi:10.1111/j.1365-246X.2004.02410.x

Guo JY, Duan XJ, Shum CK (2010) Non-isotropic Gaussian smoothing and leakage reduction for determining mass changes over land and ocean using GRACE data. Geophys J Int 181:290–302, doi:10.111Vj.1365-246x.2010.04534.x

Han SC, Shum CK, Jekeli CJ, Kuo CY, Wilson CR, Seo KW (2005a) Non-isotropic filtering of GRACE temporal gravity for geophysical signal enhancement. Geophys J Int 163:18–25. doi:10.1111/j.1365-246X.2005.02756.x

Han SC, Shum CK, Jekeli CJ, Alsdof DA (2005b) Improved estimation of terrestrial water storage changes from GRACE. Geophys Res Letts 32:L07302. doi:10.1029/2005GL022382
Jekeli C (1981) Alternative methods to smooth the Earth's gravity field. Rep. 327, Dept. of Geod. Sci. and Surv., Ohio State Univ., Columbus
Klees R, Revtova EA, Gunter BC, Ditmar P, Oudman E, Winsemius HC, Savenije HHG (2008) The design of an optimal filter for monthly GRACE gravity models. Geophys J Int 175:417–432. doi:10.1111/j.1365-246X.2008.03922.x
Kumar KK, Rajagopalan B, Cane MA (1999) On the weakening relationship between the Indian monsoon and ENSO. Science 284(5423):2156–2159. doi:10.1126/science.284.5423.2156
Kusche J (2007) Approximate decorrelation and non-isotropic smoothing of time-variable GRACE-type gravity field models. J Geod 81:733–749. doi:10.1007/s00190-007-0143-3
Misra AK, Saxena A, Yaduvanshi M, Mishra A, Bhauduriya Y, Takur A (2007) Proposed river linking project of India: boon or bane to nature? Environ Geol 51(8):1361–1376. doi:10.1007/s00254-006-0434-7
Nahar N, Hossain F, Hossain MD (2008) Health and socio-economic effects of groundwater arsenic contamination in rural Bangladesh: evidence from field surveys. Int Perspect J Environ Health 70(9):42–47
Peltier WR (2004) Global glacial isostasy and the surface of the ice-age earth: the ICE-5G(VM2) model and GRACE. Ann Rev Earth Planet Sci 32:111–149
Rodell M, Houser PR, Jambor U, Gottschalck J, Mitchell K, Meng CJ, Arsenault K, Cosgrove B, Radakovich J, Bosilovich M, Entin JK, Walker JP, Lohmann D, Toll D (2004) The global land data assimilation system. Bull Amer Meteor Soc 85(3):381–394
Rodell M, Velicogna I, Famiglietti JS (2009) Satellite-based estimate of groundwater depletion in India, Nature, doi:10.1038/nature08238
Schmidt M, Han S, Kusche J, Sanchez L, Shum C (2006) Regional high-resolution spatio-temporal gravity modeling from GRACE data using spherical wavelets. Geophys Res Lett 33:L08403. doi:10.1029/2005GL025509
Schmidt M, Seitz F, Shum CK (2008) Regional four-dimensional hydrological mass variations from GRACE, atmospheric flux convergence, and river gauge data. J Geophys Res 113:B10402. doi:10.1029/2008JB005575
Schrama EJO, Wouters B, Lavall´ee DA (2007) Signal and noise in Gravity Recovery and Climate Experiment (GRACE) observed surface mass variations. J Geophys Res 112:B08407. doi:10.1029/2006JB004882
Seitz F, Schmidt M, Shum C (2008) Signals of extreme weather conditions in Central Europe in GRACE 4D hydrological mass variations. Earth Planet Sci Lett 268(1–2):165–170. doi:10.1016/j.epsl.2008.01.001
Swenson S, Wahr J (2006) Postprocessing removal of correlated errors in GRACE data. Geophys Res Lett 33:L08402, doi:10.1029/2005GL025285
Syed TH, Famiglietti JS, Chen J, Rodell M, Seneviratne SI, Viterbo P, Wilson CR (2005) Total basin discharge for the Amazon and Mississippi River basins from GRACE and a land-atmosphere water balance. Geophys Res Lett 32:L24404. doi:10.1029/2005GL024851
Tapley BD, Bettadpur S, Watkins M, Reigber C (2004a) The gravity recovery and climate experiment: mission overview and early results. Geophys Res Lett 31:L09607. doi:10.1029/2004GL019920
Tapley BD, Bettadpur S, Ries J, Thompson P, Watkins M (2004b) GRACE measurements of mass variability in the Earth system. Science 305:503–505
Vorosmarty CJ, Green P, Salisbury J, Lammers RB (2000) Global water resources: vulnerability from climate change and population growth. Science 289(5477):284–288. doi:10.1126/science.289.5477.284
Wahr J, Molenaar M, Bryan F (1998) Time variablility of the Earth's gravity field: hydrological and oceanic effects and their possible detection using GRACE. J Geophys Res 103(B12):30205–30229
Wahr J, Swenson S, Velicogna I (2006) Accuracy of GRACE mass estimates. Geophys Res Lett 33:L06401. doi:10.1029/2005GL025305
Wouters B, Schrama EJO (2007) Improved accuracy of GRACE gravity solutions through empirical orthogonal function filtering of spherical harmonics. Geophys Res Lett 34:L23711. doi:10.1029/2007GL032098

Chapter 7
Observation and Geophysical Causes of Present-Day Sea-Level Rise

C.K. Shum and Chung-Yen Kuo

Contents

Abstract The 2007 IPCC Fourth Assessment Report (FAR) sea-level assessment has significantly narrowed the gap between the observations and the geophysical causes of sea-level rise than the 2001 IPCC Third Assessment Report (TAR). The observed present-day (1900–current) sea-level rise is approximately 1.8–2.2 mm/year. The unexplained discrepancy (observed compared with the sum of all known geophysical contributions to sea-level rise) dropped from 1.83 to 1.29 mm/year. A post-2007 IPCC FAR sea-level assessment study covering modern satellite measurement data span (2003–2008) indicates significant narrowing of the sea-level budget disagreement over IPCC TAR, to 0.44 mm/year. However, a review of more recent studies including the mountain glacier and ice-sheet mass balance estimates and the estimated sea-level fall from human impoundment of water in reservoirs reveal that the discrepancy is now up to 1.42 mm/year, drastically larger than the current assessment (0.44 mm/year). The unexplained sea-level signal represents 71% of the

C.K. Shum (✉)
Division of Geodetic Sciences, School of Earth Sciences, Ohio State University, 275 Mendenhall Lab, 125 S. Oval Mall, Columbus, OH 43210, USA
e-mail: ckshum@osu.edu

C.-Y. Kuo
Department of Geomatics, National Cheng Kung University, No.1, Ta-Hsueh Road, Taiwan 701, Taiwan
e-mail: kuo70@mail.ncku.edu.tw

R. Lal et al. (eds.), *Climate Change and Food Security in South Asia*,
DOI 10.1007/978-90-481-9516-9_7,

observed sea-level rise (~2.0 mm/year). Major geophysical contributors to sea-level rise identified which potentially have the largest errors include the ice-sheet mass balance, the knowledge of glacial isostatic adjustment forward models underneath the ice-sheets and the ocean, mountain glaciers and ice caps, and the anthropogenic effect of human impoundment of water in reservoirs and dams. Integrated analysis and interpretation using modern satellite and in situ measurements could narrow the uncertainty between the observations and the explained contributions from each of the geophysical sources to sea-level rise.

Keywords Sea-level rise • Global climate change • Intergovernmental Panel for Climate Change

Abbreviations

FAR	Fourth Assessment Report
TAR	Third Assessment Report
UNEP	United Nations Environment Program
WMO	World Meteorological Organization
GIA	glacial isostatic adjustment
GRACE	Gravity Recovery and Climate Experiment
MBT	mechanical bathythermographs
PMSL	permanent service for mean sea-level
LGM	last glacial melt

7.1 Introduction

The Earth's Quaternary climate, driven by Milankovich cycles which resulted in ice ages, can be characterized on time-scales on the order of 100,000 years with interlinked changes in temperature, greenhouse gases which are dominated by CO_2, and natural water reservoirs including the ice-sheets, glaciers and ice caps, hydrosphere and ocean (e.g., Shum et al. 2008). The global sea-level during the Last Ice Age, or at the Last Glacial Maximum (LGM) 20,000 years ago, is ~130 m lower than the present (Lambeck et al. 2002). During the twentieth century and since the onset of the Industrial Revolution, anthropogenic effects from greenhouse gases have led to global warming, resulting in accelerated ice-sheet melt and an increased rate of sea-level rise (Solomon et al. 2007). Paleoclimate studies of past ice-sheets from previous Ice Ages indicate a distinct potential of future accelerated ice-sheet melt and the corresponding sea-level rise due to anthropogenic climate change (Overpeck et al. 2006). From a science perspective, it is critical to understand the complicated processes of greenhouse gas forced warming to the changes in the Earth's natural reservoirs (ice-sheet, hydrosphere and ocean) contained by the solid-Earth and their feedbacks, with the present-day global sea-level rise as one of the consequences.

More than 70% of our planet is ocean. Approximately half of the world's population or 3.2 billion people lives within 200 km of coastlines (Hinrichsen 2009, http://peopleandplanet.net), and 30% of US population lives near the coastal regions (Crowell et al. 2007). Sea-level rise, widely recognized as one of consequences resulting from anthropogenic climate change, has a substantial social and economic impact, and is a timely scientific, societal and cross-disciplinary problem. The present-day (twentieth century and to the present) sea-level rise is a measurable signal using tide gauges over the last century and a half, and using Earth-orbiting satellite measurements over the past decade and a half. Quantifying sea-level change remains a complex interdisciplinary research problem, primarily because of the small magnitude of the sea-level rise signal: at ~1–2 mm/year over the last century. However, the signal has a very long or near-planetary spatial scale, allowing averaging of measurements over large ocean basins or globally and it has already been demonstrated that this small signal can be measured with adequate accuracy using tide gauges and using satellite altimetry (Douglas 2001; Cazenave and Nerem 2004). The small rate of the twentieth century and contemporary sea-level rise could only be *partially* explained, at present, by a number of competing geophysical processes, each of which is a complex process within the Earth-atmosphere-ocean-cryosphere-hydrosphere system. Improved quantification, understanding and future projection of sea-level change remains a challenge. The understanding of geophysical and anthropogenic processes leading to sea-level rise, towards improving its future projection, is a significant contemporary geoscience and societal problem.

7.2 Intergovernmental Panel on Climate Change Assessments on Sea-Level Rise

Recognizing the problem of global climate change, the World Meteorological Organization (WMO) and the United Nations Environment Programme (UNEP) established the Intergovernmental Panel on Climate Change (IPCC) in 1988. IPCC's First Assessment Report provides an estimate of global sea-level rise of 1.0–2.0 mm/year during the twentieth century with a large uncertainty (Warrick and Oerlemans 1990). The IPCC Second Assessment Report in 1996 indicates a consensus estimate of the twentieth century sea-level rise based on tide gauges at ~2 mm/year, attributing half of the rise to glacier ablation and the other half to thermal expansion of the ocean, with a conclusion that ice-sheets contributed little to sea-level rise (Warrick et al. 1996). The IPCC Third Assessment Report (TAR) in 2001 stated that the estimated twentieth century sea-level rise is at 1–2 mm/year (Church et al. 2001), and established more certainty on the tide gauge observed rate of 1.84 ± 0.35 mm/year (Douglas 2001). The observed thermal expansion of the ocean during the last 5 decades is estimated to be 0.55 mm/year (Levitus et al. 2000), which is significantly less than the anticipated value of ~1 mm/year from the 1996 IPCC Assessment. The discrepancy between the observed and explained geophysical causes is almost as large as the observed rate of sea-level rise (1.8 mm/year) in IPCC TAR (Church et al. 2001), and the problem remains an enigma (Munk 2002). The 2007 IPCC Fourth Assessment Report (FAR), Working Group I, *The Physical Science Basis*, concluded that the warming of the climate system is unequivocal, and with high certainty that the effect of human activities since 1750 has resulted in global warming (Solomon et al. 2007). Observational evidence confirms the anthropogenic increase of average air, land and ocean temperature, melting of snow and ice, and global sea-level rise. Compared with IPCC TAR, the 2007 IPCC Fourth Assessment Report (FAR) sea-level Assessment (Bindoff et al. 2007; Lemke et al. 2007) significantly closed the gap between observations and geophysical explanations.

Figure 7.1 (updated from Shum et al. 2008) shows the current knowledge of estimated, observed and projected sea-level rise for the past two centuries and the next century (1800–2100), characterizing the "*Hockey Stick*" *of Sea-Level Rise,* indicating a significant sea-level acceleration since ~1900 (Donnelly et al. 2004; Gehrels et al. 2006; Jevrejeva et al. 2008), which coincide with the Industrial Revolution. The pre-1900 estimate is based on geological interpretation at 0.1–0.2 mm/year; Lambeck et al. 2002), the tide gauge observed sea-level rise (1900–2005) (red, uncertainty with yellow shade) and satellite altimetry observed sea-level rise (1985–2005) (blue) (Cazenave and Nerem 2004; Church et al. 2004; Kuo 2006; Shum et al. 2009). The projected twenty-first century sea-level rise (IPCC FAR), based on the natural forcing plus greenhouse gases (ALL250 or

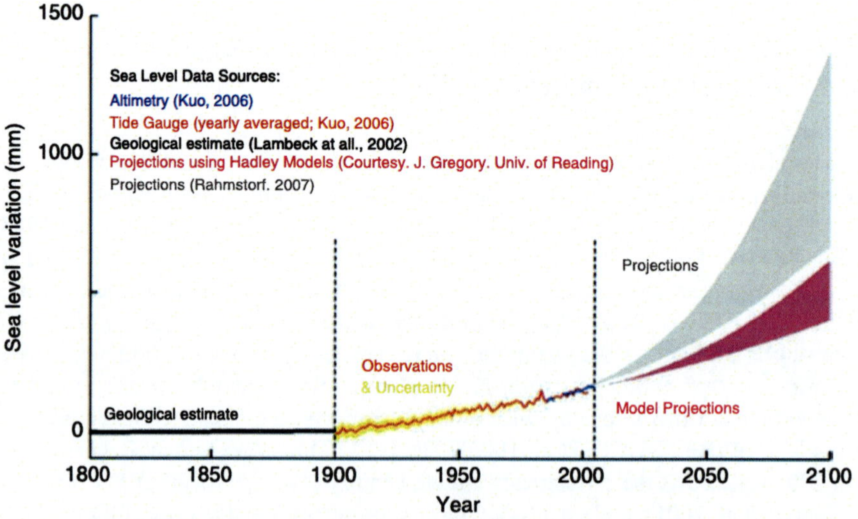

Fig. 7.1 Current knowledge of the estimated, observed and projected global sea-level rise over the past two centuries and the next century (1800–2100) (Figure updated from Shum et al. 2008)

"worse" scenario) using the HadCM3 climate model (Gregory et al. 2006) (pink envelop) is 20–60 cm. However, Rahmstorf (2007) which used empirical methods, show a much higher projected sea-level rise (70–130 cm) than the IPCC FAR projection. Siddall et al. (2009) used a climate model with ice-sheet feedbacks and assuming a maximum warming of 6.4°C and projects an end of twenty-first century sea-level rise of 82 cm (not shown here), which is in good agreement with the IPCC FAR projection. The current discrepancy of sea-level projections remains controversial and the likely improvement of model predictions depends on our enhanced understanding of the complex geophysical processes causing sea-level rise, and our ability to more accurately measure and to explain various geophysical contributions resulting in sea-level rise, including ice-sheet and glacier melt rates, oceanic thermal expansion.

7.3 Sea-Level Budget

Table 7.1 (2nd through 4th columns) summarized that the 2007 IPCC Fourth Assessment Report (FAR) sea-level assessment (Bindoff et al. 2007; Solomon et al. 2007) has significantly narrowed the gap between the observations and the geophysical causes of sea-level rise than the 2001 IPCC Third Assessment Report (TAR) (Church et al. 2001). The unexplained discrepancy (observed minus the sum of geophysical contributions) dropped from 1.83 to 1.29 mm/year, 2001 TAR to 2007 FAR assessment (Table 7.1, 2nd and 4th column), respectively. A post-2007 IPCC FAR sea-level budget assessment study by (Cazenave and Shum 2009) covering modern satellite measurement data span (2003–2008) indicates significant narrowing of the sea-level budget disagreement over IPCC TAR, to 0.44 mm/year (Table 7.1, 5th column). Here we define the sea-level trend covering time span of the twentieth century and present-day (2000–present) and provided a review of published estimates of various geophysical causes of sea-level rise and observations (Table 7.1, 6th column). We make the assumption that individual (geophysical or observed sea-level) trend estimates has mitigated long-period signals and represents "true" long-term trends for ice mass-balance or sea-level. The definition would not preclude a potential acceleration signal in the sea-level rise. The estimated rates or trends of geophysical signals (including ice-sheet mass balance) have the same definition. The result now reveals a larger discrepancy in the sea-level budget (Table 7.1, 6th column): the maximum difference between observed and explained contributions to sea-level rise is at 1.42 mm/year, which is much larger than the assessment by (Cazenave 2009) of 0.44 mm/year (Table 7.1, 5th column), and in worse agreement than the 2007 IPCC FAR assessment of 1.29 mm/year (Table 7.1, 4th column).

The unexplained sea-level signal of 1.42 mm/year represents up to 71% of the observed sea-level rise (~1.8–2.2 mm/year). This significant unraveling of our current knowledge of the sea-level budget is due primarily to a number of critical geophysical sources providing published estimates of their respective contributions

Table 7.1 Contemporary global sea-level budget (observations versus contributing geophysical sources)

Geophysical sources of sea-level rise	2001 IPCC TAR 1900–1990[a](mm/year)	2007 IPCC FAR (mm/year) (Bindoff et al. 2007) 1961–2003	2007 IPCC FAR (mm/year) (Bindoff et al. 2007) 1993–2003	Post–FAR study[c] (mm/year) 2003–2008	This study (mm/year) 1900–2008
Thermal expansion	0.5 ± 0.2	0.42 ± 0.12	1.6 ± 0.5	0.31 ± 0.15	0.4 ± 0.2
	(0.3 → 0.7)	(0.3 → 0.54)	(1.1 → 2.1)	(0.16 → 0.46)	(0.2 → 0.6)[1]
Glaciers and Ice caps	0.3 ± 0.1	0.50 ± 0.18	0.77 ± 0.22	1.1 ± 0.24	0.96 ± 0.44
	(0.2 → 0.4)	(0.32 → 0.68)	(0.55 → 0.99)	(0.86 →1.34)	(0.52 → 1.4)[2]
Greenland Ice sheet	0.3 ± 0.3	0.05 ± 0.12	0.21 ± 0.07		0.3 ± 0.33
	(0.0 → 0.6)[a]	(−0.07 → 0.17)	(0.14 → 0.28)	1.0 ± 0.15	(−0.03 → 0.63)[3]
Antarctic Ice sheet	−0.1 ± 0.1	0.14 ± 0.41	0.21 ± 0.35	(0.85 → 1.15)	0.14 ± 0.26
	(−0.2 → 0.0)[a]	(−0.27 → 0.55)	(−0.14 → 0.56)		(−0.12 → 0.4)[4]
Terrestrial hydrology		–	–	0.17 ± 0.1	0.17 ± 0.1
	−0.35 ± 0.75			(0.07 → 0.27)	(0.07 → 0.27)[5]
Water impoundment in reservoirs	(−1.1 → 0.4)	–	–	–	−0.55[6]
Sum of geophysical contributions	0.65 ± 0.84	1.1 ± 0.5	2.8 ± 0.7	2.58 ± 0.34	1.42 ± 0.82
	(−0.8 → 2.1)	(0.6 → 1.6)	(2.1 → 3.5)	(1.94 → 3.22)	(0.6 → 2.24)
Observed total sea-level rise	1.5 ± 0.5	1.8 ± 0.5	3.1 ± 0.7	2.5 ± 0.4	2.0 ± 0.2
	(1.0 → 2.0)	(1.3 → 2.3)	(2.4 → 3.8)	(2.1 → 2.9)[b]	(1.8 → 2.2)
Difference (observed–explained)	0.85 ± 0.98	0.7 ± 0.7	0.30 ± 0.99	−0.08 ± 0.52	0.58 ± 0.84
	(−0.13 → 1.83)	(0.0 → 1.4)	(−0.69 → 1.29)	(−0.60 → 0.44)	(−0.26 → 1.42)

[a]Included natural and "climate" contributions from ice-sheets (Church et al. 2001)
[b]Observed by satellite altimetry
[c]Cazenave (2009), similar studies by Peltier (2009), Leuliette and Miller (2009), Willis et al. (2008)
[1]Antonov et al. (2005), Ishii and Kimoto (2009), Gouretski and Koltermann (2007), Domingues et al. (2008)
[2]Arendt et al. (2002), Dyurgerov and Meier (2005), Kaser et al. (2006), Meier et al. (2007), (Cogley 2009)
[3]Abdalati et al. (2001), Rignot and Thomas (2002), Krabill et al. (2004), Zwally et al. (2005), Johannessen et al. (2005), Rignot and Kanagaratnam (2006), Ramillien et al. (2006), Chen et al. (2006b), Luthcke et al. (2006), Velicogna and Wahr (2006a), Cazenave (2009), Shepherd and Wingham (2007), Slobbe et al. (2009)
[4]Thomas et al. (2004), Davis et al. (2005), Wingham et al. (2006), Zwally et al. (2005), Rignot et al. (2008), Cazenave (2009), Shepherd and Wingham (2007), Shum et al. (2008). GRACE estimâtes (Ramillien et al. 2006; Chen et al. 2006a, Velicogna and Wahr 2006b, Shum et al. 2008): 0.14–0.40 mm/year
[5]Milly et al. (2003), Ngo-duc et al. (2005), Ramillien et al. (2008)
[6]Chao et al. (2008), no uncertainties provided

to sea-level rise with large discrepancies, or there are still large uncertainties for these estimates. These geophysical sources include the ice-sheet mass balance estimates, the land water contributions (including terrestrial hydrology, glaciers and ice caps and human-impoundment of water in reservoir and dams), and the effect of glacial isostatic adjustment (GIA) process, or solid Earth's viscoelastic rebound due to deglaciation of ancient ice-sheets from the Last Ice Age since the LGM. The ultimate objective is to improve the individual estimates quantifying of each of these geophysical sources causing present-day sea-level rise.

7.3.1 Ice-Sheet Mass Balance Estimates

There are significant discrepancies between the various ice-sheet mass balance estimates (Table 7.1, 6th column), from –0.03 to 0.63 mm/year and from –0.12 to 0.4 mm/year (equivalent sea-level) for Greenland and for Antarctica, respectively, using data from airborne altimetry (Krabill et al. 2004) and satellite altimetry (Cazenave and Nerem 2004), from synthetic aperture radar interferometry (InSAR) (Rignot and Thomas 2002) and from Gravity Recovery and Climate Experiment (GRACE) satellite mission (Tapley et al. 2004). The discrepancy are primarily due to (1) poor knowledge of glacial isostatic adjustment (GIA) (Peltier 2004) over the ice-sheets, which is in particular critical when GRACE is used (Velicogna and Wahr 2006b): the choice of the GIA models significantly affects Antarctica mass balance estimates, adding 0.25–0.45 mm/year of equivalent sea-level rise (Shum et al. 2008), (2) firn-compaction and ice column density variations when (airborne and satellite) altimetry are used (Helsen et al. 2008), (3) short data spans (GRACE) or lack of finer than seasonal sampling (InSAR, airborne altimetry) which may bias the trend estimate, and (4) significant differences in GRACE mass balance estimates including results with different spatial resolutions and with land-sea signal leakage (Lettenmaier and Milly 2009; Guo et al. 2010). Figure 7.2 shows the satellite radar altimetry and GRACE estimated Antarctica mass balance. Notable *discrepancies* between the estimates from altimetry and GRACE are in the Antarctic Peninsula, JJ′ (altimetry shows no large mass loss); E. Antarctica, AA″ & A″B; Siple Dome, E′E″, D″D; Enderby Land, A″B, Oakes Coast, D″D′. However, there are good agreements between satellite altimetry and GRACE observed ice-sheet mass balance, for example, over the regions including: Basin GH (Amudsen Sea), W. Antarctica, and Basin BC (Lambert Glacier/Emery Ice Shelf), E. Antarctica (Fig. 7.2). In summary, the discrepancy of the ice-sheet mass balance estimates for Greenland ranges from –0.03 to 0.63 mm/year equivalent sea-level rise, and for Antarctica ranges from –0.14 to 0.40 mm/year, respectively (Table 7.1, 6th column), representing one of the largest uncertainties contributing to the current discrepancy of the sea-level budget.

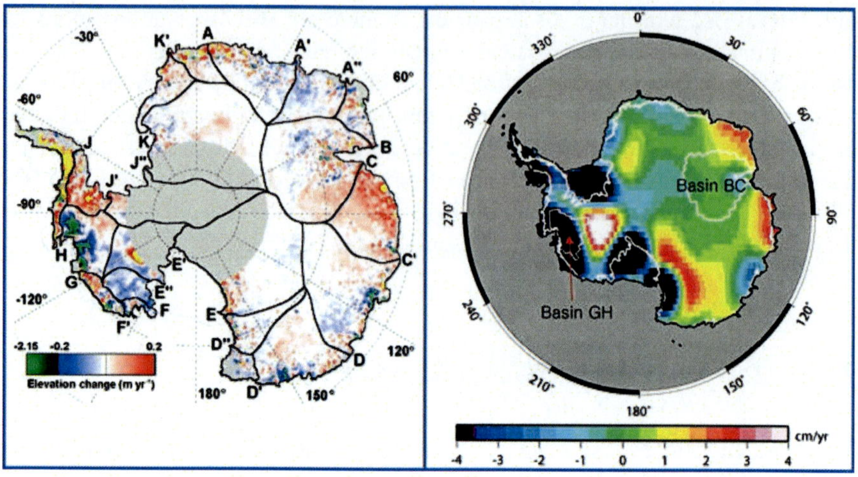

Fig. 7.2 *Left*: ERS-1–2 radar altimetry determined Antarctica ice elevation change (m/year, 1992–2003) (Figure and data courtesy, D. Wingham et al. 2006). *Right*: GRACE observed ice mass balance (cm/year, 4/2002–3/2009), GIA correction using ICE-5G (VM4) (Peltier 2004), destriped/smoothed using a 200-km radius non-isotropic Gaussian filter, with geocenter and land signal leakage corrections (Shum et al. 2008; Guo et al. 2010; Duan et al. 2009)

7.3.2 Land Water Contribution to Sea-Level

Chao et al. (2008) estimated that the sea-level fall resulting from human impoundment of water in reservoirs and dams over the last 50 years is equivalent to –0.55 mm/year. However, Lettenmaier and Milly (2009) evaluated a number of smaller land contributions to sea-level rise, and argued that since most of the large manmade reservoirs were constructed since about the 1950 and one may no longer be assumed that they contributed negatively to sea-level rise during the current time period.

The contribution of mountain glaciers and ice caps to sea-level rise is by far the largest and with a wide range of estimates: 0.52–1.4 mm/year (Arendt et al. 2002; Dyurgerov and Meier 2005; Kaser et al. 2006; Meier et al. 2007; Cogley 2009) (Table 7.1, 6th column). The latest estimate of 1.1–1.4 mm/year (Cogley 2009) is due to the revised modelling of tide water glaciers. Like Meier et al. (2007), Cogley (2009) based the estimates entirely on in situ measurements and extrapolated globally, and the estimate did not include any satellite observations (e.g., of ice cap thinning). Of the 0.8–1.1 mm/year contribution of glacier loss estimated by Meier et al. (2007) (and probably by Cogley (2009)) and the IPCC FAR study, errors are thought to be of the order of 14–23%, respectively (Milly et al. 2009). Therefore, it is critical to narrow the uncertainty of this estimate.

The contributions of terrestrial hydrologic imbalance, potentially due to global warming, to sea-level rise has been estimated to be 0.07–0.27 mm/year (equivalent sea-level rise) based on forward hydrologic models and using GRACE (Milly et al. 2003; Ngo-duc et al. 2005; Ramillien et al. 2008). This particular geophysical contribution is among the least known and further study is warranted.

7.3.3 Glacial Isostatic Adjustment

Glacial isostatic adjustment process (Peltier 2001, 2004) produces a significant signal in the solid Earth where tide gauges are located (Douglas 2001), underneath the ice-sheets and the ocean where the mass balance signal and the sea-level signal has opposite signs, respectively.

Therefore, in addition to the need of more accurate forward GIA modeling to correct tide gauge sea-level record, GIA corrections for GRACE mass change estimates over ice-sheets, and for ocean mass variations due to exchange of water fluxes from land to ocean, is critically needed. In addition the magnitude of the GIA model in terms of water thickness change over the ocean, is on the order of 1–2 mm/year (depending on whether the Paulson (2006) or the ICE-5G (Peltier 2004) GIA model is used, respectively), which has the same magnitude as the expected signal (GRACE ocean mass variations, ~2 mm/year) (Cazenave 2009; Peltier 2009; Leuliette and Miller 2009; Willis et al. 2008). On the other hand, land-water storage, according to GRACE, shows very little trend (dashed line) over 2002–2008 (Fig. 7.3, from Lettenmaier and Milly 2009). The total land water mass variations excluding Greenland and Antarctica observed by GRACE is shown here (Fig. 7.3, the shading indicates the range of estimates from the various GRACE solutions from different processing centers (data courtesy, D. Chambers), with the thick black line the mean of the three solutions). The right-hand axis shows the sea-level anomaly, which corresponds to the land equivalent water-depth anomaly. The estimated rate of land water change, ~0.3 mm/year (Cazenave 2009) (into the ocean) is "balanced" by the observed oceanic mass (~2 mm/year), *if* the GIA forward model correction (1–2 mm/year) is *perfect*. Figure 7.4 shows the ocean geoid change (mm/year) predicted by the ICE-5G(VM4) GIA model (Peltier 2004). This quantity if expressed in the form of mass variations (or water thickness change) by including self-gravitational or the elastic loading effect (e.g., Wahr et al. 1995),

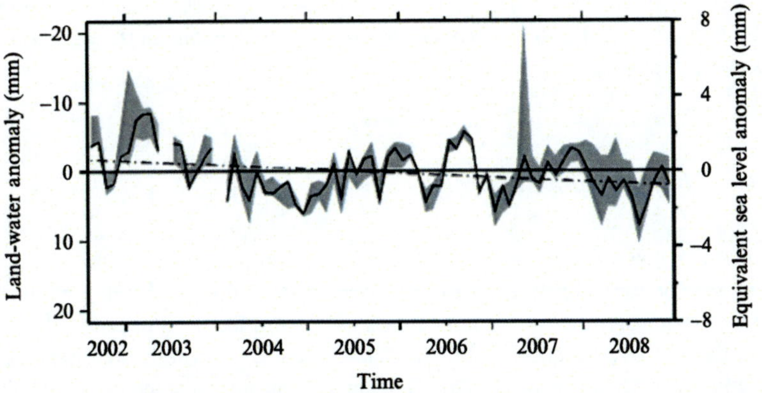

Fig. 7.3 GRACE equivalent water-depth anomalies (departures from time mean) over the global land area with exception of Greenland and Antarctica (Figure from Lettenmaier and Milly 2009)

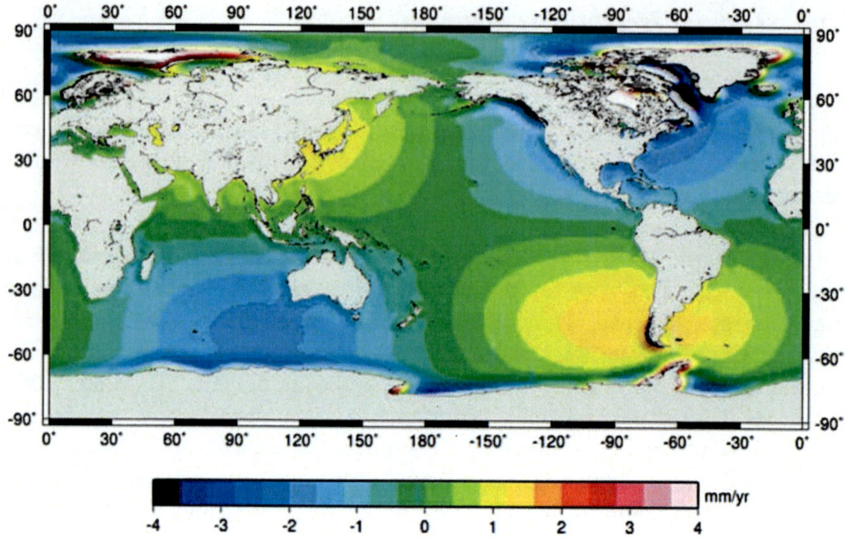

Fig. 7.4 The geoid change in the ocean predicted by the ICE-5G (VM4) glacial isostatic adjustment model (Peltier 2004)

would be amplified approximately five times of the ocean geoid change values, i.e., ~2 mm/year when averaged over the ocean (Peltier 2009). It is highly unlikely that the GIA forward model is accurate enough that an unambiguous claim is possible at present that GRACE ocean mass variations is 'balanced' by the observed land water fluxes flow into the ocean (Cazenave 2009; Peltier 2009; Leuliette and Miller 2009; Willis et al. 2008). Figure 7.4 also shows a see-saw positive and negative pattern in the ocean geoid change, indicative of an effect due to Earth rotational feedback. This pattern seems to be too large to be realistic and possibly this component of the modeling is in error. However, this error probably does not necessarily increase significantly the current uncertainty of predictive GIA models. In summary, there is a critical emphasis to improve the constraints of or to improve the GIA model itself, for improved estimation of oceanic mass variations and ice-sheet mass balance.

7.3.4 Thermal Expansion

Thermal expansion of the ocean, and to a less extent, the contraction of the ocean due to salt or the salinity (halosteric) effect affects the sea-level. Levitus et al. (2000) reported significant ocean heat transport from the deep ocean to the surface during the last five decades, and estimated that the thermosteric sea-level rising at 0.55 mm/year (Levitus et al. 2000). Since then, the discovery of significant instrument biases in the old mechanical bathythermographs (MBTs), the more

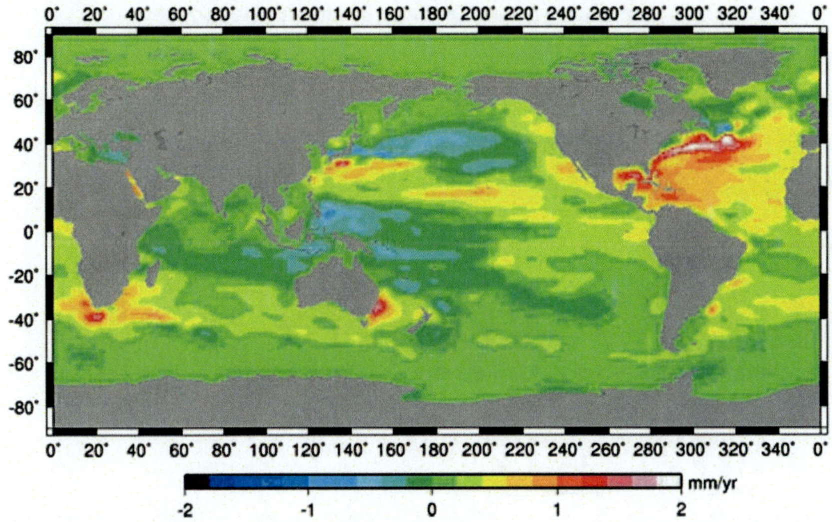

Fig. 7.5 Thermosteric sea-level trend (1955–2005): 0.24 mm/year, data based on Ishii and Kimoto (2009) data

modern expandable bathythermographs (XBTs), and the modern Argo arrays (http://sio-argo.ucsd.edu) (Willis et al. 2007), have caused several revised studies of the thermosteric sea-level rise (Antonov et al. 2005; Ishii and Kimoto 2009; Gouretski and Koltermann 2007; Domingues et al. 2008; Wijffels et al. 2008). However, the estimates still have a wide range, 0.24–0.6 mm/year (Table 7.1, 6th column). Figure 7.5 shows the thermosteric sea-level trend, 1955–2005, estimated to be 0.24 mm/year, based on the Ishii and Kimoto (2009) objectively analyzed data integrating the sea surface from 0 to 700 m depth. It is evident that more studies are needed to validate the contribution of thermal expansion to the global sea-level, and additional measurements are needed to quantify the deeper ocean (>700 m) thermosteric sea-level rise.

7.4 Sea-Level Measurements

Contemporary observations of sea-level rise (1900–2005) (WCRP Sea-Level Workshop Summary 2006) used long-term tide gauges (Douglas 2001; Miller and Douglas 2006; Holgate 2007) from Permanent Service for Mean Sea-Level (PSMSL) RLR data records (Woodworth and Player 2003), and more recently (since 1992) used satellite radar altimetry from TOPEX/POSEIDON (T/P) (Cazenave and Nerem 2004; Church et al. 2004; Kuo 2006). Tide gauges are sparsely distributed globally but have long records (>70 years). They are located near islands and continental margins, measures relative sea-level, and are susceptible to uncertainty of land motion, e.g., due to glacial isostatic adjustment (GIA).

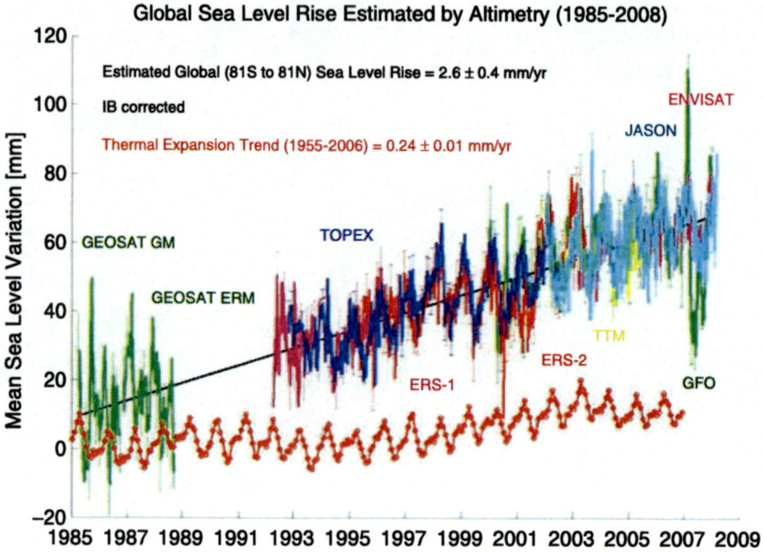

Fig. 7.6 Multiple-mission satellite radar altimetry (GEOSAT, TOPEX/POSEIDON, ERS-1/-2, JASON-1, ENVISAT and GFO, 1985–2008) observed sea-level trend (2.6 mm/year, 2.9 mm/year when GIA geoid change correction is applied using ICE-5G(VM4) model [Peltier 2004]). There is a data gap between 1989 and 1991

Satellite altimetry measures geocentric sea-level and unaffected by land motion (to a less extent, affected by changing shape of the ocean basin) and have globally coverage (Shum et al. 1995). However, they have much shorter records (10 years) than tide gauges, and require absolute calibration or monitoring for potential instrument drifts (e.g., Shum et al. 2003).

Figure 7.6 shows the multiple-mission satellite radar altimetry (GEOSAT, TOPEX/POSEIDON, ERS-1/-2, JASON-1, ENVISAT and GFO, 1985–2008) observed sea-level trend (2.6 mm/year, 2.9 mm/year when the GIA geoid correction is applied using the ICE-5G(VM4) model (Peltier 2004)). There is a data gap between 1989 and 1991. The inverted barometer correction (IB) has been applied. Also plotted is the thermosteric sea-level during the altimetry data span (data from Ishii and Kimoto, 2009). The seasonal variations are not removed. Figure 7.7 shows the sea-level trend observed by individual tide gauges (1900–2006, trend color coded and show as circles) with a global average of 1.65 ± 0.4 mm/year. Note that tide gauges sample only about a few % of the global ocean surface. The background is the satellite altimetry observed short-term (1985–2008) trend of 2.6 ± 0.4 mm/year (2.9 mm/year after corrected for the GIA effect), when averaged globally. Satellite altimetry with its global sampling reveals that the rate of sea-level rise is not uniform globally (Fig. 7.7) and that the estimated trend is potentially dominated by interannual or longer variations in the ocean. In some regions (e.g., Western Pacific), rates of sea-level rise are faster by a factor up to 3 than the global mean rate. In other regions rates are slower than the global mean (e.g., eastern Pacific). The estimated trend from altimetry (2.9 mm/year)

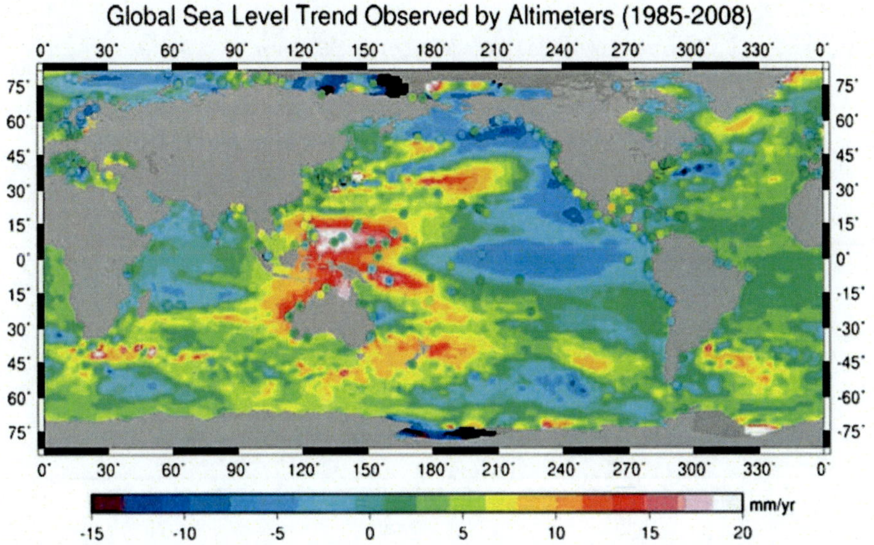

Fig. 7.7 Sea-level trend observed by individual tide gauges (1900–2006, trend color coded and show as circles) with a global average of 1.6 ± 0.4 mm/year. The background is the satellite altimetry observed short-term (1985–2008) trend of 2.6 ± 0.4 mm/year (2.9 mm/year after GIA effect is corrected), when averaged globally

which is much larger than the trend estimated by tide gauges (1.65 mm/year) may or may not necessarily indicate a recent (1990s) acceleration of sea-level rise (Woodworth et al. 2009; Merrifield and Merrifield 2009).

Figure 7.8 illustrates the differences of these two sea-level measurement types. Figure 7.8 shows the globally averaged 500 long-term (>30 years) tide gauge sea-level time series, 1900–2004 (corrected for GIA using ICE-5G model). The monthly averaged time series (grey) is shown to have significant 'false' (large) amplitudes in the seasonal signal, while the yearly averaged time series (blue) agrees between with average of satellite altimetry sea-level measurements (monthly average in green and yearly average in red). The estimated trend from tide gauges is 1.65 mm/year, compared with altimetry observed sea-level (1984–2008) trend of 2.9 mm/year (after GIA uplift correction). Figure 7.8 shows that global sampling is required to average out the variability in the sea-level trend (Fig. 7.7), and that the trend estimates using short-data span from satellite altimetry may be contaminated interannual or longer variations oceanic signals.

The Gravity Recovery and Climate Experiment (GRACE) twin-satellites measure *mass change* of the Earth (Tapley et al. 2004) (in the form of water thickness change) inferring from its temporal gravity field solutions, using the known harmonic (Stokes) coefficients, $\dot{C}_{nm}$ and $\dot{S}_{nm}$ (Wahr et al. 1998) (n is degree and m is order, equation not shown here), with monthly sampling (or finer) and spatial resolutions, ranging from 400 km (half-wavelength) (e.g., Velicogna and Wahr 2006b) to 200 km using *mascons* (e.g., Luthcke et al. 2006), from regional solutions

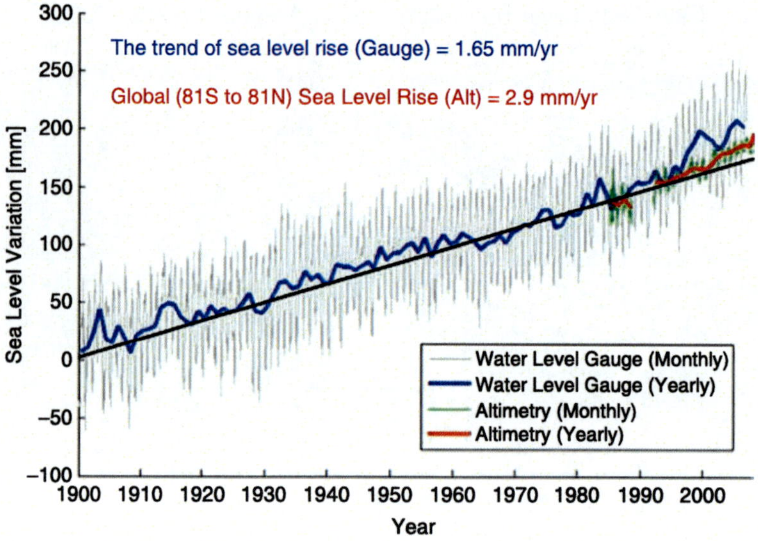

Fig. 7.8 Global sea-level trend observed by tide Gauges (1900–2006) and Altimetry (1985–2008)

including the use of spherical wavelets (Han et al. 2005; Schmidt et al. 2006), and recently achieving 200 km resolution via data post-processing (decorrelation and latitude-dependent non-isotropic filtering) (Duan et al. 2009; Guo and Shum 2009; Guo et al. 2010). The computation of mass changes involves elastic loading in the form of load Love number, k_n. GRACE is not sensitive to $n = 1$ terms (or geocenter) which has noticeable effects, e.g., on ice-sheet mass balance estimates (Chambers et al. 2007). Geocenter corrected used could be solutions from satellite laser ranging to Lageos satellites (J. Ries, pers. comm.). The smoothing or filtering and 'destripping' of Stokes coefficients representing GRACE's temporal gravity field solutions (Wahr et al. 2006) is necessary to mitigate sampling/observability problems (Swenson and Wahr 2006; Duan et al. 2009; Guo et al. 2010), and various correction algorithm for land signal leakage due to filtering near land-ocean boundaries (Velicogna and Wahr 2006b; Chen et al. 2006b; Guo et al. 2010). Figure 7.9 shows a global ocean mass (or the ocean bottom pressure) trend map estimated by GRACE. The combination of thermosteric sea-level (Fig. 7.5) and satellite altimetry (Fig. 7.7) yields the ocean mass variations (Fig. 7.9) (e.g., Kuo et al. 2008; Chambers 2006a, b).

The combination of these modern satellite and in situ data is anticipated to provide more accurate measurement of geophysical sources contributing to present-day sea-level rise, and the sea-level signal directly. In addition, the integrated analysis using these data sets allows sea-level budget studies treating each of the geophysical sources (ice-sheet, land water, ocean) separately, improving the chance to provide better quantification of the respective contributions from each of these sources to present-day sea-level rise.

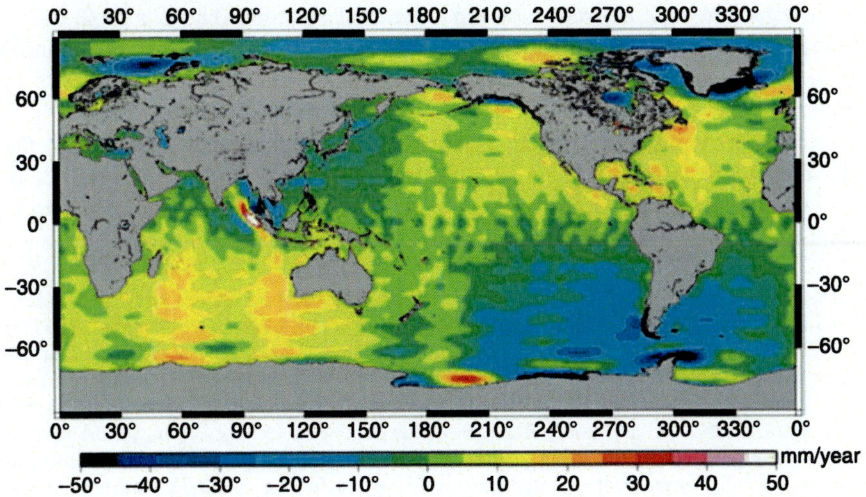

Fig. 7.9 GRACE observed global ocean botom pressure interannual variation, 2003–2008 (mm/year). CSR RL04 Level 2 data product used, water thickness change, de-striping and smoothed at 200 by 300 km by an non-isotropic filter, ICE5G(VM4) GIA geoid and seasonal signals removed, land leakage signal repaired. (Duan et al. 2009; Guo et al. 2010; Guo and Shum 2009)

7.5 Conclusions

At present, the observed and explained geophysical causes of present-day global sea-level rise appears to be worse than the assessment published by the 2007 Intergovernmental Panel for Climate Change (IPCC) Working Group I, Fourth Assessment Report (FAR). Several of the geophysical causes have been identified, including the contributions to present-day sea-level rise from ice-sheet mass balance, mountain glacier and ice caps, land water including human-impoundment of water in reservoirs, and the effect of the geodynamic process, the glacial isostatic adjustment which affects measurements of sea-level and ice-sheet mass balance using tide gauges, satellite altimetry and satellite gravimetry, GRACE. Integrated analysis and interpretation using these satellite measurements and in situ measurements including tide gauges and hydrographic data will narrow the uncertainty between the observations and the explained contributors of sea-level rise from each of these geophysical sources.

Sea-level rise is a major threat for many low-lying, highly populated coastal regions of the world (about 3.2 billion people live presently within 200 km of the coastal area (Hinrichsen 2009, http://peopleandplanet.net)). Sea-level rise exaggerates the effect of land erosion, wetland loss, storm surges associated with typhoons or hurricanes, sediment loading, land subsidence (Dixon et al. 2006) due to natural or anthropogenic (ground water pumping, oil extraction, urbanization) effects, rising water table and salt-water intrusion in freshwater aquifers (Nicholls 2002, 2007). In response to sea-level rise, in particular and for example, in Bangladesh, significant considerations should be given to the science and engineering aspects to mitigate relative sea-level rise (absolute sea-level rise adding the effect of land subsidence or vertical motion due to natural and anthropogenic effects) from the above-mentioned phenomena. In addition, the mitigation and adaptation requires an integrated approach including policy change based on scientific and engineering assessment of the risk from sea-level rise.

Acknowledgments We acknowledge Philip Woodworth, UK's Permanent Service for Mean Sea-Level (PSMSL), for providing the tide gauge records, NASA and CNES for the TOPEX/POSEIDON, Jason-1 radar altimetry data, ESA for the ERS-1/-2 and Envisat altimetry data, US Navy via NOAA's Laboratory for Satellite Altimetry for the Geosat and GFO altimetry data, NASA and GFZ for the GRACE data via University of Texas Center for Space Research and JPL-PODAAC, M. Ishii at Frontier Research Center for Global Change, Japan, for the thermal sea-level data, D. Peliter, H. Wang, and P. Wu for providing the glacial isostatic adjustment models, and D. Wingham at University College London for the ERS-1/-2 altimetry derived Antarctic ice-sheet elevation data. This research is supported by grants from NASA and from the Ohio State University's Climate, Water and Carbon Program. Chung-yen Kuo is supported by grants from the National Cheng Kung University, and from the National Science Council, Taiwan.

References

Abdalati W, Krabill W, Frederick E, Manizade S, Martin C, Sonntag J, Swift R, Thomas R, Wright W, Yungel J (2001) Outlet glacier and margin elevation changes: near-coastal thinning of the Greenland ice sheet, J Geophys. Res 106(D24): 33, 729–33, 742

Antonov J, Levitus S, Boyer T (2005) Steric variability of the world ocean, 1955–2003. Geophys Res Lett 32(12):L12602. doi:10.1029/2005GL023112

Arendt A, Echelmeyer K, Harrison W, Lingle C, Valentine V (2002) Rapid wastage of Alaska glaciers and their contribution to rising sea-level. Science 297:383–386

Bindoff N, Willebrand J (Coordinating Lead Authors) Artable V, Cazenave A, Gregory J, Gulev S, Hanawa K, Le Quere C, Levitus S, Nojiri Y, Shum CK, Talley L, Unnikrishnan A (Lead Authors) and 50 contributing authors (2007) Chapter 5: observations: oceanic climate change and sea-level, Intergovernmental Panel Climate Committee (IPCC) Working Group 1 (WG1) Fourth Assessment Report

Cazenave A, Nerem R (2004) Present-day sea-level change: observations and causes. Rev Geophys 42:RG3001. doi:10.1029/2003RG000139

Cazenave A (2009) Sea-level budget after IPCC AR4: a reevaluation from satellite altimetry, GRACE and Argo data over 2003–2008, AAAS Annual Meeting: Global Sea-Level Rise: Observation, Causes, and Prediction, Chicago, Illinois, Feb. 12–16

Cazenave A, Shum C (2009) Sea-level budget after IPCC AR4: a reevaluation from satellite altimetry, GRACE and Argo data over 2003–2008, Joint IPCC-WCRP-IGBP Workshop: New Science Directions and Activities Relevant to the IPCC AR5, Hawaii, March 3–6

Chambers D (2006a) Observing seasonal steric sea-level variations with GRACE and satellite altimetry. J Geophys Res 111:C03010. doi:10.1029/2005JC002914

Chambers D (2006b) Evaluation of new GRACE time-variable gravity data over the ocean. Geophys Res Letts 33:L17603. doi:10, 1029/2006GL027296

Chambers DP, Tamisiea ME, Nerem RS, Ries JC (2007) Effects of ice melting on GRACE observations of ocean mass trends. Geophys Res Lett 34:L05610. doi:10.1029/2006GL029171

Chao B, Wu Y, Li Y (2008) Impact of artificial reservoir water impoundment on global sea-level. Science. doi:10.1126/science.1154580

Chen J, Wilson C, Blankenship D, Tapley B (2006a) Antarctic mass rates form GRACE. Geophys Res Lett 33:L11502. doi:10.1029/2006GL026369

Chen J, Wilson C, Tapley B (2006b) Satellite gravity measurements confirm accelerated melting of Greenland Ice Sheet. Science 313:1958. doi:10.1126/science.1129007

Church J, Gregory J, Huybrechts P, Kuhn M, Lambeck K, Nhun M, Qin D, Woodworth P, 26 others (2001) Chapter 11: change in sea-level, Intergovernmental Panel Climate Committee (IPCC) Working Group 1 (WG1) Third Assessment Report

Church J, White N, Coleman R, Lambeck K, Mitrovica J (2004) Estimates of the regional distribution of sea-level rise over the 1950–2000 period. J Climate 17:2609–2625

Cogley JG (2009) Geodetic and direct mass-balance measurements: comparison and joint analysis. Ann Glaciol 50:96–100

Crowell M, Edelman S, Coulton K, McAfee S (2007) How many people live in coastal areas? J Coast Res 23:5. doi:10.2112/07A-0017.1

Davis C, Li Y, McConnell J, Frey M, Hanna E (2005) Snowfall-driven growth in east Antarctic Ice Sheet mitigates recent sea-level rise. Science 10.1126/science.1110662

Dixon T, Amelung F, Ferretti A, Novali F, Rocca F, Dokka R, Sella G, Kim SW, Wdowinski S, Whitman D (2006) Subsidence and flooding in New Orleans, Nature, doi:10.1038/441587a

Domingues C, Church J, White N, Gleckler P, Wijffels S, Barker P, Dunn J (2008) Upper-ocean warming and sea-level rise, Nature 453, doi:10.1038/nature07080

Donnelly JP, Cleary P, Newby P, Ettinger R (2004) Coupling instrumental and geological records of sea-level change: evidence from southern New England of an increase in the rate of sea-level rise in the late 19th century. Geophys Res Lett 31:L05203. doi:10.1029/2003GL018933

Douglas B (2001) In: Douglas B, Kearney M, Leatherman S (eds) Sea-level rise: history and consequences, sea-level change in the era of the recording tide gauge. Academic, San Diego

Duan XJ, Guo JY, Shum CK, van der Wal W (2009) On the post-processing removal of correlated errors in GRACE temporal gravity field solutions. J Geodesy. doi:10.1007/s00190-009-0327-0

Dyurgerov M, Meier M (2005) Glaciers and changing Earth system: a 2004 snapshot. INSTAAR, Boulder

Gehrels W, Marshall M, Larsen G, Kirby J, Eiríksson J, Heinemeier J, Shimield T (2006) Rapid sea-level rise in the North Atlantic Ocean since the first half of the nineteenth century. The Holocene 16:949–965

Gouretski V, Koltermann K (2007) How much is the ocean really warming? Geophys Res Lett 34:L01610. doi:10.1029/2006GL027834

Gregory J, Lowe J, Tett S (2006) Simulated global-mean sea-level changes over the last half-millennium. J Climate 19:4,576–4,591. doi:10.1175/JCLI3881.1

Guo JY, Shum C (2009) Application of the cos-Fourier expansion to data transformation between different latitude-longitude grids. Comput Geosci 35:1439–1444

Guo JY, Duan XJ, Shum CK (2010) Non-isotropic filtering and leakage reduction for determining mass changes over land and ocean using GRACE data. Geophys J Int 181: 290–302. doi: 10.111/j.1365-246X.2010.04534.x

Han S, Shum C, Jekeli C, Kuo C, Wilson C, Seo K (2005) Non-isotropic filtering of GRACE temporal gravity for geophysical signal enhancement. Geophys J Int 163(1):18–25. doi:10.1111/j.1365-246X.2005.02756.x

Helsen M, van den Broeke M, van de Wal R, van de Berg WJ, Meijaard E, Davis C, Li YH, Goodwin I (2008) Elevation changes in Antarctica mainly determined by accumulation variability. Sciencexpress 10.1126/science.1153894, 1–4, May

Hinrichsen, D (2009) Ocean planet in decline, peopleandplantet.net: People and coasts and oceans, http://www.peopleandplanet.net/doc.php%3Fid=429&section=6.html

Holgate S (2007) On the decadal rates of sea-level change during the twentieth century. Geophys Res Lett 34:L01602. doi:10.1029/2006GL028492

IPCC (2007) Intergovernmental Panel on Climate Change, Climate Change 2007: The Physical Science Basis, Summary for Policymakers

Ishii M, Kimoto M (2009) Reevaluation of historical ocean heat content variations with time-varying XBT and MBT depth bias corrections. J Oceanogr 65:287–299

Jevrejeva S, Moore JC, Grinsted A, Woodworth PL (2008) Recent global sea-level acceleration started over 200 years ago? Geophys Res Lett 35:L08715. doi:10.1029/2008GL03611

Johannessen O, Khvorostovsky K, Miles M, Bobylev L (2005) Recent ice-sheet growth in the interior of Greenland. Science 310:1013–1016

Kaser G, Cogley J, Dyurgerov M, Meier M, Ohmura A (2006) Mass balance of glaciers and ice caps: consensus estimates for 1961–2004. Geophys Res Lett 33:L19501. doi:10.1029/2006GL027511

Krabill W, Hanna E, Huybrechts P, Abdalati W, Cappelen J, Csatho B, Frederick E, Manizade S, Martin C, Sonntage J, Swift R, Thomsa R, Yungel J (2004) Greenland ice sheet: increased coastal thinning. Geophys Res Lett 31:L24402. doi:10.1029/2004GL021533

Kuo C (2006) Determination and characterization of 20th century global sea-level rise, OSU Report 478, ix + 158 pp

Kuo C, Shum C, Braun A, Cheng K, Yi Y (2008) Vertical motion determined using satellite altimetry and tide gauges, Special Issue: satellite altimetry over land and coastal zones: challenges and applications. Terr Atmos Ocean Sci 19(1–2):21–35. doi:10.3319/TAO.2008.19.1-2.21(SA

Lambeck K, Esat T, Potter E (2002) Links between climate and sea-levels for the past three million years. Nature 419:199–206

Lemke P, Ren J, Alley R, Allison I, Carrasco J, Flato G, Fujii Y, Kaser G, Mote P, Thomas R, Zhang TJ, and 44 contributing authors (2007) Chapter 4; Observations: changes in snow, ice and frozen ground, IPCC WG1 Fourth Assessment Report

Lettenmaier D, Milly C (2009) Land waters and sea-level. Nat Geosci 2:452–454

Leuliette EW, Miller L (2009) Closing the sea-level rise budget with altimetry, Argo, and GRACE. Geophys Res Lett 36:L04608. doi:10.1029/2008GL036010

Levitus S, Antonov I, Boyer T, Stephens C (2000) Warming of the world ocean. Science 287:2225–2229

Luthcke S, Zwally H, Abdalati W, Rowlands D, Ray R, Nerem R, Lemoine F, McCarthy J, Chinn D (2006) Recent ice sheet mass loss by drainage system from satellite gravity observations. Science. doi:10.1126/science.1130776

Meier M, Dyurgerov M, Rick U, O'Neel S, Pfeffer W, Anderson R, Anderson S, Glazovsky A (2007) Glaciers dominate eustatic sea-level rise in the 21st century, Science, doi: 10.1126/science.1143906

Merrifield MA, Merrifield ST (2009) An anomalous recent acceleration of global sea-level rise. J Climate, doi:10.1175/2009JCL12985.1, 2009

Miller L, Douglas B (2006) On the rate and causes of twentieth century sea-level rise. Phil Trans R Soc A, doi:10.1098/rsta.2006.1738

Milly P, Cazenave A, Gennero M (2003) Contribution of climate-driven change in continental water storage to recent sea-level rise. Proc Natl Acad Sci 100(23):13158–13161

Milly PCD, Cazenave A, Famiglietti J, Gornitz V, Laval K, Lettenmaier D, Sahagian D, Wahr J, Wilson C (2009) Terrestrial water storage contributions to sea-level rise and variability, Proceedings of the WCRP workshop 'Understanding sea-level rise and variability'. In: Church J, Woodworth P, Aarup T, Wilson S, et al., Blackwell Publishing, Inc., New York

Munk W (2002) Twentieth century sea-level: an enigma. Proc Natl Acad Sci 99(10):6550–6555

Ngo-Duc T, Laval K, Polcher Y, Lombard A, Cazenave A (2005) Effects of land water storage on the global mean sea-level over the last half century. Geophys Res Lett 32:L09704. doi:10.1029/2005GL022719

Nicholls RJ (2002) Rising sea-level: potential impacts and responses. In: Hester RE, Harrison RM (eds) Issues in environmental science and technology; global environmental change, vol 17., pp 83–107

Nicholls RJ (2007) The impacts of sea-level rise. Ocean Challenge 15(1):13–17

Overpeck J, Otto-Bliesner B, Miller GH, Muhs DR, Alley RB, Kiehl JT (2006) Paleoclimatic evidence for future ice-sheet instability and rapid sea-level rise. Science 311:1747–1750

Paulson A (2006) Inference of the Earth's mantle viscosity from post-glacial rebound, PhD Dissertation, University of Colorado

Peltier W (2001) Global glacial isostatic adjustment and modern instrumental records of relative sea-level history. In: Douglas B, Kearney M, Leatherman S (eds)., Sea-level rise: history and consequences, Chapter 4. Academic Press, New York, 65–93

Peltier W (2004) Global glacial isostasy and the surface of the ice-age Earth: The ICE-5G (VM2) model and GRACE. Annu Rev Earth Planet Sci 32:111–149

Peltier W (2009) Closure of the budget of global sea-level rise over the GRACE era: the importance and magnitudes of the required corrections for global glacial isostatic adjustment, Quatern Sci Rev, 28(17–18), 1658–1674

Rahmstorf S (2007) A semi-empirical approach to projecting future sea-level rise. Science 315:368–370

Ramillien G, Lombard A, Cazenave A, Ivins E, Llubes M, Remy F, Biancale R (2006) Interannual variations of the mass balance of the Antarctica and Greenland ice sheets from GRACE. Global Planet Change 53(3):198–208

Ramillien G, Bouhours S, Lombard A, Cazenave A, Flechtner F, Schmidt R (2008) Land water contributions from GRACE to sea-level rise over 2002–2006. Global Planet Change 60:381–392

Rignot E, Thomas R (2002) Mass balance of polar ice sheets. Science 297:1502–1506

Rignot E, Kanagaratnam P (2006) Changes in the velocity structure of the Greenland Ice Sheet. Science 311:986–990

Rignot E, Bamber JL, van den Broeke MR, Davis C, Li YH, van de Berg WJ, Meijaard EV (2008) Recent Antarctic ice mass loss from radar interferometry and regional climate modeling. Nature 1:106–110

Schmidt M, Han C, Kusche J, Sanchez L, Shum C (2006) Regional high-resolution spatio-temporal gravity modeling from GRACE data using spherical wavelets. Geophys Res Lett 33:L08403. doi:10.1029/2005GL025509

Shepherd A, Wingham D (2007) Recent sea-level contributions of the Antarctic and Greenland ice sheets. Science 315:1529–1532. doi:10.1126/science.1136776

Shum C, Ries J, Tapley B (1995) The accuracy and applications of satellite altimetry. Geophys J Int 121:321–336

Shum C, Yi Y, Cheng K, Kuo C, Braun A, Calmant S, Chamber D (2003) Calibration of Jason-1 altimeter over Lake Erie. Marine Geodesy 26:335–354. doi:10.1080/01490410390253487

Shum C, Kuo C, Guo J (2008) Role of Antarctic ice mass balances in present-day sea-level change. Polar Sci 2:149–161

Shum C, Cazenave A, Kuo CY (2009) Quantifying geophysical causes of present-day sea-level rise. Joint IPCC-WCRP-IGBP Workshop: New Science Directions and Activities Relevant to the IPCC AR5, Hawaii, March 3–6

Siddall M, Stocker TS, Clark PU (2009) Constraints on future sea-level rise from past sea-level change. Nat Geosci, doi:10.1038/NGE0587

Slobbe DC, Ditmar P, Lindenbergh RC (2009) Estimating the rates of mass change, ice volume change and snow volume change in Greenland from ICESat and GRACE data. Geophys J Int 176:95–106

Solomon S, Qin D, Manning M, Alley RB, Berntsen T, Bindoff NL, Chen Z, Chidthaisong A, Gregory JM, Hegerl GC, Heimann M, Hewitson B, Hoskins BJ, Joos F, Jouzel J, Kattsov V, Lohmann U, Matsuno T, Molina M, Nicholls N, Overpeck J, Raga G, Ramaswamy V, Ren J, Rusticucci M, Somerville R, Stocker TF, Whetton P, Wood RA, Wratt D (2007) Technical summary. In: Solomon S, Qin D, Manning M, Chen Z, Marquis M, Averyt K, Tignor M, Miller H (eds) Climate change 2007: the physical science basis. Contribution of Working Group I to the fourth assessment report of the Intergovernmental Panel on Climate Change. Cambridge University Press, New York, pp 21–84

Swenson S, Wahr J (2006) Post-processing removal of correlated errors in GRACE data. Geophys Res Lett 33:L08402. doi:10.1029/2005GL025285

Tapley BD, Bettadpur S, Ries J, Thompson P, Watkins M (2004) GRACE measurements of mass variability in the Earth system. Science 305:503–505

Thomas R, Rignot E, Casassa G, Kanagaratnam P, Acuna C, Akins T, Brecher H, Frederrick E, Gogineni P, Krabill W, Manizade S, Ramamoorthy H, Rivera A, Russell R, Sonntag J, Swift R, Yungel J, Zwally J (2004) Accelerated sea-level rise from West Antarctica. Science 306(5694):255–258

Velicogna I, Wahr J (2006a) Acceleration of Greenland ice mass loss in spring 2004. Nature 443:329–331

Velicogna I, Wahr J (2006b) Measurements of time-variable gravity show mass loss in Antarctica. Science 311:1754–1756

Wahr J, Han H, Trupin A (1995) Predictions of vertical uplift caused by changing polar ice volumes on a viscoelastic Earth. Geophys Res Lett 22(8):977–980

Wahr J, Molenaar M, Bryan F (1998) Time variability of the Earth's gravity field: hydrological and oceanic effects and their possible detection using GRACE. J Geophys Res 103(B12):30,205–30,229

Wahr J, Swenson S, Velicogna I (2006) Using GRACE to estimate changes in land water storage: present limitations and future potential, Proc. Understanding Sea-level Rise and Variability Workshop, Paris, France, 6–9 June

Warrick R, Oerlemans J (1990) Sea-level rise. In: Houghton J, Jenkins G, Ephraums J (eds) Climate Change: the IPPC scientific assessment. Cambridge University Press, Cambridge, pp 257–281

Warrick RA, Provost Le C, Meier M, Oerlemans J, Warrick RA, Provost Le C, Meier M, Oerlemans J, Woodworth P (1996) Changes in sea-level. In: In Climate change 1995, Intergovernmental Panel on Climate Change. Cambridge University Press, Cambridge

Wijffels SE, Willis J, Domingues CM, Barker P, White NJ, Gronell A, Ridgway K, Church JA (2008) Changing expendable bathythermograph fall rates and their impact on estimates of thermosteric sea-level rise. J Climate 21:5657–5672

Willis JK, Lyman JM, Johnson GC, Gilson J (2007) Correction to "Recent cooling of the upper ocean". Geophys Res Lett 34:L16601. doi:10.1029/2007GL030323

Willis JK, Chambers DP, Nerem RS (2008) Assessing the globally averaged sea-level budget on seasonal to interannual timescales. J Geophys Res 113:C06015. doi:10.1029/2007JC004517

Wingham D, Shepherd A, Muir A, Marshall G (2006) Mass balance of the Antarctic ice sheet. Phil Trans R Soc A 364:627–1635. doi:10.1098/rsta.2006.1792

Woodworth P, Player R (2003) The permanent service for mean sea-level: an update to the 21st century. J Coastal Res 19:287–295

Woodworth P, Jevrejeva S, Holgate S, Church J, White N, Gehrels R (2009) A review of the evidence for the recent accelerations of sea-level on multi-decade and century timescales. Int. J. Climatol. 29: 777–789

World Climate Research Programme (WCRP) Workshop Summary Statement (2006 June) Understanding the sea-level rise and variability. UNESCO, Paris, France

Zwally H, Giovinetto M, Li J, Cornejo H, Beckley M, Brenner A, Saba J, Yi D (2005) Mass changes of the Greenland and Antarctic ice sheets and shelves and contributions to sea-level rise: 1992–2002, J Glaciology, 51 (175): 509–527

Part III
Climate Change and The Environment

Chapter 8
Selenium: A Right Choice to Treat Arsenicosis in Bangladesh

Abdul Momin

Contents

Abstract One half of the Bangladeshi population has been drinking arsenic contaminated water, drawn from the ground by tube-wells since 1993. More than 38,000 arsenicosis cases are reported. The present knowledge of the management of arsenicosis is limited, and specific treatment of chronic poisoning has not yet been identified.

With the approval of the ethical review board of the Bangladesh Medical Research Council (BMRC) double blind, randomized, placebo controlled trial with selenium intervention was carried out on 174 arsenicosis patients, irrespective of age and sex, for 12 months in a hyper-contaminated rural area of Bangladesh. In this study, melanosis decreased in 76% ($n = 67$, $p < 0.00$) of selenium treated patients and palmo-planter keratosis in 81% ($n = 67$, $p < 0.00$). For the selenium group arsenic content was decreased 38.2% ($p < 0.01$) in hair and 37.2% ($p < 0.00$) in nails. Overall symptoms improved 68% ($p < 0.00$) in the selenium treated group. There was no observed toxicity in a heart, kidney and liver function test. It was found that a dose of 100 μg of selenium as selenomethionine per day along with use of arsenic safe water for 12 months in chronic arsenic toxicity is a safe and effective treatment for arsenicosis.

A. Momin (✉)
Department of Dermatology, Dhaka Medical College, Dhaka 1000, Bangladesh
e-mail: amominderma@hotmail.com

R. Lal et al. (eds.), *Climate Change and Food Security in South Asia*,
DOI 10.1007/978-90-481-9516-9_8,

Keywords Arsenicosis • Selenium • Bangladesh

Abbreviations

BMRC	Bangladesh Medical Research Council
ECG	Electrocardiogram
HG-AAS	Hydride Generation Atomic Absorption Spectrometer
NAMIC	National Arsenic Mitigation information Center
SGPT	Serum Glutamic Pyruvate Transaminase
SPSS	Statistical Package for Social Sciences
WHO	World Health Organization

8.1 Introduction

Safe drinking water, sanitation and good hygiene are fundamental to health, survival, growth and development. Unfortunately, these basic needs are still a luxury for many of the poor people in a country like Bangladesh. From 1993, half of Bangladesh's population was drinking arsenic contaminated water drawn from the ground by tube-wells. More than 38,000 cases of arsenicosis have already been reported (Douglas 1999; Elizabeth 2000; Ganapati 2000; NAMIC 2005). Tondel et al. (1999) found that arsenic concentration ranged from 10 to 2,040 μg/L in tube-well, and crude prevalence rate was 29/100 Bangladeshi people. There was a significant dose response relationship ($p < 0.05$), regardless of sex (Tondel 1999). In econometric analysis, it was found that the burden of arsenicosis is on low-income, poor people (Milton et al. 2004; Anwar 2002). There is no known antidote. Avoidance of arsenic contaminated drinking water, chelating agents, retinoid, vitamins, balanced diet, spirulina, all these have tried with varying therapeutic results (Sato et al. 2000; Guha et al. 1998; Guha et al. 2001; Boyd et al. 1989; Thianprasit 1984; Khan and Ahmad 2002; Khan et al. 2001; Ahmad et al. 1998; Kosnett 1998; Huq et al. 2000; Misbahuddin et al. 2006)

Evidence from both laboratory and epidemiological studies suggests that arsenic obviously has an inhibitory effect on the antioxidant enzymes containing selenium by reacting with -SH group. This inhibitory impact has been observed in increased arsenic accumulation (Lin and Chiang 2000; Bates et al. 1992). Inorganic arsenic, once ingested, may lead to the production of reactive oxygen species (ROS) that can induce DNA damage, including single and double-strand breaks and nucleotide base modifications (Lynn et al. 2000; Shi et al. 2004; Liu and Jan 2000; Hei et al. 1998; Amundson et al. 1999). Selenium might be a suitable agent to reduce arsenic accumulation after chronic exposure for a number of reasons, i.e., there are a number of possible points and mechanisms for metabolic interaction between arsenic and selenium which include competition for the methyl donor, S-adenosylmethionine, competition for glutathione (GSH) and inhibition of glutathione reductase by a number of arseno-glutathione complexes. Moreover, both selenium and arsenic interact extensively with sulfhydral (-SH) groups in tissues; it is possible that arsenic elimination is delayed in Se-deficiency because there could be more target -SH groups for arsenic to react with because Se intake is low (Kenyon et al. 1997). There is evidence that low Selenium intake may influence the development of arsenicosis (Islam et al. 2004). In an animal study, it was observed that the cytotoxic effect of arsenic can be prevented through dietary supplementation by selenium in mice, which is of significance in protecting against the widespread toxicity observed in humans (Biswas et al. 1999; Nasir et al. 2004), but multicentral trial of selenium in humans is lacking. In 1989 a clinical trial in China among the smelter workers exposed to arsenic showed that the chromosomal aberration rate of cultured lymphocytes in workers was lowered by 46.1% after treatment with selenium (150 μg/day for 21 days; Hu 1989). There were no soil selenium maps of Bangladesh though Spallholz et al. (1978) had measured in 25 samples from Jessore district of Bangladesh and found less amount of selenium in soil, which automatically produces low selenium containing crops. In a developing country and largely agrarian Bangladesh villages, animal protein intake is low, with fruits and vegetables often proving to be poor sources of selenium where arsenicosis is occurring and excessive selenium excretion owing to selenium/arsenic complexion may add to the likehood of arsenic being more toxic and carcinogenic over time (Spallholz et al. 1978, 2004). So, interest was found to conduct a clinical trial with selenium intervention among arsenicosis patients in Bangladesh.

8.2 Materials and Methods

Arsenicosis patients were selected from 11 villages having a population of 19,000 of Shahpur union, under Chatkhil Upazilla, of District Noakhali. The area is a hyper-contaminated arsenic zone (British Geological Survey 1999). From a total population 560 probable arsenicosis patients, screened with the help of local elites and a field team, were invited to a health camp. The patients were selected clinically by the dermatologist and confirmed by positive spot urine examination by a digital arsenitor (positive indicates an arsenic concentration 1 ppb or more).

All adults, male and female, with a history of exposure to arsenic contaminated drinking water from shallow tube-well for more than 6 months and presenting signs/symptoms of arsenicosis were included in the study. Exclusion criteria included patients not exposed to arsenic, no clinical feature of arsenicosis, patient refusal to give consent, patients not having arsenic in urine, patients having concurrent illness like malaria, tuberculosis, or history of smoking (Yes or No), alcoholics (drinking more than 2 pegs) or taking hepatotoxic drug. Pregnant and lactating mothers were also excluded. Finally, 174 patients, irrespective of age and sex, were selected. An identification number was assigned to one of the three treatment groups “A” or “B” randomly following a computer generated number.

The study was approved by the ethical review board of the Bangladesh Medical Research Council (BMRC). Prior to entering the study each patient signed an informed consent form and were assured of their right to withdraw from participation. The information were kept strictly confidential and used only for research purposes. The study was a part of research work leading to a PhD.

A baseline survey was conducted which included information about age, gender, height, weight, socio-economic status, water use data, drinking water source, cooking water source, number of tube wells, duration of use and skin manifestation of arsenicosis with duration. Samples of water from the tube well were collected for arsenic estimation. In 10% randomly selected cases, 24 h food recall surveys were conducted for the estimation of arsenic in consumed food. Urine and blood samples were collected before and at the end of 4, 8 and 12 months. Only scalp hair and finger nail samples were collected at the beginning and at the end of study period. All the collected samples were labeled properly by including the identification number, date of collection and nature of the specimen. All the samples were transported to Dhaka within 24 h in frozen containers and were stored at –20°C until analysis.

Treatment Procedure: The arsenicosis patients were assigned randomly to group 'A' or group 'B'. Each group of participants was provided the drugs, identical in appearance and blindly coded as 'A' and 'B', respectively. The drugs were delivered to each participant in a sealed air tight plastic bottle of same color and size. Only code number was written on each bottle. The bottles were packed earlier with 15 tablets of the respective group of drugs by the pharmacists who kept the code confidential. Neither the investigator nor the patients knew the intervention groups. Each patient was instructed to visit the treatment camp monthly for the drug. While receiving the drug, the patient had to bring the previously used containers in order to check the compliance. Each patient was instructed to swallow one half of the tablet daily with a glass of water. Any failure to take the drug for 1 day was recorded and instructed to take a double dose the next day. The trained field workers regularly visited the patients at home to ensure compliance. The overall supervision, both in clinic and field, was maintained by the principal investigator over 12 months. None of the patients were allowed to drink arsenic contaminated water throughout the study period. All the study subjects drank boiled surface water throughout the study period. A local office-cum clinic was set up and a registered medical officer was recruited to monitor the treatment of patients. After the laboratory analysis was complete, the drug codes were decoded and data were analyzed along with clinical, biochemical and social data. After decoding, intervening drugs were found to be 'A' for Selenium and 'B' for Placebo.

8.2.1 The Intervention Agents

(A) Selenium commercially sold in the market as a solid tablet form, named 'Selenium' (Manufactured by Schiff and distributed by Schiff products, Salt Lake City, UT, USA 84104) packed in a sealed bottle. Each tablet containing 200 μg selenium as yeast rich L-selenomethionine, without any artificial color or preservatives.

(B) Placebo preparations containing potato starch in each as a tablet form which is identical in appearance and color to the selenium tablet (manufactured by The Acme laboratories Ltd, Dhanmondi, Dhaka, Bangladesh).

8.2.2 Analysis of Biological Samples

All biological samples of hair, nails and urine were digested in high purity acids and pretreated with reductants prior to analysis for estimation of arsenic using continuous flow hydride generation with Atomic Absorption Spectrophotometer (HG-AAS) (Buck Scientific, USA, and Model 210 VGP).

8.2.3 Other Biochemical Parameters

For other biochemical parameters, such as random blood glucose, serum alkaline phosphatase, and serum alanine transaminase, estimation has been done with a clinical chemistry analyzer (Microlab 300, The Netherlands). Urine was examined for the presence of albumin and microscopically examined for any cast and RBC which denotes as abnormal findings. Electrocardiogram (ECG) was done in selective cases when there was high blood pressure (either >160 in systolic or >90 in diastolic or in combination) or high pulse rate (>100/min) found in physical examination.

8.2.4 Statistical Analyses of the Data

Data analyses were carried out using Statistical Package for Social Science (SPSS) (version 12.0 for windows, SPSS Inc, Chicago, USA). Results were expressed as mean ± SD (standard deviation) and in graphical forms after log transformation to make the data normal. The statistical analyses Pearson Chi-square test (χ^2), and Student two-tailed paired t-test were performed. Differences were considered significant with $^*p < 0.05$.

8.2.5 Clinical Outcome Evaluation

The patients were examined in every follow-up visit by the same dermatologist, and severity of keratosis, melanosis, leucomelanosis and other clinical variables were recorded in the checklist without going through the records. Primary outcome measures for this study included changes in arsenical skin lesions (incidence of new lesions and improvement of old lesions), as assessed by clinical examination and secondary outcome measure included changes of arsenic levels in hair, nails and urine.

Skin lesions related to arsenic tend to occur on certain parts of the body (particularly palms, soles, and trunk), and the efficacy of the intervention as indicated by prognosis of severity of lesions (particularly melanosis, leucomelanosis and keratosis), was evaluated as grade l = mild, grade ll = moderate and grade lll = severe (WHO 2002) and recorded in the checklist at each follow-up visit.

8.3 Results

In this study, out of 174 patients selected, 33 patients could not complete the 1 year study periodic follow-up, so the dropout rate was 18.9%. Among the study subjects, 112 (64.3 %) were young adult female and the rest were male. The majority of the subjects had five or more family members and showed that the mean BMI was 20.3 ± 3.5 kg/m^2, 94% of male subjects and 89% of female subjects had suffered from arsenicosis for more than 4 years (Fig. 8.1).

As all our study subjects were from the same homogenous environment, it was found that am average of 45.45 μg/L of total arsenic was taken by each subject daily through their food chain, and an average of 243 μg/L through drinking contaminated shallow tube well water. About 33.3% subjects were drinking contaminated water initially, but eventually this rate dropped down to zero. More than 94% cooked their meals with surface water, which was found safe from arsenic. But more than 87% were drinking arsenic contaminated water from shallow tube well water for 21.1 ± 11.4 years (range 11.5–35.6 years) (Table 8.1).

The selenium treated group showed significant improvement of their physical and clinical signs and symptoms after intervention. Marked changes were observed in the severity scores compared to their pre-treatment values.

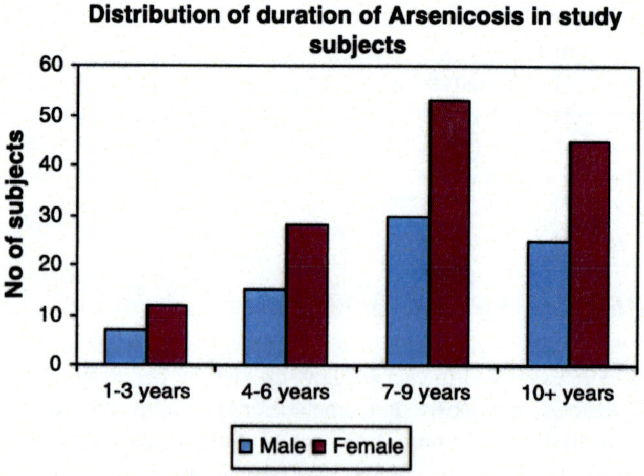

Fig. 8.1 Duration of arsenicosis in study subjects

Table 8.1 Percent distribution of source of water use

	Drinking		Cooking	
Source of water	Selenium	Placebo (%)	Selenium	Placebo (%)
Deep tube well	6.9	10.3	3.4	1.1
Shallow tube well	86.2	87.4	2.3	2.3
Pond	4.6	2.3	94.3	96.6
Filter/boiling	2.3	0.0	0.0	0.0

Table 8.2 Distribution of skin changes before and after intervention in study subjects

	Selenium (69)[a]		Placebo (72)[a]	
Skin sign	Before (%)	After (%)	Before (%)	After (%)
Grade I (mild)				
Diffuse melanosis/suspicious spotty pigmentation on trunk/limb	14.5	50.7	16.9	21.1
Mild thickening of palms/soles	19.0	49.2	21.2	21.2
Suspicious spotty depigmentation on trunk/limb	33.3	24.2	38.0	38.0
Grade II (moderate)				
Definite spotty pigmentation on trunk/limb, bilateral	30.4	36.2	26.8	22.5
Definite spotty depigmentation over trunk/limb, bilateral	48.5	50.0	39.4	35.2
Severe diffuse thickening of palms/soles	30.2	41.3	34.8	34.8
Grade III (severe)				
Definite spotty pigmentation on trunk & limb, bilateral	55.1	13.0	56.3	56.3
Definite spotty depigmentation over trunk & limb, bilateral	18.2	25.8	22.5	26.8
Mucosal pigmentation in tongue or oral mucosa	0.0	0.0	0.0	0.0
Large nodules over thickened palm/soles	50.8	9.5	43.9	43.9
Diffuse verrucous lesions with crack and fissure on soles	15.9	10.1	6.9	6.9
Non-healing ulcer, e.g., Bowen's disease, etc.	2.9	0.0	2.9	0.0
Gangrene of palm or sole	2.9	2.9	0.0	0.0

[a]Figures in parentheses are number of patients; % = percentile

The severity of melanosis and keratosis reduced to almost normal skin in many subjects after intervention. The melanosis was reduced 76% in severe grade ($n = 67$, $\chi^2 = 32.13$, $p < 0.00$) of selenium treated group after intervention, but very little change was found in Placebo group ($n = 72$, $\chi^2 = 0.59$, $p < 0.74$). The severity of keratosis in the palm and soles was reduced by 81% in severe grade of selenium intervention group ($n = 67$, $\chi^2 = 27.27$, $p < 0.00$), but no changes were observed in placebo group ($n = 72$, $\chi^2 = 0.00$, $p < 1.00$) (Table 8.2). The symptoms like anorexia, nausea, vomiting, weakness, dizziness, etc., reduced from 100% to 31.9%

(68% reduction, $p < 0.00$) in selenium intervention group but there was only 2.8% reduction in the placebo group.

The Arsenic concentration in nails reduced significantly (83.1%) after intervention in Selenium treated group ($n = 55$, $t = 2.708$, $p < 0.00$) but very little change was observed in the placebo group ($n = 61$, $t = -0.191$, $p < 0.85$). The Arsenic concentration of hair was also reduced (61.8%) significantly after intervention in the selenium treated group ($n = 60$, $t = 5.269$, $p < 0.00$), but also found a 33.7% reduction in the placebo group ($n = 67$, $t = 4.672$, $p < 0.00$) (Table 8.3).

The concentration of arsenic titer in urine also found significant reduction (42.4%) after intervention in selenium group ($t = 6.653$, $p < 0.00$), and a 41.1% reduction in the placebo group ($t = 7.493$, $p < 0.00$). The trend of arsenic excretion in urine found in selenium group was linear (Fig. 8.2).

The serum selenium concentration was found to be 1.792 ± 0.293 μg/L in study subjects, but after intervention the concentration rose to 62.4% only in the selenium treated group ($t = -2.079$, $p < 0.04$) with no change in the placebo group.

There was no substantial organopathy found in kidney function test, liver function test and random blood glucose before and after intervention in study population. Liver function test showed that there was no abnormality in serum glutamic pyruvate transaminase (SGPT) level before and after intervention in

Table 8.3 Shows distribution of concentration of Arsenic (As) in nail and hair of study subjects

Parameters	Group	Before	After	*p*-value
As concentration in nail μg/kg as mean ± SD	Selenium (55)[a]	0.314 ± 0.522	0.052 ± 0.608	<0.00
	Placebo (61)[a]	0.222 ± 0.540	0.235 ± 0.582	<0.85
As concentration in hair μg/kg as mean ± SD	Selenium (60)[a]	0.123 ± 0.442	0.047 ± 0.560	<0.00
	Placebo (67)[a]	0.608 ± 0.442	0.403 ± 0.493	<0.00

[a]Figures in parenthesis are sample numbers; μg/kg = microgram per kilogram

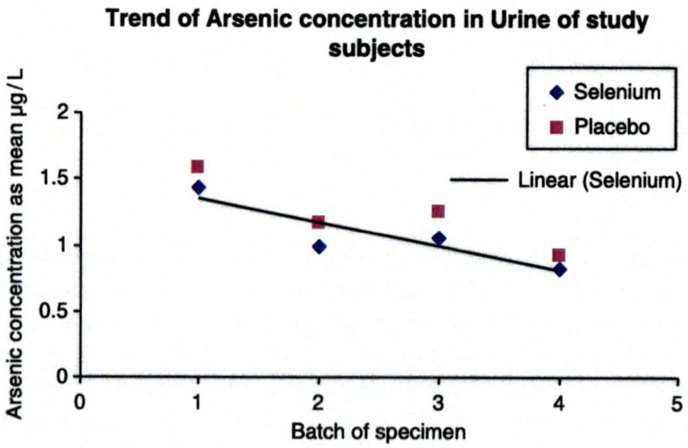

Fig 8.2 Trend of arsenic excretion in urine of selenium group

study subjects, but only 23.1% of cases had a high level of alkaline phosphatase before intervention which, after intervention, 74% reversed under the selenium group and no changes were found in the placebo group, respectively. There was no significant change in the blood glucose level of the study population after intervention ($p < 0.06$).

8.4 Discussion

In this study, it was observed that all the study subjects had multiple clinical presentation and there were gradual improvement of clinical sign and symptoms, including skin changes, after intervention. The severity of melanosis and keratosis was reduced to almost normal skin in many subjects. The severity of melanosis and keratosis was reduced by 76% and 81% respectively in selenium group after intervention but very little changed in placebo group. This study was similar to the findings by Wang et al. in China where they administered 100–200 μg selenium/day for 14 months and observed 75% and 55% reduction in clinical signs like keratosis and melanosis respectively (Wang et al. 2001). Rabbani in Bangladesh showed 63.3% improvement in planter keratosis, and a 59.7% decrease in melanosis (Rabbani et al. 2003).

A urinary concentration of total arsenic is a reliable indicator of arsenic consumption because urine is the primary route of elimination of most absorbed arsenicals (Vahter 1994). In this study, after 12 months of treatment with selenium and consumption of arsenic safe water, urinary concentration of total arsenic dropped to 50% of the pretreatment value in all groups. This drop was likely to be related to sudden cessation of arsenic consumption accomplished by taking boiled pond water. However, the peak excretion levels did not reach a constant level or a plateau at the end of the 12 month period; this indicates that treatment time might have to be prolonged to reach a maximum excretory threshold level.

In this study, the concentration of arsenic in scalp hair was significantly reduced in the selenium group. The reduction in placebo treated groups is likely to be related to the introduction of arsenic-free water. In contrast, reduction of arsenic load in hair after intervention with 200 μg/day selenium for 14 months in Mongolia found the hair arsenic content of selenium patients decreased by 73%, whereas placebo patients dropped 52% (Yang et al. 2002). In this study, the present observations support the findings by Tseng, in Taiwan, Guha in India and Yang in Mongolia that a supply of arsenic safe water alone is not sufficient to detoxify individuals exposed to arsenic for a prolonged period (Tseng et al. 1968; Guha et al. 1998; Yang et al. 2002). The findings suggest that prolonged exposure to toxic levels of inorganic arsenic may induce changes in the body that are reversible by preventing further exposure to arsenic and supplementation with selenium with a therapeutic safe dose. We found that a combination of 100 μg of selenium as selenomethionine and arsenic-safe water per day for 12 months is a safe, effective, and cheap, but needs more multicentral trials before advising mass use in combating arsenicosis in

Bangladesh. More study is also necessary for estimation of selenium in different foods and soil mapping in Bangladesh.

Acknowledgement It is a profound privilege to acknowledge a deep sense of gratitude to Professor Dr. Shah Mohammed Keramat Ali, professor of nutrition and food science, University of Dhaka, for his continuous thrust to undertake this study and necessary corrections in finalizing this manuscript. The author gratefully acknowledges the help received from the Managing Committee of Surjodayshangha, Shahpur union, Noakhali and to the staffs of National Hospital, Noakhali for their active co-operation in data collection. I express my indebtedness and gratitude to Dr. Mohammad Alauddin, chairman, Exotics Technology Center, Ashwuliya, Dhaka for his active contributions. I thank the management of The Acme Laboratories Limited, Dhaka for their immense co-operation. I am sincerely grateful to the Environmental unit of World Health Organization (WHO), Bangladesh for their assistance. Finally, thanks to my patients for their kind participation and co-operation to accomplish this study.

References

Ahmad SK, Faruquee MH, Sayeed MHSU et al (1998) Chronic Arsenicosis: management by vitamins A, E, C regimen. JOPSM 17:19–26

Amundson SA, Bittner M, Chen Y et al (1999) Florescent cDNA microarray hybridization reveals complexity and heterogenicity of cellular genotoxic stress responses. Oncogene 18:3666–3672

Anwar J (2002) The poor in Bangladesh suffers effects and treatment of arsenic poisoning. Proceedings of international workshop on arsenic mitigation, Dhaka, January 14–16

NAMIC (National Arsenic Mitigation Information Centre) (2005) Facts and Findings of the National Screening Program, News letter, Bangladesh Arsenic Mitigation Water Supply Project, May 2005

Bates MN, Smith AH, Hopenhayn-Rich C (1992) Arsenic ingestion and internal cancers: a review. Am J Epidermiol 135:462–76

Biswas S, Talukder G, Sharma A (1999) Prevention of cytotoxic effects of arsenic by short-term dietary supplementation with selenium in mice in vivo. Elsevier Science B.V. Mutant Res 441:155–160

Boyd VL, Harbell LW, O'Connor RJ et al (1989) 2, 3-Dithioerythritol, a possible new arsenic antidote. Chem Res Toxicol 2:301–306

British Geological Survey, Department of Public Health and Engineering and Mott MacDonald Limited (1999) Ground water studies for arsenic contamination in Bangadesh, Main report, Ministry of Local Government, Rural Development and Cooperatives, Government of the Peoples Republic of Bangladesh

Douglas RM (1999) Health aspects of arsenic contamination of tube well in Bangladesh. In: Final report of consultancy visit for the World Bank to Bangladesh, pp 5–26

Elizabeth MJ (2000) Arsenic 2000: an overview of the arsenic issue in Bangladesh. Water Aid Bangladesh, pp 1–63

Ganapati M (2000) Half of Bangladesh population at risk of arsenic poisoning. BMJ 320:822

Goering PL, Aposhian HV, Mass MJ et al (1999) The enigma of arsenic carcinogenesis: role of metabolism. Toxicol Sci 49:5–14

Guha MDN, Gupta JD, Santra A et al (1988) Chronic arsenic toxity in West Bengal-the worst calamity in the world. J Indian Med Assoc 96:4–7

Guha MDN, Ghoshal UC, Shaha J et al (1998) Randomized placebo controlled trial of 2, 3-dimercapto-succinic acid in therapy of chronic arsenicosis due to drinking arsenic contaminated subsoil water. J Toxicol Clin Toxicol 36:683–690

Guha MDN, De BK, Santra A, Ghosh N et al (2001) Randomized placebo-controlled trial of 2, 3-dimercapto-1-propanesulfonate (DMPS) in therapy of chronic arsenicosis due to drinking arsenic-contaminated water. J Toxicol Clin Toxicol 39:665–674

Hei TK, Liu SX, Waldren C (1998) Mutagenicity of arsenic in mammalian cells: role of reactive oxygen species. Proc Natl Acad Sci USA 95:8103–8107

Hoffman DJ, Sandrson CJ, LeCaptain LJ, Cromartie E, Pendleton GW (1992) Interactive effects of arsenate, selenium and dietary protein on survival, growth and physiology in mallard ducklings. Arch Environ Contam Toxicol 22:55–62

Hu GG (1989) Investigation of protective effect of selenium on genetic materials among workers exposed to arsenic. Zhonghua Yu Fang Yi Xue Za Zhi 23:286–288

Huq MA, Misbahuddin M, Choudhury SAR (2000) Spirulina in the treatment of chronic arsenic poisoning. Ban J Phy Pha 16:15–16

Islam L, Nabi A, Rahman M, Khan M, Kazi A (2004) Association of clinical complications with nutritional status and the prevalence of leukopenia among arsenic patients in Bangladesh. Int J Environ Res Public Health 1:74–82

Kenyon EM, Hughes MF, Levander OA (1997) Influence of dietary selenium on the disposition of arsenate in the female B6C3F1 mouse. J Toxicol Environ Health 51:279–299

Khan AW, Ahmad SA (2002) Arsenic drinking water: health effects and management. In: Research studies on health impact of arsenic exposure. Bangladesh Medical Research Council (BMRC), Mohakhali, Dhaka, pp 330–365

Khan AW, Ahmad SA, Sayed SU et al (1997) Arsenic contamination in ground water and its effect on human health with particular reference to Bangladesh. JOPSM 16:65–73

Khan MAK, Choudhury SAR, Misbahuddin M et al (2001) Effects of spirulina in the treatment of chronic arsenic poisoning in Bangladesh. Bangladesh J Med Sci 7:20–24

Kosnett MJ (1998) Clinical treatment of chronic arsenic intoxication. Proceedings of international conference on arsenic pollution of ground water in Bangladesh: causes, effects and remedies. Dhaka 8–12 Feb, pp 98–99

Levander OA (1997) Metabolic interrelationships between arsenic and selenium. Environ Health Perspect 19:159–64

Lin SM, Chiang CH (1985) Biological Trace Elements Res 811

Liu F, Jan KY (2000) DNA damage in arsenite and cadmium-treated bovine aortic endothelial cells. Free Radic Biol Med 28:55–63

Lynn S, Gurr JR, Lai HT, Jan KY (2000) NADH oxidase activation is involved in arsenic-induced oxidative DNA damage in human vascular smooth muscle cells. Circ Res 86:514–519

Milton AH, Hasan Z, Shahidullah SM et al (2004) Association between nutritional status and arsenicosis due to chronic arsenic exposure in Bangladesh. Int J Environ Health Res 14:99–108

Misbahuddin M, Islam AZM, Khandker S, Islam IAM, Anjumanara N (2006) Efficacy of spirulina extract plus zinc in patients of chronic arsenic poisoning: a randomized placebo-controlled study. Clin Toxicol 44:135–141

Nasir M, Misbahuddin M, Ali SMK (2004) Selenomethionine: a therapeutic adjunct t arsenic-free water in reducing tissue arsenic load. J Med Sci Res 1:1–5

Passwater RA (2000) New discovery expand our knowledge about selenium's importance. In: Interviews with nutritional experts. [http://www.healthworld.com]

Rabbani GH, Ali SMK, Das HH, Hossain A, Gupta SKP, Haq I, Islam MM (2003) Double blind randomized, placebo-controlled trial of antioxidant vitamins and minerals in the treatment of Chronic Arsenic Poisoning in Bangladesh. Final report, sponsored by UNICEF and Government of Bangladesh, pp 1–103

Sato Y, Kang M, Kamei T, Magara Y (2000) Performance of nanofiltration for arsenic removal. Water Res 36:3371–3377

Shi H, Hudson LG, Ding W, Wang S, Cooper KL, Liu S et al (2004) Arsenic causes DNA damage in keratinocytes via generation of hydroxyl radicals. Chem Res Toxicol 17:871–878

Spallholz JE, Collins JF, Schwarz KA (1978) A single test tube method for the fluorometric microdetermination of selenium. Bioinorg Chem 1978:453–459

Spallholz JE, Mallory-Boylan L, Rhaman MM (2004) Environmental hypothesis: is poor dietary selenium intake an underlying factor for arsenicosis and cancer in Bangladesh and West Bengal, India? Sci Total Environ 323:21–32

Styblo M, Thomas DJ (2001) Selenium modifies the metabolism and toxicity of arsenic in primary rat hepatocytes. Toxicol Appl Pharmacol 172:52–61

Thianprasit M (1984) Chronic cutaneous arsenism treated with aromatic retinoid. J Med Ass Thailand 67:93–100

Tondel M, Rahman M, Magnuson A, Chowdhury IA, Faruquee MH, Ahmad SA (1999) The relationship of arsenic levels in drinking water and the prevalence rate of skin lesions in Bangladesh. Environ Health Perspect 107:727–729

Tseng W, Chu H, How S, Fong J, Lin C, Yeh S (1968) Prevalence of skin cancer in an endemic area of chronic arsenicism in Taiwan. J Natl Cancer Inst 40:453–463

Vahter M (1994) Spieces differences in the metabolism of arsenic compounds. Appl Organomet Chem 8:175–182

Wang W, Yang L, Hous S, Tan S, Li H (2001) Prevention of endemic arsenism with selenium. Curr Sci 81:1215–1218

WHO (World Health Organization). Regional consultation on arsenicosis case – detection, management and surveillance WHO/SEARO, New Delhi. 5–9 November, 2002. Working paper 3

Yang L, Wang W, Hou S, Peterson PJ, Williams WP (2002) Effects of selenium supplementation on arsenism: an intervention trial in inner Mongolia. Environ Geochem Health 24:359–374

Chapter 9
Agricultural and Environmental Changes in Bangladesh in Response to Global Warming

S. Ahsan, M.S. Ali, M.R. Hoque, M.S. Osman, M. Rahman, M.J. Babar, S.A. Begum, D.M. Rahman, and K.R. Islam

Contents

Abstract Global climate change is a growing concern for Bangladesh. To evaluate the global climate change effects on environmental changes and agricultural production in Bangladesh, long-term data on selected climatic variables (1948–2006), agricultural production (1960–2006), and population growth (1940–2008) were collected, organized and analyzed. Results suggested that although Bangladesh emits less than 0.2% of the global carbon dioxide (CO_2), it is nevertheless facing the impact of global climate change. Average air temperature was found to be increased @ 0.7°C per decade across Bangladesh. As expected, the rainfall distribution varied regionally over time. Total average rainfall increased in the north-eastern (2.6 cm/year) region but decreased in the south-eastern region (1.4 cm/year) or remained same in the north-western and south-western regions of the country. Average sunshine duration decreased by 36 min/decade in between 1962 and 2000. While agricultural land saturation (150% cropping intensity) with increasing population

S. Ahsan
Metro State College, Denver, CO

M.R. Hoque, M.S. Osman, M. Rahman, M.J. Babar, and S.A. Begum
Department of Environment, Dhaka, Bangladesh

D.M. Rahman
Department of Soil, Water and Environmental Sciences, Dhaka University, Dhaka, Bangladesh
e-mail: dmrahman@agni.com

K.R. Islam (✉)
Ohio State University South Centers, Piketon, OH
e-mail: islam.27@osu.edu

R. Lal et al. (eds.), *Climate Change and Food Security in South Asia*,
DOI 10.1007/978-90-481-9516-9_9, © Springer Science+Business Media B.V. 2011

growth (1.8%) abounds, food production in Bangladesh under changing climates has improved over time. Total area under rice production slightly decreased, the yield, however, increased (85%). In contrast, total area under wheat (<1% to 7%), maize, potato (<1% to >2%), and oilseed production increased over time. The yield increased from <0.8 to >2 Mg/ha for wheat, 1 to >5 Mg/ha for maize, and 2.5–14 Mg/ha for potato. Total area under jute (>8 to <4%) and legume production decreased but the yield per unit of land increased over time. However, with increasing population, degraded land quality, and potential global warming, agriculture is seen as one of the major vulnerabilities facing Bangladesh in near future. More specifically, a progressive decline in sunshine duration (25%) over a period of 30 years has become a growing concern for agriculture in terms of reduced photosynthesis and food security.

Keywords Gaseous emissions • Food production • Sea level rise • Rainfall amount • Evapotranspiration

Abbreviations

SAARC	South-East Asian Association for Regional Cooperation
GHG	Greenhouse gas
MPO	Master Planning Organization

9.1 Introduction

Human activities over time have led to global climate change by emitting greenhouse gases (GHGs) especially CO_2 from fossil fuel combustion and indiscriminate land-use practices (IPCC 2007; Boden et al. 2009). Among the GHGs, CO_2 is responsible for 72% of the global climate change followed by CH_4 (18%), NOx (9%), and others (1%). Currently, fossil fuel consumption and land-use changes are accounted for 8 and 1.6 billion Mg of annual C emissions, respectively (IPCC 2007; Koonin 2007). The progressive accumulation of CO_2 has changed the atmospheric composition and created global warming by raising air temperature @ 0.2°C per decade since 1950, and consequently affected our environment (IPCC 2007). It is expected that global warming will affect most of the countries of the world to some extent (IPCC 2007). However, there is a growing concern that the impact of global warming will be a major threat to topographically low-lying countries like Bangladesh (Huq 2001; Quader et al. 2004; Ahmed 2006). The UN Climate Panel has already listed Bangladesh as one of the countries of the world most vulnerable to climate change effects (IPCC 2007).

Bangladesh is located in South Asia, between 20°34′8″–26°38′8″ N latitudes, and between 88°01′8″–92°12′8″ E longitudes with an area of about 147,570 km^2 (Rashid 1991). It is a low-lying delta of the transboundary Ganges, Brahmaputra,

Meghna and Jamuna rivers. The floodplains occupy 80% of the country (Rashid 1991). Even much before the climate change realities began to unfold; Bangladesh became prone to natural disasters (Rashid 1991; Mirza et al. 2001; Mirza 2002). It's the geography and climate (monsoon) which makes Bangladesh vulnerable to the effects of global warming (Rashid 1991; Mirza et al. 2001; Mirza 2002). Since, the landscape of Bangladesh is only a few meters above the mean sea level, it resides in a water paradox in that it receives massive volumes of water both from trans-boundary rivers and from rains (average 2.3 m annually) during monsoon, but seasonal water shortages are common during dry season (Karim et al. 1990a; Rashid 1991; Mirza et al. 2001; Mirza 2002; Agarwala et al. 2003). Among the 57 transboundary rivers, the Ganges, the Brahmaputra, the Meghna, and the Jamuna, have highest peak discharge of water in the world (Rahman et al. 1990; Mirza et al. 2001; Mirza 2002; Agarwala et al. 2003). About 1,100 km^3 of water enters Bangladesh annually through these rivers, mostly during the monsoon. To put this in perspective, it is equivalent to covering the entire country in water to a depth of 7.5 m. As a result, the country is subjected to frequent water surges and floods (Rahman et al. 1990; Mirza et al. 2001; Mirza 2002).

Bangladesh currently has a population of 150M and will add another 30M by 2025 (http://www.census.gov/ipc, http://www.who.int/research/en). About 20% of the total population living in and around highly productive but most vulnerable coastal areas and islands in the Bay of Bengal (Huq et al. 1998; Faisal and Parveen 2004). Bangladesh has 8.8 M ha of agricultural land, of which more than 85% is cultivated, so there is a limited scope for horizontal expansion of the cultivated area for food production (Karim et al. 1998; Faisal and Parveen 2004). The higher population growth has triggered agricultural intensification to meet the need for increased food production. To meet the increasing food demand for growing population, Bangladesh has no other alternatives to increase the cropping intensity. As a result, the intensive double (59%) and triple (22%) cropping increases the effective crop production by 150% with an associated degradation of land quality from continuous rice monoculture, inadequate soil–water conservation measures, unbalanced fertilization, and indiscriminate use of chemical protection (Karim et al. 1990a, b; Habibullah et al. 1998; Karim et al. 1998; Faisal and Parveen 2004). With increasing population, degraded land quality, and potential global warming, agriculture is seen as one of the major vulnerabilities facing Bangladesh today. The objectives of the paper were to discuss the effects of global warming on climate change and agricultural production, and the anticipated effects of climate change and increasing population growth on environmental degradation and food security of Bangladesh.

9.2 Materials and Methods

Long-term climatic data (1940–2004) collected by Bangladesh Meteorological Department 2008 (http://www.bmd.gov.bd/Mrain.php) and the Department of Environment 2008 (http://www.doe-bd.org) were used. The CO_2 emissions data (2008) were collected from Carbon Dioxide Information Analysis Center (Boden et al. 2009). The data on population statistics (1940–2008) were collected from the International Data Base Entry – U.S. Census Bureau 2008 (http://www.census.gov/ipc/www/idb/informationGateway.php) and World Health Organization 2008 (http://www.who.int/research/en). Bangladesh agricultural information (1960–2004) on total production area and yield of rice, maize, wheat, legumes, fibers (e.g. jute), tubers (e.g. potato), and oilseeds were collected from Food and Agriculture Organization of the United Nations 2008 (http://faostat.fao.org). The collected data were organized, processed, and presented in tables and figures. Excel® and SigmaPlot® software were used to convert the data and for making graphs. Regression analyses were performed to detect trends on climatic variables, agricultural production, and population growth over time using SigmaPlot®.

9.3 Results and Discussion

9.3.1 Carbon Dioxide Emissions, Global Warming, and Bangladesh Environmental Changes

Among all the countries of the world, Bangladesh contributed only 0.14% of the total global CO_2 emissions compared to 19.1% and 20.2% by USA and China, respectively (Table 9.1). Total amount of CO_2 emitted by Bangladesh is about 11.3 Tg/year. In terms of CO_2 emission, Bangladesh ranked 67th globally. Among the countries of the South-East Asian Association for Regional Cooperation (SAARC), Bangladesh ranked 3rd in terms of total amount CO_2 emissions followed by Pakistan (2nd) and India (1st). Further analysis of data showed that Bangladesh ranked 175th in terms of per capita CO_2 emissions globally. Bangladesh was ranked 5th in terms of per capita

Table 9.1 Total CO_2 emissions by Bangladesh as compared with developed and South-East Asian Association for Regional Cooperation (SAARC) countries in 2008 (Boden et al. 2009)

Country	Amount (Tg/year)	% Contribution	Worldwide rank	SAARC rank
Global	8,230	–	–	–
China	1,660	20.2	1	–
USA	1,570	19.1	2	–
India	427	5.2	4	1
Pakistan	38.9	0.47	34	2
Bangladesh	**11.3**	**0.14**	**67**	**3**
Sri Lanka	3.2	0.04	88	4
Nepal	0.88	0.011	126	5
Maldives	0.24	0.003	162	6
Afghanistan	0.19	0.002	163	7
Bhutan	0.10	0.001	175	8

Table 9.2 Per capita CO_2 emissions by Bangladesh as compared with developed and South-East Asian Association for Regional Cooperation (SAARC) countries in 2008 (Boden et al. 2009)

Country	Amount (Mg)	SAARC rank	Worldwide rank
Qatar	21.6	–	1
USA	5.6	–	9
Maldives	0.7	1	109
India	0.34	2	129
Pakistan	0.23	3	152
Sri Lanka	0.16	4	161
Bangladesh	**0.07**	**5**	**175**
Bhutan	0.05	6	182
Nepal	0.03	7	193
Afghanistan	0.01	8	205

CO_2 emissions among the SAARC countries. The per capita emission of CO_2 in Bangladesh was only 70 kg compared with 5.6 and 21.6 Mg/year in US (2nd) and Qatar (1st), respectively (Table 9.2). However, due to rapid industrialization, economic development, and increasing population growth, the use of commercial fuels in Bangladesh has increased sharply, and as a result, the amount of CO_2 emissions was also increased rapidly (Azad et al. 2006). Bangladesh has few indigenous renewable energy sources, and the country is heavily dependent on the imported fossil fuels. Total amount of CO_2 released from petroleum products, natural gas, and coal were 50%, 44%, and 6% CO_2, respectively (Azad et al. 2006). The analysis of energy data projected that petroleum and coal consumption in Bangladesh will be growing by more than 5%/year, however, the proportion of natural gas in total energy consumption will be increasing. In response, the government of Bangladesh has taken bold steps to reduce its future CO_2 emissions through a development of renewable energy sources and greater use of natural gas (Azad et al. 2006).

Although Bangladesh emits less than 0.2% of the global CO_2 emissions, it is nevertheless facing a climate change problem due to global warming from accelerated CO_2 emissions by developed countries (Huq 2001; Quader et al. 2004; Anonymous 2008; Boden et al. 2009). Long-term data analyses have shown that selected climatic variables such as air temperature, rainfall, relative humidity, and sunshine duration in Bangladesh have changed over time (Figs. 9.1–9.5). Average air temperature was found to be increased @ 0.7°C per decade across Bangladesh (Fig. 9.1). In other words, more than 2°C rise in air temperature had taken place from 1950 to 2004. There was a trend in average maximum temperature rising by 0.14°C per decade between 1960 and 2004 (Fig. 9.2a). Similarly, average minimum temperature increased by 0.18°C per decade (Fig. 9.2b). As expected, total annual rainfall distribution varied regionally over time (Fig. 9.3). On average, the rainfall increased by 0.73 cm/year from 1948 to 2001. When the total amount of rainfall distribution in Bangladesh was divided into four different regions, the highest amount of rainfall was found in the north-east region (Fig. 9.4a). Annual total rainfall increased in the north-east region @ 2.6 cm from 1950 to 1992. Maximum rainfall was recorded

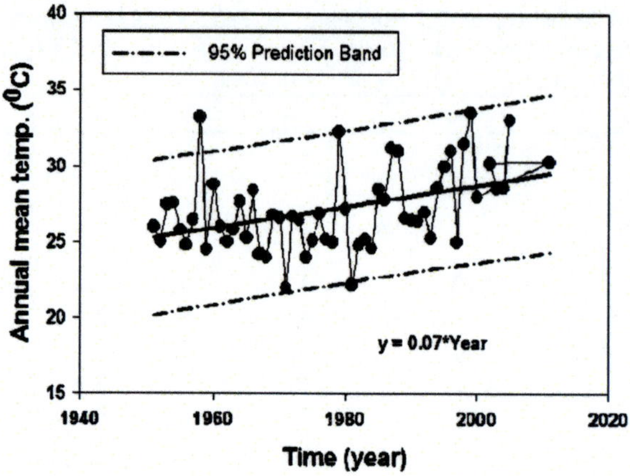

Fig. 9.1 Variations in annual average air temperature over time in Bangladesh (http://www.bmd.gov.bd/Mrain.php and http://www.doe-bd.org)

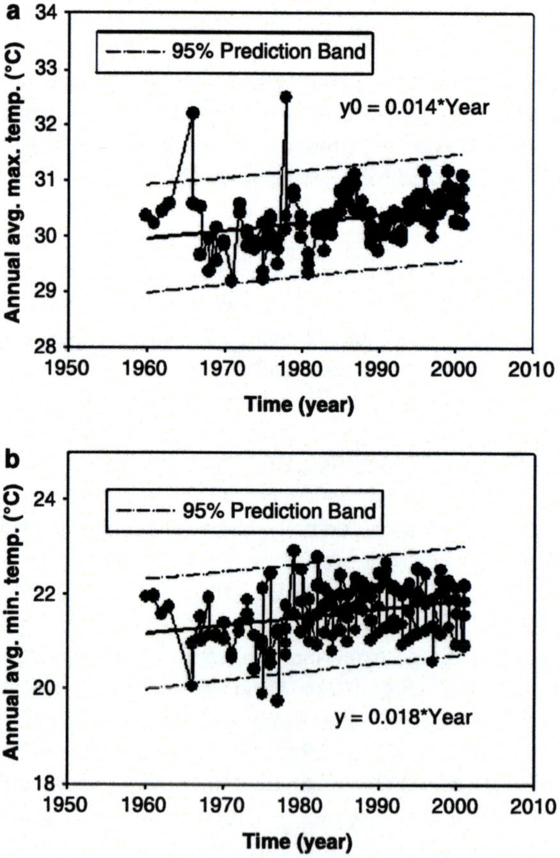

Fig. 9.2 Variations in annual average maximum (**a**) and minimum (**b**) air temperature over time in Bangladesh (http://www.bmd.gov.bd/Mrain.php and http://www.doe-bd.org)

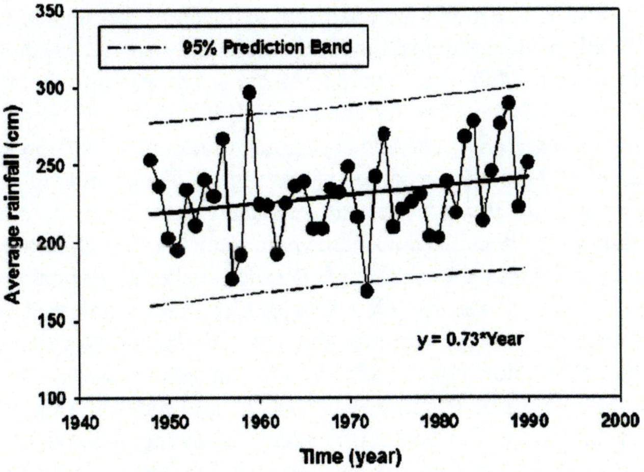

Fig. 9.3 Variations in annual average rainfall over time in Bangladesh (http://www.bmd.gov.bd/Mrain.php and http://www.doe-bd.org)

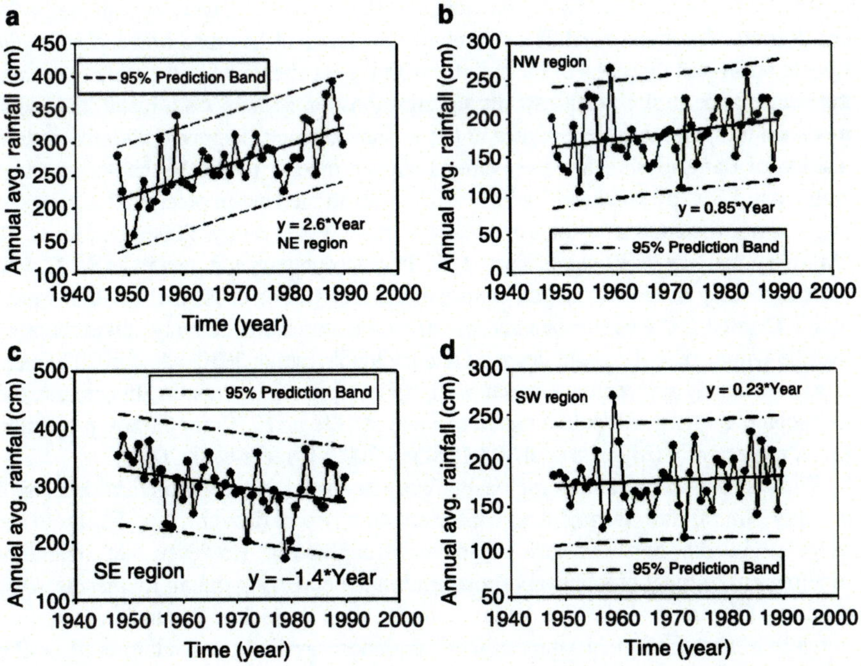

Fig. 9.4 Variations in annual average rainfall over time in north-east (**a**), north-west (**b**), south-east (**c**), and south-west (**d**) regions of Bangladesh (http://www.bmd.gov.bd/Mrain.php and http://www.doe-bd.org)

400 cm as compared to a minimum of 150 cm in the north-east region. However, the amount of total rainfall did not increase consistently in the north-west region. Highest amount of rainfall was recorded 250 cm as compared to the lowest of 100 cm/year (Fig. 9.4b). In contrast, the total rainfall decreased by 1.4 cm/year in the south-east region (Fig. 9.4c). The highest amount of rainfall of 380 cm and the minimum rainfall of 150 cm were recorded in this region. The amount of rainfall in the south-west region of Bangladesh was remained same over time (Fig. 9.4d). Maximum rainfall of about 225 cm and the minimum of 125 cm/year were recorded. Relative humidity increased by 0.81% per decade (Fig. 9.5a). Annual average maximum relative humidity was recorded 85% during 2000s compared with 75% in 1960s. However, the average sunshine duration, in general, was decreasing at an alarming rate of 36 min per decade or 3.6 min/year between 1968 and 2000 (Fig. 9.5b). The sunshine duration was 9.01 ± 0.09 h/day during the period of 1961–1975 which reduced to 5.97 ± 0.11 h/day during the period of 1991–2006.

Several studies have reported a wide spatial and temporal distribution of annual rainfall in Bangladesh (MPO 1991; Anonymous 2008). Annual rainfall reportedly ranged from 120 cm in the extreme west to over 500 cm in the east and north-east regions of the country (MPO 1991). The eastern and southern regions have been receiving more rainfall than western and northern regions of Bangladesh (Anonymous 2008). Moreover, a south to north thermal gradient in winter mean temperature was reportedly developed over time; the southern districts were 5°C warmer than the northern districts (Anonymous 2008). A temperature gradient had also oriented in southwest to north-east direction of the country with the warmer zone in the south-west and the cooler zone in the north-east during pre-monsoon season. As a result, the mean monsoon temperatures were higher in the western districts compared with the eastern districts of Bangladesh. Moreover, various climate models projected that temperature would rise 1.3°C by 2030 and 2.6°C by 2070 with an increase in precipitation, particularly during the monsoon months (Manabe et al. 1991; Mirza 2002; Quader et al. 2004). By the year 2030, the projected rise in monsoon temperature will be 0.7°C with a corresponding rise in winter temperature of 1.3–1.4°C. For 2070, the variation would be 1.7°C and 2.1°C for monsoon and winter temperatures, respectively. It was reported that the winter rainfall would decrease at a negligible rate in 2030, while in 2070 there would not be any appreciable amount of rainfall during winter months. In contrast, the monsoon precipitation would increase at a rate of 12% and 27% for the two projection years, respectively (Manabe et al. 1991; Mirza 2002; Quader et al. 2004).

With higher temperatures increasing evapotranspiration combined with a small decrease in rainfall during dry winter months, even drought, are likely to be expected in the north-western region of Bangladesh. However, net irrigation requirement to meet crop's evapotranspiration is decreasing due to increasing solar dimming (Karim et al. 1990a; Anonymous 2008). The decline in sunshine duration is really a matter of great concern for Bangladesh agriculture (Salam et al. 2003; Ramanathan et al. 2005; Anonymous 2008). A progressive increase in brown cloud coverage over south-Asia due to gradual increase in GHGs and aerosols resulting from accelerated deforestation, rapid urbanization, greater biomass use for cooking and heating, and indiscriminate land-use changes may be responsible of such an alarming rate of decrease in sunshine duration in Bangladesh (Salam et al. 2003, Ramanathan et al. 2005, Anonymous 2008).

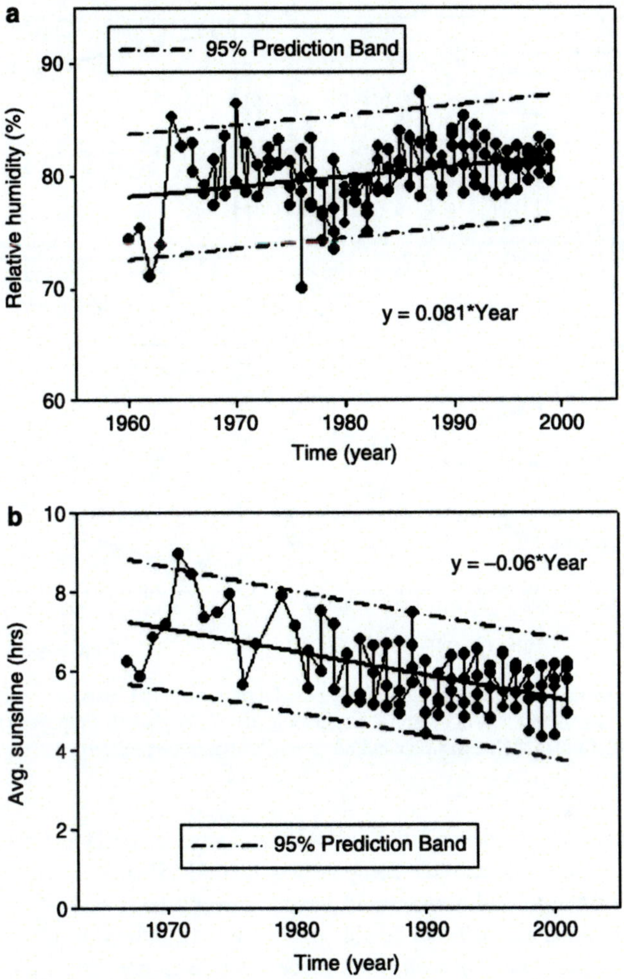

Fig. 9.5 Variations in relative humidity (**a**) and sunshine hours (**b**) over time in Bangladesh (http://www.bmd.gov.bd/Mrain.php and http://www.doe-bd.org)

9.3.2 Population Growth and Environmental Changes Impact on Agriculture

Bangladesh currently has a population approaching 150 M and will add another 100 M before stabilizing (Fig. 9.6a). Although the rate of population growth is declining (Fig. 9.6b), a high population density (>1,000 people/km^2) makes the existing agricultural land is virtually saturated, with a very limited capacity for horizontal expansion of the cultivable lands for crop production. It is projected that most of the increased population will be in urban areas, and much of Bangladesh will essentially become a city state in future (Huq et al. 1998; Faisal and Parveen 2004). The existing population places Bangladesh at great risk of reaching land

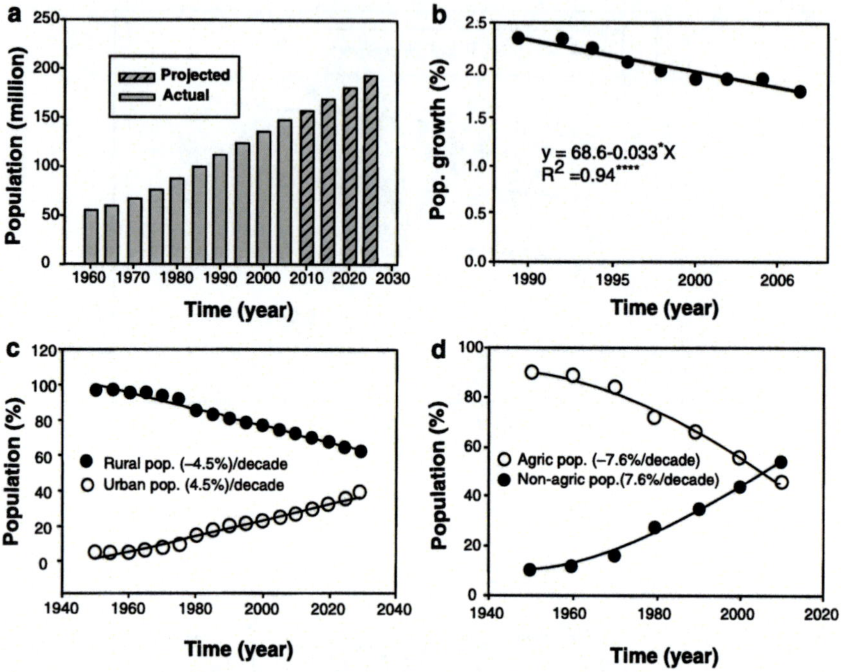

Fig. 9.6 Actual and projected population (**a**), population growth (**b**), urban and rural population (**c**), and agricultural and non-agricultural population (**d**) in Bangladesh (http://www.census.gov/ipc/www/idb/informationGateway.php and http://www.who.int/research/en)

saturation capacity to absorb further population increases into the rural labor force (Fig. 9.6c). Results showed that compared to the mid-1990s, the number of people (rural) working in agriculture sector is declining over time (Fig. 9.6d). These trends suggest that agriculture will not be able to absorb the continuing growth of population into the economically-productive labor force in future. Thus, the continuing population growth in the rural areas will have to find other sources of employment, or they will be driven into the urban areas looking for work. While agricultural land saturation by increasing population growth abounds, the country's agricultural production has improved tremendously in recent decades (http://faostat.fao.org).

Despite floods, droughts, and other problems, production of rice, wheat, maize, legumes, fiber, and tuber crops in Bangladesh has improved over time (Figs. 9.7–9.9. Although the total area under rice production slightly decreased, the yield, however, increased exponentially especially after 1990s (Fig. 9.7). The rice yield increased from <2 Mg/ha in 1960s to about 3.7 Mg/ha in 2000. In contrast, the total area under winter wheat (<1% to 7%) and maize production increased over time (Fig. 9.8ab). The wheat yield increased from <0.8 Mg/ha in 1960s to >2 Mg/ha in 2000s, i.e. four times increase in yield over time. Likewise, the maize yield increased from 1 Mg/ha in 1960s to >5 Mg/ha in 2000s. Total area under tubers especially potato increased linearly (<1% to >2%) over time (Fig. 9.8c).

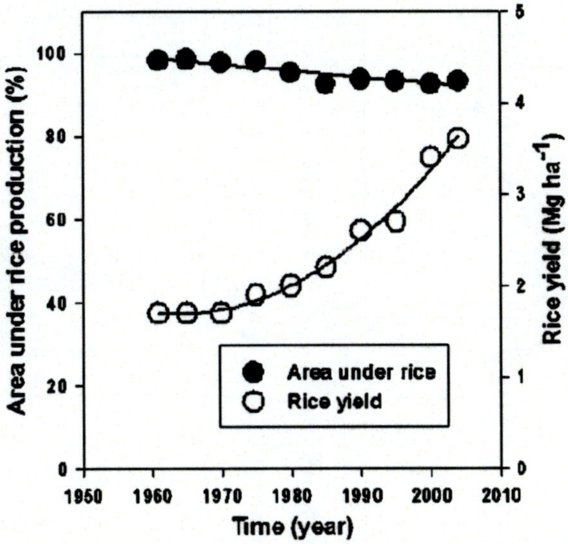

Fig. 9.7 Variations in rice yield and production area over time in Bangladesh (http://faostat.fao.org)

However, the potato yield increased exponentially (2.5–14 Mg/ha). Total area under oilseed crops production increased quadratically, however, the yield increased linearly (Fig. 9.8d). In contrast, the total area under fiber crops (e.g. jute) and legumes decreased over time (Fig. 9.9a,b). Total area under jute production decreased from 8% in 1960s to <4% during 2000s. However, the jute yield increased from 1.5 to 1.9 Mg/ha over time (Fig. 9.9a). The legume yield increased linearly (Fig. 9.9b).

A temporal increase in food production is most probably related to an increase in cropping intensity. It is reported that intensive double (59%) and triple (22%) cropping increases the crop production by 150% using irrigation, abundant chemical fertilization, and widespread use of pesticides (Karim et al. 1990b; Karim et al. 1996; Habibullah et al. 1998; Karim et al. 1998). However, Karim et al. (1996) reported that other than intensive agricultural practices, crop yields potentially responded to atmospheric CO_2 fertilization and a slight increase in air temperature from global climate change impacts. However, as the CO_2 fertilization saturates, yields could decrease. The increase in temperature may have a positive impact on crop production especially rice yields (Karim et al. 1998; Anonymous 2008). In the range of temperatures between 10°C and 32°C, a slight increase in temperatures is considered to be beneficial for crops especially rice. However, with increasing population, degraded land quality, and potential global warming, agriculture is seen as one of the major vulnerabilities facing Bangladesh. The effects of global climate change pose potential risks for Bangladesh, yet the core elements of its vulnerability are primarily contextual. Several projected climate change impacts would in fact reinforce many of these baseline stresses that already pose a serious impediment to food production and agroecosystems functionality in Bangladesh (Huq 2001; Karim et al. 1998; Agarwala et al. 2003; Faisal and Parveen 2004; Quader et al. 2004).

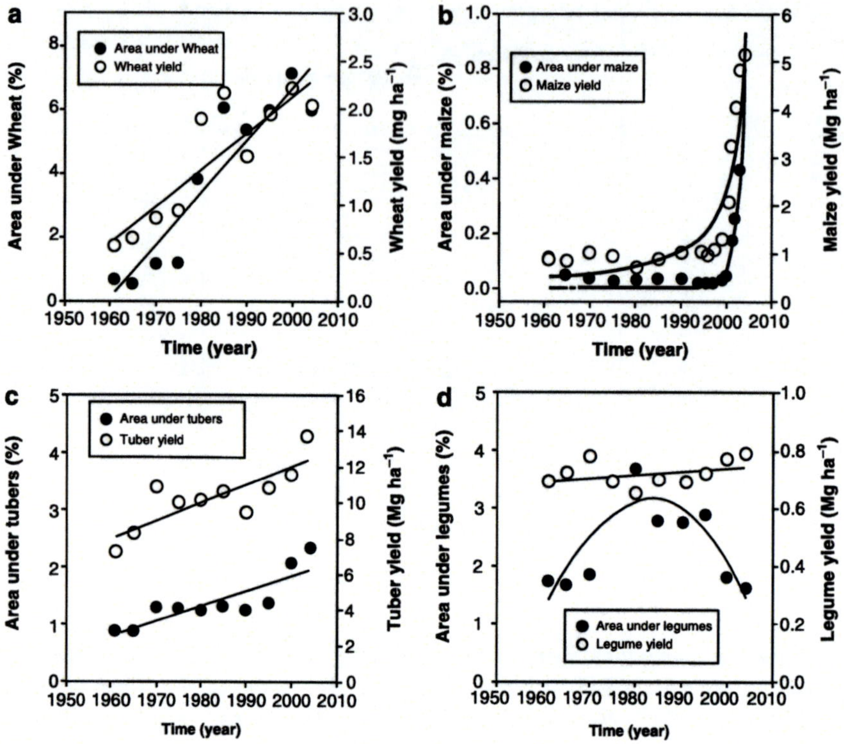

Fig. 9.8 Variations in wheat (**a**), maize (**b**), tubers (**c**), and legume (**d**) yields and production area over time in Bangladesh (http://faostat.fao.org)

Long-term data analysis suggested that the most direct and adverse effect of global warming on crop production across Bangladesh may come from an alarming rate of increase in solar dimming. The optimum day length for growing food crops in Bangladesh is 8–10 h. Due to reduction in sunshine duration by 25%, the light interception and CO_2 assimilation during photosynthesis may be reduced, and affected 15–20% of the crop yields in a typical day length experienced present day Bangladesh (Anonymous 2008). If the trend of decline in rainfall and sunshine duration continues with increasing temperature, catastrophic effects may occur on environmental degradation and agricultural production. Field preparation, planting and harvesting of crops, fertilization, and irrigation scheduling will be affected, and the incidence of pests and diseases will be severe to achieve food security. Other than direct effects of solar dimming from global warming, environmental degradation related to sea level rise and consequent salinity, flooding and soil erosion, and drought may indirectly affect food security in Bangladesh.

Rising sea-levels and consequent salinity may affect crop production by coastal flooding and reduction in availability of lands for growing crops both under ambient conditions, and even more so in the event of storm surges. It will also indirectly

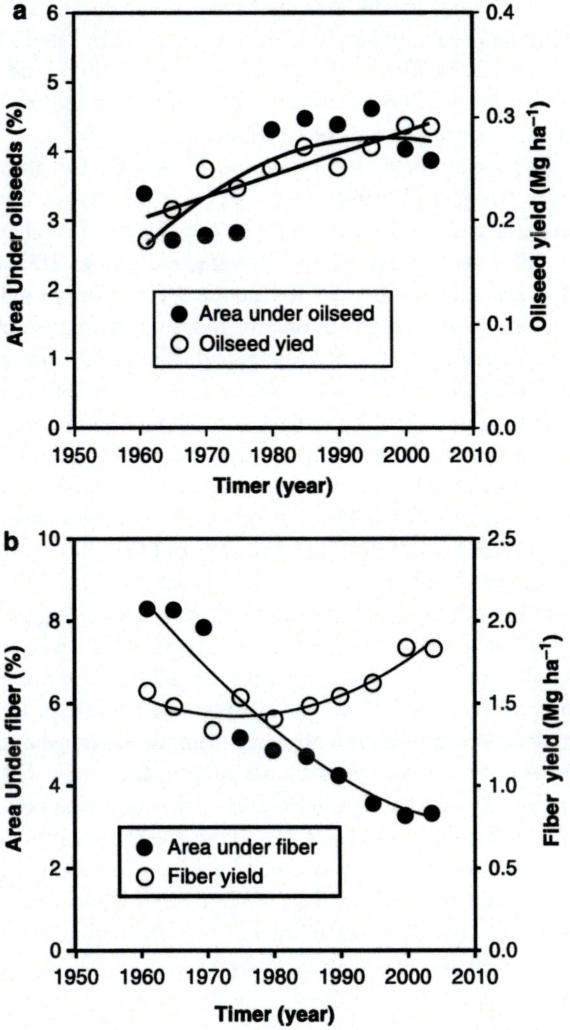

Fig. 9.9 Variations in oilseed (a) and fiber (b) crops yield and production area over time in Bangladesh (http://faostat.fao.org)

cause riverine flooding by causing more backing up of the Ganges, Brahmaputra, Meghna, and Jamuna rivers along the delta. With the expected 1 m rise in sea level, it is predicted that 20% of the southern Bangladesh will be under water, and will displace 25–30 M people (Habibullah et al. 1998; World Bank 2000; Agarwala et al. 2003; Faisal and Parveen 2004; Ahmed 2006). World Bank (2000) reported a sea level rising of about 3 mm/year in the Bay of Bengal in response to the current level of global warming. Threatening the richest and most productive region of the country, sea level rise and salinity could have dramatic consequences for crop production and the economy of Bangladeshi. It is estimated that in eastern Bangladesh alone 14,000 Mg of grain production would be lost to sea level rise in 2030 and 252,000 Mg would be lost by 2075 (Habibullah et al. 1998; WRI 2001). The impacts of soil and water salinity are already visible in the south-west region of Bangladesh (Karim et al. 1990a, b; Huq et al. 1996; Habibullah et al. 2009).

Several studies have reported that between 30–70% of the Bangladesh is normally flooded annually (Huq et al. 1996; Mirza et al. 2001; Mirza 2002). The huge amount of sediments (>2 Mg/year) brought by the Ganges, Brahmaputra, Meghna, and Jamuna coupled with a negligible flow gradient add to drainage congestion problems and exacerbate the extent of flooding. The climate models projected increased precipitation, particularly during the monsoon season which subsequently will contribute to accelerate runoff (Mirza and Dixit 1997; Mirza et al. 2001, Agarwala et al. 2003; Anonymous 2008). Mirza and Dixit (1997) have reported that a 2°C global warming with a 10% increase in rainfall would increase runoff in the Ganges, Brahmaputra, and Meghna rivers by 19%, 13%, and 11%, respectively. Satellite-image studies of the Ganges–Brahmaputra–Middle-Meghna Rivers have shown that an area of 106,300 ha lost due to flooding and riverine erosion between 1982 and 1992, while the accretion accounted to only 19,300 ha. The net erosion rate from accelerated flooding was therefore estimated at 8,700 ha per annum (Agarwala et al. 2003). Moreover, seasonal drought is a recurring problem in north-western region of Bangladesh due to upstream diversion of transboundary Ganges water by India (Karim et al. 1990a, Mirza et al. 2001). The drought problem is going to be compounded manifolds in future, more so because of the future plan to divert and withdraw greater volume of water from all the transboundary rivers by India. As a result, dry season irrigation scheduling for crops will be drastically affected. It is reported that as high as 47% area of the country is drought vulnerable where 53% of the population is currently living (Karim et al. 1990a, b; Ahmed 2006). Considering that agriculture is one of the pillars of Bangladesh economy-such global climate change impacts have the potential to adversely affect the food production in future. The impacts of increasing population and environmental degradation only exacerbate the climate change problems already facing the agriculture in Bangladesh.

9.4 Conclusions

Bangladesh is vulnerable to global climate change impacts because of its geographical location, high population growth, and greater reliance on climate-sensitive sectors particularly agriculture. Although population growth is declining, a high population density (>1,000 people/km^2) makes the existing agricultural land is virtually saturated for horizontal expansion of lands for agricultural production. Moreover, increasing temperature with less rainfall has already affected biological and physical ecosystems of Bangladesh particularly the north-western region with frequent droughts and the south-western region with increasing soil salinity. However, the most adverse effects of global warming on agriculture may come from an alarming increase (25%) in solar dimming. The decline in sunshine duration (at 36 min per decade) has become a growing concern for agriculture in terms of reduced photosynthesis and food security. Increasing population growth and environmental degradation are going to exacerbate the global climate change effects on agriculture in Bangladesh.

References

Agarwala S, Ota T, Ahmed AU, Smith J, van Aalst MK (2003) Development and climate change in Bangladesh: focus on coastal flooding and the Sunderbans. Organization for Economic Co-operation and Development, Paris, France

Ahmed AU (2006) Bangladesh climate change impacts and vulnerability – a synthesis. Climate Change Cell, Department of Environment, Comprehensive Disaster Management Program, Government of the People's Republic of Bangladesh

Anonymous (2008) Characterizing long-term changes of Bangladesh climate in context of agriculture and irrigation. Climate Change Cell, Department of Environment, Government of the People's Republic of Bangladesh. March 2008

Azad AK, Nashreen SW, Sultana J (2006) State of energy consumption and CO_2 emission in Bangladesh. Ambio 35:86–88

Bangladesh Meteorological Department (2008) Rainfall situation, Government of Bangladesh, Dhaka, Bangladesh. http://www.bmd.gov.bd/Mrain.php. Accessed 15 October 2009

Boden TA, Marland G, Andres RJ (2009) Global, regional, and national fossil-fuel CO_2 emissions. Carbon Dioxide Information Analysis Center, Oak Ridge National Laboratory (http://www.cdiac.ornl.gov/trends/tre_glob.html), US Department of Energy, Oak Ridge, TN, USA. doi 10.3334/CDIAC/00001

US Census Bureau (2008) International data base information gateway. Washington, DC. http://www.census.gov/ipc/www/idb/informationGateway.php. Accessed 15 October 2009

Department of Environment (2008) Poribesh Bhaban, E-16, Agargaon, Shere Bangla Nagar, Dhaka 1207, Bangladesh. http://www.doe-bd.org. Accessed 15 October 2009

Faisal IM, Parveen S (2004) Food security in the face of climate change, population growth and resource constraints: implications for Bangladesh. Environ Manage 34:487–498

Habibullah M, Ahmed AU, Karim Z (1998) Assessment of food grain production loss due to climate induced enhanced soil salinity. In: Huq S, Karim Z, Asaduzzaman M, Mahtab F (eds) Vulnerability and adaptation to climate change for Bangladesh. Kluwer, Dordrecht, pp 55–70

Huq S (2001) Climate change and Bangladesh. Science 294(23):1617

Huq S, Ahmed AU, Koudstaal R (1996) Vulnerability of Bangladesh to climate change and sea level rise. In: Downing TE (ed) Climate change and world food security, vol 37, NATO ASI Series I. Springer, Berlin, Heidelberg, pp 347–379

Huq S, Karim Z, Asaduzzaman M, Mahtab F (1998) Vulnerability and adaptation to climate change for Bangladesh. Kluwer, Dordrecht, p 135

IPCC (2007) Climate Change 2007: The physical science basis. In: Solomon S, Qin D, Manning M (eds) Contribution of Working Group I to the Fourth Assessment Report of the Intergovernmental Panel on Climate Change

Karim Z, Hussain SG, Ahmed M (1990a) Salinity problems and crop intensification in the coastal regions of Bangladesh. Bangladesh Agricultural Research Council, Dhaka, Bangladesh

Karim Z, Ibrahim AM, Iqbal A, Ahmed, M. (1990a) Drought in Bangladesh agriculture and irrigation schedules for major crops. Bangladesh Agricultural Research Council, Soils Publ. No. 34, p 11

Karim Z, Hussain SG, Ahmed M (1996) Assessing impacts of climate variations on food grain production in Bangladesh. Water Air Soil Pollut 92:53–62

Karim Z, Hussain SG, Ahmed AU (1998) Climate change vulnerability of crop agriculture. In: Huq S, Karim Z, Asaduzzaman M, Mahtab F (eds) Vulnerability and adaptation to climate change for Bangladesh. Kluwer, Dordrecht, pp 39–54

Koonin SE (2007) The challenge of CO_2 stabilization. Elements 4:293–294

Manabe S, Stouffer RJ, Spelman MJ, Bryan K (1991) Transient responses of a coupled ocean atmosphere model to gradual changes of atmospheric CO_2. Part I: Annual mean response. J of Climate 4:785–818

Mirza MMQ (2002) Global warming and changes in the probability of occurrence of floods in Bangladesh and implications. Global Environ Change 12:127–138

Mirza MMQ, Dixit A (1997) Climate change and water resources in the GBM basins. Water Nepal 5:71–100

Mirza MMQ, Warrick RA, Ericksen NJ, Kenny GJ (2001) Are floods getting worse in the Ganges, Brahmaputra and Meghna basin? Environ Hazards 3:38–48

MPO (1991) National Water Management Plan: Phase II, Final Report, Master Planning Organization (MPO), Ministry of Irrigation, Water Development and Flood Control, Government of the People's Republic of Bangladesh, Dhaka

Quader DA, Shrestha ML, Khan TMA, Ferdousi N, Rahman M, Mannan A (2004) Variations of surface air temperature over the land areas in and around the Bay of Bengal. Nat Hazards 31:561–584

Rahman AA, Huq S, Conway GR (1990) Environmental aspects of surface water systems of Bangladesh: an introduction. In: Rahman AA, Huq S, Conway GR (eds) Environmental aspects of surface water systems of Bangladesh. The University Press Ltd., Dhaka, Bangladesh

Ramanathan V, Chung C, Kim D, Bettge T, Buja L, Kiehl JT, Washington WM, Fu Q, Sikka DR, Wild M (2005) Atmospheric brown clouds: impacts on South Asian climate and hydrological cycle. PNAS 102:5326–5333

Rashid HE (1991) Geography of Bangladesh, 2nd edn. The University Press Ltd., Dhaka, Bangladesh

Salam A, Bauer H, Kassin K, Ullah SM, Puxbaum H (2003) Aerosol chemical characteristics of an island site in the Bay of Bengal (Bhola-Bangladesh). J Environ Monit 5:483–490

Food and Agricultural Organization of the United Nations (2008) Faostat, Statistics Division, Rome, Italy. http://faostat.fao.org. Accessed 15 October 2009

World Bank (2000) Bangladesh: climate change and sustainable development. Report No. 21104-BD, Rural Development Unit, South Asia Region. The World Bank (WB), Dhaka, p 95

World Health Organization (2008) Data and statistics, Geneva, Switzerland. http://www.who.int/research/en. Accessed 15 October 2009

WRI (2001) World Resources: 2000–2001. World Resources Institute, Washington, DC

Part IV
Climate Change and Soil Degradation

Chapter 10
Soil Degradation and Food Security in South Asia

Rattan Lal

> *"The great tragedy of science – the slaying of a beautiful hypothesis by an ugly fact."*
>
> *Thomas Huxley*

Contents

Abstract South Asia has diverse soils, climates, physiography and other natural resources conducive to agricultural intensification. Yet, the region is characterized by several problems including food insecurity, soil and environmental degradation, land desertification, pollution of natural waters, and loss of biodiversity. There is a widespread poverty and the number of food-insecure population is increasing because of lack of access to food and poor utilization caused by water and air pollution. Despite high rate of irrigation, crop yields are vulnerable to vagaries of monsoons. Soil degradation is caused by the widespread use of extractive farming including removal of crop residues, use of animal manure as household fuel, and low and unbalanced application of fertilizers. Crop yields have improved since 1960, but can be easily increased by 50% or more. There is a strong need for restoring degraded soils and ecosystems through improvements in soil organic carbon pool and creation of positive nutrient budgets. Adaptation to climate change necessitates

R. Lal (✉)
Carbon Management and Sequestration Center,
The Ohio State University, Columbus, OH 43210
e-mail: lal.1@osu.edu

R. Lal et al. (eds.), *Climate Change and Food Security in South Asia*,
DOI 10.1007/978-90-481-9516-9_10,

improvements in soil quality to buffer against the adverse impacts of extreme events on agronomic production.

Keywords Indian sub-continent • Land degradation • Salinization • Physical degradation • Chemical degradation • Per capita arable land area • Food security

Abbreviations

GM	Genetically modified
IGP	Indo-Gangetic plains
Mha	Million hectares
Mt	Million tonnes
RMPs	Recommended management practices
SA	South Asia
SOC	Soil organic carbon
SOM	Soil organic matter

10.1 Introduction

The South Asia (SA) region includes eight countries from Iran in the northwest to Sri Lanka in the Southeast along with Afghanistan, Pakistan, India, Nepal, Bhutan and Bangladesh. It is a world within the world, because of a great diversity in people, languages, climates, vegetation, terrain, soil types, crops, cropping systems, and social/cultural/political environments (Lal 2006). The region's population of 1.5 billion in 2002 may increase, with medium projections, to 2.13 billion by 2025 and 2.43 billion by 2050 (Table 10.1). Most populous countries include India, Pakistan and Bangladesh. The annual growth rate of population varies among countries.

The total land area of the SA is 641.7 million hectares (Mha), of which 228.0 Mha is arable, 22.2 Mha is grasslands, 13.9 Mha is permanent crops, 78.5 Mha is forest, and 93.3 Mha is permanent pasture (Table 10.2). Of the 228.0 Mha of arable land, 169.6 Mha is in India, 21.1 Mha in Pakistan, 15.1 Mha in Iran and 9.4 Mha in Bangladesh (Table 10.2). The land area under forest in the SA region is 78.5 Mha, of which 64.1 Mha is in India. Most forests are severely degraded because of indiscriminate removal of trees, including harvesting for fuel wood.

Careful analyses of the data in Tables 10.1 and 10.2 show an imbalance between people and the land resources. The SA region is characterized by an extremely high population density. Consequently, natural resources (especially the soil and water) are already under great stress. It is the shortage of natural resources vis-à-vis the high population that is the principal cause of degradation of soil, pollution and contamination of water, and perpetual food insecurity affecting about 300 million people in SA. Poverty and malnutrition are widespread problems in the SA region

Table 10.1 Population, land and water resources of South Asia (recalculated from Gardner-Outlaw and Engelman 1997; Engelman and Le Roy 1995)

Country	Arable land (10^6 ha)	Total freshwater (km^3)	Population (10^6)					
			1950	1960	1990	1995	2025	2050
Afghanistan	8.05	50	8.96	10.77	15.05	19.66	45.26	61.37
Bangladesh	9.44	2,357	41.8	51.41	108.1	118.2	180.0	218.2
Bhutan	0.13	95	0.73	0.85	1.54	1.77	3.65	5.18
India	169.6	2,085	357.5	442.3	850.6	929.0	1,438.9	1,532.7
Iran	15.1	118	16.9	21.55	58.9	68.4	128.3	170.3
Nepal	2.7	170	7.86	9.43	19.25	21.46	40.55	53.62
Pakistan	21.1	468	39.5	49.96	121.9	136.3	168.9	357.4
Sri Lanka	1.9	43	7.68	9.89	17.23	17.93	23.93	27.0
Total	22.80	5,386	489.89	596.16	1,192.57	1,312.72	2,129.49	2,425.77

Table 10.2 Land use in South Asia (FAO 2006)

	Area (10^6 ha)						
Country	Total	Agricultural	Arable land	Permanent crops	Grass lands	Forest	Permanent pastures
Afghanistan	65.2	8.05	8.05	0.14	16.1	1.35	30.0
Bangladesh	13.0	9.02	9.44	0.44	0.061	1.33	0.6
Bhutan	4.7	0.54	0.13	0.02	0.40	3.02	0.42
India	297.3	180.80	169.6	9.22	2.62	64.1	11.06
Iran	163.6	60.5	15.1	2.28	–	0.5	44.0
Nepal	14.3	4.23	2.7	0.125	1.11	3.9	1.74
Pakistan	77.1	25.13	91.1	0.67	1.89	2.4	5.00
Sri Lanka	6.5	2.36	1.92	1.00	0.002	1.94	0.44
Total	641.7	290.63	228.0	13.90	22.18	78.54	93.3

(Srinivasan 2000) The relation between economic crisis and long-term food security (Rosegrant and Ringler 2000), and domestic policy and food security (Ninno et al. 2007) cannot be overlooked in the SA region. Therefore, the objective of this chapter is to identify strategies of sustainable management of soil and water resources of SA to advance food security while adapting to climate change and improving the environment. Specific attention is given to the confounding effect of climate change on exacerbating the problems of degradation of soils and accentuating the scarcity of water in the SA region.

10.2 Soil and Water Resources of South Asia

The Indian sub-continent lies across the climatic divide. The northwestern part of SA (Iran, Afghanistan, Pakistan and parts of Western India including Rajasthan) comprise dry zone with about 90-day long growing period (FAO 1994). In other words, the western part of SA has adequate soil moisture to support crop growth for about 90 days. Thus, supplemental irrigation is essential to successful crop production. Most of India lies in semi-arid (southern and northwestern parts) and sub-humid (eastern parts) regions. Bangladesh and northeastern India are characterized by humid climate. Physiographically, SA can be sub-divided into three regions: (i) the mountainous region comprising of the Himalayas, Hindu Kush and the highlands of Iran. Because of the steep terrain, soils of the mountainous regions are prone to severe erosion, gullying and landslides, (ii) the alluvial plains comprising of two distinct regions including the Indo-Gangetic Plains (IGP) of Indus and Gangetic river systems, and the interior basins of Iran. Alluvial soils are flat, deep, fertile and responsive to inputs, and (iii) uplands also comprising of two distinct components including the Deccan Plateaus of south-central India, and central hill massif of Sri Lanka. Soils of the uplands are old, highly weathered, prone to erosion and drought, and have low inherent fertility.

Table 10.3 Resource availability in South Asia (recalculated from Engelman and Le Roy 1993, 1995; Gardner-Outlaw and Engelman 1997)

	Per capita arable land area (ha/person)						Per capita fresh water (m^3/person)			
Country	1950	1960	1990	1995	2025	2050	1950	1995	2025	2050
Afghanistan	0.90	0.71	0.53	0.41	0.18	0.13	5,582	2,543	1,105	815
Bangladesh	0.23	0.17	0.087	0.080	0.05	0.04	54,411	19,936	13,096	10,803
Bhutan	0.18	0.11	0.084	0.073	0.04	0.025	129,428	53,672	26,056	18,326
India	0.47	0.36	0.20	0.18	0.12	0.11	5,831	2,244	1,567	1,360
Iran	0.89	0.71	0.26	0.22	0.12	0.086	6,947	1,719	916	690
Nepal	0.34	0.19	0.14	0.126	0.07	0.05	7,862	7,923	4,192	3,170
Pakistan	0.53	0.34	0.17	0.15	0.07	0.059	11,844	3,435	1,740	1,310
Sri Lanka	0.25	0.16	0.11	0.10	0.08	0.07	5,626	2,410	1,805	1,600
Average	0.47	0.38	0.19	0.17	0.10	0.093	11,014	4,103	2,529	2,201

Averages computed on the basis of total population and total resources for the region

Most of the regions have high temperatures during the summer, with soil temperature at 1 cm depth often reaching 45°C or more. The rainfall is monsoonal, and is characterized by high intensity, erosivity and energy load. Thus, rainfall effectiveness is low because of high losses caused by surface runoff and evaporation.

The per capita arable land area in the SA region is declining regressively with increase in population. The average per capita cropland area (ha/person) for the entire region was 0.47 in 1950, 0.38 in 1960, 0.19 in 1990, and is projected to be 0.10 in 2025 and 0.093 in 2050 (Table 10.3). By 2050, the per capita arable land area (ha/person) will be 0.04 in Bangladesh, 0.025 in Bhutan, 0.05 in Nepal, 0.059 in Pakistan, and 0.07 in Sri Lanka (Table 10.3). Similar to arable land, there is also scarcity of water especially in the western regions of SA. Countries with severe shortage of renewable fresh water supply are Afghanistan, Iran, Pakistan and India (Table 10.3). Despite the national average availability of fresh water supply being adequate (>1,000 m^3/person/year) there are several regions with severe water shortages in India, Pakistan and elsewhere in SA. The problem of drought stress may be exacerbated by the projected climate change with an attendant increase in frequency and intensity of extreme events. Further, there is a strong competition for water between agriculture on the one hand and industry and urbanization on the other, as is the case in rapidly industrializing India.

Soils of SA are extremely diverse (Table 10.4), and differ among ecoregions. Predominant soils are Alfisols and Vertisols for the semi-arid, Inceptisols and Entisols for the alluvial plains, Aridisols for the deserts, and Ultisols for the humid climates (Lal 2006). In terms of the land area, predominant soils are Entisols (25.93%), Aridisols (18.65%), Inceptisols (14.48%), Alfisols (12.11%), Vertisols (9.23%), Ultisols (6.38%), Rocks and Shifting Sands (5.02% each), and Mollisols (2.92%). Small areas (~0.25%) are occupied by Histosols, Gelisols and Oxisols (Table 10.4). Intensively cultivated croplands include alluvial soils (Inceptisols), Vertisols, Alfisols, Mollisols, etc.

Table 10.4 Predominant soils of South Asia (recalculated from Eswaran et al. 1999)

Soil order	Area (Mha)	% of total area
Alfisols	79.12	12.11
Aridisols	121.78	18.65
Entisols	169.28	25.93
Gelisols	0.30	0.05
Histosols	1.16	0.18
Inceptisols	94.54	14.48
Mollisols	19.06	2.92
Oxisols	0.17	0.03
Rocks	32.79	5.02
Shifting Sands	32.75	5.02
Ultisols	41.63	6.38
Vertisols	60.30	9.23
Total	652.88	100.00

10.3 Agricultural Intensification

Rapid growth in population, especially during the twentieth century, necessitated increase in food production. Indeed, the Green Revolution of the 1960s was brought about by growing input-responsive and improved varieties of wheat and rice with irrigation and application of fertilizers. Total fertilizer use in SA increased from 0.49 million tonnes (Mt) in 1961 to 21.04 Mt in 2002, by a factor of 43. Of the total fertilizer consumption, nitrogen (N) was the most intensively used (Table 10.5). Consumption of N-fertilizer in SA increased from 0.35 Mt in 1961 to 14.25 Mt in 2002, by a factor of 41. There was also an increase in the use of phosphorus (P) from 0.084 Mt in 1961 to 4.91 Mt in 2002, by a factor of 58. In comparison, consumption of potash (K) increased from 0.057 Mt in 1961 to 1.88 Mt in 2002, by a factor of 33. Relatively more consumption of N compared with P and K, because of subsidized price of N, has created nutrient imbalance with widespread deficiency of P and K along with those of micro-nutrients (e.g., Zn, B, Fe). Yet, use of fertilizer per ha of cropland area in SA is low and is merely 92 kg/ha (NPK). The rate of fertilizer use is especially low compared with the consumption in North America and China, and may have to be doubled within the next 25 years (by 2035) to meet the food demands.

Similar to intensive use of fertilizers, there has also been a very rapid increase in cropland area under irrigation (Table 10.6), which increased from 43.4 Mha in 1961 to 91.1 Mha in 2000, by a factor of 2.1. Despite the rapid expansion, <40% of the cropland area in SA is under irrigation. Future expansion in irrigated land area is difficult because of the scarcity of water resources and excessive withdrawal of the ground water (Kerr 2009). The rate of fall of the ground water level in Punjab, India has been 4–5 m for the decade ending in 1994 (Table 10.7). The rate of the fall of water table increased drastically during 2000s (Kerr 2009). Satellite-based estimates of ground water depletion in north-western India (Punjab, Haryana, Rajasthan) have indicated that the ground water

Table 10.5 Fertilizer consumption in the South Asia region (Adapted from IFDC 2004)

	Nutrients used (10^3 Mg/year)			
Year	Nitrogen	Phosphorus	Potassium	Total
1961	353.5	83.6	56.7	493.7
1965	736.5	163.8	116.7	1,017.0
1970	1,914.2	637.3	283.3	2,834.8
1975	3,457.4	662.4	319.1	4,438.9
1980	5,002.7	1,653.2	712.2	7,368.1
1985	7,565.1	2,664.2	993.1	11,222.4
1990	9,892.5	3,812.8	1,503.0	15,208.3
1995	13,058.1	3,662.9	1,345.3	18,066.3
2000	14,559.9	5,223.3	1,732.5	21,515.7
2002	14,250.6	4,909.4	1,875.3	21,035.3

Table 10.6 Irrigated land area in South Asia (recalculated from FAO Stat 2006)

	Irrigated cropland area (10^6 ha)									
Country	1961	1965	1970	1975	1980	1985	1990	1995	2000	2003
Afghanistan	2.38	2.38	2.39	2.43	2.51	2.59	2.72	2.72	2.72	2.72
Bangladesh	0.43	0.57	1.06	1.44	1.52	2.07	2.79	3.75	4.19	4.73
Bhutan	0.008	0.01	0.018	0.022	0.026	0.03	0.04	0.04	0.04	0.04
India	24.69	26.51	30.44	33.73	38.48	42.15	46.70	53.0	56.76	57.28
Iran	4.70	4.90	5.20	5.90	4.95	6.80	7.0	7.26	7.58	7.65
Nepal	0.07	0.086	0.117	0.230	0.52	0.76	0.985	1.13	1.14	1.17
Pakistan	10.75	11.47	12.95	13.63	14.68	15.76	15.82	17.20	18.09	18.23
Sri Lanka	0.34	0.34	0.465	4.30	0.46	0.50	0.52	0.57	0.67	0.74
Total	43.37	46.27	52.64	61.82	63.15	70.66	76.58	85.67	91.19	92.56

Table 10.7 Changes in water table in June 1984–94 in Punjab (Joshi and Tyagi 1994; Aggarwal et al. 2004)

District in Punjab	Average fall of ground water (m)
Amritsar	2.3
Jallandhar	2.5
Ludhiana	1.9
Ferozpur	4.5
Kapurthala	1.8
Patiala	4.8
Sangrur	5.1
Bhatinda	1.9
Faridkot	4.5
Fatehgarh Sahib	2.7

is being depleted at a mean rate of 4.0 ± 1.0 cm/year equivalent height of water or about 17.7 ± 4.5 km^3/year (Rodell et al. 2009). Efficiency of irrigation is extremely low (~30%), and must be at least doubled by adoption of modern techniques such as sprinkler, and micro-irrigation including drip sub-irrigation and condensation irrigation. Evaporation losses must be reduced by use of conservation farming, and application of mulch including the use of bioplastic films and zeolites (NRC 2009).

10.4 Soil Degradation

Soils of SA are prone to degradation and desertification (FAO 1994; Van Lynden and Oldeman 1997; Joshi and Tyagi 1994; Douglas 2006; Acharyo and Kafle 2009). Principal processes of soil degradation include erosion by water (81.7 Mha), erosion by wind (59 Mha), decline in soil organic matter (SOM), nutrient depletion and decline in soil fertility (42.4 Mha), salinization (33.3 Mha) and waterlogging (12.8 Mha) (Table 10.8, Fig. 10.1). The problem is exacerbated by poverty and other anthropogenic factors (Fig. 10.2), which are especially a severe problem among small landholders and resource-poor farmers of SA (Table 10.9). Soil degradation and desertification are also caused by the widespread use of extractive farming practices including: removal of crop residues for fodder and other uses, use of animal dung as household fuel rather than manure, and minimal use of fertilizers and soil amendments especially in rainfed agriculture (Lal 2007). In addition to biophysical factors, soil degradation is also caused by soil, economic and political factors. Land tenure (Niazi 2003) and land fragmentation (Niroula and Thapa 2005) are also important factors. There is also a strong link between poverty and soil degradation (Winslow et al. 2004).

Table 10.8 Extent and severity of soil degradation in South Asia (Adapted from FAO 1994)

	Soil degradation (10^6 ha)				
Country	Water erosion	Wind erosion	Nutrient depletion	Salinization	Water logging
Afghanistan	11.16	2.08	–	3.1	–
Bangladesh	1.50	0	6.37	3.0	–
Bhutan	0.04	0	–	–	–
India	32.77	10.80	29.38	7.0	8.53
Iran	26.40	35.37	–	16.0	0.55
Nepal	1.59	0	–	–	–
Pakistan	7.20	10.74	5.20	4.2	3.68
Sri Lanka	1.07	0	1.43	–	–
Total	81.73	58.99	42.38	33.3	12.76

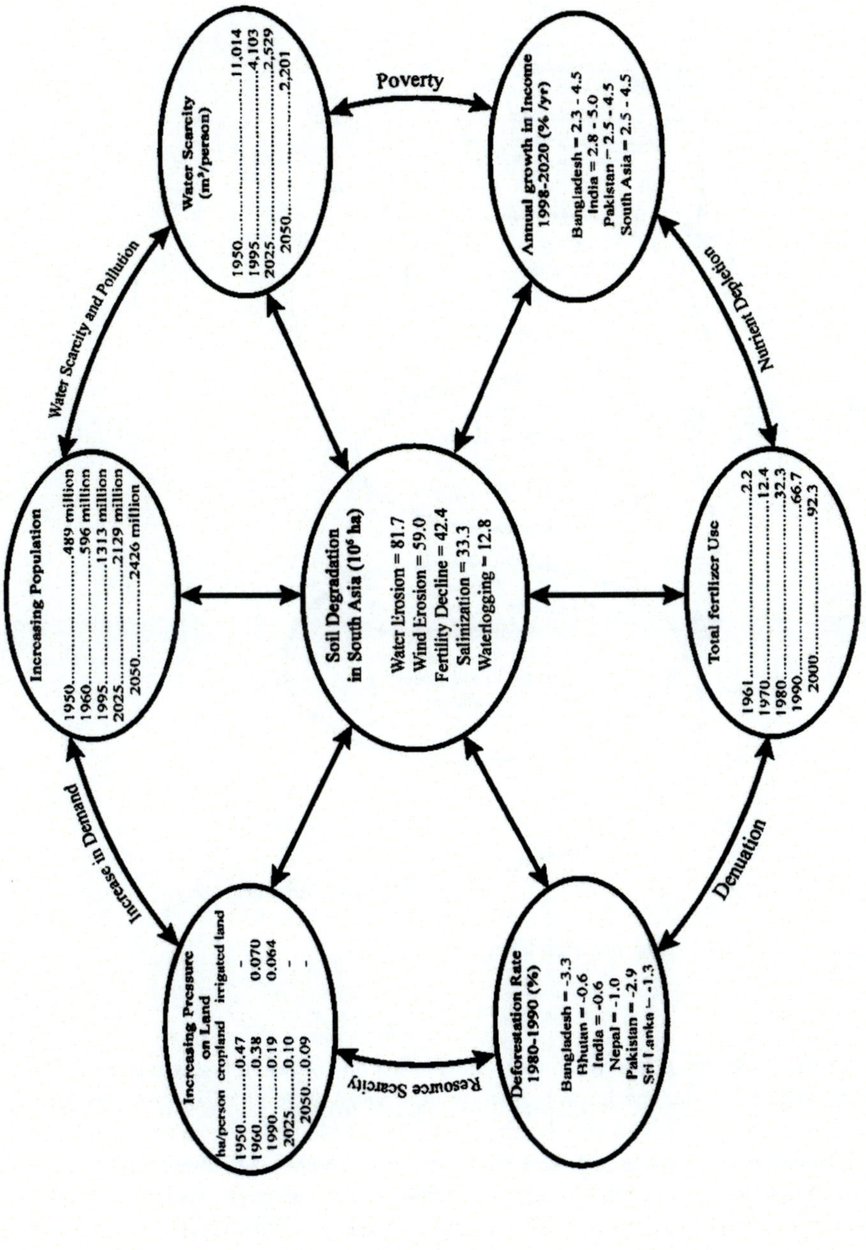

Fig. 10.1 Causes of soil degradation in South Asia (calculated from activities by FAO 1994; FAO Stats 2006; IFDC 2004; Rosegrant and Ringler 2000)

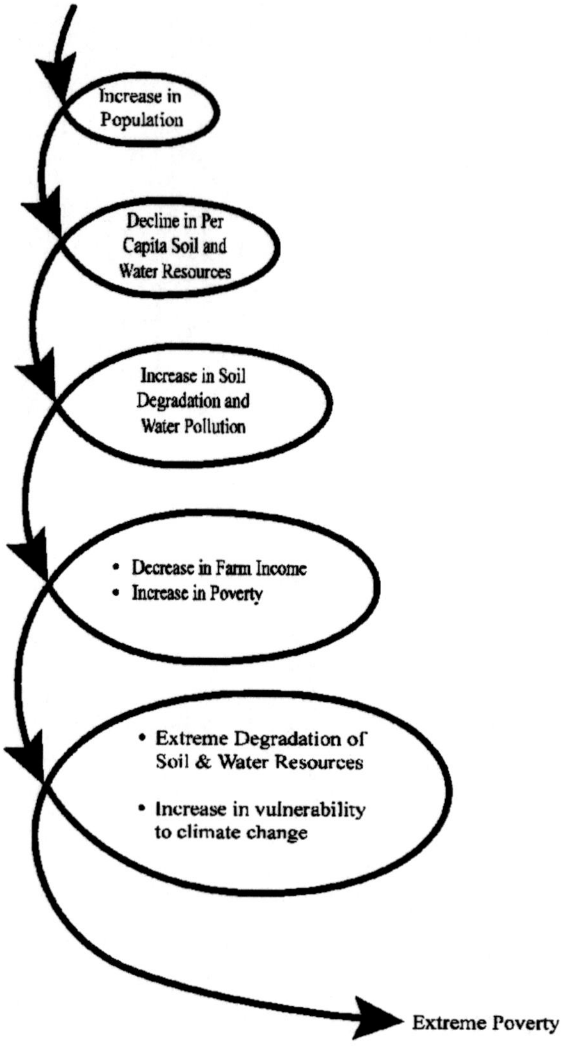

Fig. 10.2 Population driven soil degradation

There is a strong interaction between soil erosion and soil mismanagement. The latter include: (i) decline in protective ground cover by removal of crop residues, biomass burning and excessive grazing, (ii) use of little and unbalanced fertilizers, (iii) unnecessary plowing, (iv) lack of cover crops and vegetative contour hedges, and (v) none or poor establishment of a perennial vegetative cover on steep slopes and other erosion-prone regions of a watershed/landscape. The widespread problem of secondary salinization is caused by excessive and flood-type irrigation, use of poor quality (brackish) water, and no provision for drainage. Rise in ground water

Table 10.9 Poverty in South Asia region (World Bank 2005)

	Poverty headcount ratio at national poverty line (% of population)								
Country name	1991	1993	1994	1996	1999	2000	2002	2004	2005
Afghanistan	–	–	–	–	–	–	–	–	–
Bangladesh	–	–	–	51	–	48.9	–	–	40
Bhutan	–	–	–	–	–	–	–	–	–
India	–	–	36	–	–	28.6	–	–	–
Iran, Islamic Rep.	–	–	–	–	–	–	–	–	–
Maldives	–	–	–	–	–	–	–	–	–
Nepal	–	–	–	41.76	–	–	–	30.9	–
Pakistan	–	28.6	–	–	32.6	–	–	–	–
Sri Lanka	20	–	–	25	–	–	22.7	–	–

table (inundation or waterlogging), because of excessive canal-based irrigation and poor drainage, is a major cause of secondary salinization in parts of northwestern India (Punjab, Haryana, Rajasthan). There is also a close link between the drought stress and monsoon rains. Soil moisture content affects the monsoon rains (Douville et al. 2001; Kriplani and Kulkarni 2001; Maehl 1994).

Decline in soil structure, low and weak aggregation, and susceptibility to crusting and compaction are caused by depletion of SOM. The latter is exacerbated by removal of crop residues, excessive grazing and little or no application of biosolids (e.g., manure, compost, mulch, sludge, etc) as soil amendments. The soil organic carbon (SOC) concentration of most cropland soils is as low as 0.1% (1 g/kg). In contrast, the critical level of SOC for most soils of the tropics and sub-tropics, for optimal level of processes which govern the essential ecosystem services, is about 1.1% (Aune and Lal 1998). Therefore, improvements in SOC concentration (and pool) in soils of SA are essential to reversing degradation trends and restoring degraded/desertified soils and ecosystems.

10.5 Crop Yields

The Green Revolution technology increased crop yields throughout SA. The average yield of wheat (mean across the entire region) increased from 901 kg/ha in 1961 to 2,044 kg/ha in 2005, by a factor of 2.27 (Table 10.10). The largest relative increase in yield of wheat occurred in India, from 851 kg/ha in 1961 to 2,602 kg/ha in 2005, by a factor of 3.06. In comparison with Europe and East Asia, grain yield of wheat can yet be easily increased by 50% and more. Similar to wheat, there was also a substantial increase in grain yield of rice (Table 10.11). The average

Table 10.10 Grain yields of wheat in South Asia (kg/ha) (FAO Stat 2006)

Country	1961	1965	1970	1975	1980	1985	1990	1995	2000	2005
Afghanistan	1,022	972	956	1,213	1,255	1,228	1,051	1,000	724	1,816
Bangladesh	574	647	874	926	1,899	2,165	1,503	1,948	2,210	1,748
Bhutan	1,020	1,019	1,016	1,014	1,000	1,000	768	1,300	948	1,368
India	851	913	1,209	1,338	1,436	1,870	2,121	2,559	2,778	2,602
Iran	797	776	800	919	983	1,070	1,276	1,710	1,586	2,058
Nepal	1,227	1,260	1,173	1,138	1,199	1,181	1,415	1,442	1,793	2,134
Pakistan	822	863	1,171	1,320	1,568	1,612	1,825	2,081	2,491	2,586
All Countries	901	921	1,028	1,124	1,334	1,447	1,423	1,720	1,790	2,044

Table 10.11 Grain yield of rice in South Asia (kg/ha) (FAO Stat 2006)

Country	1961	1965	1970	1975	1980	1985	1990	1995	2000	2005
Afghanistan	1,519	1,727	1,812	2,071	2,173	2,248	1,903	2,294	2,000	2,540
Bangladesh	1,701	1,683	1,686	1,853	2,020	2,168	2,566	2,653	3,484	3,781
Bhutan	2,000	2,000	2,000	2,000	2,000	1,938	1,654	1,667	1,723	2,670
India	1,542	1,294	1,684	1,858	2,000	2,329	2,612	2,697	2,849	3,152
Iran	2,143	2,839	2,794	3,102	2,836	3,711	3,779	4,068	3,690	4,357
Nepal	1,938	1,987	1,949	2,074	1,932	2,016	2,407	2,391	2,703	2,782
Pakistan	1,392	17,417	2,194	2,296	2,423	2,395	2,315	2,752	3,031	3,174
Sri Lanka	1,863	1,770	2,248	1,932	2,590	3,078	3,064	3,159	3,437	3,547
All countries	2,014	2,102	2,338	2,455	2,567	2,834	2,900	3,097	3,273	3,715

yield of rice (mean across the entire region) increased from 2,014 kg/ha in 1961 to 3,715 kg/ha in 2005, by a factor of 1.84. At national level, high rice grain yields are obtained in Iran, Bangladesh, and Sri Lanka. National average grain yields are low in Nepal, India and Pakistan. Rice grain yield can be easily doubled through proper management of water and plant nutrients. Water management in rice is especially critical. Growing aerobic rice may produce a satisfactory yield of 4–5 t/ha, while saving a considerable amount of water. Therefore, growing aerobic rice is an important strategy, provided that appropriate variety, weed control and nutrient management technologies can be established for diverse soils and ecoregions (Kerye et al. 2009; Bouman et al. 2007).

10.6 Innovative Technologies

Restoring soil quality, identifying alternate crops (improved varieties) and cropping systems, and improving use efficiency of inputs (nutrients, water, energy, chemicals) are important to enhancing and sustaining high production. Adapting agricultural systems to changing climate also necessitate formulation and implementation of appropriate policies (Aggarwal et al. 2004). In this regards, the significance of increasing SOC (SOM) pool cannot be over-emphasized, especially with regards to improving soil structure, increasing total and macro porosity, improving water infiltration rate, enhancing activity and species diversity of soil fauna, and reducing losses of water and nutrients from the ecosystems. The goal is to create positive C and nutrient budgets by strengthening nutrient recycling, and reducing losses. Some of the new and innovative technologies for water and soil management are listed in Table 10.12. Important technologies for water management include micro-irrigation to replace flood irrigation, water harvesting and recycling, using waste water (city water) for irrigation, improving soil–water retention capacity. Growing aerobic rather than flooded rice is an important strategy to save water, and produce more crop per drop of water. Rice and shrimp production can be a profitable system (Ali 2006).

Soil management options include conservation agriculture with mulching and use of cover crops in the rotation cycle, application of organic amendments, and biosolids to enhance SOC pool by creating a positive C budget, integrated nutrient management to create positive nutrient budget while minimizing dependence on chemical fertilizers, and improving biotic activity of macro and micro fauna. Watershed management (Tiwari et al. 2008), and soil C sequestration (Upadhyay et al. 2005) must be appropriately emphasized.

There are also crop management options, using GM crops and complex cropping systems. The goal is to identify new crops which are adapted to changing climate, especially crops with deep and prolific root systems. Forages must be grown in rotation with food crops so that crop residues are returned or mulch, and energy plantations must be established so that animal dung is use as a compost rather than a household fuel.

Table 10.12 Innovative technologies for sustainable management of soil and water resources in South Asia (Adapted from NRC 2009)

Resource	Innovative technologies
I. Water management	(i) Micro-irrigation primarily sub-surface drip irrigation (SDI) (ii) Improved soil management (e.g., balanced nutrient application, effective weed control) and aerobic rice (iii) Water harvesting, storage and recycling including managed underground storage, and use of aquaculture (iv) Desalination and using nano materials for water purification (v) Waste water (city water/grey water) reclamation and use (vi) Improving soil–water storage through bioplastic (degradable) mulch and zeolites
II. Soil management	(i) Improving SOC/SOM concentrations and pools by creating positive C budget (ii) Enhancing soil fertility by using NIM, and using balanced fertilizers, and using remote sensing of plant physiology for nutrient management and soil quality, development of transgenic N fixation in non-legumes, manipulating micro-organisms in the rhizosphere, using phytostimulators, and microbial enhancement of P uptake by crops (iii) Using zeolites and synthesized nano-materials (iv) Promoting microbial communities to create disease-suppressive soils
III. Crop management	(i) Improved varieties adapted to high temperatures, extreme events, and other biotic and abiotic stresses, and GM crops with better root systems (ii) Rice varieties which can produce high and stable yields under aerobic rather than flooded conditions (iii) New crops which can adapt to changing climate and produce high yields under adverse climatic conditions (iv) Complex crop rotations and diversified farming systems (v) Crops with deep root systems (vi) Forages which can be grown in rotation with food crops so that crop residues are returned to the soil (vii) Fuel wood plantations to provide alternative to cow dung as cooking fuel, and agroforestry systems

10.7 Conclusions

South Asia is endowed with diverse soil, water, climate, terrain and physiographic conditions. Thus, it has a vast potential to grow a wide range of crops. Replacement of traditional and extractive farming practices with scientifically proven and modern innovations can enhance food production and eliminate food-insecurity. The vicious cycle linking poverty with soil degradation can be broken through adaptation of recommended management practices (RMPs). The latter can be facilitated through payments for ecosystem services including soil C sequestration, improving water quality, enhancing biodiversity etc. In this regard, commoditization of soil C is a win–win strategy. Identifying new opportunities, which restore soil and water resources, to enhance farm income is important to increase access to food. Important among these are aerobic rice, new crops and cropping systems, aquaculture, agroforestry, biofuel plantations to provide clean cooking fuel etc. With improvement in agriculture the SA region can be an engine of economic development for the world.

References

Acharya AK, Kafle N (2009) Land degradation issues in Nepal and its management through agroforestry. Agri Environ 10:115–123

Aggarwal PK, Joshi PK, Ingram JSI et al (2004) Adapting food systems of the Indo-Gangetic Plains to global environmental change: key information needs to improve policy formulation. Environ Sci Policy 7:487–498

Ali AMS (2006) Rice to shrimp: land use/land cover changes and soil degradation in southwestern Bangladesh. Land Use Policy 23:421–435

Aune J, Lal R (1998) Agricultural productivity in the tropics and critical limits of properties of Oxisols, Ultisols and Alfisols. Trop Agric (Trinidad) 74:96–103

Bouman BMM, Lampayan RM, Toung TP (2007) Water management in irrigated rice: coping with water scarcity. IRRI, Los Barros, Philippines, p 54

Douglas I (2006) The local drivers of land degradation in South-East Asia. Geogr Res 44: 123–134

Douville H, Chauvin F, Broqua H (2001) Influence of soil moisture on the Asian and African monsoons. Part 1: monsoon and daily precipitation. J Climate 14:2381–2404

Engelman R, Le Roy P (1993) Sustaining water: population and the future of renewable water supplies. Population Action International, Washington, DC, p 56

Engelman R, Le Roy P (1995) Conserving land: population and sustainable food production. Population Action International, Washington, DC, p 48

Eswaran H, Reich PE, Beinroth FH et al (1999) Major land resource stresses in relation to sustainable agriculture in South Asia and Pacific. Ext. Bull 474. Food & Fert Techn Center, Taipei, Taiwan

FAO (1960–2007) FAO Stat, several years, FAO, Rome, Italy

FAO (1994) Land degradation in South Asia: its severity, causes and effects upon the people. World Soil Resources 78. FAO, Rome, Italy, p 100

Gardner-Outlaw T, Engelman R (1997) Sustaining water, easing scarcity: a second update. Population Action International, Washington, DC

IFDC (2004) Global and regional data on fertilizer production and consumption 1961/62–2002/03. IFDC, Muscle Shoals, AL, p 73

Joshi PK, Tyagi NK (1994) Salt-affected and waterlogged soils in India. In: Svendsen M, Gulati A (eds) Strategic change in Indian irrigation. IFPRI/ICAR, New Delhi, India

Kerr RA (2009) Northern India's ground water is going, going, going. Science 325:798

Kreye C, Bouman BAM, Castaneda AR et al (2009) Possible causes of yield failure in tropical aerobic rice. Field Crops Res 111:197–206

Kriplani RH, Kulkarni A (2001) Monsoon rainfall variations and teleconnections over South and East Asia. Int J Climatol 21:603–616

Lal R (2006) Dryland farming in South Asia. In: Dryland Agriculture, 2nd edn. ASA, Madison, WI, pp 527–575

Lal R (2007) Soil degradation and environmental quality in South Asia. Int J Ecol Environ Sci 33:91–103

Maehl GA (1994) Coupled land-ocean-atmosphere processes and South Asian Monsoon variability. Science 266:263–267

Niazi T (2003) Land tenure, land use and land degradation: a case for sustainable development in Pakistan. J Environ Dev 12:275–294
del Ninno C, Dorosh PA, Subbarao K (2007) Food aid, domestic policy and food security: contrasting experiences from South Asia and Sub-Saharan Africa. Food Policy 32:431–435
Niroula GS, Thapa GB (2005) Impacts and causes of land fragmentation, and lessons learned from land consolidation in South Asia. Land Use Policy 22:358–372
NRC (2009) Emerging technologies to benefit farmers in sub-Saharan Africa and South Asia. National Academy Press, Washington, DC, p 262
Rodell M, Velicogna I, Famiglietti JS (2009) Satellite-based estimates of groundwater depletion in India. Nature. doi:10.1038/nature 08238) Sept 2009 p 4
Rosegrant MW, Ringler C (2000) Asian economic crisis and the long-term global food situation. Food Policy 25:234–254
Srinivasan TN (2000) Poverty and under nutrition in South Asia. Food Policy 25:269–282
Tiwari KR, Bajracharya RM, Sitaula BK (2008) Natural resource and watershed management in South Asia: a comparative evaluation with special reference to Nepal. J Agric Env 9:72–89
Upadhyay TP, Sankhayan PL, Solberg B (2005) A review of carbon sequestration dynamics in the Himalayan region as a function of land use change and forest/soil degradation with special reference to Nepal. Agric Ecosyst Environ 105:449–465
Van Lynden GWJ, Oldeman LR (1997) The assessment of the status of human-induced soil degradation in South and Southeast Asia. ISRIC Wageningen, Holland, p 35
Winslow M, Shapiro B, Thomas R et al (2004) Desertification, drought, poverty and agriculture. Research lessons and opportunities. ICRISAT, Patancheru, AP, India
World Bank (2009) Poverty index. World Bank, Washington, DC

Chapter 11
Restoration of Degraded and Desertified Lands: Experience from Iceland

Sveinn Runólfsson and Anna María Ágústsdóttir

Contents

Abstract Land degradation and desertification are a real threat to the future of human civilization in addition to being large contributors to the risks of climate change. Iceland has a long history of desertification and land degradation since its settlement by humans about 1,100 years ago. Organized battle against destruction of woodlands and soil erosion in Iceland began with a law that was set in 1907, establishing the Icelandic Soil Conservation Service (ISCS). Experience from the last 100 years shows that the key to success is reaching people through involvement and education using participatory approaches to soil conservation. The current goals of the ISCS are mitigation of land degradation, revegetation of eroded land, and attaining sustainable land use.

Climate mitigation through carbon sequestration in soil and vegetation with land restoration and revegetation must give full consideration to multiple goals, including those of the conventions of combating desertification and conserving biological diversity. In Iceland carbon sequestration is regarded as an added benefit of land restoration efforts, but not a goal in itself. Ecosystem restoration and carbon sequestration through revegetation demonstrates the synergic effects of land degradation and desertification on other environmental goals.

Keywords Desertification • Land degradation • Carbon sequestration • Land restoration • Land management • Afforestation

S. Runólfsson (✉) and A.M. Ágústsdóttir
Icelandic Soil Conservation Service, Gunnarsholt, IS-851 Hella, Iceland
e-mail: sveinn@land.is; annamaria@land.is

R. Lal et al. (eds.), *Climate Change and Food Security in South Asia*,
DOI 10.1007/978-90-481-9516-9_11, © Springer Science+Business Media B.V. 2011

Abbreviations

IPCC	Intergovernment Panel on Climate Change
ISCS	Icelandic Soil Conservation Service
NIR	National Inventory Report

11.1 Introduction

Land degradation and desertification are a real threat to the future of human civilization. The Millennium Ecosystem Assessment has ranked land degradation among the world's greatest environmental challenges that is hampering efforts to achieve the Millennium Development Goals in all ecoregions of the world. Vegetation and soil is being lost at an alarming rate in many parts of the world, leading to a wide range of environmental and socio- economic degradation and hazards. Close to 95% of food production is soil based, which therefore makes soil the most precious resource on Earth. With the interaction of current soil losses and projections on population growth, how can enough food be secured in the coming years? The same applies to many of the various services provided by the interlinked ecosystems of the globe. Water storage is dependent on the health of watersheds, but a growing proportion of the world's population is already facing water shortages. A land area of about two billion ha is prone to degradation processes and increases annually by 5–10 million ha (Oldeman 1994).

Desertification is a global problem, not only in its consequences, but also in its extent. It is far from being confined to land degradation in arid, semi-arid and dry sub-humid areas. The world's forests and woodlands are being reduced at an alarming rate in many parts of the world, and large areas are being overgrazed. The weakening of the vegetative cover can lead to a chain of ecosystem disturbances, further reducing the resilience of the ecosystems towards greater degradation. If prolonged, this can lead to desertification in a wide range of moisture regimes.

Land degradation and desertification are also large contributors to the risks of climate change. Land use changes contribute about 25% of agents causing climate change (IPCC 2007a). Equally important is that soil organic matter is the second biggest carbon pool in the planet after the oceans. Climate change, via changes in rainfall patterns and increases in average temperatures, will put further pressure on soil quality and will increase the risk of desertification.

The more carbon we keep in or add to the soil, the less carbon dioxide we will have in the atmosphere. This will not only aid in mitigating global warming, it will also diminish desertification risks, thereby sustaining agricultural production and allowing us to keep feeding the ever growing world population. The more affluent countries of the world must increase their efforts in assisting countries in need. Our common future depends on common solutions.

11.2 Experience from Iceland

Iceland has a long history of desertification and land degradation since its settlement by humans about 1,100 years ago. Organized battle against destruction of woodlands and soil erosion in Iceland began with a law that was set in 1907. Therefore Iceland has possibly the world's oldest Soil Conservation Service (ISCS), established 1907. The main goals of the current law for the ISCS are mitigation of land degradation and desertification, revegetation of eroded land, and attaining sustainable land use.

Iceland is located just beneath the Arctic Circle, on the Mid-Atlantic ridge, one side belongs to the European-, the other to the American tectonic-plate. Its size is 103,000 km^2 and population of about 315,000 people in 2008. The climate is cold temperate to alpine, mean temperature in the south ranges from around −1°C to +11°C.

Iceland was settled by Scandinavian Vikings around 874 AD. They came to a vacant country, although some Irish monks may have had some dwellings there. The Saga period, the first few centuries, was prosperous. The foundation for the initial wealth of the Icelanders was the fertility of the land. About 60% of the country was covered with lush vegetation when it was first settled in 874 (Bjarnason 1942), but today, it is left with only about 35% vegetative cover (Guðjónsson and Gíslason 1998). The forest cover, mostly Downy Birch, *Betula pubescens*, was significantly reduced from 25% since settlement (Bjarnason 1942), to 1.1% in recent decades (Guðjónsson and Gíslason 1998). There are several indications that land decline began soon after settlement. The woodlands were cut for fuel and timber or burned to get space for agriculture and grazing. Regeneration was hampered by heavy grazing and the woodlands began to reduce in size. Some of the Sagas, written about 200 years after these episodes, already then spoke of woodlands *of the past* in some areas. With the reduction of the woodland cover, sensitive soils lost their shelter. Unsustainable land use, interacting with frequent volcanic eruptions and climatic fluctuations, marked the beginning of dramatic ecosystem destruction that still lasts.

For about 1,000 years, Iceland was a country of self subsistence, to a large extent based on hay- and grazing-based livestock production in a harsh environment. Winter survival of livestock was the main determinant of population size until the late nineteenth century. There were many severe winters. If the sheep starved and died, so did the people.

Keeping the nation alive took its toll. The pressure exceeded the ecological capacity, and catastrophic soil erosion and desertification devastated large parts of the country. Much of remaining vegetation is severely degraded. A national survey of the nature and extent of soil erosion was completed in 1997, revealing that serious soil erosion is still occurring in about 40% of Iceland (Fig. 11.1, Arnalds et al. 2001).

There is much at stake for the Icelandic nation in restoring soil fertility. Not only soil and vegetation in extensive areas have been lost. Biological diversity has been greatly reduced, land fertility diminished, hydrology altered and local climate changed. Immense amounts of carbon have been lost (Figs. 11.2–11.5). Since settlement, around 20 million tons of carbon stored in vegetation may have been lost (Jónsson and Óskarsson 1996), and 120–500 million tons of soil organic carbon (Óskarsson et al. 2004). Iceland may have lost in total the equivalence of at least

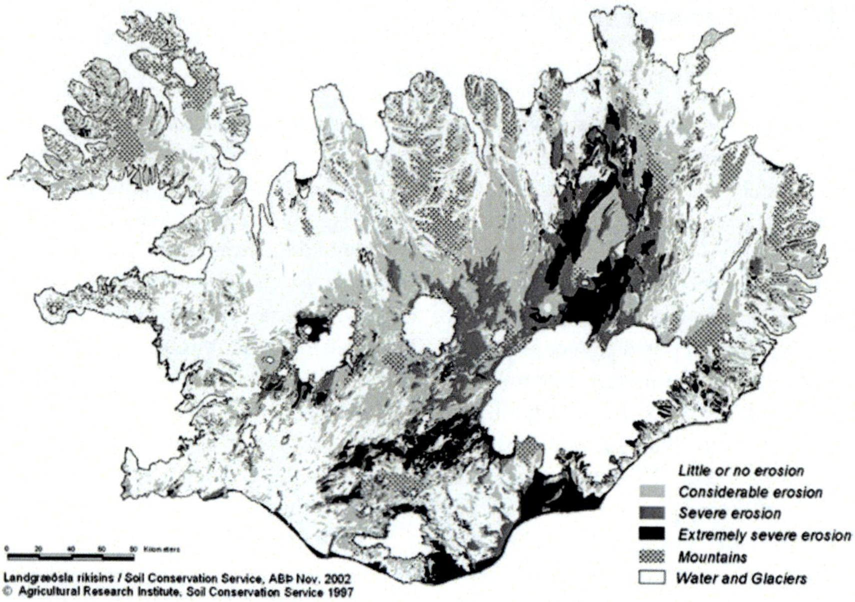

Fig. 11.1 Soil erosion in Iceland. Of the total area of Iceland of 103,000 km², little or no erosion covers an area of 37.3%, considerable erosion 22.5%, severe erosion 11%, and extremely severe erosion covers 6.2% (Arnalds et al. 2001)

Fig. 11.2 Erosion escarpments at Arnardrangur in W-Skaftafellssýsla county, show vividly the magnitude of soil erosion. The erosion has resulted in the complete removal of the soil profile down to bedrock or the lava surfaces leaving isolated islands or areas with some remaining soil and degraded vegetation. The complete lack of vegetation on the eroded surfaces shows that erosion in this site is still ongoing and severe. (Photo, Björn Barkarson ISCS)

Fig. 11.3 Þjórsárdalur valley in Árnessýsla county. Hardy grass species are used for reclamation in order to stabilize the surface of vast areas that was a source of sand drift prior to reclamation. Fertilization is required for 2–4 years to begin with in order to facilitate and speed up natural plant succession. Seeded species regress as fertilization effects wear off and natural succession continues the process. Area to the right on the photo shows the condition of the land prior to action, a deserted area with no vegetation. (Photo, Sigşrúður Jónsdóttir ISCS)

1.6 billion tons of CO_2 since 874. This is 500 times more than emissions in the baseline year of 1990. These figures are likely to be underestimates. It is difficult to assess the fate of the carbon. Much of the finer and the more valuable soil is blown long distances, much of it being lost to the ocean. Some is redistributed making remaining soils thicker, and some is carried by water either for resettlement or lost to lakes and the ocean.

Fighting catastrophic erosion is a legal responsibility of the government. During the last 100 years the key to success has usually been found by reaching people through involvement and education. Since 1990, there has been an increase in participatory approaches to soil conservation. A number of policies and practices have been set into action to arrest land degradation and to restore land and soil quality.

The highly successful "Farmers heal the land" project includes a "cost sharing" partnership with farmers, with conservation work jointly funded by government and farmers. The evolving "Better farms" project combines the forces of soil conservation, forestry, extension and nature conservation in aiding land users to produce their own property plans. The farmers have the highest stakes, as their income depends directly on the quality of the land. They play a key role in carbon sequestration. They own most of the land to be restored and with their machinery and knowhow they have an enormous capacity. About 600 farmers (25% of the total

Fig. 11.4 Beach Wildrye (*Leymus arenarius*) has been used to stabilize drifting sands and eroding fronts in Iceland for more than a century. It is the only plant species that is capable of tolerating and halting sand drift. It does not tolerate competition well and a decline of vigour in *L. arenarius* is commonly observed on sand dunes in Iceland that have become stabilized and colonized by other vascular plant species (Photo, Elín Fjóla Þórarinsdóttir ISCS)

number of farmers) participate with the Icelandic Soil Conservation Service in "Farmers heal the land".

A major step in conservation was taken with the current contract on agricultural support between sheep producers and the government. It has a cross-compliance clause, i.e. about a third of the support is dependent on quality of land use. Starting in 2003/2004, farmers must verify sustainability of their operation to the ISCS in order to obtain a full subsidy. Farmers not meeting standards must submit a conservation and land improvement plan for ISCS approval.

A Parliament agreed program gives the ISCS an operational framework for the period 2003–2014. This forms Iceland's equivalence of a National Action Plan for Combating Desertification. This 12 year program emphasizes mitigating land degradation and desertification, reclamation, sustainable land use, research, extension and awareness-raising. It includes provisions for strengthening cooperation with non-governmental organizations (NGOs), land users and other interest groups. Measures will be taken so that soil conservation will be in line with policy on development and nature conservation. The strategy also aims to fulfill international agreements on desertification and nature conservation that Iceland is a party to.

The 2008 government budget of the ISCS is US $6 million. With a population of 315,000 this is equivalent to about US $19 per capita.

Fig. 11.5 Remnants of lush birch (*Betula pubescens*) shrubs and forest still remain in eroded sites, shown here in Hraunteigur S-Iceland. These small isolated forest remnants bear evidence to the extensive woodlands and prolific ecosystems that were in place in the past (around 1850) in all of the northern part of Rangárvallasýsla county in S-Iceland. After extensive erosion the remaining vegetation was highly degraded such as the birch in this photo. (Photo, Sveinn Runólfsson ISCS)

11.3 Carbon Sequestration and Climate Change

Warming of the climate system is unequivocal, both carbon dioxide and global temperatures are rising (IPCC 2007b). The greenhouse effects add on natural fluctuations, making them more extreme. Climatic change might have more profound effects in Iceland than most of its inhabitants realize, both on its sensitive and vulnerable nature and on the active lifestyle of the Icelanders. However, these effects are hard to predict, and they could turn out to become either highly beneficial or catastrophic, with a wide range in between.

Iceland may be regarded as a cool country, and it is on the borderline for crop agriculture. A slight increase in the mean annual temperature could make Iceland the nicest spot on Earth, with a very comfortable climate. Iceland is in a warm period, possibly with the highest mean temperatures since before year 1200. As a result, glaciers that cover 10% of the island are retreating. Some estimates even predict that much of the glaciers will have disappeared after about 200 years at the current rate. Iceland would then lose one of its strongest characters and sources of beauty.

Iceland is blessed with plentiful hydrological- and geothermal power. Over 99% of electricity production and almost 80% of total energy production comes from hydropower and geothermal energy. More than 80% of houses in Iceland are heated with geothermal hot water, and most of the rapidly growing large scale industry is based in hydrological power. No other nation uses such a high proportion of renewable energy resources as Iceland. The composition of sources of emission of greenhouse gases in the baseline year of 1990 therefore may be unusual, with industry, transportation and fishing roughly emitting close to a third each. The transformation to sustainable energy took place long before 1990. Iceland therefore has more limited options than most other nations in reducing greenhouse gas emissions. This uniqueness was met in the Kyoto protocol by allowing Iceland to increase its emissions by 10% above the 1990 baseline during the first commitment period. A special clause to the Protocol has also been approved that allows countries with small overall emission and where single projects have a large effect, to exclude emissions from projects above a certain size up to a total of 1.6 million tons of CO_2 per nation. Some call this the "Icelandic clause".

Left with limited options for reducing greenhouse gas emissions, carbon sequestration is a significant tool in the Icelandic Climate Change Action Program in meeting commitments for the Kyoto Protocol. This is based on the fact that reducing emissions, preventing degradation of soil and vegetation, and carbon sequestration are all important tools in meeting the goals of conserving climate. With regard to land fertility, CO_2 may be regarded as a misplaced resource that vegetation can convert back to organic matter, to be stored in biota and soil.

Encouraged by the success of mitigation and restoration work for 100 years, the Icelandic government decided to use carbon sequestration in soil and vegetation to meet emission targets for year 2000 and established a special action program for 1997–2000. This led to some fund increase for halting soil erosion, revegetation and reforestation, compared to previous years.

A data base is being built up for carbon sequestration bookkeeping purposes for both soil conservation and forestry. Research so far indicates an average carbon sequestration rate for revegetation of 0.75 tons carbon/hectare per year, or 2.75 tons of CO_2 equivalents (NIR 2007). Revegetation in Iceland often involves converting previously unvegetated area ('other land') into grassland. Much of this carbon is stored in the soil. This may be regarded as a rather high rate, but volcanic soils as those in Iceland can store high amounts of carbon. In re- or afforestation the annual C removal factor of 1.2 t C ha^{-1} is used in the UNFCCC inventory as a precautionary estimate of data from Icelandic Forest Research, including both surface biomass and below ground biomass of coarse roots in living biomass (NIR 2007). The rate ranges widely whether it is from forestry in reclamation- and other poor condition or under better conditions using introduced species. Carbon sequestration in Iceland due to revegetation in 1990–2005 amounts to more than −244 Gg CO_2 (NIR 2007). There is no shortage of land to be reclaimed in Iceland, and it is of vital importance for the Icelanders to improve living conditions and increase food security for the future.

In carrying out projects of carbon sequestration for climate conservation purposes, full consideration must be given to multiple goals, including those of the conventions of combating desertification and conserving biological diversity. In Iceland the carbon sequestration is regarded as an added benefit, but not a goal in itself. A misplaced resource is being returned back to the land for a variety of purposes benefiting both current and future generations. Carbon sequestration by restoration of land health should not be regarded as an "escape route" from reducing emissions. It is an additional tool, a true win–win opportunity for both preventing climate change and meeting food requirements and other needs of the worlds growing populations. Ecosystem restoration and carbon sequestration through revegetation also demonstrates the synergic effects of mitigating land degradation and desertification on other environmental goals.

References

Arnalds O, Thórarinsdóttir EF, Metúsalemsson SM, Jónsson Á, Grétarsson E, Árnason A (2001) Soil erosion in Iceland. English translation, originally published in Icelandic 1997. Soil Conservation Service, Agriculture Research Institute, Reykjavík, Iceland, p 121

Bjarnason H (1942) Ábúð og örtröð. Ársrit Skógræktarfélags Íslands 1942: 8–40 (In Icelandic)

Guðjónsson G, Gíslason E (1998) Vegetation map of Iceland. 1: 500.000. General overview, 1st edn. Icelandic Institute of Natural History, Reykjavík, Iceland

IPCC (2007a) Climate Change 2007: the physical science basis. In: Solomon S, Qin D, Manning M, Chen Z, Marquis M, Averyt KB, Tignor M, Miller HL (eds) Contribution of Working Group I to the Fourth Assessment Report of the Intergovernmental Panel on Climate Change. Cambridge University Press, Cambridge, p 996

IPCC (2007b) Climate Change 2007: synthesis report. Contribution of Working Groups I, II and III to the Fourth Assessment Report of the Intergovernmental Panel on Climate Change. In: Team CW, Pachauri RK, Reisinger A (eds) IPCC. Geneva, Switzerland, p 104

Jónsson ÞH, Óskarsson Ú (1996) Skógrækt og landgræðsla til að nema koltvísýring úr andrúmslofti. Skógræktarritið 1996: 65–80 (In Icelandic)

NIR (2007) National Inventory Report. Iceland 2007. Submitted under the United Nations Framework Convention on Climate Change. Ministry for the Environment, Reykjavík, June 2007, pp 185

Oldeman LR (1994) The global extent of soil degradation. In: Greenland DJ, Szaboles I (eds) Soil resilience and sustainable land use. CAB International, Wallingford, UK, pp 99–118

Óskarsson H, Arnalds Ó, Gudmundsson J, Gudbergsson G (2004) Organic carbon in Icelandic Andosols: geographical variation and impact of erosion. Catena 56: 225–238

Chapter 12
Organic Carbon Cycling During Himalayan Erosion: Processes, Fluxes and Consequences for the Global Carbon Cycle

Valier Galy, Christian France-Lanord, Olivier Beyssac, Bruno Lartiges, and Mustafizur Rhaman

Contents

Abstract The net effect of organic carbon cycling during continental erosion depends on the balance between rock-derived organic carbon oxidation and biospheric organic carbon burial in sediments. Himalayan erosion is dominated by

V. Galy (✉)
Department of Marine Chemistry and Geochemistry, Woods Hole Oceanographic Institution, 360 Woods Hole Road, Woods Hole MA 02543
and
CRPG UPR 2300 CNRS/INSU, Université de Lorraine, BP 20, 54501 Vandœuvre-lès-Nancy, France
e-mail: vgaly@whoi.edu

C. France-Lanord
CRPG UPR 2300 CNRS/INSU, Université de Lorraine, BP 20, 54501 Vandœuvre-lès-Nancy, France
e-mail: cfl@crpg.cnrs-nancy.fr

O. Beyssac
Laboratoire de Géologie, Ecole Normale Supérieure, CNRS- UMR 8538, Paris

B. Lartiges
LEM UMR 7569 CNRS/INSU, Nancy Université, BP 40, 54501 Vandœuvre-lès-Nancy, France

M. Rhaman
Department of Soil, Water and Environment, Dhaka University, Dhaka, Bangladesh
e-mail: dmrahman@agni.com

R. Lal et al. (eds.), *Climate Change and Food Security in South Asia*,
DOI 10.1007/978-90-481-9516-9_12,

physical transport and each year up to two billion tons of sediments eroded from the Himalaya are delivered to the Bengal Fan through the Ganga–Brahmaputra (G–B) fluvial system.

We developed a sampling protocol that allows the heterogeneity of the sediment load to be accounted for. In the channel of large rivers, the total organic carbon content (TOC) is variable and decreases towards depth. TOC is positively correlated to Al/Si ratio, which characterizes the mineral and grain size sorting. In the delta of Bangladesh, sediments from Ganga, Brahmaputra and Lower Meghna have similar organic carbon loading.

Coupling Raman Micro-spectroscopy and High Resolution Transmitted Electron Microscopy allows the unambiguous detection and characterization of petrogenic (rock-derived) carbon. Comparison of Himalayan rivers and G–B in Bangladesh indicates that the most graphitised forms are selectively preserved and delivered to the Bay of Bengal. Radiocarbon characterization of sediments along depth profiles yields values for the absolute concentration of petrogenic carbon in rivers sediments. Comparison of Himalayan rocks and G–B sediments in Bangladesh shows that 40% (±10) of the organic carbon contained in the Himalayan rocks is preserved and delivered to the ocean.

The evolution of stable isotopic composition ($\delta^{13}C$) from the outflow of the Himalayan range to the delta of Bangladesh shows that during the Gangetic floodplain transit, more than 50% of organic carbon derived from the Himalaya is oxidized and replaced by organic carbon derived from the floodplain.

The organic carbon loading of recent Bengal Fan sediments is comparable to that of G–B river sediments. Biomarker abundance and $\delta^{13}C$ values show that organic carbon is dominated by terrestrial inputs. The terrestrial organic carbon burial efficiency is thus close to 100%. This strongly contrasts with other large deltaic system on earth, where ~70% of terrestrial organic carbon is oxidized prior to burial. This extreme burial efficiency is sustained by high erosion rate in Himalaya that generates high sedimentation rate and low oxygen availability in the Bay of Bengal.

The balance between biospheric organic carbon burial and petrogenic carbon oxidation indicates a net CO_2 consumption of $3.2 \pm 0.8 \times 10^{11}$ mol/year. Atmospheric CO_2 consumption through organic carbon cycling during Himalayan erosion is thus an order of magnitude higher than the CO_2 consumption through silicate weathering in the Himalayan basin (6.4×10^{10} mol/year). Efficient burial of organic carbon is a characteristic of high physical erosion typical of active orogenic systems. Enhanced physical erosion and consequent organic carbon burial buffer atmospheric CO_2 thereby exerting a negative feedback on the long-term climate.

Keywords Erosional processes • Fate of soil organic carbon • Carbon burial • Sedimentation • Fluvial transport

Abbreviations

ADCP	Acoustic Doppler current profiler
G–B	Ganga–Brahmaputra
HRTEM	High Resolution Transmitted Electron Microscopy
OC	Organic carbon
RM	Raman microspectroscopy
TOC	Total organic carbon content
WARPO	Water Resources Planning Organization

12.1 Introduction

Over geologic timescales, atmospheric CO_2 concentration is controlled by the balance between processes liberating and consuming CO_2. The main sources of CO_2 are: (1) volcanism (e.g. Berner 2003; Marty and Tolstikhin 1998), (2) metamorphic decarbonation (e.g. Becker et al. 2008; Evans et al. 2008; Perrier et al. 2009), and (3) oxidation of organic carbon contained in rocks (e.g. Berner 2003; Hayes and Waldbauer 2006). On the other hand, there are two main CO_2 sinks: (1) silicate weathering followed by carbonate precipitation in the ocean (Ebelmen 1845; Rubey 1951; Garrels et al. 1976; Walker et al. 1981), and (2) organic carbon burial in marine sediments (e.g. Berner 2003; Hayes and Waldbauer 2006; Galy et al. 2007). Silicate weathering retained most of the attention during the past three decades or so. Nevertheless, since the emergence of higher (vascular) plants on the continents, organic carbon cycling (i.e. burial and oxidation) likely exerted at least at some periods of Earth history a dominant control on atmospheric CO_2 content. Organic carbon burial and oxidation thus play a major role in the long-term climate regulation and are intimately linked to the development and maintenance of life on earth. Physical erosion of the continents leads to the export of biospheric organic carbon from the continent to the ocean. But it also promotes liberation, transport and oxidation of refractory organic carbon contained in rocks. Moreover, extensive oxidation of organic carbon in the ocean prior to its burial has been documented (e.g. Ludwig et al. 1996; Hedges et al. 1997; Schlünz and Schneider 2000; Burdige 2005; Burdige 2007). The net effect of organic carbon cycling during continental erosion thus depends on the balance between rock-derived organic carbon (petrogenic organic carbon) oxidation and biospheric organic carbon burial in sediments. Hence, determining the net effect of organic carbon cycling during continental erosion (i.e. carbon source or sink) requires to determine not only the flux of organic carbon exported by rivers to the ocean but also its composition (e.g. biospheric vs. petrogenic) as well as its fate in the ocean.

Since the 90s, small mountainous catchments have been the focus of many studies because they were considered the most efficient pathways of terrestrial OC export and burial, resulting in a disproportionately large influence on the OC cycle (Milliman and Syvitski 1992; Scott et al. 2006; Hilton et al. 2008a, b). Carbon cycling in these small mountainous rivers stands in sharp contrast to many large river basins with extensive floodplains, such as the Amazon basin. The latter typically act as "reactors", efficiently recycling labile terrestrial organic material and associated nutrients, and resulting in extensive loss of organic carbon prior to its escape to the marine sedimentary sink (e.g. Mayorga et al. 2005). However, hitherto very little was known about the flux and fate of petrogenic organic carbon and the factors controlling the preservation of terrestrial biospheric organic carbon in the oceanic system. An early study of organic carbon burial over the Cenozoic in the Bengal fan turbiditic system, the sedimentary repository of Himalayan erosion, suggested that organic carbon cycling during Himalayan erosion has been a globally significant atmospheric CO_2 sink (France-Lanord and Derry 1997). Contrary to the early 90s hypothesis that Himalayan erosion was acting as a CO_2 consumer through enhanced silicate weathering (e.g. Raymo and Ruddiman 1992), this suggests that organic carbon cycling has been the dominant process able to consume atmospheric CO_2. During the past few years, we have thus been conducting a detailed study of organic carbon cycling associated to current erosion of Himalaya and sedimentary transfer in the Ganga–Brahmaputra river system (Galy et al. 2007; Galy et al. 2008a, b). Here we present a synthesis of these studies in the scope of the global organic carbon cycle. We first explore the different processes affecting organic carbon composition and flux in the Ganga–Brahmaputra system (Fig. 12.1). Then we address the question of the fate of organic carbon delivered by the G–B system to the Bay of Bengal.

12.2 Fluvial Transport in the G–B System

In large rivers, river detritus experience strong hydrodynamic forcing during transport, which results in mineral and grain size sorting. These sorting and mixing processes are in turn responsible for very large sediment heterogeneity in the river channel. Quartz and other coarse, rounded and dense particles tend to increase with water depth and in the bedload, whereas fine and tabular phyllosilicates concentrate towards the surface and are more rapidly transported downstream. The organic carbon content of the river sediments is also variable. In the river channel, total organic carbon (TOC) concentrations are depth-dependent with surface suspended particles having the highest values and bed sediments the lowest (Fig. 12.2). Taking into account the heterogeneity of the detrital load is thus essential for deriving accurate flux estimates as well as to accurately define the composition of the material delivered by rivers to the ocean. Consequently, we have applied a sampling protocol to develop a more representative picture of sediment transport in river systems. This approach, developed during detailed surveys of the Ganga–Brahmaputra and also successfully applied to the Amazon system, involves collection of suspended and bed sediments along vertical depth profiles from the surface

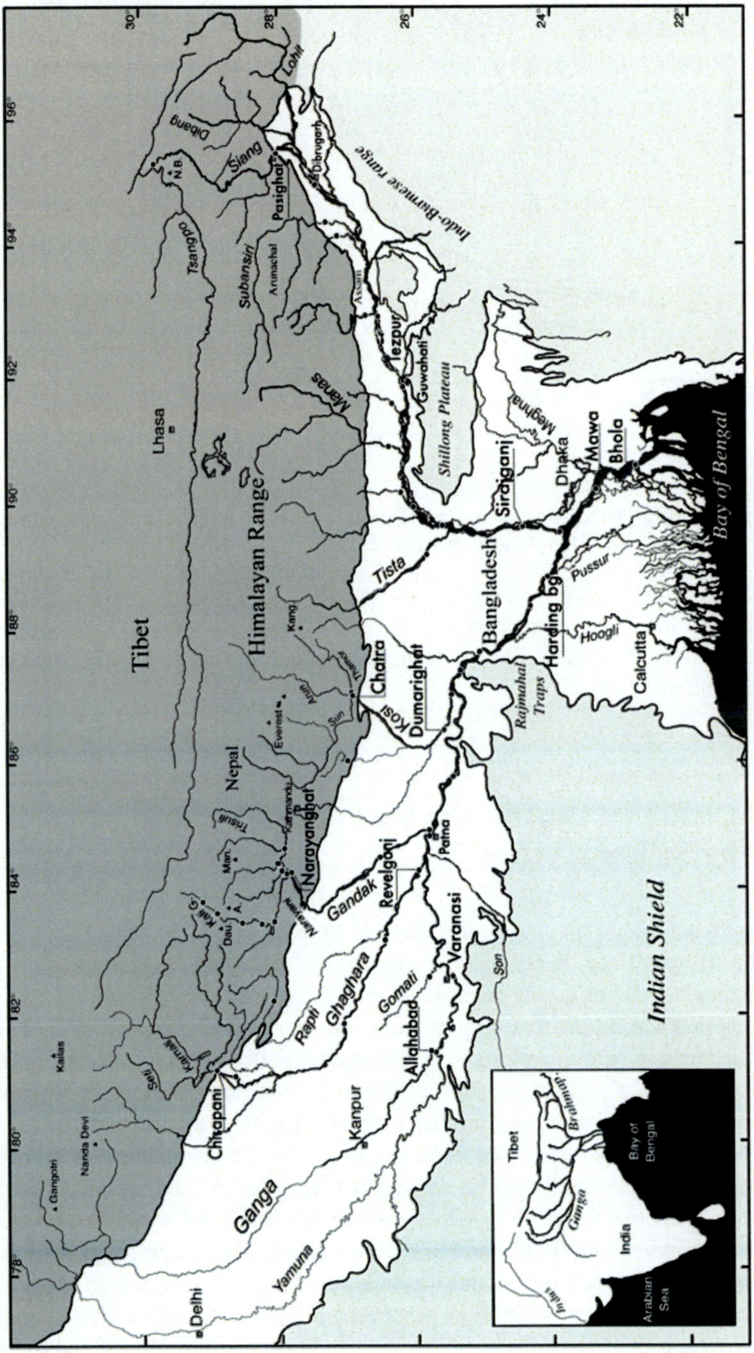

Fig. 12.1 Map of the Ganga–Brahmaputra basin showing the different physiographic units and the position of the sampling location (*black dots*)

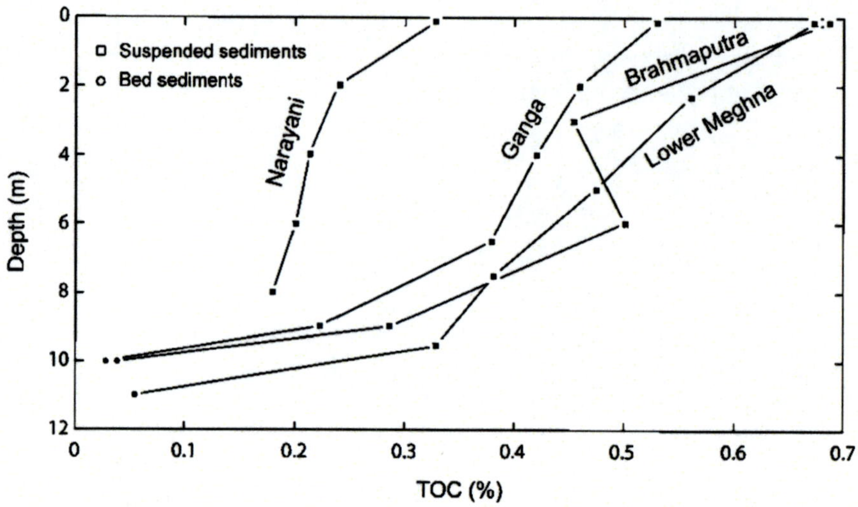

Fig. 12.2 Evolution of TOC in river sediments with sampling depth. Floodplain rivers (Ganga, Brahmaputra and Lower Meghna) as well as a large Himalayan river (Narayani) show a progressive decrease of the TOC from the surface suspended sediment to the bed sediment

to the bed of the river, coupled with flow velocity measurements via an Acoustic Doppler Current Profiler (ADCP), enabling computation of the mean composition of the sedimentary flux.

12.3 Control of Organic Carbon Content by Sediment Properties

The simultaneous and monotonic evolution of organic carbon content and sediment properties from surface suspended sediments to bed sediments suggests that organic carbon content is (1) primarily controlled by sediment properties, and/or (2) affected by mixing and sorting processes in the same way as mineral particles. In G–B river sediments, organic carbon appears to be present under two different species: organo-mineral associations and free organic particles. Detailed observations of suspended sediments collected along depth profiles indicates that free organic particles tend to be segregated as a function of their size, in a similar way as minerals: coarse organic debris tend to concentrate in coarse sediments while fine grained sediments contain mostly tiny free organic particles. While we found several good correlations between sediment properties and TOC, the most robust relationship is obtained with Al/Si ratio (Fig. 12.3). This ratio represents the bulk mineralogical composition of the sediments with low values indicating high proportion of quartz and high values indicating high proportions of micas and clays. The positive linear relationship defined by TOC and Al/Si thus indicates a control of the mineralogy on the formation of organo-mineral

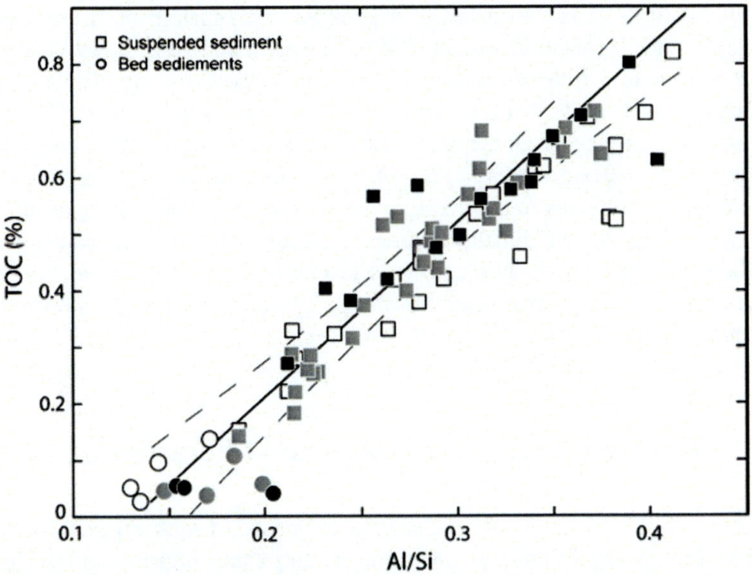

Fig. 12.3 TOC of the Brahmaputra (*grey symbols*), Ganga (*open symbols*) and Lower Meghna (*black symbols*) sediments as a function of Al/Si. Sediments from the three rivers define similar positive trend between TOC and Al/Si, indicative of similar OC loadings. Best fit and 95% confidence interval are shown for Lower Meghna sediments

associations. Microscopic observations of the sediments reveal the presence of organo-mineral aggregates. The formation of these aggregates is controlled by multiple physical parameters such as particle charge density, chemical composition (nature of exchangeable cation), crystalline structure, size and shape (e.g. Sollins et al. 1996). Specifically, our study indicates a greater affinity of organic carbon with clays and micas than with quartz, feldspar and dense minerals.

Al/Si ratio being dependent on both mineralogy and grain size it appears to adequately catch the processes controlling the repartition of both free organic particles and mineral bounded organic carbon. In the G–B system Al/Si ratio also has the advantage of being a conservative tracer because Si can be considered as an insoluble element (dissolved Si flux is three orders of magnitude lower than particulate Si flux). Al/Si ratio thus provides an efficient normalization parameter, allowing the definition of the organic carbon loading for each river and sampling location.

12.4 Evolution of Organic Carbon Loading During Transport

The comparison of different rivers and positions in the basin (high mountains, base of the range, delta) indicates that at the first order organic carbon loading varies little in the G–B system. Detailed study of the large Himalayan rivers at the base of the range however indicates some variability of the organic carbon loading.

This variability can be attributed to a complex combination of factors such as geomorphology, vegetation distribution, soil types and human activities. It is important to note that the main rivers in the plain – the Ganga and Brahmaputra – are characterized by very similar organic carbon loading (Fig. 12.3). The Lower Meghna – the mixing of Ganga and Brahmaputra in Bangladesh – also has very similar organic carbon loading (Fig. 12.3). Organic carbon loading in the delta furthermore corresponds to the average of the organic carbon loading of the large rivers at the base of the Himalayan range. Altogether these observations indicate that the exportation of organic carbon is primarily controlled by the capacity to export sediments out of the basin rather than by intrinsic parameters such as primary production.

12.5 Petrogenic Carbon: Detection and Characterization

The Himalayan range is almost exclusively composed of metamorphic rocks. The different Himalayan lithologies therefore contain some organic carbon that has been transformed – graphitized – during the metamorphic process. This graphitization consists in a chemical and structural transformation of the organic carbon leading to crystallized carbonaceous mater. With increasing temperature the graphitization is more intense and the final structure of carbonaceous mater closer to that of the graphite. Finally, when the metamorphic temperature reaches 550 degrees (or more) the graphitization process is complete and the organic carbon is transformed into graphite.

During erosion, this crystallized carbonaceous mater – hereafter referred to as petrogenic carbon – is liberated, transported by rivers and eventually oxidized or buried in marine sediments. Burial of petrogenic carbon in marine sediments does not represent a consumption of atmospheric CO_2 but a simple recycling of reduced carbon. On the other hand, oxidation of petrogenic carbon is a net source of CO_2 for the atmosphere. Evaluating the flux and fate of petrogenic carbon during erosion is therefore crucial to characterize the role of erosion in the global C cycle.

During erosion and fluvial transport, some biospheric organic carbon is admixed to the mineral matrix and petrogenic carbon. The total organic carbon content of river sediments hence account for both petrogenic and biospheric organic carbon. During the last decade or so several techniques have been developed to detect and quantify petrogenic carbon in river and marine sediments. However, these methods are based on the refractory character of the petrogenic carbon, which is not unique to petrogenic carbon. We therefore developed a new method allowing specifically detecting and characterizing the structure of petrogenic carbon (Galy et al. 2008a). This method consists in coupling Raman Micro-spectroscopy (RM) and High Resolution Transmitted Electron Microscopy (HRTEM). RM permits to characterize the microstructure of carbonaceous particles, while HRTEM allows the characterization of its nanostructure (Beyssac et al. 2002). Thanks to minimal sample preparation (hence bias) and great specificity for crystallized carbonaceous matter,

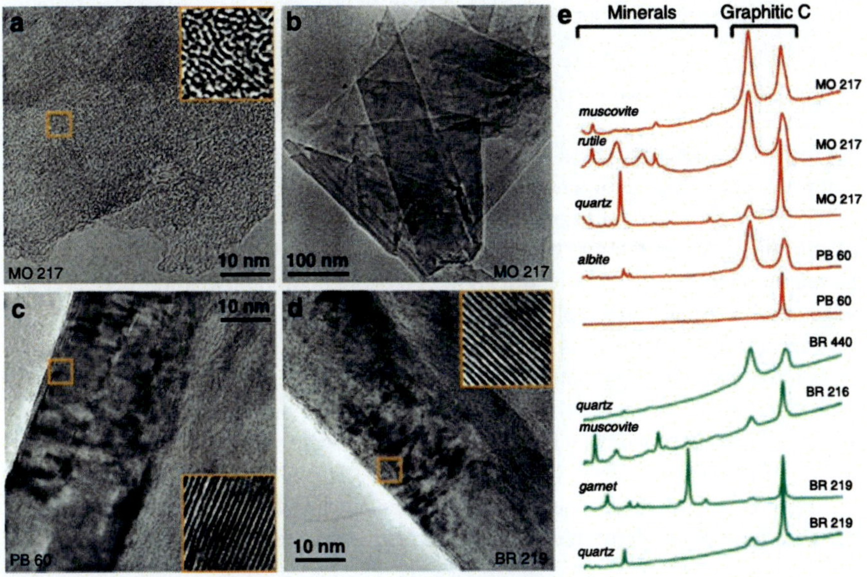

Fig. 12.4 Raman micro-spectroscopy and TEM characterization of petrogenic carbon in G–B river sediments. (**a**) TEM image of disordered and microporous petrogenic carbon in Narayani bed load. (**b**) TEM image of graphitic C in Narayani bed load. (**c**) Low magnification (LM) image of a graphite particle in Narayani suspended load. (**d**) TEM image of graphitic petrogenic carbon in Lower Meghna bed load. (**e**) Selection of representative Raman spectra from Himalayan rivers (*top panel*) and Lower Meghna (*lower panel*). These spectra were obtained from both individual petrogenic carbon particles (no mineral contribution in the spectrum) and petrogenic carbon inclusions/aggregates within minerals (mineral contribution in the spectrum as depicted)

these two methods allow the unambiguous detection and characterization of petrogenic carbon.

Using RM, petrogenic carbon was identified in various forms: (1) discrete "free" particles from a few to several tens of μm size, (2) inclusions within quartz, calcite and metamorphic minerals, (3) aggregates with minerals, mostly micas. RM shows a large structural variety of petrogenic carbon in bed and suspended loads from Himalayan rivers (Fig. 12.4). This is confirmed by TEM investigations, which reveal the presence of turbostratic disordered petrogenic carbon to perfectly crystalline graphite in these samples (Fig. 12.4). The structural variety of petrogenic carbon reflects the zoning of metamorphism in the source rocks of the Himalayan range (e.g. Beyssac et al. 2004) from low- to high-grade metamorphism. Similar structural variety is observed within the vertical depth profiles collected at the outflow of the range and in Bangladesh, although the less ordered petrogenic carbon is rarely observed. This is particularly true for the most distal sampling location in Bangladesh, where highly graphitic petrogenic carbon is largely dominant (Fig. 12.4). This implies that the less ordered petrogenic carbon is specifically oxidised during erosion and fluvial transport, while the most graphitised forms are selectively preserved and delivered to the Bay of Bengal.

12.6 Petrogenic Carbon: Quantification

Besides its detection and characterization, it is crucial to obtain a quantification of the petrogenic carbon in river sediments. We recently developed the use of vertical depth profile sampling to distinguish petrogenic and biospheric organic carbon components in river sediments on the basis of bulk ^{14}C measurements (Galy et al. 2008a). Our approach relies on a binary mixing model, which assumes the presence of two distinct populations of organic carbon in river sediments: (1) petrogenic carbon derived from rock erosion, and (2) biospheric organic carbon derived from vegetation, soil and autotrophic production in the river.

The petrogenic carbon is by definition ^{14}C free, while the biospheric organic carbon has a variable age but always contains some ^{14}C. Radiocarbon characterization of sediments along depth profiles that reflect the continuum created by sorting and mixing processes during fluvial transport yields values for the absolute concentration of petrogenic carbon in rivers sediments (Fig. 12.5).

Large Transhimalayan rivers sampled at the outflow of the range appear to carry variable amounts of petrogenic carbon. We estimate a petrogenic carbon content of

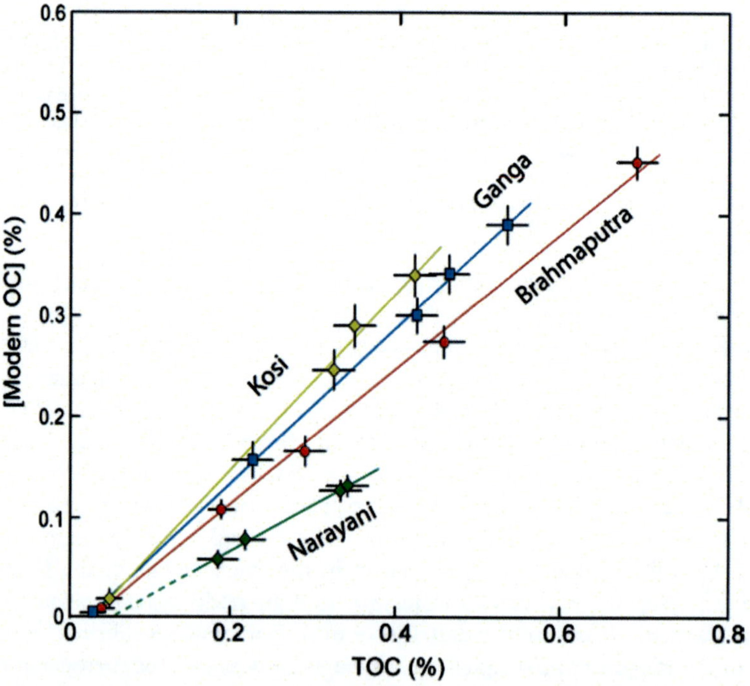

Fig. 12.5 Radiocarbon measurements in depth profile sediments collected in the Ganga–Brahmaputra system. This type of diagram shows the product of the radiocarbon composition by the TOC (or absolute modern organic carbon content) as a function of TOC. It allows calculating the absolute petrogenic carbon content for each depth profile, which is given by the intercept of the trends with the x axis

0.05% and 0.03% for the Narayani and Kosi, respectively (Fig. 12.5). This variability likely derives from intrinsic characteristics of the drainage basins. Our data consistently indicate a petrogenic carbon content of 0.02–0.03% in the sediments exported by the Brahmaputra and the Ganges sampled close to their mouth in the Bangladesh delta (Fig. 12.5).

To evaluate the fate of petrogenic carbon during erosion and fluvial transport we need to compare these concentrations to the initial mean petrogenic carbon content of the Himalayan rocks. The later has been estimated to be 0.05–0.08%, on the basis of (1) individual rock samples, and (2) composite gravels extracted from the bed of Himalayan rivers (France-Lanord and Derry 1997; Aucour et al. 2006; Galy et al. 2007b; Galy et al. 2008b). Based on these figures, at least 30% (40 ± 10) of the organic carbon contained in the Himalayan rocks appears to be preserved and delivered to the Indian Ocean.

12.7 Fate of Biospheric Organic Carbon During Fluvial Transport

Stable carbon isotopes ($\delta^{13}C$) have long been used as a biogeochemical tool to derive information on the source of organic carbon due to isotope contrasts resulting from photosynthetic C fixation mechanisms employed by different plant types (e.g. Deines 1980; Collister et al. 1994). In river sediments, the $\delta^{13}C$ of bulk organic carbon is a proxy for its provenance, because river algae, C3 and C4 plants have distinct isotope compositions. The geographic variations in proportions of C3 and C4 vegetation in the G–B drainage basin are substantial, rendering this a potentially valuable tracer, although petrogenic carbon contributions can muddy the waters, so as to speak, because of its variable and intermediate isotopic composition.

In large rivers at the outflow of the Himalayan Mountains, bulk organic carbon $\delta^{13}C$ values range between −21.2‰ and −25.7‰. Most of this variability is due to isotopic heterogeneity in organic carbon-poor bed sediments and reflects contributions from petrogenic carbon with variable $\delta^{13}C$ in the Himalayan bedrocks (Galy et al. 2008b). In suspended sediments, the $\delta^{13}C$ values display a narrower range with a mean value of ~−24‰. This isotope signature suggests only minor inputs of algal matter (which is typically significantly more ^{13}C-depleted), consistent with high sediment loads and very dynamic transport that inhibit primary production within the Himalayan rivers. In addition, this mean $\delta^{13}C$ value reveals that biospheric organic carbon is largely dominated by C3 plant inputs, consistent with vegetation patterns in the Himalayan region where forest and high altitude ecosystems dominate (e.g., Blasco et al. 1996; Dobremez et al. 1978). In summary, suspended sediments exiting the Himalayan range predominantly represent a binary mixture of organic carbon, with a portion of this comprised of petrogenic carbon from sedimentary rock erosion and the remainder from C3 vegetation that grows at lower elevations within the Himalayan range.

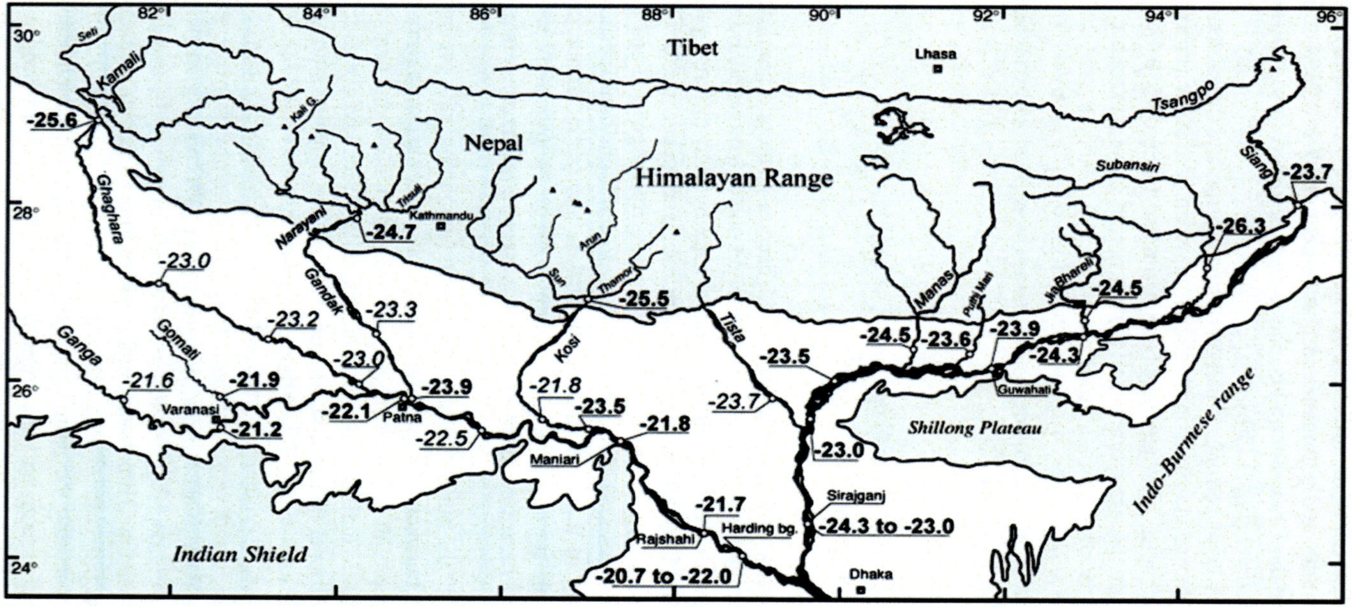

Fig. 12.6 Simplified map of the G–B system showing the evolution of bulk organic carbon $\delta^{13}C$ values from the Himalayan range to the delta. Bold numbers represent $\delta^{13}C$ values of surface suspended sediments; italic numbers represent $\delta^{13}C$ values of fine-grained bank sediments used as analogous of suspended sediments

In the floodplain, Ganga and Brahmaputra suspended sediments have distinct $\delta^{13}C$ values, ~−21‰ and −23.5‰ respectively (Fig. 12.6). The isotope composition of organic carbon in Brahmaputra suspended sediments is comparable to that of the Himalayan rivers at the outflow of the range implying that inputs are similarly dominated by C3 plants (Fig. 12.6). However, organic carbon in lower Ganga suspended sediments has $\delta^{13}C$ values that are on average 3‰ higher than that of Himalayan rivers suspended sediments (Fig. 12.6). This isotopic offset is interpreted as a consequence of significant contributions from ^{13}C-enriched C4-plant organic carbon to the Ganga suspended sediments. As the organic carbon loading in suspended sediments is similar at the outflow of the range and in the floodplain, the $\delta^{13}C$ shift is considered to reflect removal, and subsequent replacement of C3-derived organic carbon with C4 plant organic carbon during transit through the Ganga floodplain (Keil et al. 1997; Galy et al. 2008).

In the Brahmaputra floodplain, C3 plants dominate the vegetation. Hence, while this "loss-and-replacement" process may also be operative, there is insufficient isotope contrast in bulk $\delta^{13}C$ values between Himalayan and Brahmaputra floodplain vegetation to establish whether or not this is the case. While it is certainly plausible that similar processes occur along the course of the Brahmaputra River, there are differences other than vegetation type that may influence the efficacy of organic carbon turnover in the floodplain. Notably, the clay mineralogy of soils within the Ganga and Brahmaputra is quite distinct, with the former being dominated by smectite and the latter by illite. This difference manifests itself in mineral surface area measurements whereby smectite-rich sediments have high specific surface areas compared to those containing illite. The role of expandable clays as templates for organic matter attachment and transformation in soils and sediments is well known, and hence the higher proportion of these clays in the Ganga may be an important factor in dictating the exchange of organic carbon derived from different vegetation sources (e.g., Satterberg et al. 2003). Other parameters likely to limit organic carbon replacement in the Brahmaputra floodplain are the mode of fluvial transport and the size of the floodplain. While the Ganga in the plain is a meandering river and has a very long course, the Brahmaputra river is a braided river and has a much shorter course in its narrow floodplain. These geomorphologic differences further support a lower extent of organic carbon replacement in the Brahmaputra than in the Ganga floodplain.

12.8 Flux of Biospheric Carbon Delivered to the Ocean

A consequence of the huge variability of TOC along depth profiles is that estimates of organic carbon fluxes based on surface suspended sediment data are overestimated. Thus, assessing the organic carbon flux requires the mean sediment composition transported by the river to be calculated. Thanks to our sampling protocol coupling depth profiling and ADCP current profiling, the integrated suspended sediment TOC can be estimated by integrating TOC gradient, sediment concentration and flow velocity over the whole river depth. For the Lower Meghna prior to its discharge into the Bay of Bengal (Fig. 12.1) we calculated an integrated TOC of 0.41% ± 0.04%. In comparison, bed sediments have low TOC with an average value of 0.05%. Furthermore, based on radiocarbon measurements, we previously estimated that both bed and suspended sediments delivered to the Bay of Bengal contain 0.02–0.03% of petrogenic carbon. Hence the biospheric organic carbon content in mean suspended and bed sediments of the Lower Meghna is respectively 0.39% ± 0.04% and ca. 0.025%.

Suspended sediment fluxes have been measured for the Ganga and the Brahmaputra and the average total flux is around 1.15×10^9 t/year (RSP 1996). The flux of bed sediment is not directly measured but geochemical mass balance implies that bed sediment flux plus floodplain sequestration are almost equal to the suspended sediment flux (Galy and France-Lanord 2001). Taking into account these sediment fluxes, we estimate that the G–B system delivers $3.9 \pm 0.5 \times 10^{11}$ mol/year of biospheric organic carbon to the Bay of Bengal.

12.8.1 Organic Carbon Burial Efficiency in the Bengal Fan

The G–B river system supplies sediment to the Bengal Fan, Earth's largest active sedimentary system. Over the last glacial-interglacial cycle, the sediment source has remained stable and is largely dominated by Himalayan inputs (Pierson-Wickmann et al. 2001; Galy et al. 2008c). As in the rivers, characteristics of Fan sediments (grain size and mineralogy) are highly variable due to sorting during the transport within the depositional system. Nevertheless, while TOC contents are also variable, as for the river sediments they are tightly correlated with Al/Si, and thus a linear function of the proportion of clays and fine-grained minerals.

Importantly, the relationship between TOC and Al/Si defined by Bengal Fan sediments is statistically identical to that of the G–B river sediments (Fig. 12.7). This relationship implies an equivalent level of organic carbon loading in G–B river sediments and in modern Bengal Fan sediments. Several biomarker and isotope studies have shown that the organic carbon in both modern and ancient Bengal Fan sediments is overwhelmingly dominated by terrestrial inputs with negligible marine organic carbon (Cochran et al. 1989; Poynter and Eglinton 1990; Meyers and Dickens 1992; France-Lanord and Derry 1994; Freeman and Colarusso 2001; Galy et al. 2007; Galy et al. 2008c). Together, these observations imply that terrestrial organic carbon preservation in Bengal Fan sediments is exceptionally high, with the proportion of organic carbon exported by the river that is buried in fan sediments (the burial efficiency) approaching 100%! Although high terrestrial organic carbon export efficiencies have been reported for small mountainous rivers (e.g., Goni et al. 2006), this situation stands in sharp contrast with other large deltaic systems such as those of the Amazon and Mississippi rivers where terrestrial OC burial efficiencies typically do not exceed 30% (e.g. Burdige 2005; Burdige 2007; Hedges et al. 1997).

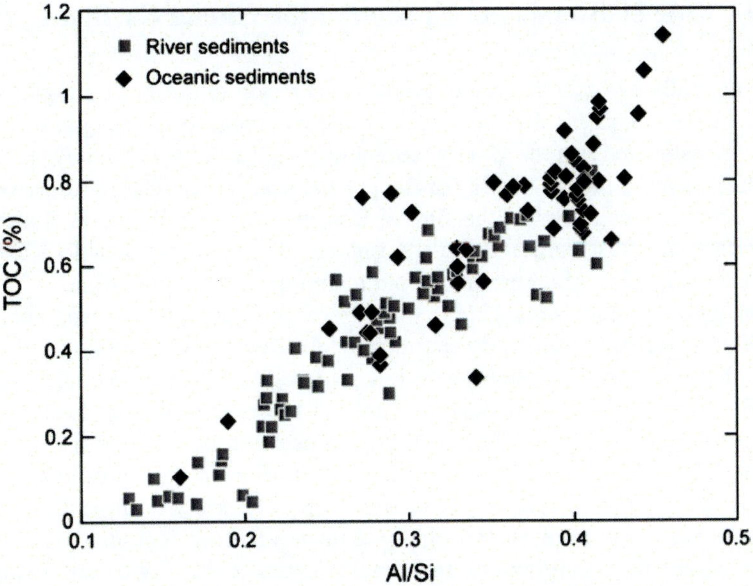

Fig. 12.7 TOC of the Bengal Fan sediments (*black diamond*) as a function of Al/Si with best fit and 95% confidence interval. Fan sediments define a trend comparable to that defined by river sediments (*grey squares*), suggesting they have similar organic carbon loadings

Very high organic carbon burial efficiency in the Bengal Fan likely derives from the transport dynamic, with rapid sediment accumulation on the delta shelf (up to 30 cm/year (Suckow et al. 2001). In shallow-water, where organic carbon oxidation is theoretically most efficient, organic carbon is protected from oxidation by quick burial under a thick layer of fresh sediments. Typical O_2 penetration depth in such sediments is only a few centimeters (Cai and Sayles 1996), i.e. the same order of magnitude as the Bengal shelf annual accumulation. Therefore, organic carbon exposure time to O_2 is only few years or so, insufficient for effective organic carbon oxidation. Sediments are finally transferred to the deep-water zone by turbiditic current and deposited in channel-levee systems. Identical organic carbon loading in the shelf, active channel-levee and mid-deep fan sediments indicates that organic carbon oxidation during sediment transfer from shelf to deep-ocean is negligible. In addition, river discharge and precipitation are responsible for a negative salinity anomaly and maintain stratified waters in the Bay of Bengal (Berner et al. 2003; Broecker et al. 1980). With high productivity in surface water sustained by high nutrient flux delivered by the G–B, the stratification of the Bay of Bengal waters favors an intense respiration of marine organic carbon in the surface waters. This process consumes O_2 and generates an extended O_2 minimum zone and generally low O_2 concentrations in the Bay of Bengal (Berner et al. 2003; Broecker et al. 1980). In Bengal Fan sediments, terrestrial organic carbon oxidation is thus limited by both short exposure to and low availability of O_2.

12.9 Role of Himalayan Erosion in the Global Carbon Cycle

We can finally evaluate the role of Himalayan erosion in the long term global carbon cycle by comparing the extent of the different process consuming or producing CO_2. We estimated that the G–B system delivers $3.9 \pm 0.5 \times 10^{11}$ mol/year of biospheric organic carbon to the Bay of Bengal. Because nearly 100% of the terrestrial organic carbon delivered to the Bay of Bengal is actually buried in Bengal fan sediments, this flux represents the atmospheric CO_2 consumption associated with biospheric organic carbon export and burial during Himalayan erosion.

On the other hand, according to the difference in petrogenic carbon between the Himalayan rocks and river sediments delivered to the Bengal fan, we estimate that Himalayan erosion generates a petrogenic carbon oxidation flux of $7 \pm 3 \times 10^{10}$ mol/year. The balance between biospheric organic carbon burial and petrogenic carbon oxidation finally indicates a net CO_2 consumption of $3.2 \pm 0.8 \times 10^{11}$ mol/year. Atmospheric CO_2 consumption through organic carbon cycling during Himalayan erosion is thus an order of magnitude higher than the CO_2 consumption through silicate weathering (coupled with subsequent carbonate precipitation) in Himalaya (6.4×10^{10} mol/year (Galy and France-Lanord 1999)). Although metamorphic outgassing represents a potentially significant source of CO_2 in Himalaya, the net effect of Himalayan erosion is to consume atmospheric CO_2, mostly through organic carbon export and burial. The efficiency of this process appears to be mostly determined by the extent of organic carbon preservation in the marine system, which is enhanced by intense orogenesis and subsequent rapid physical erosion and sediment export. The creation of new, tectonically active mountain ranges therefore acts as a driver of the global carbon cycle, hence long-term climate, through enhanced organic carbon cycling and corresponding atmospheric CO_2 consumption.

12.10 Conclusion

The Himalayan erosion generates one of the most intense physical erosion flux of the planet. The consequence is that the continental transfer to the ocean is remarkable by its particle flux. During transport in the floodplain, particle flux fixes and carries organic carbon derived from soil degradation. In spite of the relatively modest organic carbon loading of these sediments, the rapid accumulation in the ocean and relatively anoxic conditions in the Bay of Bengal concur to protect organic carbon from decay during transport. Comparatively, silicate-weathering uptake of CO_2 appears modest. This is in part linked to the lithological composition of the Himalaya that contains mostly metamorphosed recycled sediment and are therefore silicate crust already depleted in mobile elements such as Ca and Mg. As a consequence, the characteristic of the Himalayan orogeny is to promote an erosion cycle that impact the carbon cycle mostly through the fixation and burial of recent organic carbon rather than throughout the silicate weathering-biogenic carbonate precipitation pathway.

References

Aucour A-M, France-Lanord C, Pedoja K, Pierson-Wickmann A-C, Sheppard SMF (2006) Fluxes and sources of particulate organic carbon in the Ganga–Brahmaputra river system. Glob Biogeochem Cycles 20:1–12

Becker JA, Bickle MJ, Galy A, Holland TJB (2008) Himalayan metamorphic CO_2 fluxes: quantitative constraints from hydrothermal springs. Earth Planet Sci Lett 265:616–629

Berner RA (2003) The long-term carbon cycle, fossil fuels and atmospheric composition. Nature 426:323–326

Berner U, Poggenburg J, Faber E, Quadfasel D, Frische A (2003) Methane in ocean waters of the Bay of Bengal: its sources and exchange with the atmosphere. Deep Sea Res II 50:925–950

Beyssac O, Goffé B, Chopin C, Rouzaud J-N (2002) Raman spectra of carbonaceous material in metasediments: a new geothermometer. J Metamorph Geol 20:859–871

Beyssac O, Bollinger L, Avouac J-P, Goffe B (2004) Thermal metamorphism in the lesser Himalaya of Nepal determined from Raman spectroscopy of carbonaceous material. Earth Planet Sci Lett 225:233–241

Blasco F, Bellan MF, Aizpuru M (1996) A vegetation map of tropical continental Asia at scale 1.5 million. J Veg Sci 7:623–634

Broecker WS, Toggweiler JR, Takahashi T (1980) The Bay of Bengal – a major nutrient source for the deep Indian Ocean. Earth Planet Sci Lett 49:506–512

Burdige DJ (2005) Burial of terrestrial organic matter in marine sediments: a re-assessment. Glob Biogeochem Cycles 19:4011

Burdige DJ (2007) Preservation of organic matter in marine sediments: controls, mechanisms and an imbalance in sediment organic carbon budgets? Chem Rev 107:467–485

Cai W-J, Sayles FL (1996) Oxygen penetration depths and fluxes in marine sediments. Mar Chem 52:123–131

Cochran JR, Stow DAV, Auroux C, Amano K, Balson PS, Boulegue JJ, Brass GW, Corrigan J, Gartner S, Hall S, Iaccarino S, Ishizuka T (1989) Leg 116 distal Bengal fan. Ocean Drilling Program, College Station, TX

Collister JW, Rieley G, Stern B, Eglinton G, Fry B (1994) Compound-specific $\delta^{13}C$ analyses of leaf lipids from plants with differing carbon dioxide metabolisms. Org Geochem 21:619–627

Deines P (1980) The isotopic composition of reduced organic carbon. In: Fritz P, Fontes JC (eds) Handbook of environmental isotope geochemistry. Elsevier, New York

Dobremez JF et al (1978) Carte écologique du Népal 1/250 000. University of Grenoble, Grenoble

Ebelmen JJ (1845) Sur les produits de la décomposition des espèces minérales de la famille des silicates. Ann Mines 7:3–66

Evans MJ, Derry LA, France-Lanord C (2008) Degassing of metamorphic carbon dioxide from the Nepal Himalaya. Geochem Geophys Geosyst 9:Q04021

France-Lanord C, Derry LA (1994) $\delta^{13}C$ of organic carbon in the Bengal Fan: source evolution and transport of C3 and C4 plant carbon to marine sediments. Geochim Cosmochim Acta 58:4809–4814

France-Lanord C, Derry LA (1997) Organic carbon burial forcing of the carbon cycle from Himalayan erosion. Nature 390:65–67

Freeman KH, Colarusso LA (2001) Molecular and isotopic records of C4 grassland expansion in the late Miocene. Geochim Cosmochim Acta 65:1439–1454

Galy A, France-Lanord C (1999) Weathering processes in the Ganges–Brahmaputra basin and the riverine alkalinity budget. Chem Geol 159:31–60

Galy A, France-Lanord C (2001) Higher erosion rates in the Himalaya: geochemical constraints on riverine fluxes. Geology 29:23–26

Galy V, France-Lanord C, Beyssac O, Faure P, Kudrass H, Palhol F (2007) Efficient organic carbon burial in the Bengal fan sustained by the Himalayan erosional system. Nature 450:407–410

Galy V, Beyssac O, France-Lanord C, Eglinton TI (2008a) Recycling of graphite during Himalayan erosion: a geological stabilization of carbon in the crust. Science 322:943–945

Galy V, France-Lanord C, Lartiges B (2008b) Loading and fate of particulate organic carbon from the Himalaya to the Ganga–Brahmaputra delta. Geochim Cosmochim Acta 72:1767–1787

Galy V, François L, France-Lanord C, Faure P, Kudrass H, Palhol F, Singh SK (2008c) C4 plants decline in the Himalayan basin since the Last Glacial Maximum. Quatern Sci Rev 27:1396–1409

Garrels RM, Lerman A, Mackenzie FT (1976) Controls of atmospheric O_2 and CO_2: past, present and future. Am Sci 64:306–315

Goni MA, Monacci N, Gisewhite R, Ogston A, Crockett J, Nittrouer C (2006) Distribution and sources of particulate organic matter in the water column and sediments of the Fly River Delta, Gulf of Papua (Papua New Guinea). Estuar Coast Shelf Sci 69:225–245

Hayes JM, Waldbauer JR (2006) The carbon cycle and associated redox processes through time. Philos Trans R Soc Lond B Biol Sci 361:931–950

Hedges JI, Keil RG, Benner R (1997) What happens to terrestrial organic matter in the ocean? Org Geochem 27:195–212

Hilton RG, Galy A, Hovius N (2008a) Riverine particulate organic carbon from an active mountain belt: the importance of landslides. Glob Biogeochem Cycles 22:1–12

Hilton RG, Galy A, Hovius N, Chen M-C, Horng M-J, Chen H (2008b) Tropical-cyclone-driven erosion of the terrestrial biosphere from mountains. Nat Geosci 1:759–762

Keil RG, Mayer LM, Quay PD, Richey JE, Hedges JI (1997) Loss of organic matter from riverine particules in deltas. Geochim Cosmochim Acta 61:1507–1511

Ludwig W, Probst J-L, Kempe S (1996) Predicting the oceanic input of organic carbon by continental erosion. Glob Biogeochem Cycles 10:23–41

Marty B, Tolstikhin IN (1998) CO_2 fluxes from mid-ocean ridges, arcs and plumes. Chem Geol 145:233–248

Mayorga E, Aufdenkampe AK, Masiello CA, Krusche AV, Hedges JI, Quay PD, Richey JE, Brown TA (2005) Young organic matter as a source of carbon dioxide outgassing from Amazonian rivers. Nature 436:538–541

Meyers PA, Dickens GR (1992) Accumulation of organic matter in sediments of the Indian Ocean: a synthesis of results from scientific deep sea drilling. In: Duncan RA, Rea DK, Kidd RB, von Rad U, Weissel JK (eds) Synthesis of results from scientific drilling in the Indian Ocean. American Geophysical Union

Milliman JD, Syvitski PM (1992) Geomorphic/tectonic control of sediment discharge to the ocean: the importance of small mountainous rivers. J Geol 100:525–544

Perrier F, Richonb P, Rajaured S, France-Lanord C, Revil A (2009) A direct evidence for high carbon dioxide and radon-222 discharge in Central Nepal. Earth Planet Sci Lett 278:198–207

Pierson-Wickmann A-C, Reisberg L, France-Lanord C, Kudrass H (2001) Os-Sr-Nd results from sediments in the Bay of Bengal: implications for sediment transport and the marine Os record. Paleoceanography 16:435–444

Poynter J, Eglinton G (1990) Molecular composition of three sediments from Hole 717c: the Bengal Fan. In: Cochran JR, Stow DAV (eds) Proceedings of the ODP, science research. Ocean Drilling Program, College station, TX

Raymo ME, Ruddiman WF (1992) Tectonic forcing of late Cenozoic climate. Nature 359:117–122

RSP (1996) Spatial representation and analysis of hydraulic and morphological data. Water Resources Planning Organization (WARPO)

Rubey WW (1951) Geologic history of sea water. Bull Geol Soc Am 62:1111–1148

Satterberg J, Arnarson TS, Lessard EJ, Keil RG (2003) Sorption of organic matter from four phytoplankton species to montmorillonite, chlorite and kaolinite in seawater. Mar Chem 81:11–18

Schlünz B, Schneider RR (2000) Transport of terrestrial organic carbon to the oceans by rivers: re-estimating flux and burial rates. Int J Earth Sci 88:599–606

Scott DT, Baisden WT, Davies-Colley R, Gomez B, Hicks DM, Page MJ, Preston NJ, Trustrum NA, Tate KR, Woods RA (2006) Localized erosion affects national carbon budget. Geophys Res Lett 33:1–4

Sollins P, Homann P, Caldwell BA (1996) Sabilization and destabilization of soil organic matter: mechanisms and controls. Geoderma 74:65–105

Suckow A, Morgenstern U, Kudrass H-R (2001) Abslolute dating of recent sediments in the cyclone-influenced shelf area off Bangladesh: comparison of Gamma spectrometric (^{137}Cs, ^{210}Pb, ^{228}Ra), radiocarbon, and ^{32}Si ages. Radiocarbon 43:917–927

Walker JCG, Hays PB, Kasting JF (1981) A negative feed back mechanism for the long-term stabilization of Earth's surface temperature. J Geophys Res 86:9776–9782

Part V
Climate Change and Food Security

Chapter 13
World Food Security: The Challenges of Climate Change and Bioenergy

Ramasamy Selvaraju

Contents

R. Selvaraju (✉)
Climate, Energy and Tenure Division (NRC), Food and Agriculture Organization of the United Nations, Viale delle Terme di Caracalla, Rome 00100, Italy
e-mail: Selvaraju.ramasamy@fao.org

R. Lal et al. (eds.), *Climate Change and Food Security in South Asia*,
DOI 10.1007/978-90-481-9516-9_13,

Abstract Securing world food security in light of the impact of climate change may be one of the biggest challenges we face in this century. Rising temperatures, more intense droughts, floods, and greater weather variability all mean productivity losses to crops and livestock. Greater numbers of crop failures, livestock deaths, forest fires and damage to fishery infrastructure are already imposing economic losses and undermining food security.

Impacts of climate change can influence the whole food supply chain and all four dimensions of food security, namely: availability of food from domestic production and imports, access to resources for producing or buying food, stability of food supply, utilization of food. In parallel, growing energy demand is adding additional string and could compete with food security objectives.

As importance to bioenergy is growing, the challenges and opportunities posed by biofuels, in view of the world's food security, energy and sustainable development needs to be addressed. In this paper the challenges of climate change and bioenergy on food security are discussed together with potential adaptation and mitigation options. The key messages from the FAO's high level conference on "World Food Security: The Challenges of Climate Change and Bioenergy" are presented.

Keywords Extreme climate events • Population growth • Water resources • Pests and diseases • Climate risk management • Biofuel • Biodiversity • Adaptation • Sustainability

13.1 Introduction

One sixth of humanity is undernourished, more than ever before (FAO 2009b). Changing climatic conditions are projected to affect food security through their impact on local food systems. More frequent and intense extreme weather and climate events, increasing uncertainties in rainy season patterns and rising sea levels are already having significant impacts on food production, food distribution infrastructure, food emergencies, livelihood assets and human health, in both rural and urban areas (FAO 2008a).

Climate change will generate significant costs to both developing and developed countries. Such costs will include increased frequency and intensity of severe weather events such as floods, tornados and hurricanes; increased drought in some regions; loss of coastal areas and water shortages; and changes in the incidence of disease. Projected changes in the frequency and severity of extreme climate events have significant consequences for food and forestry production, and food insecurity, in addition to impacts of projected mean climate (Easterling et al. 2007).

Higher growing season temperature can have significant impacts on agriculture productivity, farm incomes, and food security (Battisti and Naylor 2009).

In mid-to high-latitude regions, moderate warming benefits cereal crop and pasture yields, but even slight warming decreases yields in seasonally dry and tropical regions (IPCC 2007b) and will increase risk of hunger in poor developing countries. These countries are highly vulnerable to climate change as under nourishment is already prevalent among considerable proportion of the population.

Many of the world's small-scale farmers work on marginal land in the tropics. The adaptive capacity of the smallholder farmers are constrained, will experience the negative effects on yields of low latitude crops, combined with a high vulnerability to extreme events. Climate change is likely to affect the suitability of land for different types of crops, livestock, fish and pasture. It would also have an impact on the health and productivity of forests, the incidence of pests and diseases, biodiversity and ecosystems. Regional changes in distribution and productivity of particular fish species are expected due to continued warming.

Growing support to the production of biofuels such as ethanol and biodiesel from crops adds another dimension to the factors influencing food security. Demand for biofuels could place additional pressure on the natural resource base, with particularly harmful environmental and social consequences for people who already lack access to energy, food, land and water (FAO 2007). On the one hand, a stronger link between agriculture and the demand for energy could result in higher agricultural prices, output and gross domestic product (GDP) and long-term improvements in food security. On the other hand, there is a risk that unsustainable biofuel development could threaten the food security of the world's poorest people (FAO 2008c). This paper presents the challenges of world food security posed by climate change and bioenergy with potential adaptation and mitigation options.

13.2 Trends of Agriculture Growth and Food Consumption

13.2.1 Population Growth and Demand for Food

World population is expected to grow by over a third (or 2.3 billion people) between 2009 and 2050 (FAO 2009b). Nearly all the further population increases will be occurring in countries several of which even in 2050 may still have inadequate food consumption levels, hence significant possibility for further increases in demand. World's average per capita agriculture GDP has grown from US$328 in 1980 to US$472 in 2004 (FAO 2007) and is much less than overall per capita GDP. Continued growth of world agriculture is expected even after the end of world population growth. Feeding a world population of 9.1 billion people in 2050 would require raising overall food production by some

70% over the period from 2005/07 to 2050 (FAO 2009a). The combined effects of climate change, land degradation, crop land losses, water scarcity and species infestations may cause projected yields to be 5–25% short of demand by 2050 (Nellemann et al. 2009).

To compensate this demand and to facilitate growth in agriculture output, another 185 million ha of rainfed-crop land (+19%) and 60 million ha of irrigated land (+30%) will have to be brought into production (IPCC 2007b). On the one hand, the entire agricultural land expansion need to take place in developing countries especially in sub-Saharan Africa and Latin America (Cassman et al. 2003). On the other hand, sub-Saharan Africa, Asia and Latin America, with high rates of population growth and natural resource degradation, are likely to continue to have high rate of poverty and food insecurity (Alexandratos 2005). Climate change will add an additional challenge to the dual challenge of meeting food demand while at the same time protecting natural resources.

13.2.2 Trends of Food Consumption

The historical trend towards increased food consumption per capita as a world average and particularly in the developing countries will likely continue, but at slower rates than in the past. The average of the developing countries, that rose from 2,111 kcal/person/day 35 years ago to the present 2,650 kcal, may rise further to 3,070 kcal by 2050 (Fig. 13.1). However, not all countries may achieve food consumption levels constant with requirements for good nutrition (FAO 2006). Potential exists for several of these countries to make gains by assigning priority to the development of local food production. But in countries that have limited agricultural potential, the problem of production constrained food insecurity and increase in undernourishment may persist.

For example, the total food production in South Asia has increased threefolds from 117 million tonnes in 1961 to 348 million tonnes in 2006 (Fig. 13.2), but the dietary energy consumption has increased only marginally and the current dietary energy consumption is 2,364 kcal/person/day. It requires additional efforts in yield improvement, given the fact that there is limited scope for expanding area under cultivation and area under irrigation. The above facts and figures underline the importance of putting in place effective poverty reduction strategies, and sustainable agriculture development programs.

13.3 Climate Change and Food Security

Agriculture has been described as the most weather-dependent of all human activities (Oram 1989). Importantly, agriculture in its many different forms and locations remains highly sensitive to climate variations, the dominant source of the

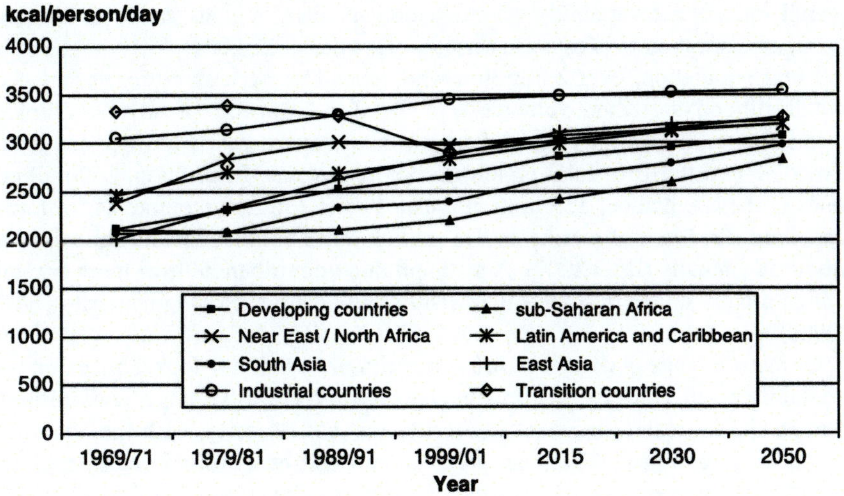

Fig 13.1 Per capita food consumption (kcal/person/day) (Data from FAO 2006)

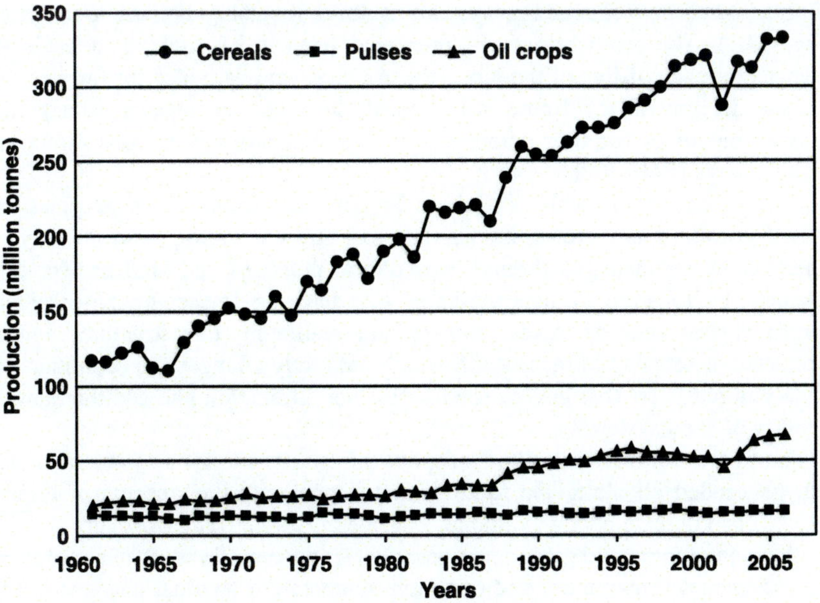

Fig 13.2 Food production in South Asia (Source: FAOSTAT)

overall interannual variability of production in many regions and a continuing source of disruption to ecosystem services (Howden et al. 2007).

Rural population, over 2.6 billion who depend on agriculture for sustenance and livelihood are often vulnerable to the direct impacts of adverse climate events. In total 2.2 billion tonnes of cereals are produced yearly for food and feed, providing two-thirds of total protein intake by humans. In addition, 260 million tones of meat and about 139 million tonnes of fish (36 million tones of freshwater; 72 million tones of marine and 31 million tones of other aquatic animals) are produced annually (FAO 2007). Aquatic products contribute 50% or more of total animal protein intake in some small islands and other developing countries (FAO 2008d).

Climate change will affect food security in all of its four dimensions – availability, accessibility, utilization and stability. Negative impacts of climate change and increasing climate variability on food security, with the potential of reducing food production, access to and utilization of food in many regions already vulnerable today, are expected. In particular, stability of food supply is likely to be disrupted by more frequent and severe climate extremes and associated variability of agricultural production across all areas. Utilization of food may be affected negatively by increases in crop, livestock and human pests and diseases, as well as by reduced water availability and water quality. The result could be a substantial decline in labour productivity and increase in poverty and mortality rates.

The conceptual framework presents a simplified description of the dynamics of potential climate change impacts and feedback loops in a holistic food system (Fig. 13.3). The implications are presented linearly by looking at projected changes for each of five of the most important climate variables for food system process. In general the framework presents how climate change affects food security outcomes for four components of food security in various direct and indirect ways (FAO 2008b).

Climate change variables influence biophysical factors, such as plant and animal growth, water cycles, biodiversity and nutrient cycling, and the ways in which these are managed through agricultural practices and land use for food production. However, climate variables also have an impact on physical and human capital such as roads, storage and marketing infrastructure, houses, productive assets and human health which indirectly changes the economic and socio-political facts that govern food access and utilization and can threaten the stability of food systems.

The framework illustrates how adaptive adjustments to food system activities will be needed all along the food chain to cope with the impacts of climate change. The climate change variables considered in the framework are: the CO_2 fertilization of increased greenhouse gas (GHG) concentrations in the atmosphere, increasing mean, maximum and minimum temperature, gradual changes in precipitation, increase in the frequency and intensity of extreme events and greater seasonal weather variability and changes in growing season determinants.

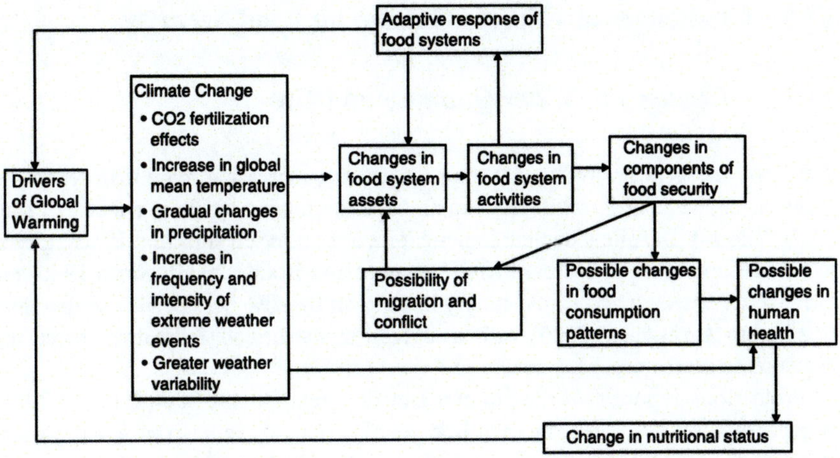

Fig. 13.3 The conceptual framework of climate change and food security (Source: FAO 2008b)

13.4 Agriculture and Global Climate Change

Agriculture is not only at risk from climate change, it is a major driver of environmental and climate change itself. Agriculture land occupied 4,974 Mha in 2000 and most of the area is under pasture (3,442 Mha), arable crops (1,397 Mha) and permanent crop occupied 135 Mha (FAO 2007). Agriculture gained almost 500 Mha during the last four decades from other land uses, a change driven largely by increasing demands for food from a growing population (Smith et al. 2007). At the global level, 3,952 Mha of land are forested (Nabuurs et al. 2007). Agricultural lands (lands used for agricultural production, consisting of cropland, managed grassland and permanent crops including agro-forestry and bioenergy crops) occupy about 40–50% of the Earth's land surface (IPCC 2007c).

Agriculture is a notable source of the three major greenhouse gases: carbon dioxide, methane and nitrous oxide. Agricultural activities and land-use changes contribute about one-third of the total carbon dioxide emissions and are the largest sources of methane from livestock and flooded rice production and nitrous oxide primarily from application of inorganic nitrogenous fertilizer. Globally, agriculture accounted for an estimated emission of 5.1–6.1 $GtCO_2$-eq/year in 2005 (accounts for 10–12% of the total anthropogenic emissions of GHGs) (IPCC 2007c). Of the global atmospheric emissions of GHGs agriculture accounts for 58% of N_2O and about 47% of CH_4 (Smith et al. 2007). Forestry (including deforestation) accounted for 17.4% of total greenhouse gas emissions in 2004, with emissions from intensive crop and livestock production contributing another 13.5% (IPCC 2007c).

13.5 Challenges of Climate Change on Food Security

13.5.1 Changes in Water Quantity and Use

Water use has grown rapidly over the past century, increasing more than sevenfold between 1900 and 2000 while the human population grew by about a factor of four (UNDP 2006). Despite a decline in per capita consumption since the 1980s, global water use continues to increase (Shiklomanov and Rodda 2003). Since 1960, the ratio of water use to accessible supply has grown by 20% per decade (Millennium Ecosystem Assessment 2005) with an overall upward trend, indicating increasing pressure on freshwater resources.

Projections reported in the Human Development Report 2006 (UNDP 2006) suggest that, by 2025, over three billion people are likely to be experiencing water stress and 14 additional countries might be classified as water-scarce. An additional 1.8 billion people could be living in a water scarce environment by 2080 (UNDP 2008). Climate change will affect rainfall and fresh water availability for agriculture and other uses. Adverse effects of climate on freshwater systems aggravate the impacts of other stresses, such as population growth, changing economic activity, land-use change, and urbanization (Kundzewicz et al. 2007).

Changing climate patterns will have important implications for water availability especially mean run-off changes (Milly et al. 2005), glacier melt during summer season, ground water recharge and floods (Kleinen and Petschel-Held 2007) and geomorphic processes including erosion, slope stability, channel changes, and sediment transport (Dennis et al. 2003). Climate models projected increased precipitation in high latitudes and parts of the tropics, and decrease in some sub-tropical and lower mid-latitude regions (Bates et al. 2008).

Agriculture accounts for about 70% of all water use worldwide and up to 95% in many developing countries and thus influences both the quantity and quality of water available for other human uses (FAO 2007). Over 277 million hectares are classified as irrigated land (FAO 2007) and of all sectoral water demands, the irrigation sector will be affected most strongly by climate change, as well as by changes in the effectiveness of irrigation methods (Kundzewicz et al. 2007). Given the dominant role of irrigated agriculture in global water use, management practices that increase the productivity of irrigation water use can greatly increase the availability of water for other human and environmental uses.

13.5.2 Distribution, Incidence and Intensity of Pests and Diseases

Changing weather patterns may help spread crop pests and diseases in the future. Climate change and increasing climate variability is creating favourable conditions for animal and plant pests and diseases. The movement of plant pests, animal diseases and invasive aquatic organisms across physical and political boundaries threatens food security. For example, Bluetongue, a sheep disease is moving north into more temperate zones of Europe (Van Wuijckhuise et al. 2006). Fleming and Tatchell (1995) predicted that over the next 50 years, aphids will appear at least 8 days earlier in the spring. This may increase their severity as pests, will also depend partly on how the phenologies of their host plants change (Harrington et al. 2007). Increased plant density will tend to increase leaf surface wetness and leaf surface wetness duration, and so make infection by foliar pathogens more likely (Huber and Gillespie 1992). Changes in plant architecture may affect microclimate and thus risks of infection.

Altered weather patterns can increase crop vulnerability to infection, pest infestations, and weeds. Range of crop weeds, insects, and diseases are projected to expand to higher latitudes (Rosenzweig et al. 2001). In particular, CO_2-temperature and CO_2-precipitation interactions are recognised as the key factors in determining plant damage from pests in future (Easterling et al. 2007; Zvereva and Kozlov 2006). New vectors, selection and recombination of disease may occur when animal species and breeds and plant species mix or when insect pests and vectors are introduced without their natural enemies. Climate change may result in changes in species composition and interactions that will augment the emergence of new diseases and pests. Climate change would impact vector-borne diseases and may also result in new transmission pathways and different host species (Garrett et al. 2006).

Entry and establishment, emergence and outbreaks of animal and plant pests and diseases have historically resulted in major food problems either directly through yield reductions of food crops and losses in animal production, or indirectly through yield reduction of cash crops (e.g. rinderpest, potato blight, locusts). Climate change would result in a higher volatility and, therefore, is likely to cause additional crises in agricultural production, in particular for small farmers and those involved in subsistence agriculture.

Redistribution of plant pests and changes in pest incidence and intensity may result in additional and inappropriate pesticide use. Consequently, there may be higher levels of pesticide and veterinary drugs in food. Changes in rainfall, temperature and relative humidity may favour the growth of fungi that produce mycotoxins and thus may make food such as groundnuts, wheat, maize, rice and coffee unsuitable for human and animal consumption.

13.5.3 Impacts of Extreme Weather/Climate Events

Changes in extreme weather and climate events have significant impacts and are among the most serious challenges to society in coping with a changing climate (Karl et al. 2008). Natural disasters are increasing, with more frequent and more severe occurrences fuelled by global warming. Extreme events drive changes in natural and human systems much more than average climate. FAO/GIEWS data indicate that sudden-onset disasters – especially floods – have increased from 14% of all natural disasters in the 1980s to 20% in the 1990s and 27% since 2000 (FAO 2008e).

There is growing evidence that a warming world will be accompanied by changes in the intensity, duration, frequency and geographic extent of weather and climate extremes. The frequency of heavy precipitation events will be very likely to increase over most areas during the twenty-first century, with consequences for the risk of floods. At the same time, the proportion of land surface in extreme drought at any one time is projected to increase, in addition to a tendency for drying during summer, especially in the sub-tropics, low and mid latitudes (Bates et al. 2008). Changes in extremes are already observed to be having impacts on social, economic and natural systems, and future changes associated with continued warming will present additional challenges (Karl et al. 2008). The damage and loss due to hydro-meteorological disasters in terms of economic value has increased dramatically over the past few decades. The damage trends have increased significantly despite ongoing adaptation efforts that have been taking place (Mills 2005).

Worldwide, flood occurrence has risen from about 50 floods per year in the mid-1980s to more than 200 today (CRED 2008). As sudden-onset emergencies leave much less time for planning and response than slow-onset ones, these trends have important implications for mobilization of resources needed to prepare for, and respond to, emergencies in order to save lives and protect livelihood systems.

13.5.4 Increased Vulnerability of Fishery-Dependent Communities

Fisheries employ some 150 million people in developing countries. Fish products provide more than 2.8 billion people with about 20% of their average per capita intake of animal protein (FAO 2008d). Fish contributes to, or exceeds, 50% of total animal protein intake in some small island developing states. Fisheries and aquaculture are threatened by climate change due to higher water temperatures, rising sea levels, melting glaciers, changes in ocean salinity and acidity, more cyclones in some areas, less rain in others, shifting patterns and abundance of fish stocks.

Fishery-dependent communities may also face increased vulnerability in terms of less stable livelihoods, decrease in availability of fish, and safety risks due to fishing in harsher weather conditions and further from their landing sites (Hall-Spencer et al. 2008; McClanahan et al. 2008). Direct effects of increasing temperature on marine and freshwater ecosystems are already evident, with rapid poleward shifts in regions and changes in distribution and production (Drinkwater 2005). Evidence from the Pacific and the Atlantic suggests that nutrient supply to the upper productive layer of the Ocean is declining due to reductions in the maridional overturning circulation and up welling (Curry and Mauritzen 2005). Most of the large global marine-capture fisheries are affected by regional climate variability associated with El-Nino/Southern Oscillation

(ENSO), Pacific Decadel Oscillation (Lehodey et al. 2003) and North Atlantic Oscillation (Drinkwater et al. 2003).

Vulnerability of fishery-dependent communities in Small Island Developing States (SIDS) will stem from their resource dependency and exposure to extreme weather events. Combined effect of predicted warming, the relative importance of fisheries to national economies and diets, and limited societal capacity to adapt to potential impacts are prioritised as reasons of vulnerability of countries to climate change (Allison et al. 2009). Climatic changes could increase physiological stress on cultured stock and affect productivity and increase vulnerability to diseases and, in turn, impose higher risks. Extreme weather events could result in escapes of farmed stock and contribute to reductions in genetic diversity of the wild stock, affecting biodiversity. However, new opportunities and positive impacts emerging from such areas as changes in species and new markets also could be part of future changes.

Stability of supply will be impacted by changes in seasonality, increased variance of ecosystem productivity, increased supply risks. Access to fish for food will be affected by changes in distribution of fish species and in livelihoods combined with impacts transferred from other sectors such as increases in prices of substitute food products and competition for supply. Utilization of the nutrients and the nutritional value of fishery products will be affected by changing supply quality and market chain disruptions (FAO 2008d).

13.5.5 Impact on Agriculture Biodiversity

The biodiversity associated with agricultural ecosystems is generally regarded as the multitude of plants, animals and micro-organisms at genetic, species and ecosystem levels. Agriculture biodiversity plays sustaining key functions for food production and food and livelihood security (FAO 2007) and is the outcome of the interactions among the environment, genetic resources and the management systems and practices used by farmers.

Biodiversity and climate change are closely linked. Biodiversity is threatened by climate change, but biodiversity resources can reduce the impacts of climate change on population and ecosystems. Observed impacts of climate change on vulnerable systems such as polar and high mountain ecosystems have showed greater vulnerability due to temperature increase. IPCC fourth assessment report (IPCC 2007b) projected an increasing risk of species extinction with higher confidence as warming proceeds. The report elaborated that approximately 20–30% of plant and animal species assessed so far are likely to be at increased risk of extinction if increases in global average temperature exceed 1.5–2.5°C over 1980–1999 levels. Increases in sea surface temperature of about 1–3°C are projected to result in more frequent coral bleaching events and widespread mortality, unless there is thermal adaptation or acclimatisation by corals.

Concerns have been raised in recent years over the loss of agricultural biodiversity due to expansion of cropland into low productive rainfed lands (Nellemann et al. 2009) and homogenization of agricultural production systems. Two major concerns are: increasing levels of genetic vulnerability and genetic erosion (FAO 1997). Genetic vulnerability occurs where a widely used crop or livestock variety is susceptible to changing climatic conditions. Genetic erosion is the loss of genetic resources through the extinction of a livestock variety or crop. Climate change and increased climate variability may increase genetic vulnerability and intensify genetic erosion. Without proper management of agricultural biodiversity, some key functions of the agro-ecosystem may be lost, such as maintenance of nutrient and water cycles, pest and disease regulation, pollination and land erosion control.

13.6 Expanding Biofuel Production and Food Security

Biofuel crops, increasingly an important source of energy is being promoted in the context of their critical role in adaptation to climate change and mitigation of carbon emissions. Biofuels currently account for 0.2% of total global energy consumption, 1.5% of total road transport fuels, 2% of global crop land, 7% of global coarse grain use and 9% of global vegetable oil use (FAO 2009d). These shares are projected to rise over the coming decades. Climate change and expanding biofuel production is likely to lead to competition for access to land. For the millions of farmers, pastoralists, fisher folk and forest dwellers with no formal land tenure rights, this increased competition poses a threat to their livelihoods.

Investments in biofuels have grown rapidly since beginning of this century. Steenblik (2007) estimated US corn-based ethanol production at roughly 18 billion litres in 2006 followed by Brazil at 17 billion litres of ethanol from sugarcane. A growing number of developing countries are beginning to invest in feedstocks for the production of ethanol and biodiesel.

Unsustainable biofuel production will negatively affect the food security of low income food deficit countries. The combined effect of rising oil and food prices has stressed many developing economies and poor households, because many of the low-income food-deficit countries are also net importers (Runge and Senauer 2007). The high demand for energy and the potential of biofuels are no guarantee that small farmers and poor people in developing countries will receive the benefits.

13.7 Adapting to Climate Change in Agriculture

13.7.1 Cost of Adaptation

Immediate action is vital to increase the resilience of rural people to climate change and help them adapt to new conditions. The Stern's report (Stern 2007) emphasized that taking action early on climate change would greatly reduce estimated costs. It is estimated that the cost of removing most of the climate change risks to an acceptable level would be around 1% of global Gross Domestic Product (GDP) by 2050, if decisive action starts now.

A range of methods exists for costing adaptation in agriculture (Parry et al. 2009). The top-down approach used for costing by McCarl (2007)) for UNFCC report on "adaptation options for agriculture, forestry and fisheries" estimated an additional funding of 12.9 billion without mitigation and US$11.3 billion with mitigation in the year 2030. Interpretation of the study of Fischer et al. (2007) provides a cost of US$8 billion for adapting crop and irrigation systems to climate change by 2030. A different approach by Cline (2007), using simple crop growth models, provides an estimate of US$14.5 billion for the year 2030 for reduction in the value of global crop outputs due to climate change. Recent estimates by Nelson et al (2009) for IFPRI put the aggressive agriculture productivity investments of US$7.1–7.3 billion to raise calorie consumption enough to offset the negative impacts of climate change on the health and well being of children.

Global warming is already underway and adaptation strategies are now a matter of urgency, especially for the most vulnerable poor countries, which are even now being disproportionately affected. It is essential to facilitate adaptation, notably through better risk management and safety nets to protect the most vulnerable (World Bank, 2009). Many adaptation measures need to focus on strengthening measures that already exist, such as sustainable natural resource management, water management and enhanced water productivity, climate risk management, early warning systems, disaster risk management, rural investments to reduce the impacts of climate variability on food security, through crop insurance and incentives that encourage farmers to adopt better agricultural and land use practices. This chapter provides a brief background on the potential adaptation strategies towards enhanced food security.

13.7.2 Adaptation Strategies to Enhance Food Security Under Changing Climate

13.7.2.1 Sustainable Natural Resource Management

Adaptation efforts must be strengthened to enhance the production in a sustainable way to ensure that a growing world population has access to sufficient, safe and nutritious food under changing climate. Disruption or decline in global and local food supplies due to climate change can be avoided through more efficient natural resource management, improved crop varieties, improved land cultivation, farm and livestock management and the development of crop varieties and breeds that posses potential of adapting to changing climatic conditions.

The specific action to be implemented include: prioritizing location specific adaptation measures considering existing natural resources; enterprise diversification;

water resource management; increasing water productivity; and diversified livelihood systems; institutional strengthening; integration of traditional farmer practices and gender perspectives into adaptation strategies. Integrated watershed management, land use and land cover planning depending on the available natural resources and socio-economic needs of the local communities are considered sustainable.

Adaptation measures range from temporal and spatial variations in production systems. Better protection against temperature changes, changing rainfall variability and patterns, salinization though sea level rise, and pest attacks require adjusting planting or fishing dates, rotations, multiple cropping/species diversification, crop-livestock pisciculture systems and agroforestry. Investing in soil, water and biodiversity conservation and development include options such as building soil biomass, restoring degraded lands, rehabilitating rangelands, harvesting and recycling water, developing adapted cultivars and breeds, protecting aquatic ecosystems in order to maintain long-term productivity.

13.7.2.2 Water Management and Enhanced Water Productivity

Reducing Uncertainty in Water Availability As a consequence of climate change, farmers will face growing unpredictability and variability in water supplies and increasing frequency of droughts and floods. Open access or loose property rights on water resources and irrigation systems lead to the overexploitation of aquifers and unsustainable irrigation practices that exhaust, contaminate and increase irrigation costs. Land degradation is also an outcome of inefficient use of water resources and inadequate irrigation management practices, resulting in productivity reductions and increasing losses of cropland.

Current water management practices may not be robust enough to cope with the impacts of climate change on water resources and water supply reliability in agriculture. In many locations, water management cannot satisfactorily cope even with current climate variability. As a first step, improved incorporation of information about current climate variability into water-related management would assist adaptation to longer-term climate change impacts (Bates et al. 2008).

Water Harvesting Water harvesting refers to a number of technologies, traditional and modern, that either harvest surface runoff or increase water infiltration. These include water channels and dams to catch and convey water, techniques to increase soil moisture content, and reservoirs for irrigation and household use and to reduce flood peaks. Pretty et al. (2006) confirmed that these practices also provide a notable improvement in water productivity, especially for rainfed agricultural systems Water-harvesting practices are known to enhance the surface and subsurface water resources.

Investment in Irrigation In Africa, less than 5% of cropland is irrigated. Large benefits could accrue to small farmers by expansion of irrigated land to increase and stabilize the level of production, while also minimizing the role of rainfall uncertainty in agriculture. Irrigation investment projects have high rates of return, estimated as exceeding 15% and even reaching 30% in sub-Saharan Africa (World Bank 2007; FAO 2008e). Significant gains in terms of welfare improvements are also expected from expanding irrigation investment. Increasing investment in irrigation by 1% has been estimated as having reduced poverty by nearly 5% in Kenya (Thurlow et al. 2007).

Soil and Water Conservation Numerous studies have established the positive impact of soil and water conservation practices in enhancing the water productivity. No-till systems are found to reduce water runoff, increase water infiltration, and reduce soil erosion (Lal 2007; Humberto Blanco-Canqui and Lal 2008). Conservation tillage combined with residue management and proper fertilizer use can help to preserve soil moisture, maximum water infiltration, increase carbon storage, minimise nutrient run-off and increase crop yields. Rotational grazing, improved livestock distribution and increased tree cover on pastures have been found to improve water recharge.

Enhancing Water Productivity The ability to produce more food for a growing world population has improved significantly in recent decades as a result of expansion in irrigated cropland (FAO 2008a). Increasing the proportion of irrigated agricultural land has provided a solid base for boosting productivity and reducing the volatility of agricultural yields. With demand for water rising and climate change imposing further restrictions, efficiency in the management of available water resources becomes necessary for productivity increases in agriculture and for food security.

13.7.2.3 Biodiversity Conservation

The conservation of crop and livestock genetic diversity may be ensured either ex situ or in situ. Ex situ methods include seed and gene banks, while in situ conservation takes place in farmers' fields, ponds or forests. The two approaches are complementary; the ex situ collections preserve a static set of genetic resources, while in situ efforts preserve a dynamic process of evolution, as genetic resources adapt to changing pressures from natural and human selection.

The necessary measures for biodiversity conservation in agriculture depend on the type of biodiversity to be conserved on production systems and location. The three main ways in which farmers service institutions and policy-makers can contribute to biodiversity conservation are: (i) reducing agricultural expansion into biodiversity-rich lands; (ii) adopting agricultural production systems that support the joint production of biodiversity conservation and agricultural products; and (iii) conserving agricultural biodiversity.

A wide range of methods exist for conserving agricultural biodiversity, depending on the specific component that is focused upon. Methods differ in terms of the degree of human intervention in the natural system, ranging from highly managed ex situ gene and seed banks to maintaining wild relatives of cultivated species in wilderness areas. Measures also include the on-farm conservation and utilization of traditional varieties of crops and livestock, which are often highly adapted to their local environments. Diversity can be promoted by providing incentives to maintain a heterogeneous set of crop varieties in production, particularly traditional varieties, or by managing field margins to encourage pest-suppressing natural enemies and pollinators.

Agricultural biodiversity is directly linked to agricultural production, working within agricultural market channels to provide incentives to farmers to conserve agricultural diversity is an important strategy. In recent years, the international community has provided support to farmers for conserving agricultural biodiversity in situ. These programmes seek to increase the availability and productivity of diversity in production systems, or enhance the returns to maintaining diverse systems. Increasing the demand for diverse products through the establishment of labelling, certification or origin schemes and increasing the diversity of agricultural seed supply systems (FAO 2006) are the applicable strategies.

There is an urgent need to determine the distribution of biodiversity for food and agriculture both in the wild and in the fields and assess its vulnerability to climate change. Matching biodiversity distribution mapping with different climate change scenarios is a basic requirement for countries to develop conservation strategies.

13.7.2.4 Climate Risk Management

The risk of more complex, frequent, intense and unpredictable climate-related extreme events associated with global temperature increase, changing precipitation patterns and sea-level rise coupled with gradual changes, suggests the need for a renewed focus on the climate risk management. Climate risk management need to play an increasingly key role in dealing with the impacts of climate change on agriculture and food security.

Climate risk management require extensive high quality data and information on climate, and on agricultural, environmental and social systems affected by climate, with a view to carrying out realistic vulnerability and risk assessments and looking towards the knowledge on areas of concern. Vulnerability assessment observes impacts of variability and changes in mean climate (inter-annual and intra-seasonal variability) on agricultural systems given the knowledge on biophysical and socio-economic context and available adaptation options. However, agricultural production systems have their own dynamics and adaptation has a particular emphasis on future agriculture and the potential benefits of adaptation depend on sustained policy support.

Climate risk management contribute to facilitate adaptation to climate variability and change, including: (i) a historical climate data archive; an archive on climate impacts on agriculture; (ii) monitoring tools using systematic meteorological observations; (iii) climate data analysis (to determine the patterns of inter-annual and intra-seasonal variability and extremes); (iv) information on the characteristics of system vulnerability and adaptation effectiveness such as resilience, critical thresholds and coping mechanisms; and (v) crop weather insurance indices to reduce the risk of climate impacts for lower-income farmers.

Provision of advance climate information and integration with operational crop models to develop alternative scenarios for operational decision making, and capacity building is part of the climate risk management framework. Climate forecasts and early warning can be a useful part of the decision making process for adaptation of food systems to climate change (Challinor 2009). Decision makers must therefore prepare for the range of possibilities, and often employ risk management strategies that reduce negative impacts of climate extremes and inefficient use of natural resources (Hansen and Sivakumar 2006). Adjustment of crop management practices to climate fluctuations are not new, but proactive management taking advantage of recent advances in climate prediction has raised the prospects to manage the climate risks and opportunities in agriculture. Several case studies have demonstrated the benefits of localized adaptation strategies conditioned by reliable climate information to improve the production and income of smallholders (Sivakumar and Hansen 2007).

Agricultural policies, national extension services and national meteorological services must therefore develop synergies to ensure that planning and the management of crops, livestock, forests and fisheries benefits from weather and climate-based advice. Farmers, herders, and fishers need the benefit of weather and climate-based advisories because food systems are expanding more and more into marginal and vulnerable areas. With the modern and low cost information and communication technologies (ICTs), climate information can now be systematically collected in real-time from villages, analyzed centrally and management options can be prioritized and communicated to farmers. Climate information for decision making have the potential to optimize the economic return from farm, livestock and fishery activities; optimize the use of land, water, fertilizer, pesticides etc., in the light of reducing short-term risk.

13.7.2.5 Pest and Diseases Management

Increased climate extremes may trigger plant diseases and pest outbreaks (Gan 2004). Control and management of new diseases and pests are emerging challenges under changing climate. Pest and disease management programmes need to strengthen national animal and plant health services as a top priority and need to focus on taxonomy, modelling, population ecology and epidemiology. These efforts should also consider how to better consolidate and organise national animal and plant health services. Investments in early control and detection systems will be key to avoid the higher costs of eradication and management.

New research efforts should focus on identification of resistance gene in hosts against pest and diseases, which is insensitive to temperature increase. New heat stable antimicrobial compounds derived from plant sources have the potential to control plant diseases.

Monitoring pest incidences need to be strengthened and new thresholds has to be followed to decide on the pest and disease management methods. Efforts should be initiated in developing new knowledge systems and Integrated Pest management

(IPM) technologies to counter new pests or the intensification of new ones (SP-IPM 2008). Integrated pest management approach can be further elaborated and strengthened taking into consideration of climate related risks. The aim to minimise the amount of pesticides by employing biological, cultural and chemical methods enhances food safety.

13.7.2.6 Fisheries and Aquaculture

Better Use of Production Aquaculture production to just maintain the current dietary production of fish by 2050 will require a 56% increase as well as new alternatives to wild fisheries for the supply of aquaculture feed (Nellemann et al. 2009). Losses caused by spoilage amount to about 10–12 million tonnes per year and an estimated 20 million tonnes of fish a year are discarded at sea (FAO 2008d). Reducing post-harvest losses and increasing the percentage of use for direct human consumption by creating better storage facilities offers opportunities for better use of production.

Ecosystem Approach for Adaptation Climate change compromises the sustainability and productivity of key economic and environmental resources, but it also presents opportunities, especially in aquaculture. Adaptation strategies should be based on an "ecosystem approach", defined as comprehensive and holistic processes to understanding and anticipating ecological change, and developing appropriate management responses (FAO 2008d).

13.8 Agriculture and Climate Change Mitigation

The mitigation efforts over the next two to three decades will determine to a large extent the long-term global mean temperature increase and the corresponding climate change impacts that can be avoided (IPCC 2007a). Agriculture plays an important role as a carbon "sink" through its capacity to sequester and store greenhouse gases, especially as carbon in soils and in plants and trees.

Agriculture practices collectively can make a significant contribution at low cost to increasing soil carbon sinks, to GHG emission reduction, and by contributing biomass feedstock for energy use. Agriculture has the technical potential to mitigate between 5.5 and 6.0 Gt of CO_2 per year by 2030, mainly through soil carbon sequestration (89%) (IPCC 2007c). Additionally, several agriculture-based mitigation options generate significant co-benefits for both food security and climate change adaptation (FAO 2009c). The actions that have large mitigation potential and high co-benefits are increasing soil carbon sequestration through improved crop and grazing land management (e.g., improved agronomic practices, conservation tillage, and crop residue management), forestry and agro-forestry initiatives, improving efficiency of nutrient management and restoration of organic soils and degraded lands. Mitigation is also possible with improved water and rice management and improved livestock and manure management.

In the forestry sector, the mitigation practices include reduced emission from deforestation and forest degradation (REDD), sustainable forest management and forest restoration, including afforestation and reforestation. About 65% of the total mitigation potential is located in the tropics and about 50% of the total could be achieved by reducing emission from deforestation (IPCC 2007c). The adoption of agroforestry, rehabilitation of degraded forests and establishment of forest plantation and silvopastoral systems count among the many land-use changes that can generate above-ground carbon sequestration (Palm et al. 2005). Co-benefits of REDD, for instance alleviating poverty, improving governance, conserving biodiversity and providing other environmental services have been part of the debate and greatly enhanced (Campbell 2009).

Besides carbon sequestration in the above ground biomass, significant potential exists for sequestration of carbon in soils. Total global soil carbon sink capacity, approximately equal to the historic carbon loss of 78 ± 12 Pg, can be filled at the potential maximum rate of about 1 Pg C/year (Lal and Follett 2009). Strategies to increase the soil carbon pool include soil restoration and woodland regeneration, no-tillage farming, cover crops, nutrient management, manuring and sledge application, improved grazing, water conservation, and harvesting, efficient irrigation, agroforestry and growing energy crops on spare lands (Lal 2004). Conservation agriculture is practiced in about 95 million hectares which provide significant soil carbon sequestration services (Derpsch 2005). Cropping systems could be changed to achieve substantial soil carbon sequestration. Around 30% (4.7 million km^2) of the land characterized by medium-to -high potential for carbon sequestration is located in areas where agricultural production is practiced, representing 15% of total croplands (FAO 2008a). Carbon sequestration provides multiple associated benefits as the resultant increase in root biomass and soil organic matter, enhance water and nutrient, availability and plant uptake.

Existing options that can mitigate GHG emissions from the livestock sector are discussed by Steinfeld et al. (2006). Mitigation options in the pastoral systems of the tropics are reviewed by Reid et al. (2004). Carbon can be sequestered from improved management in grasslands which include conversion of cropland to grassland, reduction in grazing intensity, avoiding biomass burning, improving degraded lands, reducing erosion and changes in species mix. Although methane emission from livestock is projected to increase (Herrero et al. 2008) in future, technical options do exist to mitigate the emission (Thornton et al. 2009). Mitigation activities have the chance of success if they build on traditional pastoral institutions and knowledge, while providing pastoralists with food security benefits at the same time.

Emission reductions due to biofuels, feedstocks and associated production technologies are estimated to be smallest (10–30%) for ethanol from maize and largest (70–90%) for ethanol from sugarcane and second-generation biofuels (FAO

2009c). These emission reductions will be smaller to the extent that increased biofuel production accelerates conversion of forests or grasslands to cropland.

13.9 Integrated Adaptation and Mitigation Strategies

The strong trends in climate change already evident, the likelihood of further changes occurring, and the increasing scale of potential climate impacts give urgency to addressing agricultural adaptation more coherently (Howden et al. 2007). Integrated strategies involving adaptation, mitigation and short and long-term approaches are required to overcome the multiple threats of climate change on food security. Effective implementation of these strategies will require increased investment in agricultural development and natural resources management at all levels. Adaptation interventions confirms the need for multiple and integrated pathways across sectors to improve adaptive response of local communities. Short-term and long-term adaptive measures in agriculture linked with focus on future anticipated risks need to be integrated into cross sectoral planning.

The community based adaptation interventions promote integrated strategies to enhance the community resilience and livelihood assets (FAO 2008f). These include, for example: (i) undertaking physical adaptive measures, such as link canals, irrigation, storage facilities for efficient water conveyance and drainage; (ii) adjusting existing agricultural practices to match future anticipated risks, such as adjustment of cropping pattern, selection of adapted crop varieties, diversification of cropping and/or farming systems, better storage of seeds and fodder, more efficient use of irrigation water on rice paddies, more efficient use of nitrogen application on cultivated fields, and improved water management including water harvesting; (iii) introducing alternative enterprises and farming systems such as drought tolerant tree species, goat rearing and poultry production and agroforestry; (iv) making socio-economic adjustments, e.g. livelihood diversification or market facilitation; (v) strengthening local institutions and self-help capacities together with risk insurance, risk sharing mechanisms; (vi) strengthening formal institutions and community based organizations; (vii) formulating policy to catalyze enhancement of adaptive livelihood opportunities; and (viii) promoting awareness and knowledge sharing.

Climate change response strategies need to be integrated into the overall development agenda. A successful climate strategy will motivate rapid reductions in emissions as well as major investments in adaptation. Such a strategy needs to address the warming climate's connection to food production. Hallegatte (2009) examined integrated strategies that involved: (i) no-regret strategies that yields benefits even in absence of climate change; (ii) favoring reversible and flexible options; (iii) buying safety margins in new investments; (iv) promoting soft adaptation strategies, including long-term prospective; and (v) reducing decision time horizons. Adaptation-mitigation strategies also call for integrated design and assessment of adaptation and mitigation policies.

13.10 Capacity Building

The capacity to identify, collect and share data, use information and build knowledge relevant for climate change adaptation, mitigation and food security is critical because of rapidly changing climatic, environmental and socio-economic conditions. Extension services need to be strengthened substantially in order to address adaptation and mitigation if it will have to provide an efficient interface between policy-makers and the farming community.

Technical capacity to assess the impacts of climate change and apply adaptation and mitigation measures in agriculture, forestry and fisheries needs strengthening at national and local levels. Further, to implement national climate change and food security policies, there is need for in-depth knowledge of appropriate methods and tools as well as awareness of available funding mechanisms, such as the carbon market and adaptation funds.

13.11 Agriculture Research

Agricultural research need to provide several new location specific adaptation and mitigation options. However, research for a rapidly changing climatic condition is different from research for managing past and current climate risks. New investment in research and development (R&D) is the most productive way to support agriculture under changing climate. The priority areas of research are improving data collection, dissemination and analysis, developing varieties tolerant to drought, water stagnation and salinity, enhancing water productivity and pest and disease resistance. Significant public and private investments in research is required if agriculture is to benefit from the use of new technologies and techniques (FAO 2009a). Adaptation and mitigation in agriculture requires participatory and practical learning and action research to develop and replicate innovative technologies jointly with farmers, extension services and research institutions.

Traditional and indigenous knowledge and local biodiversity are one of the many suitable entry points, but likely to be insufficient in changing climate conditions. In addition, methodologies, farm management practices, crops and crop varieties need to be developed for future conditions. Crop and livestock productivity-enhancing research, including biotechnology, will be essential to help overcome stresses due to climate change (Nelson et al. 2009). Research results need to be public in an enabling environment in which methods, germplasm, crop varieties and animal breeds are accessible for use by most vulnerable people and introduction in adaptation and mitigation programs.

13.12 Sustainable Bioenergy

Bioenergy presents both opportunities and risks for food security. Bioenergy meets approximately 10% of global energy demand, around 80% of it as solid biomass for heating and cooking. Liquid biofuels account for less than 2% of road transport fuels worldwide: this is projected to rise to nearly 5% by 2030 (FAO 2008g). This expansion could revitalise the agriculture sector, foster rural development and alleviate poverty. But if not managed sustainably, it could seriously threaten food security. Policy makers have a major role in ensuring that bioenergy is developed sustainably, safeguarding food security and ensuring other benefits such as market and technology promotion, participatory processes and social protection (FAO 2008g).

Sustainable bioenergy need greater attention at this point of time when agriculture has to play a major role to supply food, fodder, fiber and recently energy is added to the list. Without an understanding of these new and complex interactions, it will be difficult to make the most fundamental policy decisions (FAO 2008c). Bioenergy can reduce emissions by substituting for transport fuels and replacing fossil fuels such as coal for power and heat generation. But, bioenergy development can have impacts on water use, soil erosion and biodiversity conservation. A major problem with current patterns of biomass use for energy, particularly for traditional bioenergy systems in developing countries, is its low conversion efficiency, frequently as low as 10% (Kaltschmitt and Hartmann 2001), and related degradation of carbon stocks. Improving bioenergy efficiency is a fairly straightforward means of reducing carbon emissions and it represents a large potential source of carbon payments for those countries that currently depend on traditional bioenergy (Jürgens et al. 2006).

Access to energy is a critical factor in development and poverty alleviation. Sustainable energy supply responses must be complemented with policy measures that manage energy demand growth. Bioenergy, including liquid biofuels for transport, can play a role in ensuring sustainable energy supply, but its contribution will be limited. While traditional biomass use will dominate developing country energy systems for many years to come, modern bioenergy systems will only meet a small share of total energy demand. Other sources of bioenergy besides liquid biofuel (biodiesel, bioethanol) can be used for transport, including biogas and woody biomass transformed into electricity. These entail less risk but more technological issues to sort out but it is feasible now or in a relatively short term.

As a new source for agricultural investment biofuel demand can stimulate agricultural growth and poverty alleviation by creating income and employment, but only if it is produced sustainably with participation in growing markets. The best ways to reduce the competition between food and fuel is to develop better integrated food-energy systems, including: (i) intercropping food and fuel crops,

(ii) agroforestry, (iii) using by products from agro processing units as feedstock to produce bioenergy. Liquid biofuels can be a sustainable form of energy, but only if safeguard measures are adopted to ensure that environmental and social risks are managed appropriately. However, an international approach is needed to agree and implement sustainability standards for biofuels without creating new trade barriers for developing countries (FAO 2008h).

13.13 FAO's High Level Conference

In order to put agriculture, forestry, fisheries and food security on the international climate change agenda, the Food and Agriculture Organization of the United Nations (FAO), in cooperation with the Consultative Group on International Agricultural Research (CGIAR), the International Fund for Agricultural Development (IFAD) and the World Food Programme (WFP), has organized a High-Level Conference on "World Food Security: The Challenges of Climate Change and Bioenergy" held at FAO Headquarters in Rome, Italy on 3–5 June 2008.

In preparation for the Conference, eight expert meetings were held to assemble the best available knowledge through expert meetings on: (i) Climate change adaptation and mitigation in agriculture, forestry and fisheries; (ii) Climate change, water and food security; (iii) Climate-related trans-boundary pests and diseases, including relevant aquatic species; (iv) Climate change and disaster risk management; (v) Bioenergy policy, markets and trade and food security; (vi) Global perspectives and food and fuel security; (viii) Climate change and fisheries and aquaculture and (viii) Climate change and biodiversity for food and agriculture. Each of these expert meetings examined the opportunities and constraints for the agriculture, forestry and fisheries sectors, including cross-sectoral linkages between food security, rural development and the environment. Additionally, stakeholder consultations for civil society and the private sector were organized to bring in a broader range of views and experiences and to identify areas for collaboration.

13.13.1 Prioritized Policy Options

The key policy recommendations derived from the expert meetings and stakeholder consultations are presented below under different thematic areas:

13.13.1.1 Climate Change Adaptation and Mitigation

- Adaptation measures need to focus on: climate change "hot spots" analysis, early warning systems, disaster risk management, crop insurance, incentives to adopt better agricultural and land use practices
- Building capacity and awareness on climate change adaptation

- Strengthening data collection, monitoring, analysis and dissemination within the national extension and research services
- Promoting soil carbon sequestration as a potential option for mitigation in agriculture

13.13.1.2 Climate Change and Water

- Integration of adaptation and mitigation measures for agricultural water management in national development plans
- Promoting management measures to improve the water use efficiency in rainfed and irrigated agriculture
- Developing knowledge on climate change and water, and share good practices among countries and regions
- Integration of risk management in national policies and promoting better monitoring networks
- Accessing adaptation funds to meet the challenges of water and food security

13.13.1.3 Climate Change and Disaster Risk Management

- Enhance understanding of climate change impacts at local level
- Diversifying livelihoods suitable for local conditions
- Improving weather and climate forecasting and early warning systems
- Risk management and contingency plans taking into consideration new and evolving risk scenarios
- Adjustment of land use plans considering climate change projections
- Cost/benefit analysis on structural mitigation measures

13.13.1.4 Climate-Related Trans-boundary Pests and Diseases

- Strengthening national animal and plant health services
- Focusing on basic sciences – taxonomy, modeling, population ecology and epidemiology
- Consolidation and organizing national animal and plant health services
- Investing in early control and detection systems, including broader inspections

13.13.1.5 Climate Change, Fisheries and Aquaculture

- Defining and implementing adaptation strategies based on ecosystem approach
- Developing appropriate management responses to adapt to anticipated ecological change

13.13.1.6 Climate Change, Biofuels and Land

- Promoting sound land tenure policies and planning
- Ensuring land tenure security to mitigate climate change
- Encouraging investments in sustainable land use practices

13.13.1.7 Bioenergy and Food Security

- Ensuring sustainable bioenergy development
- Safeguarding food security and ensuring that benefits include market and technology promotion and encouraging participatory processes

13.13.1.8 Climate Change and Biodiversity

- Assessment and distribution of biodiversity for food and agriculture both in the wild and in the fields
- Assessment of impact of climate change on agriculture biodiversity
- Biodiversity distribution mapping with different climate change scenarios

The Civil Society Organization (CSO) and Non Governmental Organization (NGO) forum reaffirmed that food production should be given the highest priority. In that context, it is emphasized that the diversion of production to bioenergy crops should in every case be handled with caution. The Small Island Developing States (SIDS) forum stressed the need for increased attention, both human and financial resources, to assist SIDS to mitigate the impacts of climate change on their food security. The Africa forum insisted on the fact that local knowledge has always been underestimated and emphasized the need to give importance to indigenous solutions and build a bridge between traditional knowledge and technology. The need to share successful experiences of small farmers in adaptation to climate change and farmer to farmer networks were suggested.

13.13.2 Conference Declaration

Following significant discussion and negotiations, the conference concluded with the adoption by acclamation of a declaration calling on the international community to increase assistance for developing countries, in particular the least developed countries and those that are most negatively affected by high food prices. The countries agreed that the issues of food security, bioenergy and climate change are all closely linked.

The conference urged governments to assign appropriate priority to the agriculture, forestry and fisheries sectors, in order to create opportunities for the world's smallholder farmers and fishers, including indigenous people, in particular vulnerable areas, to participate in, and benefit from financial mechanisms and investment flows to support climate change adaptation, mitigation and technology development, transfer and dissemination.

On bioenergy, the conference stressed the need to address the challenges and opportunities posed by biofuels, in view of the world's food security, energy and sustainable development needs. The conference concluded that in-depth studies are necessary to ensure that production and use of biofuels is sustainable and take into account the need to achieve and maintain global food security. The conference called upon relevant inter-governmental organizations, national governments, partnerships, the private sector, and civil society, to foster a coherent, effective and results-oriented international dialogue on biofuels in the context of food security and sustainable development.

13.14 Conclusions

Agriculture is most weather dependent and climate sensitive sector. Small farmers, marginal ethnic groups, indigenous people, livestock herders, forest dependent communities and fishers living in vulnerable areas will be affected by most of the climate phenomenon manifest due to climate change. Developing countries, Small Island Developing States, Least Development Countries that already vulnerable to climate extremes, low incomes and high incidence of hunger and poverty are expected to be adversely affected by climate change.

Climate change adaptation strategies are now a matter of urgency, especially for the most vulnerable communities, which are even now being disproportionately affected. Integrated strategies involving adaptation, mitigation priorities and short and long-term approaches are required to generate significant benefits for food security, and climate change adaptation and mitigation. Increasing soil carbon sequestration through forestry and agro-forestry initiatives and tillage practices, improving efficiency of nutrient management and restoring degraded lands are examples of actions that have large mitigation potential and high co-benefits.

The practices having strong adaptation and mitigation synergies in agriculture, livestock production, fisheries and forestry may offer new opportunities for financing and benefit smallholder farmers. However, agriculture needs a strong presence in climate change negotiations. There are constraints which includes lack of sufficient data, monitoring and policy and institutional frameworks. There are innovative policy options already exist and new strategies are being developed continuously and are expected to provide necessary mechanisms to promote more resilient adaptation and mitigation practices without compromising food security.

Climate change combined with demand for alternative energy possibly from biofuels produced from food crops is capable of reducing the availability of land, and water for food production. Bioenergy can contribute to rural income, supply rural households with electricity and heat, and mitigate climate change by substituting fossil fuels. However, if biofuels are produced unsustainably, their contribution to mitigating climate change is negative. It is the challenge how best to respond to the new opportunities, while making sure people can continue to grow or buy adequate food.

References

Alexandratos (2005) Countries with rapid population growth and resources constraints: issues of food, agriculture and development. Popul Dev Rev 31:237–258

Allison EH, Perry AL et al (2009) Vulnerability of national economies to the impacts of climate change on fisheries. Fish and Fisheries 10:173–196

Bates BC, Kundzewicz ZW et al (eds) (2008) Climate Change and Water. Technical paper of the intergovernmental panel on climate change, IPCC Secretariat, Geneva, p 210

Battisti DS, Naylor RL (2009) Historical warming of future food security with unprecedented seasonal heat. Science 323:240–244

Blanco-Canqui H, Lal R (2008) No-tillage and soil-profile carbon sequestration: an on-farm assessment. Soil Sci Soc Am J 72:693–701

Campbell BM (2009) Beyond Copenhagen: REDD+, agriculture, adaptation strategies and poverty, Global Environmental Change 19:379–399

Cassman KG, Dobermann A et al (2003) Meeting cereal demand while protecting natural resources and improving environmental quality. Annu Rev Environ Resour 28:315–358

Challinor A (2009) Towards the development of adaptation options using climate and crop yield forecasting at seasonal to multi-decadel timescales. Environ Sci Policy 12:453–465

Cline WR (2007) Global warming and agriculture: estimates by country. Centre for Global Development and Peterson Institute for International Economics, Washington

CRED (2008) Centre for research on the epidemiology of disasters, Université Catholique de Louvain. 2008. Disaster data: a balanced perspective. CRED Crunch 11:1–2

Curry R, Mauritzen C (2005) Dilution of the Northern North Atlantic Ocean in recent decades. Science 308:1772–1774

Dennis IA, Macklin MG et al (2003) The impact of the October–November 2000 floods on contaminant metal dispersal on the River Swale catchment, North Yorkshire. Hydrol Process 17:1641–1657

Derpsch R (2005) The extent of conservation agriculture adoption worldwide: implications and impact. In: Proceedings of the Third World Congress on Conservation Agriculture: Linking Production, Livelihoods and Conservation, Nairobi, Kenya, 3–7 October 2005 [CD-Rom]. Harare, African Conservation Tillage Network Productions

Drinkwater KF (2005) The response ofAtlantic cod (*Gadus morhua*) to future climate change. ICES J Mar Sci 62:1327–1337

Drinkwater KF, Belgrano A et al (2003) The response of marine ecosystems to climate variability associated with the North Atlantic Oscillation. The North Atlantic oscillation: climatic significance and environmental impact. In: Hurrell JW, Kushnir Y et al (eds) Geophysical monograph. American Geophysical Union, Washington, DC, pp 211–234

Easterling WE (2007) Climate change and the adequacy of food and timber in the 21st century. Proc Natl Acad Sci USA 104(50):19679

Easterling WE, Aggarwal PK et al (2007) Food, fiber and forest products. In: Parry ML, Canziani OF et al (eds) Climate Change 2007: impacts, adaptation and vulnerability. Contribution of

Working Group II to the Fourth Assessment Report of the Intergovernmental Panel on Climate Change. Cambridge University Press, Cambridge, pp 273–313

FAO (1997) The state of the world's plant genetic resources for food and agriculture. Food and Agriculture Organization of the United Nations, Rome

FAO (2006) World agriculture: towards 2030/2050 prospects for food, nutrition, agriculture and major commodity groups, Global Perspective Studies Unit, Food and Agriculture Organization of the United Nations, Rome, June 2006 Interim report, pp 71

FAO (2007) The state of food and agriculture – paying farmers for environmental services. Food and Agriculture Organization of the United Nations, Rome

FAO (2008a) Climate change adaptation and mitigation: challenges and opportunities for food security. Information document prepared for the High level conference on World Food Security: the Challenges of Climate Change and Bioenergy, 3– 5 June 2008, Rome

FAO (2008b) Climate change and food security: a framework document. Interdepartmental working group on climate change, Food and Agriculture Organization of the United Nations, Rome

FAO (2008c) Food, energy and climate: a new equation, FAO at work 2007–2008. Food and Agriculture Organization of the United Nations, Rome

FAO (2008d) Report of the FAO expert workshop on climate change implications for fisheries and aquaculture, Food and Agriculture Organization of the United Nations, Rome

FAO (2008e) The state of food insecurity in the world 2008. High food prices and food security – threats and opportunities, Food and Agriculture Organization of the United Nations, Rome, FAO Fisheries Report No. 870. Rome, 7–9 April 2008

FAO (2008f) Community based adaptation in action a case study from Bangladesh, project summary report (Phase I) improved adaptive capacity to climate change for sustainable livelihoods in the agriculture sector. Food and Agriculture Organization of the United Nations, Rome

FAO (2008g) Bioenergy, food security and sustainability – towards an international framework. High level conference on world food security: the challenges of Climate Change and Bioenergy, Rome, 3–5 June 2008, Food and Agriculture Organization of the United Nations, Rome

FAO (2008h) The state of food and agriculture - Biofuels: prospects, risks and opportunities. Food and Agriculture Organization of the United Nations, Rome

FAO (2009a) How to feed the world in 2050. Food and Agriculture Organization of the United Nations, Rome

FAO (2009b) More people than ever are victims of hunger. FAO food outlook and background note on hunger. Food and Agriculture Organization of the United Nations, Rome

FAO (2009c) Background note on Global agriculture towards 2050, High level expert forum, 12–13 October 2009. Food and Agriculture Organization of the United Nations, Rome

FAO (2009d) Background note on climate change and bioenergy challenges for food and agriculture. Food and Agriculture Organization of the United Nations, Rome

Fischer G, Tubrello FN et al (2007) Climate change impacts on Irrigation water requirements: effects of mitigation 1990–2080. Technol Forecast Soc Change 74:1083–1107

Fleming RA, Tatchell GM (1995) Shifts in the flight periods of British aphids: a response to climate warming? In: Harrington R, Stork N (eds) Insects in a changing environment. Academic, New York, pp 505–508

Gan J (2004) Risk and damage of southern pine beetle outbreaks under global climate change. Forest Ecol Manag 191:61–71

Garrett KA, Dendy SP et al (2006) Climate change effects on plant diseases: genomes to ecosystems. Annu Rev Phytopathol 44:489–509

Hallegatte S (2009) Strategies to adapt to an uncertain climate change. Glob Environ Change 19:240–247

Hall-Spencer JM, Rodolfo-Metalpa R et al (2008) Volcanic carbon dioxide vents show ecosystem effects of ocean acidification. Nature 454:96–99

Hansen JW, Sivakumar MVK (2006) Advances in applying climate prediction to agriculture. Climate Res 33(1):1–2

Harrington R, Clark SJ et al (2007) Environmental change and the phenology of European aphids. Glob Change Biol 13:1550–1564

Herrero M, Thornton PK et al (2008) The spatial distribution of methane emissions from African domestic ruminants to 2030. Agric Ecosyst Environ 16:122–137

Howden SM, Soussana JF et al (2007) Adapting agriculture to climate change. Proc Natl Acad Sci USA 104(50):19691–19696

Huber L, Gillespie TJ (1992) Modelling leaf wetness in relation to plant disease epidemiology. Annu Rev Phytopathol 30:553–577

IPCC (2007a) Climate change 2007: synthesis report. In: Pachauri RK, Reisinger, A (eds) Contribution of working group I, II and III to the Fourth Assessment Report of the Intergovernmental Panel on Climate Change IPCC, Geneva, Switzerland, pp 104

IPCC (2007b) Climate change 2007: impacts, adaptation and vulnerability. In: Parry ML, Canziani OF et al (eds) Contribution of Working Group II to the Fourth Assessment Report of the Intergovernmental Panel on Climate Change. Cambridge University Press, Cambridge, UK, p 976

IPCC (2007c) Climate Change 2007: mitigation. In: Metz B, Davidson OR (eds) Contribution Of Working Group III to the Fourth Assessment Report of the Intergovernmental Panel on Climate Change. Cambridge University Press, Cambridge, UK and New York, USA

Jürgens I, Schlamadinger B et al (2006) Bioenergy and the CDM in the emerging market for carbon credits. Mitigation and Adaptation Strategies for Global Change 6, 11(5–6), 1051–1081

Kaltschmitt M, Hartmann H (2001) Energie aus Biomasse. Grundlagen, Techniken und Verfahren. Springer, Berlin

Karl TR, Meehl GA, et al (2008) Weather and climate extremes in a changing climate. Synthesis and assessment product 3.3 Report by the US Climate Change Science Programme and the subcommittee on global change research. pp 162

Kleinen T, Petschel-Held G (2007) Integrated assessment of changes in flooding probabilities due to climate change. Clim Change 81:283–312

Kundzewicz ZW, Mata LJ et al (2007) Freshwater resources and their management. In: Parry ML, Canziani OF et al (eds) Climate Change 2007: Impacts. Adaptation and Vulnerability. Contribution of Working Group II to the Fourth Assessment Report of the Intergovernmental Panel on Climate Change, Cambridge University Press, Cambridge, UK, pp 173–210

Lal R (2004) Soil carbon sequestration impacts on global climate change and food security. Science 304:162–1627

Lal R (2007) Constraints to adopting no-till farming in developing countries. Soil Tillage Res 94(1):1–3

Lal R, Follett RF (2009) Soil carbon sequestration and greenhouse effect, 2nd edn. Soil Science Society of America, Inc, Madison, WI, pp 410

Lehodey P, Chai F et al (2003) Modelling climate-related variability of tuna populations from a coupled ocean biogeochemical-populations dynamics model. Fish Oceanogr 12:483–494

McCarl BA (2007) Adaptation options for agriculture, forestry and fisheries, a report to the UNFCCC Secretariat Financial and Technical Support Division

McClanahan TR, Cinner JE et al (2008) Conservation action in a changing climate. Conserv Lett 1:53–59

Millennium Ecosystem Assessment (2005) Ecosystems and human well-being synthesis. Island Press, Washington, DC

Mills E (2005) Insurance in a climate of change. Science 309:1040–1044

Milly PCD, Dunne KA et al (2005) Global pattern of trends in stream flow and water availability in a changing climate. Nature 438:347–350

Nabuurs GJ, Masera O et al (2007) Forestry. In: Metz B, Davidson OR et al (eds) Climate Change 2007: mitigation. Contribution of Working Group III to the Fourth Assessment Report of the

Intergovernmental Panel on Climate Change Cambridge University Press, Cambridge, United Kingdom and New York, NY, USA
Nellemann C, MacDevette Manders et al (eds) (2009) The environmental food crisis – the environment's role in averting future food crises. A UNEP rapid response assessment. United Nations Environment Programme, p 101
Nelson GC, Rosegrant MW et al (2009) Climate change, impact on agriculture and costs of adaptation. International Food Policy Research Institute, Washington, DC, p 19
Oram PA (1989) Sensitivity of agricultural production to climate change: an update. In: Climate and Food Security, IRRI, Manila. The Philippines, pp 25–44
Palm CA, van Noordwijk M et al (2005) Carbon losses and sequestration following land use change in the humid tropics. In: Palm CA, Vosti SA et al (eds) Slash and burn: the search for alternatives. Columbia University Press, New York, pp 41–63
Parry M, Arnell N et al (2009) Assessing the costs of adaptation to climate change: a review of the UNFCCC and other recent estimates. International Institute for Environment and Development and Grantham Institute for Climate Change, London
Pretty JN, Noble AD et al (2006) Resource-conserving agriculture increases yields in developing countries. Environ Sci Technol 40(4):1114–1119
Reid RS, Thornton PK et al (2004) Is it possible to mitigate greenhouse gas emissions in pastoral ecosystems of the tropics? Environ Dev Sustain 6:91–109
Rosenzweig C, Iglesias A et al (2001) Climate change and extreme weather events. Implications for food production, plant diseases, and pests. Global Change Human Health 2:90–104
Runge CF, Senauer B (2007) How biofuels could starve the poor, Foreign Affairs 83(3) May–June
Shiklomanov IA, Rodda JC (2003) World water resources at the beginning of the Twenty-First Century, Cambridge University Press, Cambridge, UK
Sivakumar MVK, Hansen J (2007) Climate prediction and agriculture: advances and challenges. Springer, Berlin, p 306
Smith P, Martino D et al (2007) Agriculture. In Climate Change 2007: mitigation. In: Metz B, Davidson OR et al (eds) Contribution of Working Group III to the Fourth Assessment Report of the Intergovernmental Panel on Climate Change Cambridge University Press. USA, Cambridge
SP-IPM (2008) The role of integrated pest management. How IPM contributes to the CGIAR system priorities and the Millennium Development Goals. IPM Research Brief No. 5. SP-IPM Secretariat, International Institute of Tropical Agriculture (IITA), Ibadan, Nigeria
Steenblik R (2007) Biofuels – at what cost? Government support for ethanol and biodiesel in selected OECD countries. A synthesis of reports addressing subsidies for biofuels in Australia, Canada, the European Union, Switzerland and the United States. The Global subsidies initiative of the International Institute for Sustainable Development, Geneva, Switzerland
Steinfeld H, Gerber P et al (2006) Livestock's long shadow: environmental issues and options. FAO, Rome, Italy
Stern N (2007) The economics of climate change: stern review. Cambridge University Press, Cambridge
Thornton PK, van de Steeg J et al (2009) The impacts of climate change on livestock and livestock systems in developing countries: a review of what we know and what we need to know Agricultural Systems 101(2009):113–127
Thurlow J, Kiringai J et al (2007) Rural investments to accelerate growth and poverty reduction in Kenya. Discussion Paper No. 723, Washington, DC, IFPRI
UNDP (2006) Human Development Report 2006, Beyond scarcity: power, poverty and the global water crisis, United Nations Development Programme (UNDP), pp 48
UNDP (2008) Human Development Report 2007/2008: fighting climate change: human solidarity in a divided world. United Nations Development Programme (UNDP), pp 31
van Wuijckhuise L, Dercksen D et al (2006) Bluetongue in the Netherlands; description of the first clinical cases and differential diagnosis; Common symptoms just a little different and in too many herds. Tijdschr Diergeneesk 131:649–654

World Bank (2007) World Development Report 2008: Agriculture for Development. Washington, DC

World Bank (2009) Development and climate change: World Development Report 2010. The World Bank, Washington, DC

Zvereva EL, Kozlov MV (2006) Consequences of simultaneous elevation of carbon dioxide and temperature for plant–herbivore interactions: a meta-analysis. Glob Change Biol 12:27–41

Chapter 14
Implications of Climate Change on Agriculture and Food Security in South Asia

M. Ad Spijkers

Contents

Abstract It is estimated that the current food production in South Asia has increased three times from 117 million tonnes in 1961 to 348 million tonnes in 2006, but the dietary energy consumption has improved not enough emphasising the fact that the growth rate is not sufficient to tackle the emerging challenges in addition to population pressure. Ensuring food security in the future requires a great deal of additional efforts in yield improvement, with limited scope for expanding area under cultivation.

FAO has revealed recently (2008) that the Boro rice output in Bangladesh is estimated at record 17.54 million tonnes, increased by some 17.2% from the previous year and 29.3% above the 5-year average. This increase of production was mainly due to favourable weather conditions and extra efforts made by farmers and Government in response to the high food prices and production loss of 1.4 million tonnes in 2007 Aman season following severe flood and Cyclone Sidr.

We in FAO strongly feel that it is not enough; adaptation and mitigation requires socio-institutional learning process and participatory community based actions for technology refinement and transfer. Location-specific technologies and good practices need to be built upon an improved understanding of the links between climate change and food provision, while promoting socio-economic development and limiting further environmental degradation.

M. Ad Spijkers (✉)
FAO Representation, FAO, House #37, Road #8, Dhanmondi Residential Area, Dhaka, 1205, Bangladesh
e-mail: ad.spijkers@fao.org

R. Lal et al. (eds.), *Climate Change and Food Security in South Asia*,
DOI 10.1007/978-90-481-9516-9_14,

Keywords Gross domestic product • Poverty line • Fragile ecosystems • Glacier melting • Hydrologic cycle • Disaster risk management

14.1 Introduction

Securing world food security in light of the impact of climate change may be one of the biggest challenges we face in this century. An estimated 1,020 million people in the world today suffer from hunger. Of those, more than 95% live in developing countries, the countries expected to be most affected by climate change. Projected population and socio-economic growth will double current food demand by 2050. To meet this challenge in developing countries, cereal yields need to increase by 40%, net irrigation water requirements by 40–50%, and 100–200 million ha of additional land may be needed, largely in Asia, sub-Saharan Africa and Latin America.

The fourth assessment report of the Intergovernmental Panel on Climate Change (IPCC) states that warming of the climate system is unequivocal, as is now evident from observations of increase in global average air and ocean temperatures, widespread melting of snow and ice and rising global average sea level (IPCC 2007). Observational evidences from all continents and most oceans show that many natural systems are being affected by regional climate changes, particularly temperature increase. In terrestrial ecosystems, earlier timing of spring events and pole ward and upward shifts in plant and animal ranges are with very high confidence linked to recent warming. In some marine and freshwater systems, shifts in ranges and changes in algal, plankton and fish abundance are with high confidence associated with rising water temperatures, as well as related changes in ice cover, salinity, oxygen levels and circulation. All these changes are already affecting agriculture and allied sector and food security in many parts of the world.

South Asia (as defined in this paper comprises eight countries; Afghanistan, Bangladesh, Bhutan, India, Maldives, Nepal, Pakistan and Sri Lanka,) faces key development challenges such as population growth, high incidence of poverty, urbanization and the degradation of the environment. Climate change could make this region more vulnerable and reverse country's efforts towards achieving the Millennium Development Goals (MDGs) by causing reduction in agricultural yields, increased water stress due to changes in precipitation patterns and glacier melting, decline in fishery products, degradation of natural grasslands and impacts on forest products. South Asia lags on most human development goals, although it will likely meet the poverty reduction MDG (World Bank 2008).

The climatic conditions, combined with socio-economic situations make South Asia, one of the most vulnerable regions in the world with regard to climate change. The climate of South Asia is characterised by extremes and natural hazards like cyclones, wind storms, droughts and heat waves, floods, Glacial Lake Outburst Floods (GLOS), land slides, pest and disease outbreaks etc.. The assessments of the impacts of natural disasters revealed that South Asia accounts for almost 80% of the total population affected and 86% of total damage due to drought in Asia.

Similarly, this sub-region account for 35% of the total population affected and 28% of the total damage due to floods in Asia (CRED 2008). Moreover, the productive sectors which include agriculture accounted for over half of associated damages and losses. Climate change will superimpose itself on these existing trends, significantly increasing production risk and rural vulnerability, particularly in regions that already suffer from poverty and hunger.

14.2 Current Agriculture Production and Food Security Status in South Asia

The majority of the south Asian countries share similar economic and sustainable development challenges. The most prominent similarities are frequent occurrence and susceptibility to natural hazards, excessive dependence on agriculture, widespread poverty and vulnerability to climate change. Five out of eight countries in the region are characterised as Least Developed Countries (LDC), and of these five LDCs, three are Land-Locked Least Developed Countries (LLDC) possessing low-income, weak human resources and economic vulnerability. The South Asia region contains a population of 1,507.5 million (2007) and projections anticipate 1,727 million by 2015. Agriculture represents a high share of GDP and approximately 150 million households, with 751 million people classified as agriculture dependent. Since 1990, millions more people are chronically hungry in sub-Saharan Africa and in Southern Asia, where half the children under age 5 are malnourished (United Nations 2005).

South Asia possesses diverse farming systems ranging from intensive rice-wheat systems to sparse arid regions and mountains. Large parts of the region face severe environmental constraints like erratic and uneven distribution of rainfall, water stress risk, low soil suitability, steep slopes and mountains, severe land degradation and low to medium climate production potential (Dixon et al. 2001). The area under arable land and permanent crops is estimated at 213 million ha (FAO 2008a) and expected to show only a marginal increase by 2030; the region's irrigated land area will grow from 85 million ha to about 95 million ha in 2030.

The current total food production in South Asia has increased threefolds from 117 million tonnes in 1961 to 348 million tonnes in 2006 (FAO 2008b), but the dietary energy consumption has increased only marginally. The current dietary energy consumption is 2,364 kcal/person/day and is expected to increase to 2,790 by 2015 and 3,040 by 2030. It requires additional efforts in yield improvement, given the fact that there is limited scope for expanding area under cultivation and area under irrigation.

South Asia is home to the largest concentration of poverty and undernourished population. FAO estimates that 312 million (21%) people are still undernourished (FAO 2008c) and 26.4% of the population is below poverty line (ADB 2008). Indicators of other dimensions of poverty, such as female illiteracy (59%), child mortality (89 per 1,000 in children <5 years), and child malnutrition (51%) also point to extensive poverty. Nearly 40% of the world's poor earning less than a dollar a day live in the region (Dixon et al. 2001).

14.3 Climate Change and Its Implications for Agriculture and Food Security in the Sub-region

Changes in climate and other important environmental factors pose a major concern to food security in the region. This is because such changes not only directly threaten the production of food from land and sea for local consumption, but also threaten revenue generation at farm scale. The adverse impacts of climate change are a major barrier to food security and achievement of sustainable development goals in South Asia. They are anticipated to exacerbate the impact of existing development challenges such as loss of market and declining value of traditional exports, declining domestic food production and increasing imports; and environmental degradation.

Arable land, water resources and biodiversity are already under pressure and are expected to be stressed by changes in precipitation patterns. With climate change, negative impacts on agriculture are predicted; coral reefs and mangroves will be threatened by increased sea surface temperatures, and sea-level rise. Predicted impacts of climate change in the region include extended inundation of arable land, salinity intrusion and reduced fresh water availability. For example, in India freshwater availability is predicted to decrease by 47% in 2025 due to climate change and population growth.

The fragile ecosystems vulnerable to climate change impacts are: mountain/Himalayan ecosystems (e.g. Nepal, India, Bhutan), mangroves, salt marshes and coral reefs (e.g. India, Bangladesh and Sri Lanka), semi-arid and arid resource poor dry lands (e.g. India and Pakistan). The low lying coastal regions would be affected due to Sea Level Rise (SLR) and/or increase in extreme climate events (e.g. Maldives and Bangladesh).

Semi-arid tropics are vulnerable due to reduced rainfall and increased evapotranspiration and drought (e.g. central and peninsular India; Sindh and Balochistan of Pakistan; North West Bangladesh), while small islands are extremely vulnerable due to high exposure of population and agricultural infrastructure to sea level rise (e.g. Maldives) and increased storm surge. The Magna basin and north eastern hoar region of Bangladesh is vulnerable to flash floods.

Fresh water availability in South Asia is projected to decrease and temporal and spatial changes in precipitation and associated droughts have major implications for agriculture. Temperatures are projected to increase by as much as 3–4°C towards the end of the twenty-first century. In India, climate change may aggravate the current problems of sustainability and profitability of agriculture in many regions. Studies on socio-economic impact of climate change indicate that the loss in farm-level net revenue may range between 9% and 25% for a temperature rise between 2.0°C and 3.5°C (MoEF-GOI 2004).

In Pakistan, studies indicate that fourteen crops (eight field crops, three vegetables and three fruits) have shown some degree of vulnerability to heat stress under a climate change scenario of a rise in temperature of 0.3°C per decade. Under the scenario where rainfall decreases by 6%, net irrigation water requirements could increase by 29%. Over 1.3 million farm households (30% of the total), cultivated cotton and 27% of reporting households had paddy fields which will be exposed to negative climate change impacts (MoE-GIRP 2003).

Glacier melting in the Himalayas is projected to increase flooding and affect water resources within the next two to three decades. Glacial Lake Outburst Floods (GLOF), landslides, flash flood and droughts are key hazards affecting Bhutan. Decreased water availability for crop production, increased risk of extinction of already threatened crop species (traditional crop varieties), loss of soil fertility due to erosion of top soil and runoff, loss of fields due to flash floods, landslides, crop yield loss due to hailstorms and forest fires are the key vulnerabilities. There are an estimated 2,674 glacial lakes in Bhutan out of which 562 are associated with glaciers and 24 glacial lakes are potentially dangerous (NEC-RGB 2000).

Agriculture remains Nepal's principal economic activity, employing 80% of the population. The Terrai plains constitute 43% of the total cultivated land. Recurring natural disasters undermine agricultural productivity causing poverty and food insecurity. In Nepal, the potential yield of Terrai rice is estimated to increase by about 18–21% when CO_2 increases to 580 ppm. However, with an increase of temperature beyond 4°C, the yield is projected to decrease. Similarly, temperature changes will affect the availability of forages and alter the movement of Yaks in the mountains between 3,000 and 5,000 m elevation (MoEST 2004).

It is projected that Sri Lanka's rice output would be reduced by 5.91% with a temperature increase of 0.5°C. Approximately 740,000 ha are cultivated with paddy in Sri Lanka and of this 44% is irrigated under major irrigation schemes and another 24% under minor irrigation schemes (GoSL 2000). The bulk of this land is in the dry and intermediate zones and is vulnerable to fluctuation in rainfall pattern. Increased temperatures are also expected to negatively affect high value crops such as vegetables and potatoes. The impact of salt water intrusion on low lying agriculture would be significant and loss and degradation of arable lands will significantly lower the agricultural output in coastal areas. A temperature rise of about 2°C may have substantial impacts on the distribution, growth and reproduction of fish stocks.

Over 80% of the land area of the Maldives is less than 1 m above mean sea level and is extremely vulnerable to sea level rise and beach erosion. Agriculture is a comparatively small sector in the Maldivian economy with a GDP share of about 3.5%. However, the sector is considered important because of its potential on generating employment, income opportunities and attaining local food security. A wide range of crops are grown, with a heavier concentration of root crops in the south and more field and grain crops in the north. The Maldives tuna fishery is affected by the seasonal monsoon and their associated currents.

The hydrological cycle is predicted to be more intense with increased intensity of daily rainfall during monsoon season which may lead to intensified flooding and inundation of agricultural areas. The most intense cyclones crossing the East coast of India and Bangladesh are associated with storm surges, strong winds, coral bleaching, ocean spray and inundation of land, and erosion. In South Asia, analysis of data has shown that the number of very warm days and nights is increasing. There is a trend towards an overall increase in precipitation, with prolonged dry spells having occurred over the last few decades. Climate change projections show marked increase in both rainfall and temperature over the region. Temporal and spatial changes in glacier melting and precipitation patterns; associated droughts, floods, and more intense or frequent cyclones are likely to negatively affect all agricultural sub-sectors.

The crop yields could decrease up to 30% in the region by the mid-twenty-first century and considering the population growth, the risk of hunger is projected to remain very high. Impact analysis based on statistical crop models and climate projections for 2030 from 20 general circulation models, revealed that South Asia, without sufficient adaptation measures, will likely suffer negative impacts on several crops that are important to large food-insecure human populations (Lobell et al. 2008). Other projected impacts of climate change include inundation of arable land, salinity intrusion, reduced fresh water availability and persistence of trans-boundary pest and diseases. Irrigation demand for agriculture in arid/semi-arid regions is expected to increase by 10% for temperature increase by 1°C; and increased dryness during pre-summer season may accelerate the rate of forest fire incidence and threaten rural livelihoods.

Extreme weather events can be very damaging to fisheries industry, hitting fishing gear, fishing vessels, but also coral reefs, mangroves and coastal vegetation, which act as protective barriers for the coastlines. The unprecedented increase in the movement of people, animals and goods, multiplies the pathways for the dissemination of trans-boundary animal diseases and plant pests (including insects, pathogens, and plants as pests) and aquatic species. Once introduced, climatic change combined with change in crops, landscapes and human activities may create favourable ecological conditions for the persistence of transboundary diseases and pests.

14.4 Initiatives on Climate Change Adaptation and Mitigation in the Agriculture Sector

Several regional and national initiatives have been undertaken in the recent past on climate change adaptation and mitigation. The South Asian Association for Regional Cooperation (SAARC) has prioritized regional issues, strategies, programs and projects on food security. The prioritized issues related to climate change impacts are: low and stagnating production and productivity, high pre and post harvest losses, overexploitation and degradation of natural resources.

The fourteenth SAARC Summit (New Delhi, 3–4 April 2007) expressed deep concern over global climate change; the New Delhi declaration called for pursuing a climate resilient development in South Asia. The SAARC Expert Group Meeting on Climate Change (Dhaka, 1–2 July 2008) recommended a draft SAARC Action Plan on Climate Change and stressed the need for actions relevant to agriculture, among others: adaptation of technologies and practices, sharing of best practices on sustainable forest management and sharing of good practices in disaster management.

India released its National Action Plan on Climate Change in June 2008 with a focus on harnessing renewable energy. The action plan identifies eight priority missions that will promote India's development objectives, with the "co-benefit" of tackling climate change. The missions are: solar energy, enhanced energy efficiency, sustainable habitats, water conservation, sustaining the Himalayan ecosystem, developing a 'green' India, sustainable agriculture and building a strategic knowledge platform on climate change. The Government of Bangladesh has prepared a first draft Climate Change Strategy and Action Plan (2008) for discussion. Three of the five LDCs in the region, -Bangladesh, Bhutan and Maldives-, have submitted their National Adaptation Programme of Action (NAPA), recognizing that agriculture (including livestock, fishery and forestry) is highly vulnerable to climate change and prioritizing several adaptation projects in these sectors.

FAO assists member countries in the region in identifying potential adaptation and mitigation options most applicable to their particular circumstances, in mainstreaming climate change responses in food and agricultural policies and programmes and in including adaptation measures into National Programmes for Food Security (NPFSs), Special Programmes for Food Security (SPFSs), National Forest Programmes (NFPs) and other policy and planning processes.

FAO has initiated programmes targeting food security in the sub-region: (i) pro-poor policy formulation, dialogue and implementation to reduce rural poverty through enhanced institutional capacity to analyze, formulate and implement agricultural and rural development policies, (ii) support for the preparation of a Regional Programme for Food Security (in Bangladesh, Bhutan, India, Maldives, Nepal, Pakistan and Sri Lanka) and (iii) the Initiative on Soaring Food Prices.

FAO has implemented and/or is in the process of implementing many projects in South Asia directly addressing climate change adaptation at the local level: (i) strengthening support to disaster preparedness in agriculture in Bangladesh (2004–2006), (ii) livelihood adaptation to climate change in drought-prone and coastal areas of Bangladesh (FAO/UNDP-CDMP: 2005–2009), (iii) strengthening capacities for disaster preparedness and climate risk management in the agriculture sector in Nepal (FAO-TCP: 2008–2010), (iii) enhancing capacities for disaster risk management (DRM) and climate change adaptation (CCA) for sustainable livelihoods in the agriculture sector in Nepal (FAO-UNDP). In Pakistan, technical support is being given for crop area and yield forecasting.

The donor funded projects have identified a multitude of adaptation options for South Asian countries which include engineering solutions such as sea defences, costal embankments, polders and the provision of water storage. The technological solutions in agriculture include use of more drought and salinity tolerant crops, terracing, contour farming, rain water harvesting, vegetative barriers, wind breaks and watershed management. In spite of the wide range of adaptation options, there are constraints that can limit the choices of options and their implementation such as inadequate data, lack of technical and institutional capacity and limited financial resources. FAO provides capacity building through field projects and legislative solutions such as agro-ecological zoning, land zoning around coasts, coastal forestry and updating food security policy and provides technical assistance to the focal ministries to participate in the NAPA processes.

The priority actions to protect local food supplies, assets and livelihoods against the effects of increasing weather and extreme events are: (a) vulnerability and risk analysis and risk management specific to agricultural eco-systems, (b) crop varieties and breeds adapted to changing climate conditions, (c) introducing alternatives to provide food, fodder and energy and enhance cash incomes, (d) promotion of insurance and relocation of vulnerable assets, and (e) rain water harvesting, water storage and in-situ conservation.

The priority actions to avoid disruptions or decline in food supplies due to changes in temperature and precipitation are: (a) efficient agricultural water management and drainage, (b) adjustment of planting and harvesting schedules, (c) floating agricultural systems for flooding, risk diversification in drought prone areas, (d) improving weather/climate and flood forecasting and warning, and (e) improved livestock management, altered grazing and rotation of pasture.

The eco-system management through provision of environmental services needs to be ensured by: (a) use of degraded/marginal lands for productive planted forests or biomass for alternatives fuels, (b) watershed management and prevention of land degradation, (c) regulation through planning legislation and zoning, (d) protection of coastal areas from cyclones and other coastal hazards, (e) forest fire management through altered stand layout, (f) preservation of mangroves and their contribution to coastal fisheries, and (g) biodiversity conservation.

FAO promotes options for climate change mitigation in the food and agriculture sector by reducing emissions without compromising food security. Sustainable land management practices can diminish the conversion from forested area to cultivated or grazing land, increase the efficiency of water, soil and energy use and at the same time reduce emissions from deforestation, forest degradation (REDD), cropland and pastures. For example, Bhutan has 72.5% of its total land area under forest cover and high per capita GHG sequestration potential.

In agriculture, reducing methane emissions from ruminant livestock, rice paddies, manure and nitrous oxide emissions from soil need attention. Carbon sequestration in biomass and soils can be enhanced by (a) sustainable forest management, (b) reforestation and afforestation, (c) rehabilitation and restoration of degraded grasslands, (d) rehabilitating and restoration of cultivated organic soils, (e) promoting conservation agriculture, (f) grazing land management and (g) residue retention and conservation tillage systems.

14.5 Experiences Gained and Lessons Learned

FAO is implementing programs and projects in the region on climate change adaptation at multiple levels targeting agriculture sector. The experience clearly shows that the climate change impacts exacerbate existing vulnerabilities; adaptation must be addressed in the broader context of vulnerability. Addressing current climatic risks is a suitable operational entry point to launch climate change adaptation. Adaptation is considered as a social learning process and needs to involve institutions and multiple actors within the agriculture sector and other relevant sectors.

Awareness raising and institutional capacity building are the key. Adaptation is location specific, requires demand-driven research and extension strategies for technology development and transfer. Cross-sectoral livelihoods perspectives with strengthened institutional systems are essential to capture farmers' needs and to respond to location specific demands.

Improved operational linkages between climate change adaptation, disaster risk management and development are needed. Re-strengthening agriculture research (action-oriented/adaptive research), extension services and development links are essential for continuous adaptation. Synergies between climate change adaptation and mitigation exist and need to be exploited.

Responses to climate change need to be coordinated and integrated with existing policies of socio-economic development and environmental conservation to facilitate sustainable development. There are several initiatives like regional consultations on food security and ministerial meeting on climate change organised by SAARC to implement potential adaptation measures to help increase resilience to the impacts of climate change. These initiatives and follow-up actions need to focus on strengthening of institutions, capacity building, and mainstreaming climate change issues into policy and regulations, and on field activities such as the promotion of water storage and drought and salinity tolerant crops.

Re-orientation of water resource management to take care of the impacts of climate change is very much needed as South Asia's food production depends heavily on fresh water resources including groundwater. National Agriculture Ministries, Agricultural Research Systems, Universities, FAO and several other development agencies have promoted innovative approaches of water management, and key issues still need to be addressed, such as: (i) reform of irrigation agencies, (ii) modernization of irrigation systems, (iii) changes in water governance, (iv) river basin management and (v) water policy related to agriculture sector aiming to meet food security in the region.

Community-level actions to demonstrate viable adaptation options in the drought-prone areas of Bangladesh improved the adaptive capacity of the marginal farmers against climate risks and helped to accelerate adaptation processes. The Farmers Field School (FFS) approach promoted by FAO and Danida provides excellent opportunities for integrating climate risk management strategies and practices into the FFS modules.

14.6 Knowledge Gaps, Opportunities and Key Messages

Comprehensive impact assessments on agriculture and food security and adaptation strategies for smallholder systems need be strengthened. Highly vulnerable micro-environments and their significance to the socio-economic status of the small and marginal farmers should be identified and suitable adaptation strategies have to be prioritized. Information and knowledge on the effects of elevated CO_2 and increased temperature on non-cereals, pest, diseases and weeds have to be improved to facilitate development of new technologies. Impacts of climate change on aquatic biota, coastal and mountain ecosystems have to be properly understood. Policy instruments to guide adaptation and mitigation actions in agriculture are required to speed up location-specific actions.

Adaptation requires targeted research and participatory extension strategies for technology development and transfer. Location-specific technologies and good practices are needed as well as decentralised ways of working, within the framework of coherent national and international policies. Within this context, support to the emergence of national/regional programmes for food security that are responding to location-specific needs to be associated with climate change concerns. There are several initiatives in the sub-region aimed at livelihood adaptation (e.g. LACC), cyclone risk management and rehabilitation programmes (in Bangladesh) and strengthening institutional capacity to manage risks associated with climate variability, change and natural hazards. These initiatives and good practices need to be scaled up.

Adoption of risk management practices and policies for hazard vulnerability reduction help South Asian countries in preparing better for climate change impacts. Risk transfer could occur through micro-insurance, catastrophe bonds and reduced insurance premiums as an incentive to take preventive measures. However, the lack of financial mechanisms act as an obstacle to insurance initiatives.

Policy formulation needs to be built upon an improved understanding of the links between climate change and food provision, while promoting socio-economic development and limiting further environmental degradation. Given the magnitude of the impacts of climate change, a comprehensive climate strategy for the region needs to focus on adaptation as a first priority. Mitigation activities such as soil carbon sequestration, ecosystem restoration, protecting mangroves in coastal areas and reduced emission from deforestation and forest degradation promotes ecosystem resilience and improves adaptive capacity of the communities against climate risks.

Several opportunities exist to streamline the climate change adaptation and mitigation interventions through the Regional Action Plan adopted by SAARC member states. Synergies are expected in identification, development and promotion of agricultural practices for mitigation and adaptation without compromising food security in the region. Collection, exchange, and dissemination of meteorological, hydrological data and statistics within the SAARC region would strengthen the adaptation and mitigation planning process.

The diversity of experience in the region provides an opportunity to upscale climate change adaptation and mitigation initiatives. Collaboration between countries in the sub-region in developing regional and national policies, legislation for sustainable development and institutional mechanisms would provide a common platform for sharing of information, technologies and capacity building.

The Government of Bangladesh recognised the need for regional collaboration and requested the development partners to assist in establishing an International Centre for Adaptation in Bangladesh which will provide a forum to study aspects of the vulnerability of countries to climate change, scope and constraints to adaptation, develop relevant data bases, provide a network among countries and professionals (Karim 2007). FAO has already responded to the request and submitted a technical proposal (FAO 2007) for consideration.

Acknowledgement The paper is prepared in consultation and inputs from Environment, Climate Change and Bioenergy Division, FAO, Rome.

References

ADB (2008) Basic statistics, developing member countries economics and research department, development indicators and policy research division, Asian Development Bank. May 2008 (www.adb.org/Statistics)

CRED (2008) Emergency events database. Centre for Research on Epidemiology of Disasters (CRED) (http://www.emdat.be/)

Dixon J, Gulliver A, Gibbon D (2001) Farming systems and poverty, improving farmer's livelihoods in a changing world. FAO and World Bank, Rome and Washington, DC

FAO (2007) Interactive Climate Change Adaptation Information Centre, Providing quality assured data and information for dealing with vulnerability, impacts and adaptation interventions, draft concept paper submitted to the Government of Bangladesh. December, 2007

FAO (2008a) Resource statistics. FAOSTAT domain, Rome (available at http://faostat.fao.org/site/405/default.aspx)

FAO (2008b) Production statistics. FAOSTAT domain, Rome (available at http://faostat.fao.org/site/567/default.aspx)

FAO (2008c) Food Security Statistics. FAOSTAT domain. Rome (available at http://www.fao.org/faostat/foodsecurity/index_en.htm)

GoSL (2000) Initial national communication under the United Nations framework convention on climate change. Government of Sri Lanka, October, 2000

IPCC (2007) Climate Change 2007: synthesis report, summary for policymakers. An assessment of the intergovernmental panel on climate change

Karim CS (2007) Footprints of climate change now writ large on Bangladesh, Ministries of Agriculture, Fisheries & Livestock, and Environment & Forest, Advisor's speech given before Development Partners on 25 November 2007, Dhaka

Lobell DB, Burke MB, Tebaldi C, Mastrandea MD, Falcon WP, Naylor RL (2008) Prioritizing climate change adaptation needs for food security in 2030. Science 319:607–610

MoE-GIRP (2003) Pakistan's initial national communication on climate change, Government of Islamic Republic of Pakistan Ministry of Environment, Islamabad-Pakistan, November, 2003

MoEF-GoI (2004) India's initial national communication to the United Nations framework convention on climate change, Ministry of Environment and Forests, Government of India, 2004

MoEST (2004) Initial national communication to the conference of the parties of the United Nations convention on climate change, Ministry of Environment, Science and Technology, Government of Nepal, July 2004

NEC-RGB (2000) Initial national communication under the United Nations framework convention on climate change, National Environment Commission, The Royal Government of Bhutan, September 2000

United Nations (2005) The millennium development goals report 2005, United Nations, New York, 2005

World Bank (2008) Global monitoring report 2008. MDGs and the environment: agenda for inclusive and sustainable development, The international bank for reconstruction and development/The World Bank 1818 H Street NW, Washington, DC 20433

Chapter 15
Climate Change and Food Security in India

Nabansu Chattopadhyay

Contents

Abstract Indian agriculture is extremely vulnerable to weather and climate. In recent past there was substantial loss of crop in the country due to extreme weather and unusual weather conditions. Significant variations in the weather and climatic parameters, as projected in different Global Circulation Models Climate Change experiments, are expected to have substantial impact on crop production in the country in future. In the present paper elaborate discussion was made to understand the trends of different weather parameters during the last few decades over the Indian region and the linkage of weather with the Indian agriculture is also highlighted. Future projections of weather parameters from various Global Circulation Models Climate Change experiments over the country and its implication on Indian agriculture have also been documented. Under changing climate, food security of the country might come under threat. To cope up with climate change more effectively, integrated adaptation and mitigation options for a range of agroecosystems, so as to enable a favorable policy environment for the implementation of the framework, have been identified. Several adaptation measures and mitigation strategies to reduce vulnerability to climate change by enhancing adaptive capacity

N. Chattopadhyay (✉)
Agricultural Meteorology Division, India Meteorological Department, Pune, India
e-mail: nabansu_c@yahoo.co.in

R. Lal et al. (eds.), *Climate Change and Food Security in South Asia*,
DOI 10.1007/978-90-481-9516-9_15,

and increasing resilience and also to offset the negative effect of climate change on Indian agriculture have also been mentioned here.

Keywords Food production • Historic climate data • Seasonal temperature projections • Adaptation to climate change • Mitigation

15.1 Introduction

Climate change is a threat to mankind. Since the end of the nineteenth century the earth's average surface temperature has increased by 0.3–0.6°C. Over the last 40 years, the rise has been 0.2–0.3°C. Recent years have been the warmest since 1860, the year when regular instrumental records became available. The potential impact on the global climate of increasing atmospheric concentrations of carbon dioxide (CO_2) and so called green house gases is now well documented. Many studies of climate change have shown that the Earth's atmosphere is being modified by anthropogenic and biogenic emission of carbon dioxide and other radiatively active gases. According to Raper et al. (1996), the most likely rate of future global warming over the period 1900–2100 due to combined human modification of the atmosphere is estimated to be between 0.1°C and 0.3°C decade^{-1} several times the mean rate of warming over the past 100 years.

Food security refers to the availability of food and one's access to it. Food production in any given year is affected most directly by the values of the critical climate elements (temperature, radiation, precipitation, etc.) during the year. The stability of available food supplies is governed by the inter-annual variability of these elements. Access to food supplies in different regions of the world is determined by their share of the food production, the role of cereals in the diet of the people, and the various political and market forces that act upon the global food security system (Sinha et al. 1988). Climate change will affect agricultural yield directly because of alterations in temperature and rainfall, and indirectly through changes in soil quality, pests and diseases. Temperature rises conditions will become more favorable for pests such as grasshoppers to complete a number of reproduction cycles thereby increasing their population. Extreme weather conditions such as high temperature, heavy rainfall, floods, droughts, etc. will also affect crop production (Fig. 15.1).

Global and regional weather conditions are also expected to become more variable than at present, with increases in the frequency and severity of extreme events such as cyclones, floods, hailstorms, and droughts. Changes in temperature and precipitation associated with continued emissions of greenhouse gases will bring changes in land suitability and crop yields. The Fourth Assessment Report of Intergovernmental Panel on Climate Change (IPCC 2007) concluded that 'there is high confidence that recent regional changes in temperature have had discernible impacts on many physical and biological systems'. Climate change affects

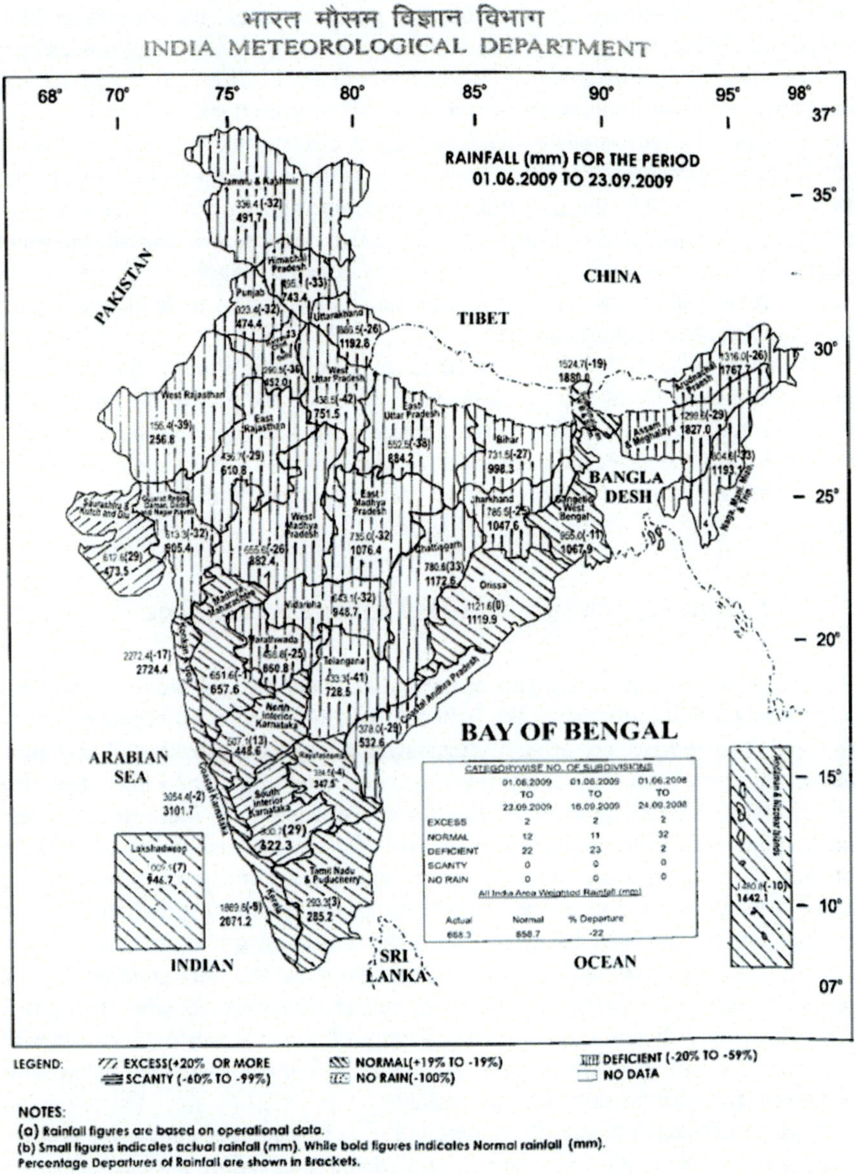

Fig. 15.1 Distribution of rainfall in different states in some selected weeks during Southwest Monsoon season 2009 (IMD 2009) of India

agriculture and food production in complex ways. It affects food production directly through changes in agro-ecological conditions and indirectly by affecting growth and distribution of incomes, and thus demand for agricultural produce. By bringing greater fluctuations in crop yields and local food supplies and higher risks of landslides and erosion damage, they can adversely affect the stability of food supplies and thus food security. The importance of the various dimensions and the overall impact of climate change on food security will differ across regions and over time and most importantly, will depend on the overall socio-economic status that a country has accomplished as the effects of climate change set in. The recent IPCC report also emphasizes that increases in daily temperature will increase the frequency of food poisoning, particularly in temperate regions. Extreme rainfall events can increase the risk of outbreaks of water-borne diseases particularly where traditional water management systems are insufficient to handle the new extremes namely food availability (i.e., production and trade), access to food, stability of food supplies, and food utilization.

The paper reviews the studies of the ranges of climatic changes that has taken place and also likely to take place in India, based on the projections of different Global Circulation Models, and also the possible impacts of these changes on agriculture. Finally the paper indicates the different adaptation and mitigation strategies to combat the negative effects of climate change on agriculture over the region.

15.2 Indian Agriculture and Its Linkage to Weather

Agriculture represents a core part of the Indian economy and provides food and livelihood activities to much of the Indian population. Factors influencing agriculture and food security are primarily increasing population, growing urbanization, decreasing crop land, continuing crop loss, declining crop production, declining bio-diversity, wide range of pests diseases and weeds etc. In India strong strides made in increasing the production in the past 50 years are mainly due to adoption of High Yielding Varieties (HYVs) of crops and other technological developments. Subsistence agriculture with small land holdings and skewed distribution of land is an important feature in Indian agriculture. Other aspect of agriculture in the country is wide variation in regional productivities. Majority of the food grain production in the country still depends on rainfed agriculture. Stagnation/decline in yields is due to the diversification, quality and quantity of water resources, incidences of pests and diseases etc. Agricultural production is frequently affected by extreme weather events such as droughts and cyclones.

Climate induced vulnerability of agriculture cause plateau in agriculture productivity in the country. Wide variation of rainfall and temperature not only affect the crops in *kharif* season, but the effects are also being manifested on *rabi* crops in winter season. As per the FAO and IPCC assessment, agricultural productivity in India would be reduced substantially in future [2020 (2.5–10%), 2050 (5–30%)]. While the magnitude of impact of climate variability and climate change varies greatly by region, climate change is expected to impact on agricultural productivity and shifting crop patterns. Study of the multi-decadal changes considering the data of past 50 years in break days during monsoon season show that number of break days are more (Table 15.1) in July as compared to August.

Table 15.1 Data of past 50 years show that the number of break days are greater in July as compared to August

	Number of break days during					
	July			August		
Period	01–10	11–20	21–31	1–10	11–20	21–31
1888–1917	46	49	53	43	84	26
1918–1947	14	36	21	55	54	25
1948–1977	22	44	64	21	33	41
1978–2003	23	32	39	6	14	37

Heat/cold wave, more variable rainfall, increased extremes weather events, erratic onset, advance and retrieval of monsoon, shift in active/break cycles, intensity and frequency of monsoon systems often affect the agricultural production in the country. Crops have to cope with increased variability of weather, extreme events, and changing climate patterns throughout the growing season. Agriculture may learn to adapt to climate change but climate variability needs to be combated. The frequency of occurrence of extreme climate conditions dictates the response of agriculture to climate variability/change.

15.3 Drought in India – 2009: Causes and Policy Implications

Although the southwest monsoon 2009 (June –September) was in time, the advancement of monsoon delayed in central and northern parts of the country resulting in delay in advancement of monsoon in other parts of the country. It was deficient by 39% as on 10th June, −54% as on 24th June which has resulted in drought like conditions in some parts of the country (IMD 2009). However, the rainfall has improved over the last week of July. At the end of July, the rainfall was normal by −19%. The cumulative rainfall from 1st June to 9th September is deficient by 20%. In the week ending August 26, the rainfall remained only 5% below normal. As many as 298 of 533 districts faced less than normal rainfall in the season as on 23rd September 2009 (Fig. 15.2). In fact, the rainfall in the month of September, 2009 has primarily helped the regions that witnessed scanty rainfall (more than 60% deficit). Deficit of 20% in southwest monsoon rainfall is expected to be worse than previous droughts in 2002, 1987 and 1979 when the shortfall was about 19% (Fig. 15.3).

Primarily the eleven states i.e. Assam, Jharkhand, Himachal Pradesh, Manipur, Nagaland, Uttar Pradesh, Bihar, Karnataka, Maharashtra, Madhya Pradesh and Rajasthan have faced drought-like situation. First drought declaration came on 25th June, from the north-eastern state of Manipur. Its neighbors, the states of Assam and Nagaland, followed on 14th and 15th July. The central Indian state of Jharkhand followed with its drought declaration on 20th July. Between 25th and 30th July the huge state of Uttar Pradesh declared drought in various districts. On 6th August its western neighbor, Himachal Pradesh, declared drought. And on 10th August its

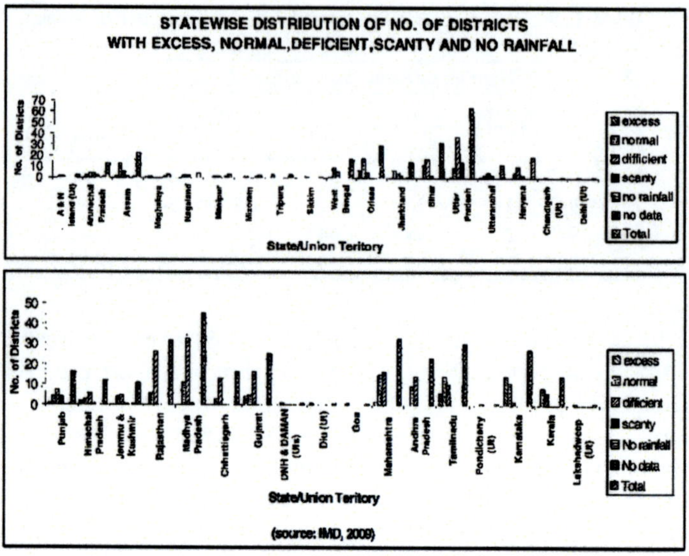

Fig. 15.2 Statewise distribution of no. of districts with excess, normal, deficient, scanty and no rainfall

eastern neighbor Bihar did so. India's monsoon rains have strengthened from end of July, easing a drought that has hit large swathes of the country, though some areas remain severely affected (Fig. 15.4).

Low rains in July and August ravaged India's rice, soybean, cane sugar and groundnut crops, and disrupted the flow of water into the main reservoirs that are vital for hydropower generation and winter irrigation. Scanty rainfall has delayed the sowing of important crops including paddy – in Uttar Pradesh. In the big western state of Maharashtra, the coverage of crop land under cereals, oilseeds has declined. Figures with the Agriculture Ministry reveal that there is a shortfall in the paddy sowing this year (Fig. 15.5).

In order to support the farmer under drought situation, policies are made at state and central level. It has been reported that the Ministry of Agriculture has provided a distress diesel subsidy – "to enable the farmers to provide supplementary irrigation through diesel pump sets in the drought and deficit rainfall affected areas to protect the standing crops; this will help in mitigating the adverse impact of drought/deficit rainfall conditions on food grain production". Widespread drought declarations prompt the central government (and affected state governments) to cut back on social sector spending. It is also reported that the instructions are given to the panchayats or village councils and the rural development department to step up relief measures in these areas and create more jobs under the National Rural Employment Guarantee Scheme. Agricultural experts and scientists in different states have decided to take up agricultural activities in each district in accordance with its climatic requirements in view of the failure of this year's monsoon and the threat of drought looming large over the State. Most of the State Governments

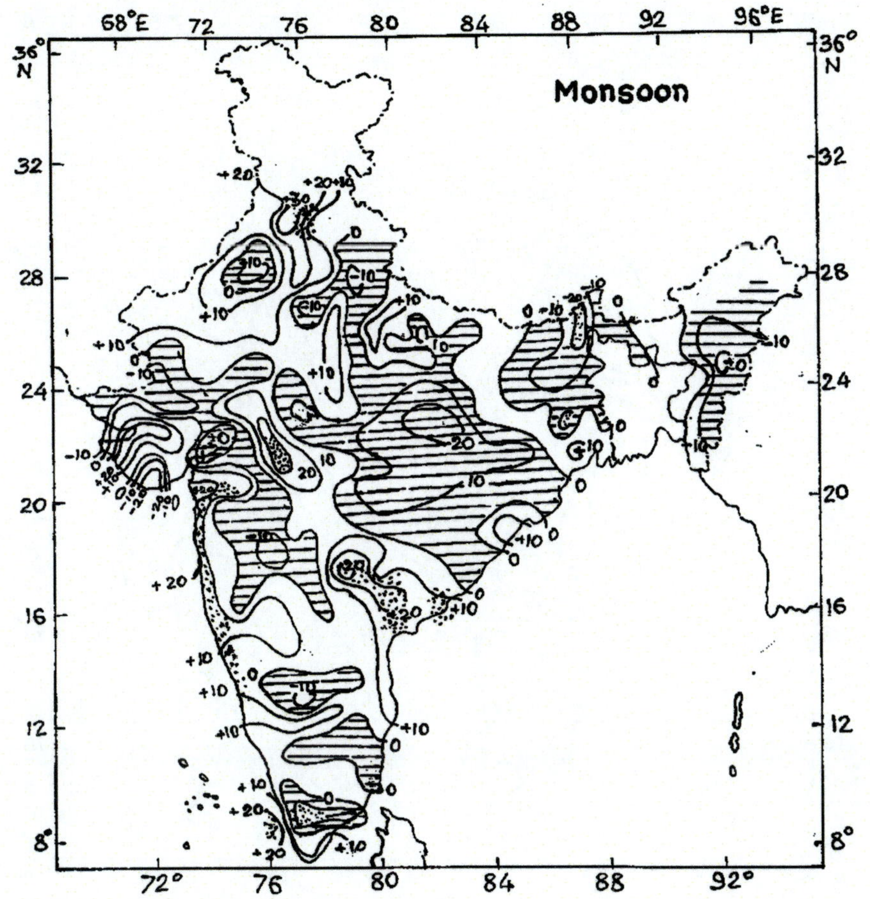

Fig. 15.3 Linear trend expressed as a percentage of normal per 100 years, 1971–1984 for monsoon rainfall. Hatched areas indicates negative trend and stipped areas indicates significance at 5% level (Rupa Kumar et al. 1992)

are fully conscious of the gravity of the drought situation in the State and taking all measures to provide relief to farmers (Fig. 15.6).

15.4 Signals of Climate Change Based on Historic Data

Global temperature has increased 0.15–0.3°C decade^{-1} for 1990 to 2005. Next 2 decades, warming of 0.2°C decade^{-1} is also projected. Indian scenario is not different. Climate change studies for India with respect to temperature and rainfall have already been made by a number of workers (Hingane et al. 1985; Thapliyal and Kulshrestha 1991; Srivastava et al. 1992; Rupakumar et al. 1994; Rupa Kumar

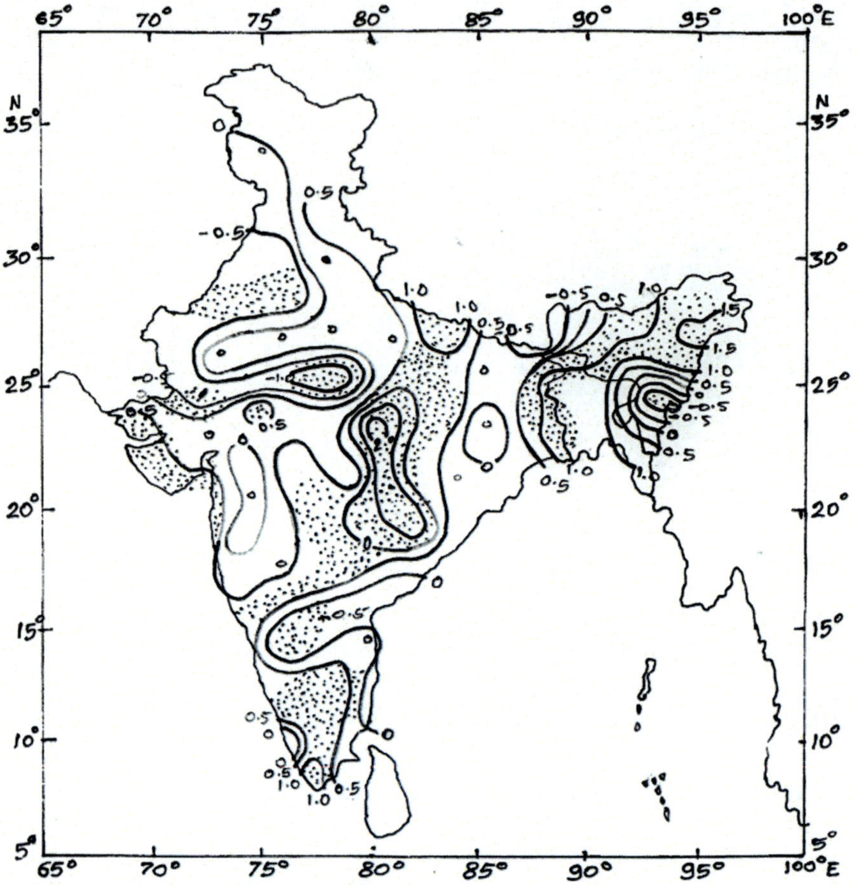

Fig. 15.4 Linear trend of mean annual temp (°C per 100 years). *Shaded areas* indicates significance at 95% level (Hingane et al. 1985)

et al. 2002; Govinda Rao et al. 1996; Kotawale and Rupa Kumar 2005; Chattopadhyay and Hulme 1997; Dash and Hunt 2007). Studies show that Indian temperatures are steadily increasing and mean annual temperature has increased by about 0.4°C in India during the past century. In general, it can be mentioned that an increasing trend in temperature has observed in southern and central India in the post monsoon season. The warming is generally been accompanied by increased diuranality. Srivastava et al. (1992) observed increasing trends of annual mean, maximum and minimum temperature south of 23°N and cooling trends north of 23°N. A number of other workers (Hingane et al. 1985; Rupa Kumar and Hingane 1988; Rupakumar et al. 1994; Govinda Rao 1993) concluded from their studies that an increasing trend in mean temperature in most parts of Indian subcontinent has been observed most strongly in post monsoon and winter seasons. Kotawale and Rupa Kumar (2005)) reported that while all India mean annual temperature has shown significant warming trend of 0.5°C10 year^{-1} during the period

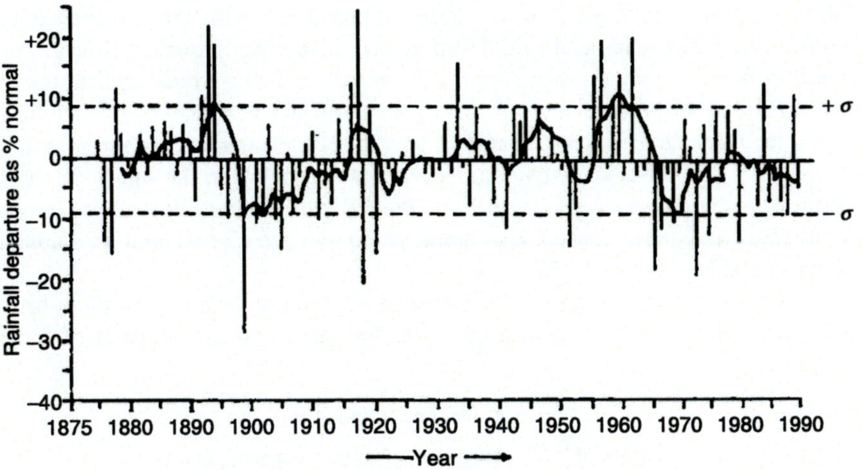

Fig. 15.5 Annual rainfall of India. Thick line-5 year running mean (Thapliyal and Kulshrestha 1991)

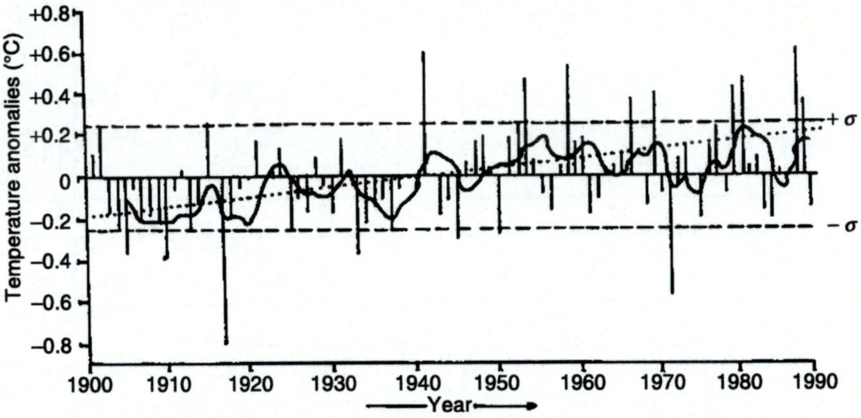

Fig. 15.6 Anomalies of average land surface maximum and minimum temperature (°C) relative to their respective means (1901–2003) values in northern and southern parts of India. The smoothed curve is obtained using 21 point binomial filter (Thapliyal and Kulshrestha 1991)

1901–2003, the recent period 1971–2003 has seen a relatively accelerated warming of 0.22°C10 year^{-1} which is largely due to the unprecedented warming during the last decade. As per the study of Dash and Hunt (2007) frequency of intense rainfall events has increased over past 53 years. Extreme rainfall events also increased over the west coast of India (based on analysis of 100 years of data; 1901–2000). Analysis of meteorological measurements in India indicates large difference in trends in the minimum temperature and cloud amounts between north and south India. There is also asymmetry in the increasing temperature trends between different seasons in a year. These observations along with the occurrence of extreme weather events lead to the importance of regional climate changes. The interplay between the aerosols, clouds and mesoscale flows around Indian mountains in global warming atmosphere may play a crucial role in the regional climate in future. According to Rupa Kumar et al. (2002) the summer monsoon rainfall during 1901–2000 has shown significant decreasing trends in the sub-division of northeast India, Orissa and East Madhya Pradesh while increasing rainfall trends in Konkan and Goa, coastal Karnataka along the west coast and in Punjab, Haryana and Delhi (Fig. 15.7).

In spite of general increase in temperature over recent decades, there has been decreased trend in Pan Evapotranspiration (Ep) in almost all the parts of India

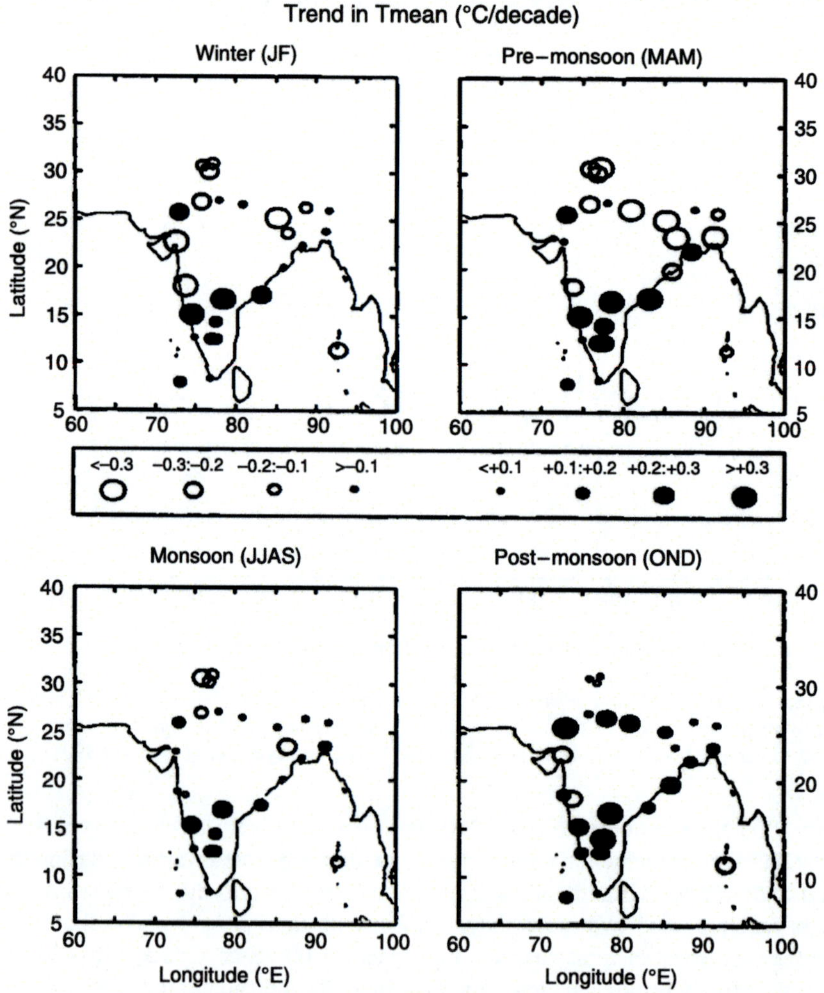

Fig. 15.7 Linear trend (°C decade^{-1}) in mean temperature for 1940–1990 for different seasons over India based on 27 stations (*dots*) Dot size is related to trend (Chattopadhyay and Hulme 1997)

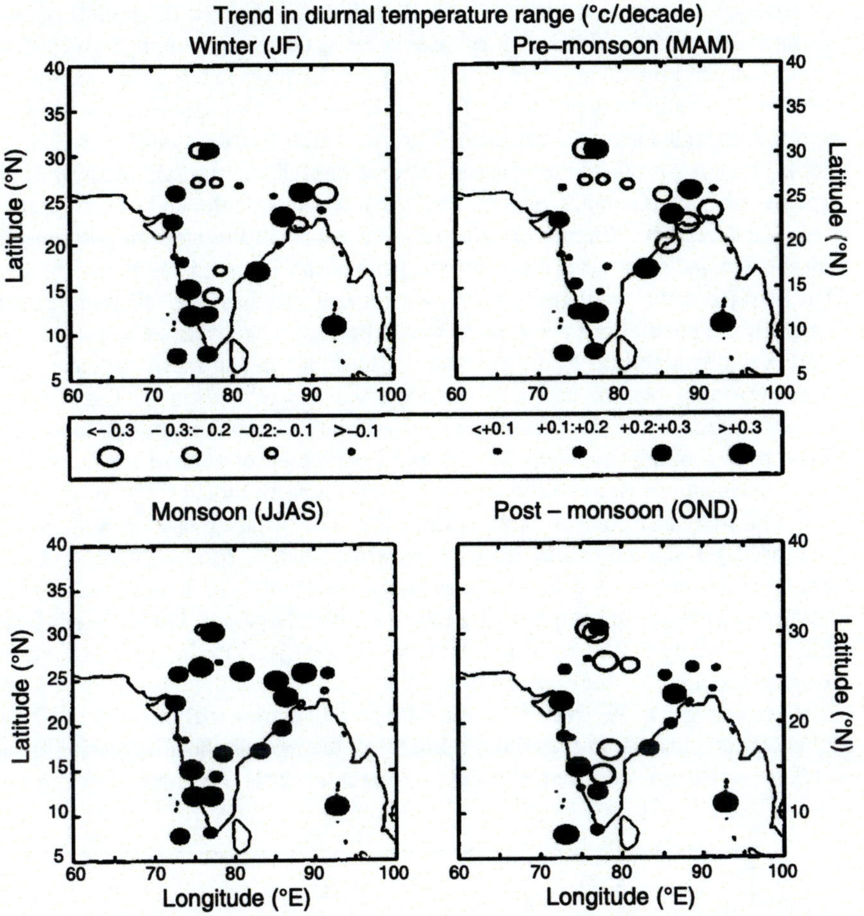

Fig. 15.8 Linear trend (°C decade^{-1}) in diurnal temperature range for 1940–1990 for different seasons over India based on 27 stations (*dots*). Dot size is related to trend (Chattopadhyay and Hulme 1997)

particularly significant in premonsoon and monsoon season (Chattopadhyay and Hulme 1997). Seasonal and spatial pattern of changes in Potential Evapotranspiration (PE) are similar to those for Ep, but magnitude of changes is less. In monsoon and post monsoon seasons PE has decreased over the whole country, whereas in the winter and pre-monsoon season the trend is less consistence (Fig. 15.8).

15.5 Projection of Climate Change in India

Using a number of Global Circulation Models different scenarios have been generated for the future climate change in India. It has been projected that average surface temperature will increase by 2–4°C during 2050s, marginal changes in monsoon rain in monsoon months (JJAS) and large changes of rainfall during non-monsoon months. Number of rainy days set to decrease by more than 15 days and intensity of rains to increase by 1–4 mm/day. Increase in frequency and intensity of cyclonic storms is projected. The hydrological cycle is predicted to be more intense, with higher annual average rainfall as well increased drought (Bhattacharya, 2006). There is a predicted increase in extreme rainfall and rainfall intensity in all three river basins (Ganga, Godavari and Krishna) towards the end of the century twenty-first century. Number of rainy days decreases in the western parts of the Ganga basin, but with increases over most parts of the Godavari and Krisha basins. Thus, surface water availability showed a general increase over all three basins though future populations projections would need to be considered to project per capita water availability. According to Lal (2001) an annual mean area-averaged surface warming over the Indian subcontinent will range between 3.5°C and 5.6°C over the region by 2080. These projections showed more warming in winter season over summer monsoon. The spatial distribution of surface warming suggests a mean annual rise in surface temperatures in north India by 3°C or more by 2050. The study also suggests that during winter the surface mean air temperature could rise by 3°C in north and central parts while it would rise by 2°C in southern parts by 2050. In case of rainfall, a marginal increase of 7–10% in annual rainfall is projected over the subcontinent by the year 2080. However, the study suggests a fall in rainfall by 5–25% in winter while it would be 10–15% increase in summer monsoon rainfall over the country (Fig. 15.9).

Future changes in PE over India and adjoining countries will increase in all the global climate models. In the winter seasons, all the models show increasing trend in PE over southern and central India up to around 25°N . In most of the model

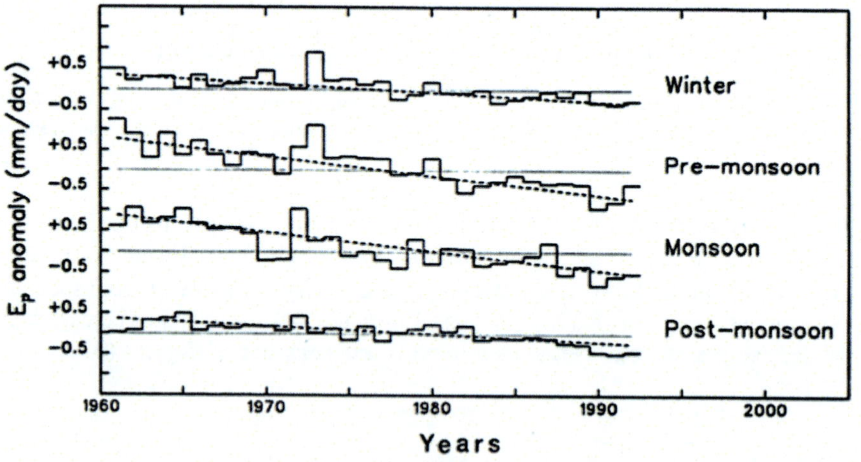

Fig. 15.9 Regionally averaged annual Ep anomalies (mm day^{-1}) for the period 1961–1992 with respect to the 1976–90 mean for different seasons over India. Number of stations is ten between 1961–1975 and 1991–1992 and 19 between 1976 and 1990. *Dashed lines* show best fit linear trend (Chattopadhyay and Hulme 1997)

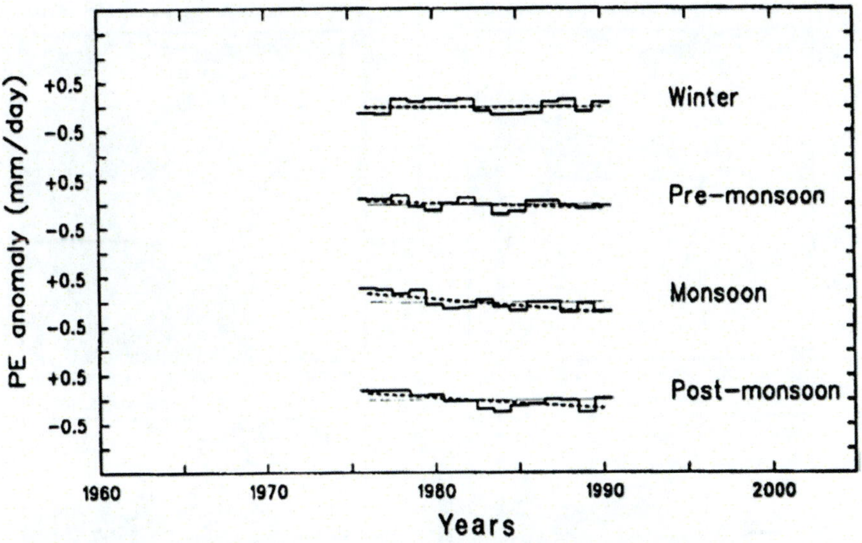

Fig. 15.10 Regionally averaged annual PE anomalies (mm day^{-1}) for the period 1976–1990 with respect to the 1976–1990 mean for different seasons over India. Number of stations averaged is ten. *Dashed lines* show best fit linear trend (Chattopadhyay and Hulme 1997)

experiments maximum winter increased in PE is of the order of 3–4% per degree Celsius of global warming and is seen in peninsular and most central parts of India. In the monsoon season maximum increased in PE over northwestern India. Inter relationship between PE and rainfall was assessed by mapping the number of GCM experiments which yield and increased the P/PE ratio for the monsoon season. A number of GCMs agree that P/PE ratio becomes more favorable over northeastern India and changes in this ratio are less favorable in post monsoon season and in the extreme south in the country (Chattopadhyay and Hulme 1997) (Fig. 15.10).

15.6 Impact of Climate Change on Agriculture

India, located in south central Asia, has great economic dependence on agriculture. A likely impact of climate change on agricultural productivity in India is causing great concern to the scientists and planners as it can hinder their attempts for achieving household food security. Any major changes in water budget and change in temperature have major consequence in hydrologic processes and agriculture and in turn economy of the country. The potential effect of climate change on agriculture in India would be the shift in the sowing time and length of growing season which would ultimately alter planting and harvesting dates of crops and varieties currently use in a particular areas. With warmer temperatures, evapotranspiration rates would rise, which would call for much greater efficiency of water use. Also weeds and insect pests could sift (Fig. 15.11).

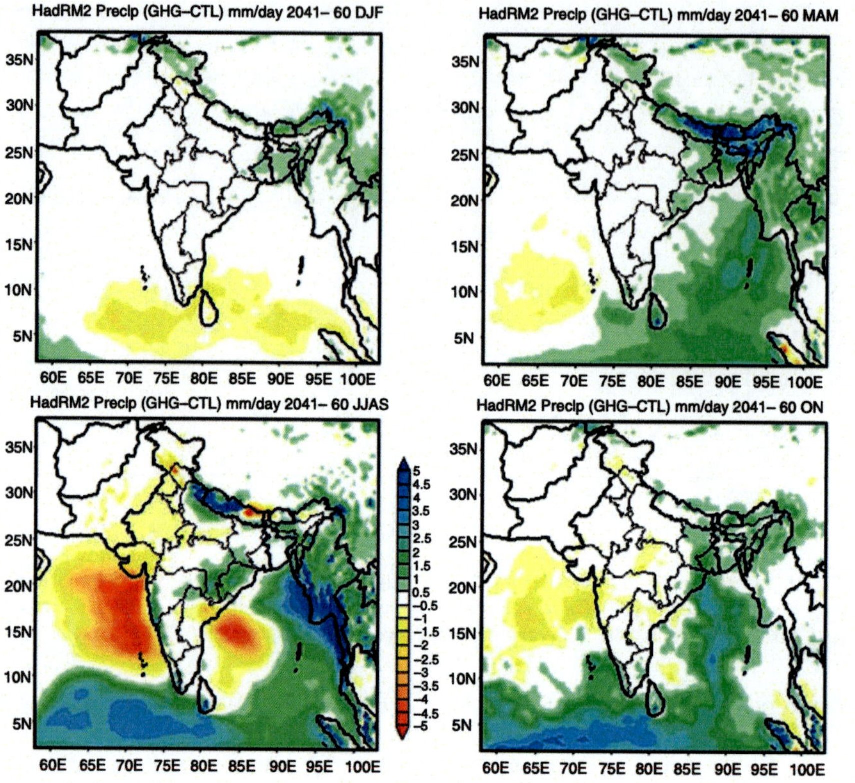

Fig. 15.11 Rainfall projections at different seasons (Bhattacharya 2006)

As per the findings in the AR4 of the IPCC, Working Groups I, II and III, there will be decrease up to 30% in south and central Asia by 2050. Sinha and Swaminathan (1991) showed that an increase of 2°C in temperature could decrease the rice yield by about 0.75 ton ha^{-1} in the high yield areas; and 0.5°C increase in winter temperature would reduce wheat yield by 0.45 ton ha^{-1}. Increased temperature is likely to reduce the wheat production particularly in north India. Morey and Sadaphal (1981) reported a decrease of wheat yield by 400 kg ha^{-1} for a unit increase of 1°C. Rao and Sinha (1994) showed that wheat yields could decrease between 28% and 68% without considering the CO_2 fertilization effects; and would range between +4% to −34% after considering CO_2 fertilization effects. Aggarwal and Sinha (1993) using WTGROWS model showed that 2°C temperature rise would decrease wheat yields in most places (Fig. 15.12). Saseendran et al. (2000) showed that for every one degree rise in temperature, the decline in rice yield would be about 6%. Decrease in yield of crops would be due to the temperature increase in different parts of India For example a 2°C increase in mean air temperature, rice yields could decrease by about 0.75 ton ha^{-1} in the high yield areas and by about 0.06 ton ha^{-1} in the low yield coastal regions. Major impacts of climate change will be on rain fed crops (other than rice and wheat), which account for nearly 60% of cropland area. In India, poorest farmers practice rainfed agriculture. The loss in farm-level net

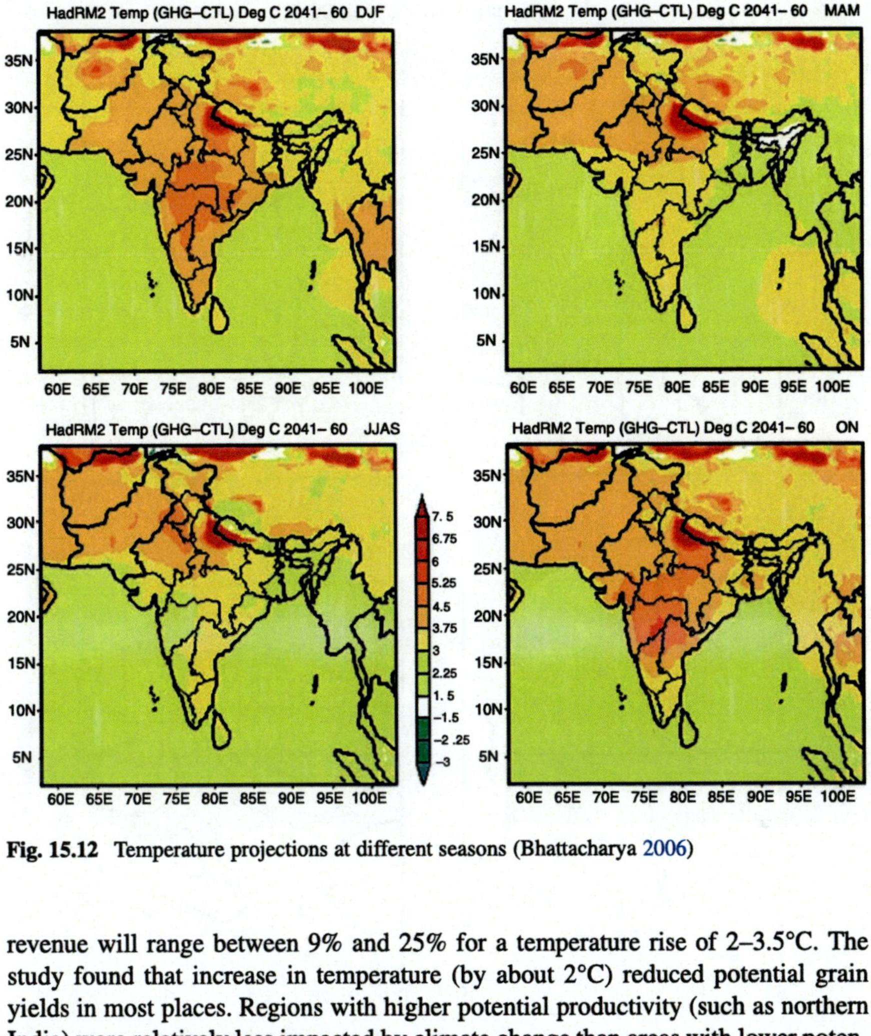

Fig. 15.12 Temperature projections at different seasons (Bhattacharya 2006)

revenue will range between 9% and 25% for a temperature rise of 2–3.5°C. The study found that increase in temperature (by about 2°C) reduced potential grain yields in most places. Regions with higher potential productivity (such as northern India) were relatively less impacted by climate change than areas with lower potential productivity (the reduction in yields was much smaller). Climate change is also predicted to lead to boundary changes in areas suitable for growing certain crops (Fig. 15.13). Reduction in yields as a result of climate change are predicted to be more pronounced for rain fed crops (as opposed to irrigated crops) and under limited water supply situations because there are no coping mechanisms for rainfall variability. The difference in yield is influenced by baseline climate. Overall temperature increases are predicted to reduce rice yields. An increase of 2–4°C is predicted to result in a reduction in yields. Eastern regions are predicted to be most impacted by increased temperatures and decreased radiation, resulting in relatively fewer grains and shorter grain filling durations. By contrast, potential reduction in yields due to increased temperatures in Northern India are predicted to be offset by higher radiation, lessening the impacts of climate change. Although additional CO_2 can benefit crops, this effect was nullified by an increase of temperature.

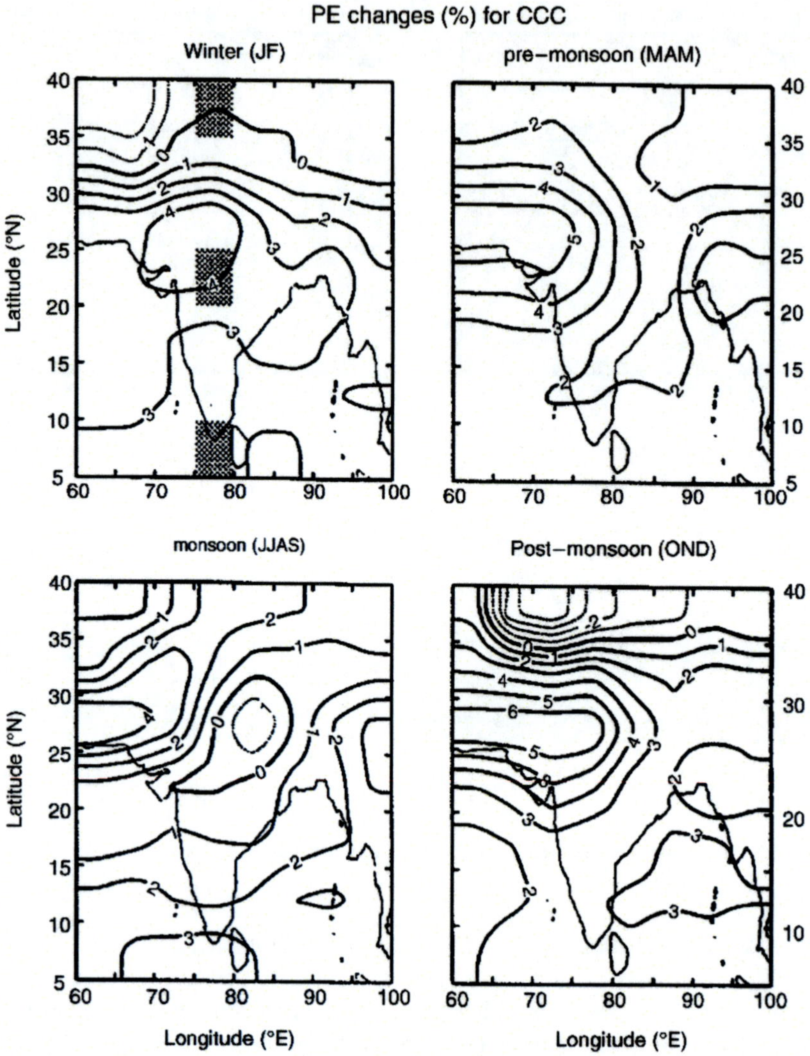

Fig. 15.13 Calculated change (%) in mean seasonal PE for 1°C of global warming for the CCC experiment (Chattopadhyay and Hulme 1997)

The Study clearly indicates that the yield of wheat, mustard, barley and chickpea show sign of stagnation or decrease following rise in temperature at all the four northern states. However, the extent of decrease was different for crops as well as there locations (Fig. 15.14).

Agriculture will be worst affected in the coastal regions of Gujarat and Maharashtra, where agriculturally fertile areas are vulnerable to inundation and salinisation. Standing crop in these regions is also more likely to be damaged due to cyclonic activity. In Rajasthan, 2°C rise in temperature was estimated to reduce production of pearl

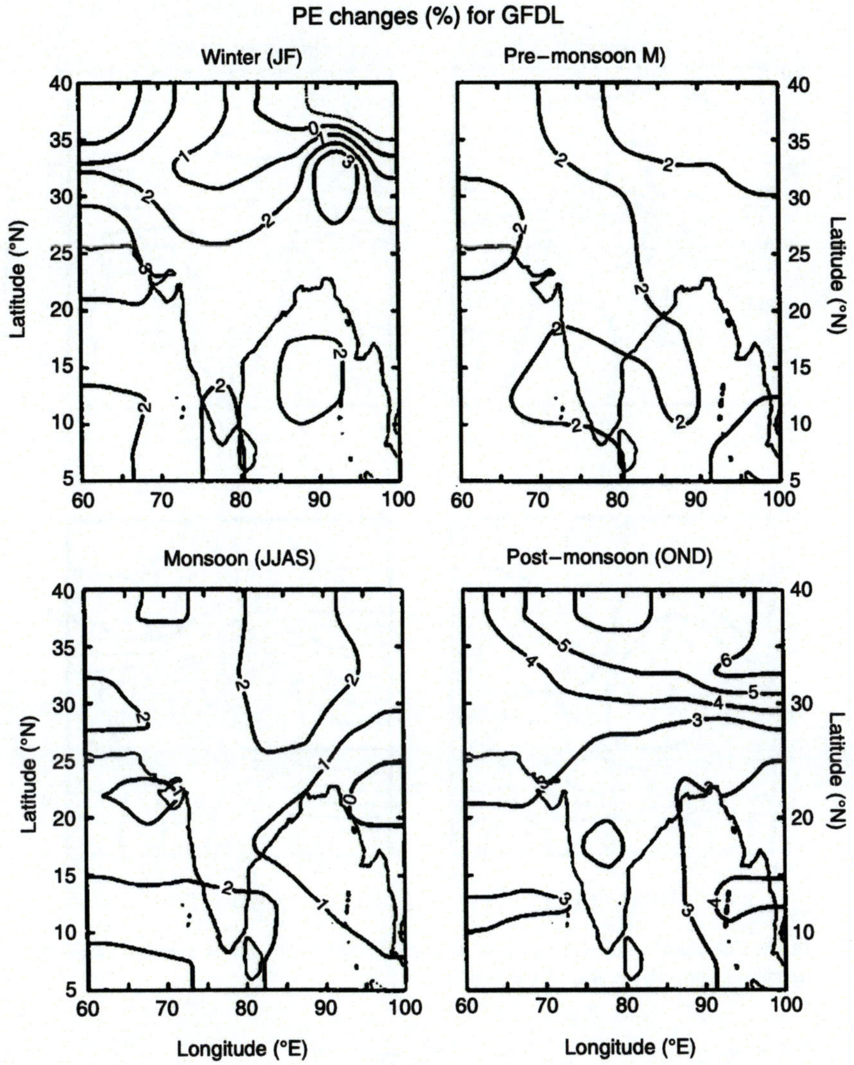

Fig. 15.14 Calculated change (%) in mean seasonal PE for 1°C of global warming for the GFDL experiment (Chattopadhyay and Hulme 1997)

millet by 10–15%. The state of Madhya Pradesh, where soybean is grown on 77% of all agricultural land, could dubiciously benefit from an increase in carbon dioxide in the atmosphere. According to some studies, soybean yield could go up to by as much as 50% if the concentration of carbon dioxide in the atmosphere doubles. However, if this increase in carbon dioxide is accompanied by an increase in temperature, as expected, then soybean yields could actually decrease. Changes in the soil, pests and weeds brought by climate change will also affect agriculture in India (Fig. 15.15).

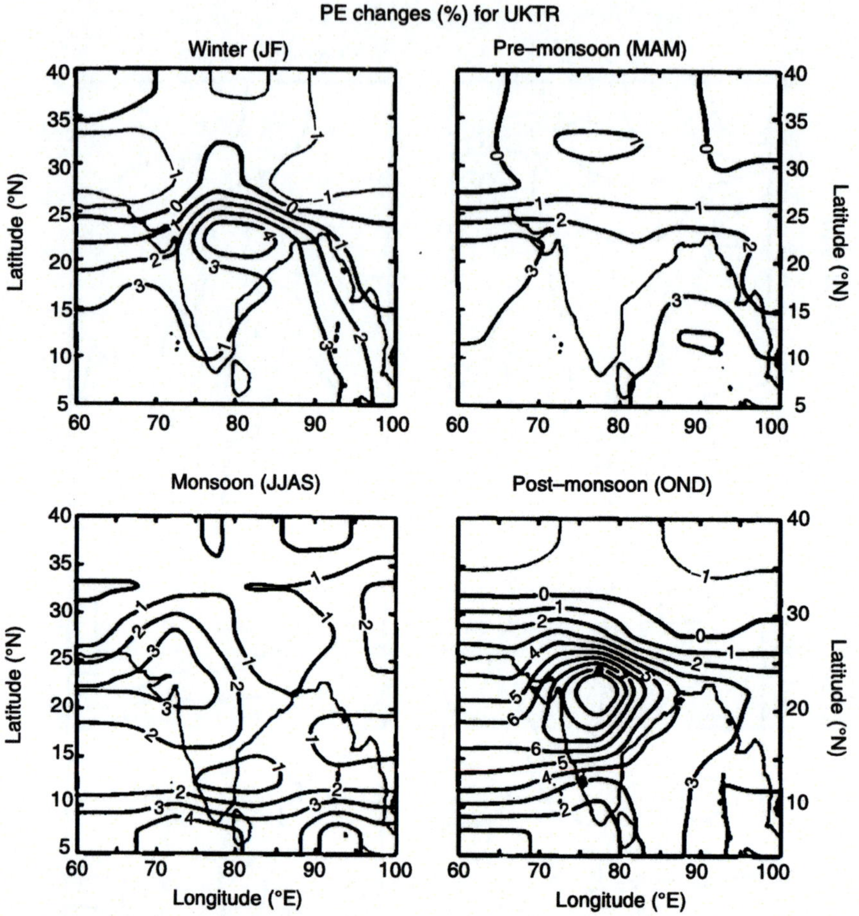

Fig. 15.15 Calculated change (%) in mean seasonal PE for 1°C of global warming for the UKTR experiment (Chattopadhyay and Hulme 1997)

15.7 Adaptation to Climate Change

Adaptation can be defined as any action that seeks to reduce the negative effects of climate change. Several adaptation measures are available to reduce vulnerability to climate change by enhancing adaptive capacity and increasing resilience. Adaptive actions may be either anticipatory or reactive in nature. An example of an anticipatory adjustment is the development of heat and drought-tolerant crop varieties. The levels of adaptation made by a region may have significant effects on how climate change will affect agriculture in that area. Farmers can adopt coping mechanisms that withstand climate variability through activities such as the use of drought-resistant or salt-resistant crop varieties, the more efficient use of water resources and improved pest management. Adjustments may include the introduction of late-maturing crop varieties, switching cropping sequences, sowing earlier, adjusting timing of field operations, conserving soil moisture through appropriate tillage methods and improving irrigation efficiency. Options such as switching crop varieties might not be expensive while others such as irrigation entail major investments. Changes in cultivation patterns can include the reduction of fertiliser use, better management of crop production, improvement of livestock diets and better management of their manure. In addition, governments have an important role to play in enforcing land-use policies which discourage slash and burn expansion, extensive livestock rearing and raising opportunities for rural employment. While changes in planting schedules or in crop varieties may be readily adopted, modifying the types of crops does not ensure equal level of food production or nutritional quality, nor can it guarantee equal profit for farmers. Expanded irrigation may lead to groundwater depletion, soil solemnization and water-logging. Increased demand for water by competing sectors may limit the viability of irrigation as an adaptation to climate change. This is particularly critical for many island atolls where the availability of water for drinking is a serious issue. Expansion of irrigation as a response to climate change will be difficult and costly, even under the best circumstances. Mounting societal pressures to reduce environmental damage from agriculture will likely foster an increase in protective regulatory policies that can further complicate the process of adaptation. Adaptation cannot be taken for granted. Improvements in agriculture have always depended on the investment made in agricultural research and infrastructure. It would help to identify, through research, the specific ways farmers adapt to present variations in climate. Success in adapting to possible future climate change will depend on a better definition of what changes will occur where and on prudent investments made in timely fashion, in adaptation strategies. To complicate matters, the availability for insurance generally and for agriculture in particular is a cause of major anxiety among many developing countries. Crop insurance sounds like a far cry in many developing countries but given the likely impacts on agriculture because of climate change, this possibility should no longer be regarded as a luxury. Indeed, insurance for crops should be an adaptation option which should be seriously pursued. The Agromet services provide a very special kind of inputs to the farmer as advisories that can make a tremendous difference to the agriculture production by taking in time actions against extreme weather events. This has a potential to change the face of India in terms of food security and poverty alleviation (Fig. 15.16).

15.8 Mitigation to Climate Change

Improved water and fertilizer management in rice paddies could reduce emission of GHGs. Improved management of livestock population and its diet could also assist in mitigation of GHGs. Approaches to increase soil carbon such as organic manures, minimal tillage and residue management should be encouraged. These have synergies wroth sustainable development as well. Use of nitrification inhibitors, such as neem coated urea and fertilizer placement practices need further consideration for GHGs mitigation. Improve the efficiency of energy use in agriculture by using better

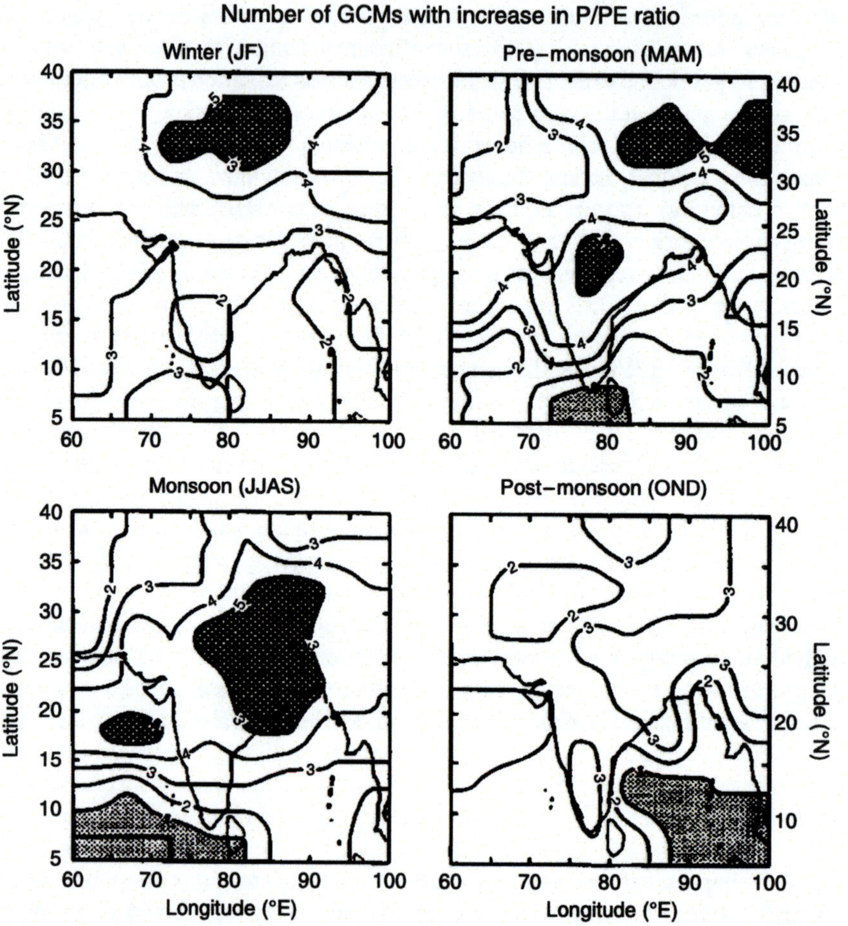

Fig. 15.16 Number of GCM experiments which yield an increase in P/PE ratio for each season. Maximum number is six. Areas of agreement in the sign of the change between all six GCMs are shaded (Chattopadhyay and Hulme 1997)

designs of machinery and by conservation practices. Changing land use by increasing area under bio fuels and agro-forestry could also mitigate GHG emissions. This, however, may have trade-offs with goal of increasing food production.

15.9 Conclusions

Climate change is a global problem and India will also feel the heat. The various studies conducted in the country have shown that the surface air temperatures in India are going up at the rate of 0.4°C per 100 years, particularly during the post-monsoon and winter season. Using models, they predict that mean winter temperatures will increase by as much as 3.2°C in the 2050s and 4.5°C by 2080s, due to Greenhouse gases. Summer temperatures will increase by 2.2°C in the 2050s and 3.2°C in the 2080s.Extreme temperatures and heat spells have already become common over Northern India, often causing loss of human life

Climate change, it appears is now underway. Nearly 700 million rural people in India directly depend on climate-sensitive sector (agriculture, forests, and fisheries) and natural resources for their subsistence and livelihood. Under changing climate, food security of the country might come under threat. In addition, the adaptive capacity of dry land farmers, forest and coastal communities is low. In increased in weather extremes like torrential rain, heat wave, cold wave, flood besides year to year variability in rainfall affects agricultural productivity significantly and lead to stagnation/ decline in production across various agro-climatic zones.

In developing countries like India, climate change could represent an additional stress on ecological and socioeconomic systems that are already facing tremendous pressures due to rapid urbanization, industrialization and economic development. With its huge and growing population, a 7,500-km long densely populated and low-lying coastline, and an economy that is closely tied to its natural resource base, India is considerably vulnerable to the impacts of climate change. Increased temperatures will impact agricultural production. Higher temperatures reduce the total duration of a crop cycle by inducing early flowering, thus shortening the 'grain fill' period. The shorter the crop cycle, the lower the yield per unit area. There is growing need to qualify the effects of rising temperature on yield of crops in different agroecologies and agri-production environments.

There is thus an urgent need to address the climate change and variability issues holistically. Climate change, energy security and food security are interlinked, and require an integrated approach. Some specific options have already been identified, tested and documented for climate change mitigation and adaptation for agriculture sector, such as sustainable land and forest management; changing varieties; more efficient water use; altering the timing or location of cropping activities; improving the effectiveness of pest, disease and weed management practices and making better use of seasonal climate forecasts to reduce production risks. If these options are widely adopted, they could have substantial potential to offset negative impacts from climate change and take advantage of positive impacts. To cope with climate change more effectively, it is necessary to identify integrated adaptation and mitigation options for a range of agroecosystems so as to enable a favorable policy environment for the implementation of the framework. The policy implications are wide-reaching, as changes in agriculture could affect food security, trade policy, livelihood activities and water conservation issues, impacting large portions of the population

References

Aggarwal PK, Sinha SK (1993) Effect of probable increase in carbon dioxide and temperature on productivity of wheat in India. J Agric Meteorol 48:811–814

Bhattacharya S (2006) Climate Change and India. International Workshop on Future International Climate Policy, August 9, 2006, University of Sao Paulo, Brazil

Chattopadhyay N, Hulme M (1997) Evaporation and potential evapotranspiration in India under conditions of recent and future climate change. Agric Forest Meteorol 87:55–73

IPCC (Intergovernmental Panel on Climate Change Working Group II) Climate Change 2007, Impacts, Adaptation and Vulnerability IPCC Working Group II

Dash SK, Hunt JC (2007) Variability of climate change in India. Curr Sci 93(6):783–782

Govinda Rao P (1993) Climate changes and trends over major river basins in India. Clim Res 2:215–223

Govinda Rao P, Kelly PM, Hulme M (1996) Climate change, greenhouse gas emissions, future climate and response strategies: the implications for India. Theor Appl Climatol 55:41–64

Hingane LS, Rupa Kumar K, Murty R (1985) Long-term trends of surface air temperature in India. J Climatol 5:521–528

IMD (2009) Press release on 25th September 2009, India Meteorological Department, Ministry of Earth Sciences, Government of India

Kotawale DR, Rupa Kumar K (2005) On the recent changes in surface temperature trend over India. Geophys Res Lett 32:L18714. doi:10.1029/2005GL023528

Lal M (2001) Future climate change: implications for Indian summer monsoon and its variability. Curr Sci 81(9):1205

Morey DK, Sadaphal MN (1981) Effect of weather elements on yield of wheat at Delhi. Punjabrao Krishi Vidyapeeth Res J 1:81–83

Rao GD, Sinha SK (1994) Impacts of climate change on simulated wheat production in India. In: Rosenzweig C, Iglesias I (eds) Implications of climate change for international agriculture: crop modeling study. EPA, USA, pp 1–10

Raper SCB, Warrick RA, Wigley TML (1996) Global sea level rise: past and future. In: Milliman JD, Haq BU (eds) Sea-level rise and coastal subsidence: causes, consequences and strategies. Kluwer, Dordrecht, Germany, p 369

Rupa Kumar K, Hingane LS (1988) Long term variations of surface air temperature at major industrial cities of India. Clim Change 13:287–307

Rupa Kumar K, Pant GB, Parthasarathy B, Sontakke NA (1992) Spatial and sub seasonal pattern of the long-term trends of Indian summer monsoon rainfall. Int J Climatol 12:257–268

Rupa Kumar K, Kriskna Kumar K, Pant GB, Srinivasan G (2002) Climate change – The Indian scenario. In: Backgroud paper prepared by FICCI, International Conference on Science and Technology Capacity Building for Climate Change, October 20–22, New Delhi, 5–17

Rupakumar K, Kriskna Kumar K, Pant GB (1994) Diurnal assymetry of surface air temperature trends ver India. Geophys Res Lett 15:677–680

Saseendran SA, Singh KK, Rathore LS, Singh SV, Sinha SK (2000) Effect of climate change on rice production in tropical humid climate of Kerla, India. Climate Change 44:495–514

Sinha SK, Swaminathan MS (1991) Deforestation, climate change and sustainable nutrition security: A case study of India. Climate Change 19:201–209

Sinha SX, Rao NH, Swaminathan MS (1988) Food security in the changing global climate. In the conference proceedings for the changing atmosphere: implications for Global Security, 27–30 June 1988, in Toronto, Canada, 167–192. WMO-No. 170. Geneva: World Meteorological Organization

Srivastava HN, Dewan BN, Dikshit SK, Prakash Rao GS, Singh SS, Rao KR (1992) Decadal trends in climate over India. Mausam 43:7–20

Thapliyal V, Kulshrestha SM (1991) Climate change and trends over India. Mausam 42:333–338

Part VI
Mitigation and Adaptation Options

Chapter 16
Global Climate Change and Food Security in South Asia: An Adaptation and Mitigation Framework

P.K. Aggarwal and Mannava V.K. Sivakumar

Contents

P.K. Aggarwal
Indian Agricultural Research Institute, New Delhi, 110012 India

M.V.K. Sivakumar (✉)
World Meteorological Organization, 7bis Avenue de la Paix, 1211, Geneva, 2, Switzerland
e-mail: msivakumar@wmo.int

R. Lal et al. (eds.), *Climate Change and Food Security in South Asia*,
DOI 10.1007/978-90-481-9516-9_16, © Springer Science+Business Media B.V. 2011

Abstract South Asia is home to nearly 22% of the world's population, including 40% of the world's poor. Agriculture plays a critical role in terms of employment and livelihood security for a large majority of people in all countries of the region. The region is prone to climatic extremes, which regularly impact agricultural production and farmers' livelihood. Himalayan glaciers, a major source of water for the rivers in the Indo-Gangetic plains, are projected to significantly recede in future that could affect food and livelihood security of millions of people in Pakistan, Nepal, Bhutan, India and Bangladesh. Climate change is further projected to cause a 10–40% loss in crop production in the region by the end of the century. The increased climatic variability in future would further increase production variability. Producing enough food for the increasing population in a background of reducing resources in a changing climate scenario, while minimizing environmental degradation is a challenging task. Simple adaptation strategies such as changes in planting dates and varieties could help in reducing impacts of climate change to a limited extent. A Regional Adaptation and Mitigation Framework for South Asia is proposed that could assist the region in increasing its adaptive capacity to climate change. This includes assisting farmers in coping with current climatic risks, intensifying food production systems, improving land, water and forests management, enabling policies and regional cooperation, and strengthening research in critical areas. South Asian agriculture is a significant source of greenhouse gases (GHG) emissions, primarily due to methane emission from rice paddies, enteric fermentation in ruminant animals, and nitrous oxides from application of manures and fertilizers to soils. While a considerable fraction of this is inevitable, some reduction in emissions could be obtained by midseason drainage or alternate drying in irrigated rice, increasing nitrogen use efficiency and soil carbon, and improvements in livestock diet. Clear inclusion of agricultural GHG mitigation options in future international agreements would lead to improved soil fertility, higher income for the farmers, and food security.

Keywords Climate change • South Asia • Food security • Adaptation • Mitigation

Abbreviations

CDM	Clean development mechanisms
COP	Conference of parties
GDP	Gross domestic product

GHG	Greenhouse gas
MDG	Millennium development goals
SAARC	South Asian Association in Regional Cooperation
UNFCCC	U.N. Framework Convention on Climate Change

16.1 Introduction

A large human population, continuing high rate of population growth, and poverty characterize South Asia. The region has nearly 1.4 billion people (22% of the world's population), including 40% of the world's poor. Agriculture continues to play a critical role in terms of employment and livelihood security in all countries of the region, although its share has now considerably declined to 20–25% of GDP. More than half of the population of the region is still engaged in agriculture. Most farms in the region are small (less than 1 ha). In addition to the marginal and small farmers there are a large number of landless households. There are more than 300 million hungry and malnourished people in the region.

South Asia is very prone to climatic extremes. Droughts and floods are common in the whole region whereas heat waves and cyclones occur regularly in some parts. Degradation of natural resources due to intensive human activities is also common in the region. There is evidence of gradual deterioration in natural resources, particularly in areas that benefited from the Green Revolution technologies. The introduction of canal irrigation in India, for example, has resulted in almost seven million ha of cultivated land becoming effected by soil salinity and waterlogging (Joshi and Tyagi 1994) while the rapid increase in the number of tubewells during the last three decades in north-western India, eastern Pakistan and other regions has resulted in overexploitation of groundwater, leading to a rapid fall in water-tables. There is now growing concern about the decline in soil fertility, changes in water-table depth, deterioration in the quality of irrigation water, and rising salinity (Sinha et al. 1998).

Despite impressive development of irrigation potential, food production and consequently economy of the region is still considerably dependent on monsoon. There have been several droughts and floods in all countries of the region, which have affected their food security.

During last few decades, however, the region as a whole has shown progress on several fronts at the macro level. The population growth rate has shown some decline, although there are significant differences among countries. The region has generally achieved food self-sufficiency and the per capita availability of food is rising. In general, there has also been an increase in per capita income, which has led to increasing demand of superior grains, and animal and horticultural products (Bhalla et al. 1999).

South Asian population is predicted to increase by almost 700 million people in the next 40 years. This, and the rising income of people is expected to result in a large demand for food. It is estimated that the food grain requirement by 2020 in the region will be almost 50% more than in 2000 (Paroda and Kumar 2000). The additional quantities will have to be produced from the same land resource, or less, due to the increase in competition for land and other resources by non-agricultural sectors. The situation may be further complicated by global climatic change that directly affects agriculture and hence food supply.

The global mean annual surface air temperature increase by the end of this century is likely to be in the range of 1.8–4.0°C (IPCC 2007a). For South Asia, the projections for increases in mean annual temperature are 0.5–1.2°C by 2020, 0.88–3.16°C by 2050 and 1.56–5.44°C by 2080 depending upon the scenario of future development (Table 16.1; IPCC 2007a). Overall, the temperature increases are likely to be much higher in the *rabi* (winter) season than in the *kharif* (monsoon) season. It is very likely that hot extremes, heat waves, and heavy precipitation events will become more frequent. The projected sea level rise by the end of this century is likely to be 0.18–0.59 m. Rupa Kumar et al. (2006) have shown similar results for the region using a regional climate model.

The absolute amount of precipitation is likely to increase in South Asia in future in all months except in the period December–February when this is likely to decrease (Table 16.1). This increase may, however, be associated by heavier precipitation events and fewer rainy days leading to increased frequency of floods and droughts in the region.

The IPCC (2007b) showed that densely populated megadeltas of the region such as those in India and Bangladesh, and islands and coastal areas such as those in Sri Lanka and Maldives, India and Bangladesh are more vulnerable to global climate change. Enhanced glacier melt in the Himalayas is projected to reduce availability of water resources in agriculturally important Indo-Gangetic plains of Pakistan, India, Nepal and Bangladesh.

The Thirteenth Conference of Parties to the UNFCCC Conference in Bali (COP-13) in 2007 saw some progress in cooperative action by all countries for meeting these challenges. Bali Action Plan calls for enhanced action on mitigation and adaptation by nationally appropriate commitments and actions, technology development and transfer, and provision of financial investments and resources to support these.

FAO organized in June 2008 a conference entitled 'World Food Security: the Challenges of Climate Change and Bioenergy'. It urged governments to explore how farmers, smallholders in particular, could adapt and assist in mitigation through the global financial mechanisms and investment flows, and technology development and transfer (FAO 2008).

Producing enough food to meet the increasing demand against the background of reducing resources in a changing climate scenario, while also minimizing further environmental degradation in South Asia, will be a challenging task. Addressing climate change is central for the region's future food security and attainment of Millennium Development Goals (MDGs), especially on poverty alleviation. The region will need to implement strategies, linked with its development plans, to enhance its adaptive capacity, mitigate emissions of greenhouse gases (GHG) and sequester more carbon.

Table 16.1 Projected changes in surface air temperature and precipitation for South Asia under SRES A1FI (highest future emission trajectory) and B1 (lowest future emission trajectory) pathways for 2020s, 2050s and 2080s (From IPCC 2007a)

	2020				2050				2080			
	Temperature (°C)		Precipitation (%)		Temperature (°C)		Precipitation (°C)		Temperature (°C)		Precipitation (%)	
Season	A1FI	B1	A1FI	B1	A1FI	B1	A1FI	B1	A1FI	B1	A1FI	B1
Dec.–Feb.	1.17	1.11	–3	4	3.16	1.97	0	0	5.44	2.93	–16	–6
Mar.–May	1.18	1.07	7	8	2.97	1.81	26	24	5.22	2.71	31	20
June–Aug.	0.54	0.55	5	7	1.71	0.88	13	11	3.14	1.56	26	15
Sept.–Nov.	0.78	0.83	1	3	2.41	1.49	8	6	4.19	2.17	26	10

The Fifteenth South Asian Association for Regional Cooperation (SAARC) Summit held in Colombo on 2 and 3 August 2008 reiterated the need for increased regional cooperation in assessing and managing risks and impacts of climate change, and to develop a people-centered short to medium term strategy to ensure region-wide food security (http://www.saarc-sec.org/data/summit15/colombostatementonfoodsecurity.htm).

Some Governments of the region have developed their *National Action Plans on Climate Change* detailing strategies they will follow for increasing their adaptive capacity and for mitigation (Government of India 2008; MOEF 2008).

In the light of these developments, a Regional Adaptation and Mitigation Framework for South Asia is being proposed in this paper which articulates the vision on technological and policy options for enhancing adaptive capacity, mitigating GHG emissions, and sequestering carbon in agricultural systems of South Asia, while keeping focus on meeting increased food production and other developmental targets.

16.2 Impacts of Climate Change on South Asian Agriculture

IPCC (2007b) indicates that crop yields could decrease up to 30% in South Asia by the end of the century even if the direct positive physiological effects of CO_2 are taken into account. Several global studies indicate a probability of 10–40% loss in crop production in India with increases in temperature by 2080–2100 (Rosenzweig and Parry 1994; Fischer et al. 2002; Parry et al. 2004; Cline 2007; Stern 2007). These long-time horizon estimates generally assume the business as usual scenario, no new technology development, and no or limited adaptation by all stakeholders. Recent studies in India and Pakistan suggest a 5–7% decline in wheat for every degree Celsius increase in temperature provided irrigation remains available in future at today's levels. In Sri Lanka, initial estimates indicate a loss of 6% rice output with an increase of 0.5°C temperature (Government of Sri Lanka 2000).

The projected negative impacts of increased temperatures on cereal yields in the low latitude regions (see Box 16.1) imply that food security of most developing countries, including the South Asian countries, is likely to become vulnerable in near future. At the same time, differential impacts of climate change on food production in different parts of the world is likely to have consequences on international food prices and trade (Easterling et al. 2007).

Analysis of the historical trends in yields of crops in the Indo-Gangetic plains using regional statistics, long-term fertility experiments, other conventional field experiments and crop simulation models has shown that rice yields during last three decades are showing a declining trend and this may be partly related to the gradual change in weather conditions during last two decades (Aggarwal et al. 2000a; Pathak et al. 2003).

All countries of South Asia frequently experience natural climatic disasters. Bangladesh, parts of India and Sri Lanka are low-lying coastal areas prone to cyclones, sea level rise, and floods. Bhutan, Nepal, India and Pakistan have several mountain ecosystems that are vulnerable to glacier melt. Semi-arid and arid areas of India, Afghanistan, Pakistan and Bangladesh frequently experience heat and drought stress. Such climatic extremes are known to negatively impact agricultural production, and farmers' livelihood. The projected increase in these events could result in greater instability in food production and threaten livelihood security of farmers. Increased production variability could be perhaps the most significant impact of global climate change on South Asian economies.

Rise in sea level can significantly affect the livelihood of coastal communities due to salinization of cultivated areas and water bodies, and associated change in land use.

Cold waves cause significant losses to crops such as mustard, mango, guava, papaya, brinjal, tomato, and potato in northern India, Nepal, Afghanistan and Pakistan. There are indications that such cold waves and frost events could decrease in future due to global warming and hence yield losses in these crops associated with frost damage are likely to decrease.

The IPCC has indicated a significant increase in runoff in many parts of the world including South Asia due to global warming (IPCC 2007b). This, however, may not be very beneficial because the increase is largely in the wet season and the extra water may not be available in the dry season, when it is most needed, unless storage infrastructure is vastly expanded. This extra water in the wet season, on the other hand, may lead to increase in frequency and duration of floods. The increased melting and recession of glaciers associated with global climate change could further change the runoff scenario. In recent decades, Himalayan glaciers have receded between 2.6 and 28 m/year (Kulkarni and Bahuguna 2002). These glaciers are a major source of water for the rivers such as the Indus, Ganga and Brahmaputra in the Indo-Gangetic plains, which are crucial to millions of people in Pakistan, Nepal, Bhutan, India and Bangladesh. Such increases in glacier melt in Himalayas could affect availability of water for irrigation especially in the Indo-Gangetic plains, and thus affect food and livelihood security of millions.

The nutritional quality of cereals may be moderately affected by climatic changes, which in turn, may have consequences for nutritional security of several developing countries where cereals are the primary diet. Research has indeed shown that the increasing CO_2 concentrations and temperature lead to a decline in grain protein content in cereals (Hocking and Meyer 1991; Ziska et al. 1997).

Crop–pest interactions will change significantly with climate change leading to impact on pest distribution and crop losses (Easterling et al. 2007). Diseases and insect populations are strongly dependent upon the temperature and humidity. Any increase in them, depending upon their base value, can significantly alter their population, which ultimately results in yield loss. With small changes, the virulence of different pests changes. The swarms of locusts produced in the Middle East usually fly eastward into Pakistan and India during summer season and they lay eggs during monsoon period. Changes in rainfall, temperature and wind speed pattern may influence the migratory behavior of locusts and other similar pests, thus threatening food and livelihood security.

Heat stress associated with global warming directly impacts forage quality, ingestion of food and feed, declines in physical activity, and ultimately reduces dairy milk yield (St-Pierre et al. 2003). Increases in air temperature and/or humidity have the potential to affect the reproductive behavior of domestic animals. Global warming would further increase water, shelter, and energy requirement of livestock for meeting the projected increased milk demands.

Increasing sea and river water temperature is likely to affect fish breeding, migration, and harvests (Sharp 2003; Easterling et al. 2007). A rise in temperature as small as 1°C could have important and rapid effects on the mortality of fish and their geographical distributions. Corals are affected by warm surface waters leading to bleaching due to losses of associated symbiotic algae. Heat stress in 1998 and 2002 caused considerable bleaching of corals in the Indian Ocean (Wilkinson et al. 1999; Kumaraguru et al. 2002). Such events are likely to become more frequent in future due to global climate change.

Box 16.1 IPCC's key observations on vulnerability of agriculture to global climate change (IPCC 2007b)

In mid- to high-latitude regions, moderate to medium local increases in temperature (1–3°C), along with associated carbon dioxide (CO_2) increase and rainfall changes can have small beneficial impacts on crop yields. In low-latitude regions, even moderate temperature increases (1–2°C) are likely to have negative yield impacts for major cereals. Further warming has increasingly negative impacts in all regions (medium to low confidence).

About 2.5–10% decrease in crop yield is projected for parts of Asia in 2020s and 5–30% decrease in 2050s compared with 1990 levels without CO_2 effects (medium confidence).

Projected changes in the frequency and severity of extreme climate events have significant consequences for food production, and food insecurity, in addition to impacts of projected mean climate (high confidence).

Smallholder and subsistence farmers, pastoralists and artisanal fisherfolk will suffer complex, localised impacts of climate change (high confidence).

Food trade is projected to increase in response to climate change, with increased dependence on food imports for most developing countries (medium to low confidence).

The marginal increase in the number of people at risk of hunger due to climate change must be viewed within the overall large reductions due to socio-economic development (medium confidence).

Projected changes in the frequency and severity of extreme climate events have significant consequences for food production, and food insecurity, in addition to impacts of projected mean climate (high confidence).

Simulations suggest rising relative benefits of adaptation with low to moderate warming (medium confidence), although adaptation stresses water

(continued)

Box 16.1 (continued)

and environmental resources as warming increases (low confidence). On average, in cereal cropping systems worldwide, adaptations such as changing varieties and planting times enable avoidance of a 10–15% reduction in yield corresponding to 1–2°C local temperature increase. Adaptive capacity in low latitudes is exceeded at 3°C local temperature increase.

16.3 A Regional Framework for Adapting Agriculture to Climate Change

In view of the projected large-scale adverse effects of global warming on food production in South Asia, we need to analyze the possible options that could assist in increasing the regions adaptive capacity. There is considerable traditional wisdom in the region for adapting to climatic risks and it is valuable even today. Sharing such experiences accumulated over centuries across South Asia could be useful at the household, community as well as regional level. The region has earlier adapted to climatic stresses by resorting to mixed cropping, changing varieties and planting times, by diversifying sources of income for farmers, and maintaining buffer stocks of food for managing periods of scarcity. These management strategies would also help in the future climate change scenarios but may not be enough in view of the increasing intensity of climatic risks and pressure on land to produce more food and also with much higher efficiency. Some of the possible adaptation options, which are relevant for adapting to current climatic risks as well as to climate change, are discussed below (see Box 16.2). These include options available with farmers (often referred to as autonomous adaptations) as well as with communities and governments (often referred to as planned adaptation). The latter are generally policy options that facilitate increases in the adaptive capacity of the agricultural systems.

Box 16.2 Key elements of the adaptation framework for South Asia

1. Assisting farmers in coping with current climatic risks
   - Improving collection and dissemination of weather related information
   - Establishing a regional early warning systems for climatic risks
   - Promoting insurance for climatic risk management
   - Facilitating establishment of community partnership in food, forage and seed banks
2. Intensifying food production systems
   - Bridging yield gaps in crops
   - Enhancing livestock productivity
   - Enhancing fisheries

(continued)

Box 16.2 (continued)

3. Improving land, water, and forests management
   - Implementing strategies for water conservation and use efficiency
   - Managing coastal ecosystems
   - Increasing the dissemination of resource conserving technologies
   - Exploiting the irrigation and nutrient supply potential of treated wastewaters
   - Improving management of forests
4. Enabling policies and regional cooperation
   - Integrating adaptation perspectives in current policy considerations
   - Providing incentives for resource conservation
   - Establishing regional food security programs
   - Securing finances and technologies for adaptation
   - Raising capacity in regional climate change assessments
5. Strengthening research for enhancing adaptive capacity
   - Assessing regional impacts on crops, livestock, fisheries, pests, and microbes
   - Evolving 'adverse climate tolerant' genotypes
   - Evaluating the biophysical and economic potential of various adaptation strategies

16.4 Assisting Farmers in Coping with Current Climatic Risks

16.4.1 Improving Collection and Dissemination of Weather Related Information

South Asia experiences considerable number of climatic extremes even today. A prerequisite for managing climatic risks, and thus increasing adaptive capacity, is timely knowledge of their spatial and temporal magnitude. A dense network of Weather Stations for standardized, real-time monitoring of rainfall and temperatures needs to be established in the whole region. The *SAARC* Meteorological Research Centre (SMRC) could be very effective for this purpose. The weather data, together with short- and medium-range weather forecasts, will enable scientists and extension workers to quickly translate the information into region-specific, value-added agricultural services for farmers such as time of crop planting, applying inputs, and timing the farm operations. Development of decision support tools for translating weather information into operational management practices will further facilitate this.

16.4.2 *Establishing a Regional Early Warning System for Climatic Risks*

A reliable and timely early warning system of impeding climatic risks could help in determining the potential food insecure areas and communities given the type of risk. Success of such a system is also dependent upon community preparedness to quickly use such warnings. To enable this, contingency plans should be developed for different types and durations of risks for various agro-ecological regions with due consideration to the required response time and available resources. Cooperation in developing such an early warning system, especially using modern tools of information and space technologies will be mutually beneficial for all countries of the region. This is especially critical for monitoring movements of insects and pathogens within South Asia. As climatic features, agricultural profiles, and pests are similar across large parts of the region, countries may not be able to address the issues related to transboundary movement of pests individually. Mechanisms do exist in some countries of the region to keep track of the movements of few pests such as locusts and avian bird flu. These early warning mechanisms, including regional cooperation for risk analysis, exchange of information and coordinated action, need considerable strengthening for a variety of key pests.

16.4.3 *Promoting Insurance for Climatic Risk Management*

The increasing probability of floods and droughts and other uncertainties in climate may seriously increase the vulnerability of resource-poor farmers of South Asia to global climate change. Policies that encourage crop/livestock insurance spread the risk and thus provide protection to the farmers if their farm production is reduced due to natural calamities. Weather insurance, coupled with standardized weather data collection, can greatly help in providing an alternate option for adapting agriculture to increased climatic risks. South Asian countries, in cooperation with other vulnerable regions such as in Africa, should endeavor to secure global adaptation and mitigation funds for improving climatic services and for implementation of weather related risk management programs.

16.4.4 *Facilitating Establishment of Community Partnership in Food, Forage and Seed Banks*

Climatic risks such as drought, cyclones and floods, often result in destruction of standing crops and loss of livestock. In such events, farmers need food for themselves, forage for their surviving livestock, and seeds to replant. Since such activities cost money and individual farmers may not be able to manage them on their own, communities of farmers could group together and establish their own stocks of food, forage and seeds. Such self-help groups will require technical and financial support from the government and other development partners, at least in the initial stages.

16.5 Intensifying Food Production Systems

16.5.1 Bridging Yield Gaps in Crops

Recent years have witnessed stagnation in food production in South Asia. The national average yields of rice and wheat crops are less than 4 t/ha today, whereas climatic factors in the region allow reasonably high yield potential of most crops. For example, in the productive Indo-Gangetic plains, potential rice and wheat yields are estimated to be 10 and 8 t/ha, respectively (Aggarwal et al. 2000b) indicating large yield gaps. Such yield gaps exist in all crops and across all ecosystems and bridging them could ensure meeting increased food demands of the future (Singh et al. 2001). Even if a fraction of these yield gaps could be bridged, food security in the region could be strengthened and the vulnerability of the food sector to climate change could be reduced. Fragile seed sector, poor technology dissemination mechanisms, lack of adequate capital for inputs, and poor markets and infrastructure are the key reasons for yield gaps (Aggarwal et al. 2004). A mission-oriented approach for key crops focusing on eliminating these constraints will enhance food security. Considering increasing competition for land and water among agriculture, urban settlements, industry, and recently biofuel plantations, such a mission approach must also ensure more yields per unit of land, water, and nutrients.

16.5.2 Enhancing Livestock Productivity

In recent past, livestock productivity has increased in South Asia due to intensification and industrialization. This will need to be further augmented to meet rapidly growing demands for milk and meat. Most livestock in the region feed primarily on crop by-products, household waste, and open grazing areas. Since productivity is the key to growth, livestock productivity will have to be raised through scientific breeding, feeding and management (Nin et al. 2007; Staal et al. 2008). Climate change is likely to impact availability of feed, water and energy for livestock. Livestock species with higher adaptation to harsher environments are available but they often have low productivity. Heat tolerance of the key livestock species needs to be enhanced by allele mining. Livestock based industry will also need to adapt to provide cooler housing for animals for alleviating their heat related distress and decline in reproduction.

This will require formulation of long-term policies by the governments and significant investments in the industry. In view of large number of low productive livestock in the region, South Asia must assess the size of the sustainable livestock population with a perspective on carbon emissions, water use, milk production and social values.

16.5.3 Enhancing Fisheries

Fisheries are a major source of nutrition and livelihood security to millions of fishers and others in South Asia (FAO 1997). This sector has seen phenomenal growth in last few decades. To meet future demands of fish, there is a need to utilize scientific and sustainable fishing practices, enhancing fish productivity, strengthening the whole chain from fish production to consumption, and imparting skills in community management and training in different aspects of fisheries and aquaculture. Climate change is likely to impact fish breeding, migration and harvest in marine as well as inland ecosystems (Nellemann et al. 2008). Being a highly perishable commodity, the quality of fish and fish products may erode due to rise in atmospheric temperature. Increase in extreme events will also have considerable impact on fisheries and aquaculture. Alternate fish species and fishing practices will need to be studied that could meet the demands for fish in the face of climate change. A comprehensive assessment of the fisheries potential of the increasing waterlogged and inundated areas as a result of higher precipitation caused by global warming should be made. Similarly, the scope for enhancing mariculture may be explored for the areas affected by sea level rise and incursion of seawater into the coastal areas.

16.6 Improving Land, Water, and Forests Management

16.6.1 Implementing Strategies for Water Conservation and Use Efficiency

Agriculture sector uses the largest amount of water in South Asia, yet a large fraction of arable land remains rainfed. There is an urgent need to develop technologies and policies for integrated storage and distribution of surface, ground, and rainwater. Technical standards, design and operation of irrigation systems should take note of the changing climatic variability. Scientific watershed management programs can assist in this endeavor and at the same time yield multiple benefits such as sustainable production, resource conservation, ground water recharge, drought moderation, employment generation and social equity (Dhyani et al. 1997).

In South Asia, there are several transboundary river basins such as Indus, Ganges and Brahmaputra and Tista. Comprehensive planning and regional collaboration from the perspective of glacier melt and water availability, its appropriate storage and integrated use could help in conserving water resources and utilizing them for alleviating droughts during periods of water stress. Appropriate measures should be put in place to manage increased silting of the irrigation systems caused by increased runoff and soil erosion associated with projected higher frequency of the intense rainfall events in the future.

16.6.2 Managing Coastal Ecosystems

Sea level rise is likely to inundate heavily populated, large coastal lands in India, Bangladesh, Maldives and Sri Lanka. This may lead to salinization of land and water bodies, and drainage congestion. Coastal lands are also vulnerable to frequent cyclones and other climatic extremes (Alam and Laurel 2005). There has been a significant increasing trend in the cyclone frequency over the Bay of Bengal during November and May, the main months for cyclone in the Bay of Bengal (SMRC 2003). Due to their limited financial resources, expensive adaptation mechanisms to prevent inundation of coastal lands may not be feasible. Salinity management, alternate land/resource use systems including salt tolerant crops, fish and fisheries, and diversification of livelihood strategies acceptable to local communities will, however, enhance their adaptive capacity.

16.6.3 Increasing the Dissemination of Resource Conserving Technologies

Recent researches have shown that surface seeding or zero-tillage establishment of upland crops after rice gives similar yields to those planted under normal conventional tillage over a diverse set of soil conditions. This reduces costs of production, allows earlier planting and thus higher yields, results in less weed growth, and above all reduces the use of fuel, and shows improvements in efficiency of water and fertilizers. Such technologies have become quite popular in eastern Pakistan and northwestern India (RWC 2007). Studies should be made to analyze and eliminate the key limitations for its accelerated diffusion in other parts of South Asia.

16.6.4 Exploiting the Irrigation and Nutrient Supply Potential of Treated Wastewaters

In the future, irrigation water availability in South Asia is likely to be less due to glacier retreat, increased variability in precipitation, and competition from rapidly growing industrial and urban sectors. In order to meet the increasing food demands in such a scenario of increasing water deficits there is a need to improve water productivity at all scales. Additionally, industrial and sewage wastewater could be evaluated as a potential irrigation source. Such effluents, once properly treated, can also be a source of nutrients for crops and potentially for fish culture. Since water serves multiple uses and users, effective inter-departmental coordination in the government is needed to develop the location specific framework of sustainable water management and cost-effective recycling of wastewater.

16.6.5 Improving Management of Forests

Forests account for about 20% of the arable land use in south Asia and are a large reserve of biodiversity and provide many services to local communities and national economy apart from protection against floods, cyclones and sea level rise. IPCC (2007b) has shown that the net primary productivity of forest ecosystems is likely to increase due to CO_2 fertilization, at least in the initial decades, and may decline in the later periods. Studies by Ravindranath et al. (2006) on forest ecosystems of India have shown that nearly 75% area is likely to experience changes in forest or vegetation type. Since forests are highly vulnerable even at moderate warming, there is a need for developing effective adaptation practices and policies. These strategies include planting of tolerant species, mixed species forestry, implementation of fire protection practices, conservation of biodiversity, halting of forests fragmentation, and promoting community forestry (Ravindranath et al. 2006).

16.7 Enabling Policies and Regional Cooperation

16.7.1 Integrating Adaptation Perspectives in Current Policy Considerations

South Asian countries have limited resources, which are targeted towards attaining food security and poverty alleviation, and other developmental targets. Addressing future climate change concerns at the cost of current priority of development cannot be an option. However, integrating perspectives on climatic risks in their current policies and programs in different sectors such as disaster management, water resources management, land use, biodiversity conservation, and agricultural development will lead to increased adaptive capacity to current as well as future climatic variability.

16.7.2 Providing Incentives for Resource Conservation

Adaptation to environmental change could be in the form of a policy where positive incentives are provided to farmers and industry for increasing the resilience of food production systems. Necessary provisions need to be included in the development plans to address the issues of attaining twin objectives of containing environmental changes and improving resource use productivity. For example, policies and incentives for appropriate organic matter management in rural areas would encourage farmers to sequester carbon in the soil and thus improve soil health, and use water and energy more efficiently. Rational pricing of surface and groundwater and incentives for their conjunctive use can arrest its injudicious use.

16.7.3 Establishing Regional Food Security Programmes

History has shown that climatic extremes do not impact the entire South Asian region at the same time. Despite being land and food scarce, several countries in the region have evolved a policy of storing food in good years to manage food scarcity in bad years as well as for providing support to relatively poor population. Assuming that climate change associated extremes are not likely to occur all over South Asia at the same time in the future as well, and considering the huge costs of storing food, SAARC countries should establish a Food Security Fund or a Food Bank to augment the regional food security that will mutually benefit countries in reducing supply fluctuations during adverse weather. Such emergency reserves should be specifically meant for supporting only the poor and needy.

16.7.4 Securing Finances and Technologies for Adaptation

Implementing various strategies for enhancing adaptive capacity would require considerable financial resources. Several global funds for adaptation are now available and some more are likely to be established in future. South Asia should attempt to secure these funds for 'climate proofing' food supplies in its vulnerable regions. Some specific programs where such funds could be used are developing and strengthening adaptation related infrastructure, implementation of weather related risk insurance programs, enhancing research capacity, and for securing 'patented' knowledge/technologies related to adaptation, including germplasm/genes from various sources.

16.7.5 Raising Capacity in Regional Climate Change Assessments

Effective handling of environmental issues in agriculture needs a close interaction between scientists, development partners, policy makers, administrators, trade and industry, farmer organizations and other stakeholders. Different types of capacity building programs need to be developed at various levels to ensure efficient management of natural resources for sustainable agricultural development. A network could be established to facilitate continuous dialogue, share experiences, and to develop strategies for implementation of desired changes. Enhancing capacities of SAARC regional centers on agriculture, disaster management, coastal zones, meteorology, and forestry, will strengthen regional cooperation and enhance adaptive capacity.

16.8 Strengthening Research for Enhancing Adaptive Capacity

16.8.1 Assessing Regional Impacts on Crops, Livestock, Fisheries, Pests, and Microbes

Developing an effective adaptation response requires a comprehensive understanding of the impacts and vulnerabilities of different agricultural commodities, and microbes and pests. A large part of current understanding of impacts on South Asia is based on generic global scale assessments since there are relatively few regional studies. Indigenous research is needed to understand the probable impacts especially on native crops such as legumes, oilseeds, and plantations. This requires special facilities such as carbon dioxide enrichment and temperature enhancement in open fields, controlled environment chambers, and validated simulation models. Regional and international cooperation for capacity enhancement in this should be sought. Special attention is needed in the region in short-term to assess the impacts of enhanced climatic variability at critical stages of crop growth and its management.

16.8.2 Evolving 'Adverse Climate Tolerant' Genotypes

Future breeding efforts will need to address multiple stresses – droughts, floods, heat, salinity, and pest load – imposed by changing global climate. There will be a need to stack several adaptive traits in a suitable agronomic background. This requires substantial breeding efforts, including collection, conservation and distribution of appropriate genetic material among breeders and other researchers. There is a need for a better understanding of wild relatives and landraces, and their distributions and sensitivity to climatic variables. In view of large biodiversity for stress tolerance in South Asia, regional cooperation in development of adapted genotypes should be encouraged.

16.8.3 Evaluating the Biophysical and Economic Potential of Various Adaptation Strategies

There are several strategies available with farmers such as using alternate crops/ varieties, livestock and fish species, and modified input management that can be employed to manage small changes in climatic parameters. Larger changes in climatic parameters may require consideration of alternate land use systems. Since the socio-economic scenario is rapidly changing, and there is considerable uncertainty in the magnitude of global climate change, a thorough quantitative assessment of the potentials and constraints of land is needed, in which scientific knowledge, socio-economic conditions and the conflicting interests of various stakeholders can be harmonized and most efficient and sustainable land use systems identified. Current availability of simulation models and other systems research tools provides an opportunity for an interdisciplinary approach for this (Roetter et al. 2007). Capacity for such research in the region could be enhanced through international collaboration.

16.9 A Framework for Mitigation of Emissions from Agriculture

The global share of average emissions of GHG from agriculture is 13.5% (IPCC 2007c). This fraction is much larger in South Asian countries due to relatively large role of agriculture in national GDP. The IPCC concluded that there has been a large growth in emissions from agriculture in South Asia in last few decades relative to the developed world due to expanding use of nitrogen fertilizers and manure to meet demands for food resulting from rapid population growth (IPCC 2007c). In the future, the percentage of emissions from agriculture in the region is likely to be smaller, and closer to the global average, due to relatively much higher growth in energy use (and hence emissions) in transport and industrial sectors.

The emissions from agriculture are primarily due to methane emission from rice paddies, enteric fermentation in ruminant animals, and nitrous oxides from application of manures and fertilizers to agricultural soils (Fig. 16.1). IPCC has identified various strategies for enhancing carbon sequestration in soils and for mitigating emissions of GHG from agricultural systems (see Box 16.3). The key ones in the context of South Asia are discussed below.

16.10 Sequestering Soil Carbon and Mitigating GHGs

Soil carbon sequestration is considered the main mechanism responsible for mitigation potential (Smith et al. 2007). There are several approaches to increase carbon sequestration in soils of South Asia (Lal 2004). These include addition of organic manures, minimal tillage, residue management, agroforestry, scientific water and nutrient management, and restoration of degraded soils. Lal (2004) estimates that such practices can lead to carbon sequestration of 25–50 Tg C/year in soils of South Asia for several decades, including 11–22 Tg C/year on croplands.

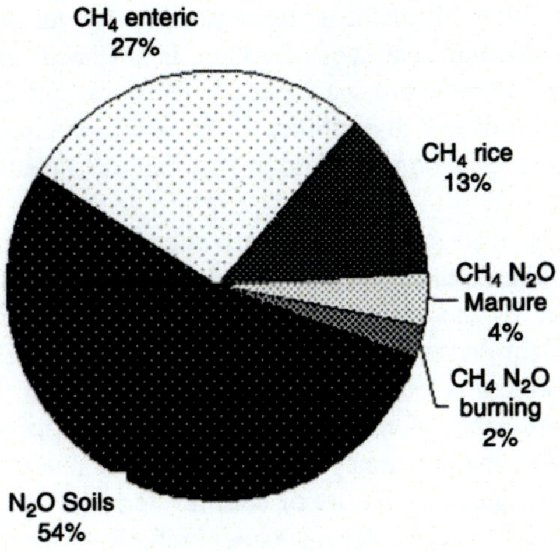

Fig. 16.1 Contribution of various agricultural sectors to GHG emissions in South Asia (Smith et al. 2007)

Box 16.3 IPCC's key observations for mitigation of GHGs and carbon sequestration in agricultural systems (IPCC 2007c)

A variety of options exist for mitigation of GHG emissions in agriculture. The most prominent options are improved crop and grazing land management (e.g., improved agronomic practices, nutrient use, tillage, and residue management), restoration of organic soils that are drained for crop production and restoration of degraded lands. Lower but still significant mitigation is possible with improved water and rice management; set-asides, land use change (e.g., conversion of cropland to grassland) and agro-forestry; as well as improved livestock and manure management.

Soil carbon sequestration (enhanced sinks) is the mechanism responsible for most of the mitigation potential (high agreement/much evidence), with an estimated 89% contribution.

Current GHG emission rates may escalate in the future due to population growth and changing diets (high agreement/medium evidence). Deployment of new mitigation practices for livestock systems and fertilizer applications will be essential to prevent an increase in emissions from agriculture after 2030.

Organic soil restoration has a host of biodiversity/environmental co-benefits but opportunity cost of crop production lost from this land; economic impact depends upon whether farmers receive payment for the GHG emission reduction.

Controlling emission of methane from paddies by midseason drainage or alternate drying instead of continuous flooding in irrigated areas has reasonable mitigation potential. Improved management of livestock diet through use of feed additives, substitution of low digestibility feeds with high digestibility ones, concentrate feeding, and changing microflora of rumen also leads to a reduction in methane emission.

Appropriate crop management practices, which lead to increase nitrogen use efficiency and yield, hold the key to reduce nitrous oxide emission. Curtailing the nitrification process by the use of nitrification inhibitors, particularly the cheap, locally available, plant-derived materials, such as neem cake, will be useful to mitigate nitrous oxide emission from soil.

Improving the efficiency of energy use in agriculture by using better designs of machinery, increasing fuel efficiency in agricultural machinery, commercialization of wind/solar power potential, and use of laser levelers also lead to mitigation.

Changing land use by increasing area under biofuels or agro-forestry could also mitigate GHG emissions (Smith et al. 2007). This, however, may have trade-offs with goal of increasing food production. South Asian countries that are short of land resources need to examine the consequences of diverting land to biofuels on their food security (Aggarwal et al. 2007).

16.11 Facilitating Mechanisms for Payments to Farmers for Carbon Sequestration

A Clean Development Mechanism (CDM) is being implemented under the UNFCCC to mitigate GHG emissions. This allows for carbon credits to be traded primarily in the energy and forestry sectors. It does not at this time specifically include carbon sequestration and mitigation in agriculture. The IPCC (2007c) estimates that the agricultural GHG mitigation options are cost-competitive with non-agricultural options. If these mitigation options could be included in future agreements on climate change, it would lead to better soil fertility and higher income for the farmers in addition to the primary goal of carbon sequestration. To facilitate this, methodologies are needed to upscale the carbon sequestered/mitigated in individual farms to a large regional scale in order to keep the transaction costs at a low level.

16.12 Conclusions

Global climate change is likely to affect food and livelihood security of the millions of poor farmers and landless in South Asia and other developing countries. Climatic changes could limit the future capacity of South Asia to remain agriculturally self-sufficient. Urgent steps are needed to increase its adaptive capacity to face current as well as the future climatic risks. Action needs to be taken now since it takes time for adaptive practices to become effective. These adaptation strategies will need to simultaneously consider the background of changing demand due to globalization, population increase and income growth, as well as the socio-economic and environmental consequences of possible adaptation options. This would require increased adaptation research, capacity building, development activities, changes in policies, and support of global adaptation and mitigation funds and other resources.

The costs of adaptation and mitigation are not clearly known today but these are expected to be high (Stern 2007; Cline 2007). A universally preferable solution is to start with such adaptation strategies that are anyway needed for sustainable development. Strategies that maximize synergies between adaptation, mitigation, food production and sustainable development would be most appropriate. For example, increasing the efficiency of fertilizer and water use will lead to higher profits for the farmers as well as sequester more carbon, which will improve soil health and lead to higher production. Such practices would become more attractive if farmers could be paid incentives for environmental services that agriculture provides. It is time that the society considers such incentives to farmers in the interest of global environment, food security, and poverty alleviation. Development partners should consider the adaptation and mitigation strategies outlined in this paper to help in poverty alleviation and enhancing food security of the vulnerable South Asia.

International climate change negotiations revolve primarily around mitigation of GHG emissions from industrial and other sources, and other related activities. Such discussions must simultaneously address poverty alleviation and the vulnerability of poor farmers of developing countries.

South Asia has a history of regional cooperation through SAARC and other mechanisms. Climate change makes the need for such cooperation even more important. Regional collaboration in many proposed areas of adaptation and mitigation would be useful due to large areas across borders with similar agro-ecological features. For example, Punjab in India and Pakistan, Bengal state of India and Bangladesh, Tarai region of Nepal and Uttar Pradesh and Bihar in India, have similar agro-ecological features. Having agreements on agriculture related trans-boundary issues such as water and pests movement would help in managing climatic risks as and when they occur. Also, such collaboration and cooperation would be necessary to raise the voice of South Asian farmers in global negotiations.

References

Aggarwal PK, Bandyopadhyay SK, Pathak H, Kalra N, Chander S, Sujith Kumar S (2000a) Analysis of the yield trends of rice-wheat system in north-western India. Outlook Agric 29:259–268

Aggarwal PK, Talukdar KK, Mall RK (2000b) Potential yields of rice-wheat system in the Indo-Gangetic plains of India. Rice-Wheat Consortium Paper Series 10. New Delhi, India. RWCIGP, CIMMYT, p 16

Aggarwal PK, Joshi PK, Ingram JS, Gupta RK (2004) Adapting food systems of the Indo-Gangetic plains to global environmental change: key information needs to improve policy formulation. Environ Sci Policy 7:487–498

Aggarwal PK, Joshi HC, Kumar S, Gupta N, Kumar S (2007) Fuel ethanol production from Indian agriculture: opportunities and constraints. Outlook Agric 36:167–174

Alam M, Laurel AM (2005) Facing up to climate change in South Asia, Gatekeeper Series, 118. International Institute for Environment and Development, London, UK

Bhalla GS, Hazell P, Kerr J (1999) Prospects for India's cereal supply and demand to 2020. International Food Policy Research Institute, USA, p 34

Cline W (2007) Global warming and agriculture: impact estimates by country. Peterson Institute for International Economics, p 250

Dhyani BL, Samra JS, Juyal GP, Ram B, Katiyar VS (1997) Socio-Economic analysis of a participatory integrated watershed management in Garhwal Himalaya (Fakot Watershed). Central Soil and Water Conservation Research and Training Institute, Dehradun

Easterling WE, Aggarwal PK, Batima P, Brander KM, Erda L, Howden SM, Kirilenko A, Morton J, Soussana J-F, Schmidhuber J, Tubiello FN (2007) Food, fibre and forest products. In: Parry ML, Canziani OF, Palutikof JP, van der Linden PJ, Hanson CE (eds) Climate Change 2007: impacts, adaptation and vulnerability. Contribution of Working Group II to the Fourth Assessment Report of the Intergovernmental Panel on Climate Change. Cambridge University Press, Cambridge, UK, pp 273–313

FAO (1997) Review of the state of world aquaculture. FAO Inland Water Resources and Aquaculture Service, Fishery Resources Division. FAO Fisheries Circular. No. 886, Rev. 1. Rome, FAO, p 163

FAO (2008) High-level conference on world food security: the challenges of climate change and bioenergy. Food and Agriculture Organization, Rome, p 51

Fischer G, Shah M, van Velthuizen H (2002) Climate change and agricultural vulnerability. International Institute for Applied Systems Analysis, Austria

Hocking PJ, Meyer CP (1991) Carbon dioxide enrichment decreases critical nitrate and nitrogen concentrations in wheat. J Plant Nutr 14:571–584

Government of India (2008) National action plan on climate change. Prime Minister's Council on Climate Change, p 49

IPCC (2007a) Climate Change 2007: the physical science basis. In: Solomon S, Qin D, Manning M, Chen Z, Marquis M, Avery KB, Tignor M, Miller HL (eds) Contribution of Working Group I to the Fourth Assessment Report of the Intergovernmental Panel on Climate Change. Cambridge University Press, Cambridge, p 996

IPCC (2007b) Climate Change 2007: impacts, adaptation and vulnerability. In: Parry ML, Canziani OF, Palutikof JP, van der Linden PJ, Hanson CE (eds) Contribution of Working Group II to the Fourth Assessment Report of the Intergovernmental Panel on Climate Change. Cambridge University Press, Cambridge, UK, p 976

IPCC (2007c) Climate Change 2007: mitigation. In: Metz B, Davidson OR, Bosch PR, Dave R, Meyer LA (eds) Contribution of Working Group III to the Fourth Assessment Report of the Intergovernmental Panel on Climate Change. Cambridge University Press, Cambridge, p 851

Joshi PK, Tyagi NK (1994) Salt affected and water logged soils in India: a review. In: Svendsen M, Gulati A (eds) Strategic change in Indian irrigation. ICAR, New Delhi, India, and IFPRI, Washington, DC, pp 237–252

Kulkarni AV, Bahuguna IM (2002) Glacial retreat in the Baspa basin, Himalaya, monitored with 17 satellite stereo data (Correspondence). J Glaciol 48:171–172

Kumaraguru AK, Jayakumar K, Ramakritinan CM (2002) Coral bleaching 2002 in the Palk Bay, southeast coast of India. Curr Sci 85:1787–1793

Lal R (2004) The potential of carbon sequestration in soils of South Asia. Paper presented at the 13th International Soil Conservation Organization Conference, Brisbane

MoEF (2008) Bangladesh Climate Change Strategy and Action Plan 2008. Ministry of Environment and Forests, Government of the People's Republic of Bangladesh, Dhaka, Bangladesh, xvi + p 68

Nellemann C, Hain S, Alder J (eds) (2008) In Dead water – merging of climate change with pollution, over-harvest, and infestations in the world's fishing grounds. United Nations Environment Programme, p 64

Nin A, Ehui S, Benin S (2007) Livestock productivity in developing countries: an assessment. Handbook Agric Econ 3:2461–2532

Paroda RS, Kumar P (2000) Food production and demand in South Asia. Agric Econ Res Rev 13:1–24

Parry ML, Rosenzweig C, Iglesias A, Livermore M, Fischer G (2004) Effects of climate change on global food production under SRES emissions and socio-economic scenarios. Glob Environ Change 14:53–67

Pathak JK, Ladha PK, Aggarwal S, Peng S, Singh Y, Singh B, Kamra SK, Mishra B, Sastri ASRAS, Aggarwal HP, Das DK, Gupta RA (2003) Trends of climatic potential and on-farm yields of rice and wheat in the Indo-Gangetic Plains. Field Crops Res 80:223–234

Ravindranath NH, Joshi NV, Sukumar R, Saxena A (2006) Impact of climate change on forest in India. Curr Sci 90:557–560

Roetter RP, van den Berg M, Laborte AJ, Hengsdijk H, Wolf J, van Ittersum M, van Keulen H, Agustin EO, Nguyen TS, Lai X (2007) Combining farm and regional level modelling for integrated resource management in East and South-East Asia. Environ Model Software 22:149–157

Rosenzweig C, Parry ML (1994) Potential impact of climate change on world food supply. Nature 367:133–138

Rupa Kumar K, Sahai AK, Krishna Kumar K, Patwardhan SK, Mishra PK, Revadekar JV, Kamala K, Pant GB (2006) High-resolution climate change scenarios for India for the 21st century. Curr Sci 90:334–345

RWC (2007) RCT induced impacts in Indo-Gangetic Plains. In: RWC Research Highlights 2006. Rice-Wheat Consortium, New Delhi, p 10

Sharp GD (2003) Future climate change and regional fisheries: a collaborative analysis. FAO Fisheries Technical Paper. No. 452. Rome, FAO, p 75

Singh P, Vijaya D, Chinh NT, Pongkanjana A, Prasad KS, Srinivas K, Wani SP (2001) Potential productivity and yield gap of selected crops in the rainfed regions of India, Thailand, and Vietnam. Natural Resource Management Program Report No. 5. Patancheru 502 324, Andhra Pradesh, India: International Crops Research Institute for the Semi-Arid Tropics, p 50

Sinha SK, Singh GB, Rai M (1998) In decline in crop productivity in Haryana and Punjab: myth or reality? Indian Council of Agricultural Research, New Delhi, p 89

Smith P, Martino D, Cai Z, Gwary D, Janzen H, Kumar P, McCarl B, O'Mara F, Rice C, Scholes B, Sirotenko O, Howden M, McAllister T, Ogle S, Pan G, Romanenkov V, Schneider U, Towprayoon S, Wattenbach M, Smith J (2007) Policy and technological constraints to implementation of greenhouse gas mitigation options in agriculture. Agric Ecosyst Environ 118:6–28

SMRC (2003) The vulnerability assessment of the SAARC coastal region due to sea level rise: Bangladesh case, SMRC-No. 3. SMRC Publication, Dhaka, Bangladesh

Government of Sri Lanka (2000) Initial national communication under the United Nations Framework Convention on Climate Change, UNFCCC, p 122

Staal SJ, Nin Pratt A, Jabbar M (2008) Dairy development for the resource poor. Part 3: Pakistan and India dairy development case studies. PPLPI Working Paper No. 44–3. ILRI, Kenya, p 78

Stern N (2007) Stern review on the economics of climate change. Cambridge University Press, Cambridge, p 692

St-Pierre NR, Cobanov B, Schnitkey G (2003) Economic losses from heat-stress by US livestock industries. J Dairy Sci 86(E. Suppl):E52–E77

Wilkinson CR, Linden O, Cesar H, Hodgson G, Rubens J, Strong AE (1999) Ecological and socioeconomic impacts of 1998 coral mortality in the Indian Ocean: an ENSO impact and a warning of future change. Ambio 28:188–196

Ziska LH, Namuco O, Moya T, Quiland J (1997) Growth and yield response of field grown tropical rice to increasing carbon dioxide and air temperature. Agron J 89:45–53

Chapter 17
Options on Land Management and Land Use for Coping with Climate Change in South Asia

Yuji Niino

Contents

Abstract The South Asia region is characterized by high population density and scarcely available land for sustainable agriculture. The region has a land area of 642 million ha (Mha) and a population of 1,587 million. The agricultural population is 786 million, 49% of the total. The area of agricultural land is 230 Mha, 36% of the total land area. The average ratio of agricultural land to agricultural population in the region is 0.33 ha per caput. Food security situation in South Asia is further affected by low productivity and prone to natural disaster. Causes of these problems are high population pressure, urbanization, increased demand for bio-fuels, building materials, and agricultural lands. The diversion of agricultural land to non-agricultural uses is also a growing concern. Fertile farm lands are being indiscriminately diverted, thus reducing the area available for food production. Vulnerability to food security is rising because of climate change which makes

Y. Niino (✉)
Land Management Officer, FAO Regional Office for Asia and the Pacific,
39 Phra Atit Road, Bangkok, 10200 Thailand
e-mail: Yuji.Niino@fao.org

R. Lal et al. (eds.), *Climate Change and Food Security in South Asia*,
DOI 10.1007/978-90-481-9516-9_17, © Springer Science+Business Media B.V. 2011

more areas, particularly coastal and low-lying areas, disaster-prone while growing population is further forcing vulnerable people to settle in risk-prone areas and crop land inundated. The problem is aggravated by factors such as skewed distribution of assets and income, degradation of the natural resource base and unsustainable management of land and water resources. Smallholder and subsistence farmers may not be able to cope with climate change effectively, due to reduced adaptive capacity and higher climate vulnerability. Under such conditions, pressure to cultivate marginal land or to adopt unsustainable cultivation practices as yields drop is likely, and may increase land degradation, water scarcity and endanger biodiversity.

To cope with climate change and increase agricultural productivity through conservation and efficient use of agricultural land and water resources with technical, policy support and environmental considerations, improved land management and land-use planning are possible adaptation methods. Those risks could be reduced through resource conservation technologies, providing information and education, and through suitable policy measures. There is a need to balance food security needs with sustainable use of land and water. Diversifying from high water consuming crops to high value horticulture and livestock products is preferable and growing. Monitoring of climate change may also mitigate the serious implications for the South Asia's agriculture. For the mitigation to climate change by controlling GHG emissions in agriculture, a variety of options exist. The most prominent options are improved crop and grazing land management, restoration of organic soils, and restoration of degraded lands. In less extent, but still significant mitigation is possible with improved water and rice management, agroforestry, and improved livestock and manure management.

Sustainable use and management of soil and water resources in South Asia is a major issue to be addressed through adaptations of technologies such as low tillage and maintenance of permanent soil cover that can increase SOM and reduce impacts from flooding, erosion, drought, heavy rain and winds. Also the potential areas being explored are conservation agriculture, organic agriculture and risk-coping production systems that incorporate crop rotations, agroforestry, crop–livestock associations, crop–fish systems and the use of hedges, vegetative buffer strips and other farm landscaping practices. Surface mulch cover protects soil from excess temperatures and evaporation losses and can reduce crop water requirements by 30%.

Keywords Land cover • Land slides • Vulnerability to food insecurity • Cropland inundation • Land degradation • Land use • Conservation agriculture • Agroforestry • Biofuel

Abbreviations

ADPC	Asian Disaster Preparedness Center
CDMP	Comprehensive Disaster Management Program
GHG	Greenhouse gas

GLCN	Global Land Cover Network
GWP	Global warming potential
HWSD	Harmonized World Soil Database
IPNM	Integrated Plant Nutrient Management
SOM	Soil organic matter
WOCAT	World overview of conservation agric. practices

17.1 Background

South Asia is extremely vulnerable to natural disasters such as earthquakes, cyclones, landslides, flooding and drought. Floods, cyclones and landslides are common hazards in the region due to the region's climatic and geographical diversity, particularly in India, Bangladesh, Pakistan and Nepal. The number of persons killed and affected between 1974 and 2003 in South Asia is more than 580 million (41.4%) out of the total number of 1.5 billion disaster victims of the world (SDMC 2009). In 8393 disaster events during 1974–2008, 40% were in Asia and 30% of those were in South Asia. The total number affected in 2007 is over 64 million and the estimated damage reaches over 4.7 billion dollars.

A significant number of landslides caused by various natural and man-made factors contribute to the high incidence of landslides each year. Natural factors include steep slopes, undercutting of their banks by incised rivers, weathered, fractured and weak rocks in the mountains, high rainfall and seismic activities. Man-made factors responsible for landslides in the country are intensive deforestation, improper agriculture and irrigation practices, overgrazing on the slopes, quarrying for construction materials, and construction of infrastructure beyond the bearing capacities of the hill slopes, Landslides frequently occur in the monsoon season following an earthquake in the previous year.

The main causes for the frequent occurrence of floods are heavy seasonal rainfall, deforestation, lack of flood protection schemes, unplanned development activities, etc. (ADPC 2007). Droughts of varying duration and impact with changing weather patterns, populations, deforestation and the resultant adverse effects on the ecological balance have been observed. Socio-economic conditions in the region including population pressures and inhabiting on marginal lands, such as flood plains, deserts, earthquake prone areas and steep slopes and environmental degradation further aggravate the situation. Lack of investments in infrastructure in hazard prone areas to reduce vulnerability and prevent potential loss of economic assets is also identified as critical.

Floods and landslides which occur mostly during the monsoon season are often interrelated and some landslides are triggered by riverbank erosion. Glacial lake outburst floods are common in the Himalayan region, and are triggered by a wide range of hydrological, geological and seismic factors (ADPC 2008). Himalayan glaciers are retreating at rates ranging from 10 to 60 m per year and many small glaciers (<0.2 km^2) have already disappeared. As glaciers retreat, glacial lakes grow, and many Himalayan basins are reporting very fast growing lakes (Bajracharya et al. 2007). Disastrous flash floods usually occur in Nepal when landslides or debris block a river for several hours and the water is then releases suddenly, inundating the adjacent area downstream in a rapid and turbulent manner. Flash floods are also caused by continuous heavy rainfall in many rivers originating in hilly regions. It may also be caused by an avalanche, snowstorm or cloudburst.

17.2 Climate Change, Disaster and Food Security

Its general physical environment is featured by climate, landform and soils which cause natural hazards of degradation. The problem is aggravated by factors such as rapidly growing population, skewed distribution of assets and income, degradation of the natural resource base and unsustainable management of land and water resources. Food security situation in South Asia is thus affected by low productivity and prone to natural disaster. Vulnerability to food security is rising because of climate change which makes more areas, particularly coastal and low-lying areas, disaster-prone while growing population is further forcing vulnerable people to settle in risk-prone areas and crop land inundated. Climate variability will result in more frequent and intensive disasters with the most severe consequences on the food security and livelihoods of agriculture dependent populations in vulnerable countries (FAO 2008a).

Agriculture in South Asia is subject to variety of risks arising from rainfall aberrations, temperature fluctuations, hailstorms, cyclones, floods and climate change. These risks are exacerbated by price fluctuations, weak rural infrastructure, imperfect markets and lack of financial services, including limited span and design of risk mitigation instruments such as credit and insurance. These factors not only endanger the farmer's livelihood and income but also undermine the viability of the agriculture sector and hence food security. Smallholder and subsistence farmers may not be able to cope with climate change effectively, due to reduced adaptive capacity and higher climate vulnerability. Under such conditions, pressure to cultivate marginal land or to adopt unsustainable cultivation practices as yields drop is likely, and may increase land degradation, water scarcity and endanger biodiversity.

17.3 Land Resources in South Asia

The South Asia region is characterized by high population density and scarcely available land for sustainable agriculture. The region has a land area of 642 Mha and a population of 1,587 million (FAO 2007a). The agricultural population is 786 million, 49% of the total. The area of agricultural land is 230 Mha, 36% of the total land area (Table 17.1). The average ratio of agricultural land to agricultural population in the

Table 17.1 Land use in South Asia (FAO 2007a)

Country	Total land (Mha)	Arable and permanent crops (Mha)	%	Permanent pasture (Mha)	%	Forests and woodlands (Mha)	%	Other land (Mha)	%
Afghanistan	65.21	8.05	12	30	46	0.87	1	26.29	40
Bangladesh	13.02	8.42	65	0.6	5	0.87	7	3.13	24
Bhutan	4.7	0.18	4	0.42	9	3.2	68	0.91	19
India	297.32	169.65	57	10.53	4	67.7	23	49.44	17
Iran	163.62	17.6	11	44	27	1,108	7	90.95	56
Maldives	0.03	0.01	33	0	3	0	3	0.02	50
Nepal	14.3	2.49	17	1.74	12	3.64	25	6.44	45
Pakistan	77.09	22.07	29	5	7	1.9	3	48.12	62
Sri Lanka	6.46	1.92	30	0.44	7	1.93	30	2.17	34
Suoth Asia Region	641.75	230.38	36	92.72	15	91.18	14	220.78	34
Asia–Pacific	3,462.86	581.31	17	1,300.16	38	724.89	22	856.51	25
World	13,009.8	1,544.62	12	3,437.6	26	3,952.03	30	4,075.55	31

region is 0.33 ha per caput which is slightly higher than the Asia–Pacific average of 0.30 ha per caput, but lower than the world average of 0.59 ha per caput. Especially for Bangladesh and Nepal, only 0.1 ha per caput is available while in India and Pakistan there are about 0.3 ha per caput where agricultural population of 566 million and 76 million which are 72% and 10% of agricultural population of the region, respectively. The region has a large growing population which put great pressure on land, but little opportunity for further expansion.

Identified environmental problems in South Asia include land degradation, water scarcity and degradation, deforestation and biodiversity loss (FAO 1994; UNEP 1997). A study estimated the severity and costs of land degradation in South Asia worth at least US$10 billion annually which is equivalent to 2% of the Gross Domestic Products of the region (FAO 1994). The 43%, 140 Mha, of the region's total agricultural land suffer from degradation. These economic losses will impact on future generations. Irrigated land in South Asia covers over 91 Mha which is the largest in the Asia and the Pacific region (FAO 2007a). Unsuitable irrigation caused land degradation in water erosion and salinization. Water in general in the region is scarce and ground water depletion has also been recognized (ESCAP 2000).

South Asia is projected to be predominantly rural in 2030 despite of its rapid growth of cities. Massive food demand to support growing population of the region has been attained by the intensification of agriculture and its expansion to marginal lands, and also with extensive and inefficient irrigation systems (ESCAP 2006). Key environmental issues identified are soil degradation (including soil salinization and loss of soil fertility), soil erosion, overgrazing, deforestation, desertification, biodiversity loss, and air and water pollution (FAO 1994; ESCAP 2000).

There is a need to balance food security needs with sustainable use of land and water. Smallholder and subsistence farmers may not be able to cope with climate change effectively, due to reduced adaptive capacity and higher climate vulnerability. Under such conditions, pressure to cultivate marginal land or to adopt unsustainable cultivation practices as yields drop is likely, and may increase land degradation, water scarcity and endanger biodiversity. This should be reduced through resource conservation technologies, providing information and education and through suitable policy measures.

17.3.1 Land Degradation

Land degradation is intensifying in many parts of the world, according to a study using data taken over a 20-year period. Defined as a long-term decline in ecosystem function and productivity, land degradation is increasing in severity and extent in many parts of the world, with more than 20% of all cultivated areas, 30% of forests and 10% of grasslands undergoing degradation. An estimated 1.5 billion people, or a quarter of the world's population, depend directly on land that is being degraded. The consequences of land degradation include reduced productivity, migration, food insecurity, damage to basic resources and ecosystems, and loss of biodiversity through changes to habitats at both species and genetic levels. Land degradation also has important implications for climate change mitigation and adaptation, as the loss of biomass and soil organic matter (SOM) releases carbon into the atmosphere and affects the quality of soil and its ability to hold water and nutrients (Fig. 17.1).

The data indicate that despite the stated determination of 193 countries that ratified the United Nations Conference to Combat Desertification in 1994, land degradation is worsening rather than improving. Some 22% of degrading land is in very arid to dry-subhumid areas, while 78% of it is in humid regions. The study found that degradation is being driven mainly by poor land management. Comparing with

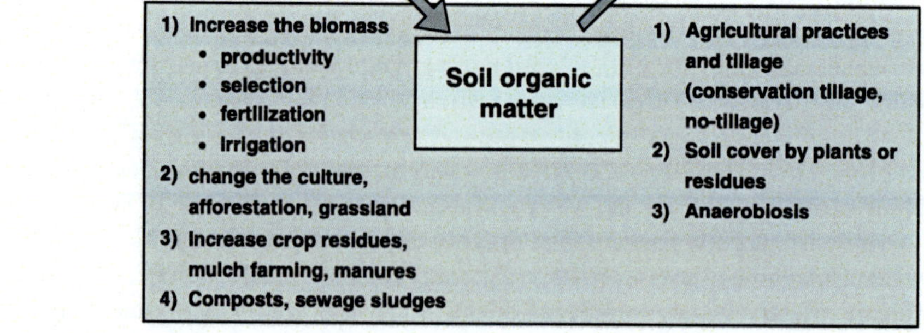

Fig. 17.1 Soil organic matter management (FAO 2001)

previous assessments, the present study shows that land degradation since 1991 has affected new areas; meanwhile some historically degraded areas were so severely affected that they are now stable having been abandoned or managed at low levels of productivity.

Global Land Assessment of Degradation (FAO 1994) showed 43% of total agricultural land was degraded with different degree of degradation. Afghanistan, Pakistan and India suffer most from land degradation and desertification. In India, about 175 Mha are degraded, of which about 141 Mha are subject to water and wind erosion and the remaining are through water logging and salinity. Soil erosion has caused sedimentation of dams and reducing their capacity to retain water, control floods and generate electricity. The soil constraints assessment (Dent 1990) for identifying the major natural constraints to agricultural production showed 76% of the soils in South Asia have some extent of problems including too dry, too steep, too salty and too clayey.

Major causes of these problems are high population pressure, urbanization, increased demand for bio-fuels, building materials, and agricultural lands. The diversion of agricultural land to non-agricultural uses is also a growing concern. Fertile farm lands are being indiscriminately diverted, thus reducing the area available for food production. In the context of food security, the need for effective land use policy for housing, infrastructure and urbanization is required to minimize the impact of this trend.

17.3.2 Climate Change and Land Degradation

Deforestation, biomass burning, conversion of natural to agricultural ecosystems, drainage of wetlands and soil cultivation are considered contributed carbon emissions through land use changes. According to a crude estimate of the carbon emitted from soil to the atmosphere based on an average loss of 30–40 megagram (Mg) C ha^{-1} due to historical land use (Fig. 17.2), soil degradation and fertility mining practices is about 5–7 Pg (Lal et al. 2004). As a result of its large-scale activities, agriculture is a significant contributor to land degradation and, in particular, a major emitter of greenhouse gases. Overall, agriculture is responsible for 25% of carbon dioxide (largely from deforestation), 50% of methane (rice and enteric fermentation), and more than 75% of N_2O (largely from fertilizer application) emitted annually by human activities. However, the assessment of degrading areas with global land cover revealed that 19% of degrading land is cropland, 24% is broadleaved forest, and 19% needle-leaved forests while cropland occupies only 12% of the land area. Thus land degradation is mainly associated with farming may be over-represented in cropland globally (Bai et al. 2008).

The depletion of soil C causes soil degradation which is exacerbated by misuse and mismanagement of land and soil resources. Soil carbon conservation and sequestration through adoption and promotion of a restorative land use and recommended management practices could contribute in reduction of the emission of CO_2 while having improved conditions for food security, agro-industries, water quality and the environment.

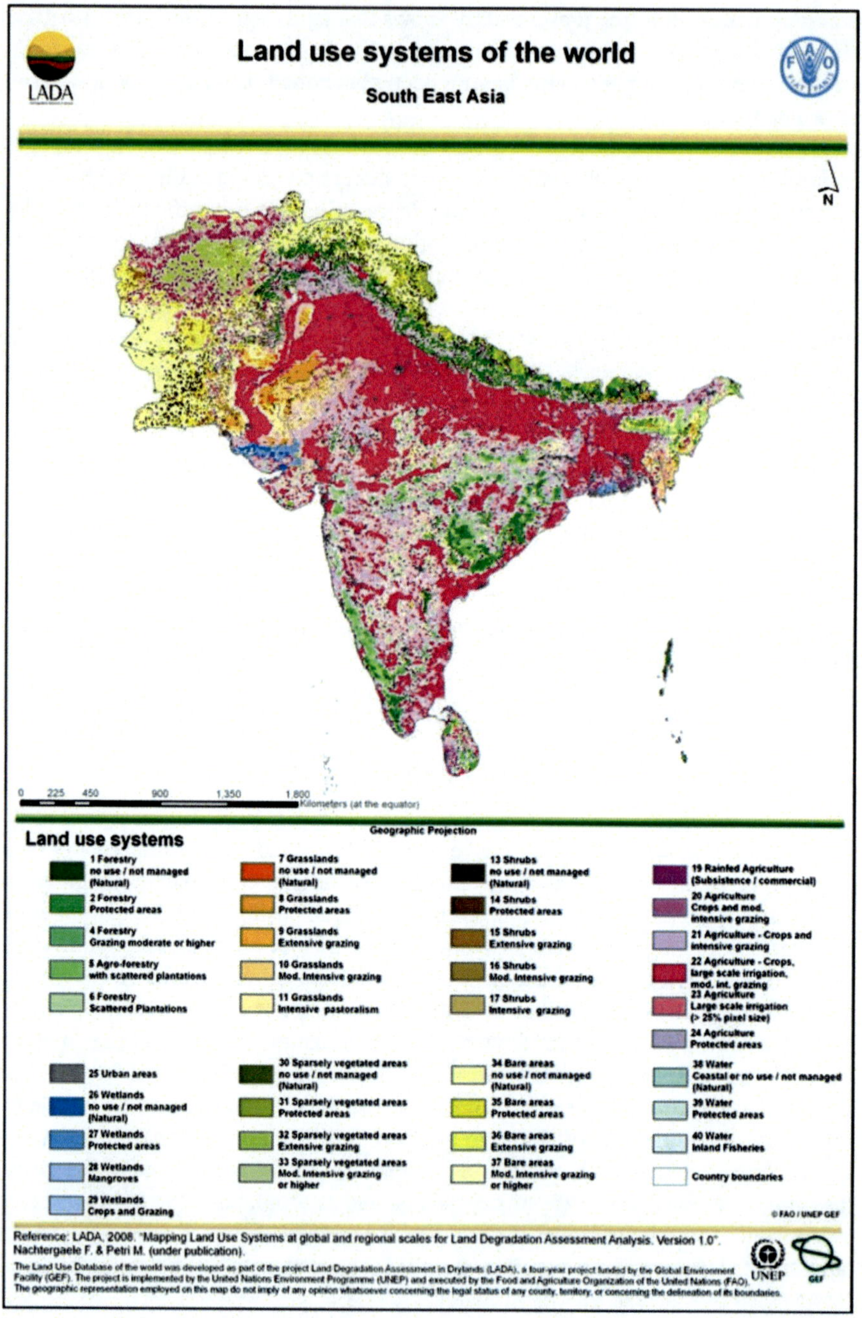

Fig. 17.2 Land use system in South Asia (Nachtergaele and Petri 2008)

It not only restores degraded soils, but also enhances biomass production and availability of surface and ground waters. Therefore soil C sequestration is considered a "win–win" strategy. Potential soil carbon restorative land use and management practices include adoption of conservation agriculture practices (minimum tillage with cover crops and crop residue mulch), integrated plant nutrient management (IPNM) with principally returning organic materials to soils, and other soil and water resources management. Sequestration of soil C may reach 50–1,000 kg ha^{-1} $year^{-1}$ and its global contribution is estimated to be 0.9 ± 0.3 Pg C $year^{-1}$ which offset one-fourth to one-third of the annual increase in atmospheric CO_2 estimated at 3.3 Pg C $year^{-1}$ (Lal 2004). However, sequestration of soil carbon is considered more a development and policy challenge than technical although it is already challenging for soils of dry and warm environment of the region.

17.4 Land Use and Management for Climate Change Adaptation and Mitigation

Adaptation to climate change is one of the approaches considered likely to reduce the impacts of long-term changes in climate variables. Adaptation is a process by which strategies to moderate and cope with the consequences of climate change, including climate variability, can be enhanced, developed and implemented, and is needed in diverse ecosystems, including agro-ecosystems (crops, livestock, grasslands), forests and woodlands, inland waters and coastal and marine ecosystems.

Climate change adaptation for agricultural cropping systems requires a higher resilience against both excess and lack of water. A key element to counter problems is SOM management (Fig. 17.1). SOM improves and stabilizes the soil structure and hence enhances absorbance and retention of water without causing surface run off, soil erosion, and downstream flooding. Depletion of SOM which was released into the atmosphere as GHGs (CO_2, CH_4 and N_2O) in South Asia is critical. Removal and burning of rice straw and wheat stubble aggravate the situation. Plowing also aggravate loss of SOM through mineralization and oxidation. It is also influenced by other factors including clay content and its mineralogy, soil temperature and soil moisture regimes (Lal et al. 2004).

Sustainable use and management of soil and water resources in South Asia is a major issue to be addressed. Soils in the warm, semi-arid and dry region of South Asia are coarse-textured and contain predominantly low-activity clay. For that environment, FAO promotes low tillage and maintenance of permanent soil cover that can increase SOM and reduce impacts from flooding, erosion, drought, heavy rain and winds (FAO 2007b). Among areas being explored are conservation agriculture, organic agriculture and risk-coping production systems that incorporate crop rotations, agroforestry, crop–livestock associations, crop–fish systems and the use of hedges, vegetative buffer strips and other farm landscaping practices. While intensive soil tillage reduces SOM through aerobic mineralization, low tillage and the maintenance of a permanent soil cover (through crops, crop residues or cover crops and the introduction of diversified crop rotations) increases SOM. A no- or low-tilled soil conserves the structure of soil for fauna and related macrospores (earthworms, termites and root channels) to serve as drainage channels for excess water. Surface mulch cover protects soil from excess temperatures and evaporation losses and can reduce crop water requirements by 30%.

17.4.1 Conservation Agriculture

Conservation agriculture and organic agriculture that combine zero or low tillage and permanent soil cover are promising adaptation options for their ability to increase soil organic carbon, reduce mineral fertilizers use and reduce on-farm energy costs. Risk-coping production systems, resilient to land and water modifications, require diversified structures in space and time such as crop rotations, agroforestry, crop–livestock associations, crop–fish systems and the use of hedges, vegetative buffer strips and other farm landscaping practices. Accomplishing this can have an enormous impact on adaptation to drought, heavy rains and winds.

Conservation agriculture systems for crops and pastures sequester carbon from the atmosphere into long-live SOM pools and contribute maintaining and increasing productivity, promote a healthy environment and strengthen rural communities. In general, 1.8 Mg CO_2 per hectare per year can be sequestered through adaptation of best conservation agriculture practices, and one-third of the current annual global emission of CO_2 from the burning of fossil fuels can be sequestered if it is practiced on five billion hectares of agricultural land. Early markets have shown that carbon offsets from conservation agriculture can be quantified, verified and traded. Carbon credit trading may provide an economic opportunity for farmers to adopt these ecologically based approaches to farming. The Conservation Agriculture Carbon Offset Consultation held on 28–30 October 2008 in Indiana, USA recommended the below priority actions (FAO 2008a).

1. Enable conservation agriculture to be recognized within any renegotiated clean development mechanism formed by the United Nations Framework for Climate Change Convention after the current expiration in 2012.
2. Promote closer interaction between government agencies, farmers, private sector, technology generators and disseminators, and nongovernment organizations in policy reform, as well as for the design and application of stewardship incentives for broad acceptance of conservation agriculture.
3. Mobilize the commercial sector to recognize the importance of ecosystem services provided by conservation agriculture, particularly its role in soil carbon sequestration.
4. Develop a science-based synthesis on how carbon might provide ecosystem services from conservation agriculture. This should include optimized measurement methodologies and determining the potential of soil carbon sequestration for defined crop management systems and ecoregions of the world.

5. Develop standardized protocols for applying science-based information to conservation agriculture projects that provide ecological goods and services using internationally accepted guidance, such as ISO 14064 for greenhouse gas emission reductions.
6. Seek government endorsement of conservation agriculture for development of national and international policies to value ecosystem services.
7. Support adoption of conservation agricultural technologies through payment for ecosystem services (e.g., soil carbon sequestration and water quality benefits).

Intensive rice-wheat cropping has caused not only reduced soil quality, but depletion of ground water in some regions. Identifying and including other possible crops such cotton, sunflower, sugarcane, maize, and vegetables in rotation would be particularly important. It is also important to identify alternative crops for rice which require more water. Improved rice farming may also reduce methane emissions. Those species should be adaptable in various cropping systems such as mixed or relay cropping to maximize resources use.

Conservation agriculture may be effective to sequester carbon into soil and biomass, but in some cases crop residues may be needed for animal feed, or leaving them on the ground may delay re-sowing. The economics of farming are harsh, especially in the developing world. So the several different farming systems, such as agroforestry, grazed grassland and arable land have also been considered (FAO 2001).

17.4.2 Agroforestry

Agroforestry, association of trees with crops or pastures, can represent a sustainable alternative to deforestation and shifting cultivation (Sanchez et al. 1999; Schroeder 1994; Sanchez 1995). It has a huge potential for carbon sequestration in croplands (Sanchez et al. 1999). Schroeder (1994) evaluated in tropical areas, carbon storage of 21 (sub-humid) to 50 Mg C ha^{-1} (humid) can be obtained with cutting cycles of 8 or 5 years, far shorter than for forests. Schroeder also estimated the land potentially available for conversion to agroforestry to be 160 Mha suitable in the tropics. The global C storage would be somewhere between 1.5 and 8 Pg.

Secondly, agroforestry systems can be established on unproductive croplands with low levels of OM and nutrients. The conversion to agroforestry would permit tripling the C stocks, from 23 to 70 Mg ha^{-1} over a 25-year period. Soil carbon sequestration for improved land management with some leguminous species (*Sesbania sesban, Tephrosia vogelii, Gliricidia sepium, Crotalaria grahamiana, Cajanus cajan*) can also supply 0.1 to 0.2 Mg N ha^{-1} $year^{-1}$. Agroforestry would be one of the interesting changes in land use related to C sequestration because of its relatively high rate of C gain (0.2 to 3.1 Mg ha^{-1} $year^{-1}$), mitigation of CO_2 emission from deforestation, and provision of a sustainable system from technical, ecological and economic points of view although there are some difficulties in social and cultural reasons.

17.4.3 Best Management Practices

The appropriate strategy to manage GHG emissions must involve intensive crop management practices that enhance nutrient use efficiency while continuing to achieve yield increase. Intensification does not necessarily increase the global warming potential (GWP) per unit of production of a cropping system (Dobermann 2007). High-yielding crops can mitigate GHG emissions through increased soil C storage, provided they are grown with Best Management Practices (BMPs) such as choice of the right combination of adapted varieties or hybrids, planting date, and plant population to maximize crop biomass production; use of tactical water and N management, including frequent N applications to achieve high N use efficiency with minimal opportunity for N_2O emissions; and use of residue management approaches that favoured a build-up of SOC as a result of large amounts of crop residues returned to the soil. Although little is known about their measured impact on GHG emissions by many fertilizer BMPs, improved crop recovery, reduced NH_3 loss and reduced NO_3-N leaching are the expectations among the other considerable potential to reduce fertilizer-associated GHG emissions and GWP.

Identified fertilizer BMPs are (1) choice of the N source that best matches the agronomic needs of the specific crop and soil system, and which minimizes risks of N loss via all loss pathways, particularly risks of N_2O emission; (2) use of appropriate N rates to optimize crop yield and minimize residuals of NO_3-N; (3) timing of N applications to coincide, as practically and logistically possible, with crop N uptake demand and to avoid losses to air or water; (4) use of balanced fertilization by supplying all required nutrients to increase crop growth, increase crop N use efficiency, and maximize crop capture of CO_2; and (5) use of the right combinations of source, rate, placement, and timing of N increase the probability of maximizing crop yields and farmer profits, while reducing the net GWP associated with the specific local crop and soil system (Snyder et al. 2007; Roberts 2007; Fixen 2007).

To enhance efficient and effective nutrient use by the cropping system or rotation, continuous system-level management planning may be needed. Intensive crop management in the system is essential to minimize nutrient losses from agro-ecosystems. Conservation practices such as crop rotation matched to specific site characteristics, effective irrigation scheduling and rates to match consumptive water and nutrient demand, cover crops to retain and recover residual inorganic N, and managed drainage and wetlands to further reduced N_2O emissions (Snyder et al. 2007).

17.4.4 Soil and Water Conservation

Climate change adaptation for agricultural cropping systems requires a higher resilience against both excess of water (due to high intensity rainfall) and lack of water (due to extended drought periods). A key element to respond to both problems is SOM that improves and stabilizes the soil structure, so that the soils can absorb higher amounts of water without causing surface runoff, which could result in soil erosion and, further downstream, in flooding. SOM also improves the water absorption capacity of the soil during extended drought. A broad range of agricultural water management practices and technologies are available to spread and buffer production risks. Enhancing residual soil moisture through land conservation techniques assists significantly at the margin of dry periods while buffer strips, mulching and zero tillage help to mitigate soil erosion risk in areas where rainfall intensities increase.

Soil and water management techniques have been developed under the FAO partnership of the World Overview of Conservation Approaches and Techniques (WOCAT), which identify which methods have proven workable under specific biophysical and socio-economic conditions. Land use planning approaches have been developed that stress participatory approaches for identifying priority areas at district and national level and identifying where investments are most needed under changing climatic conditions.

A report from Soil and Water Conservation Society (2003) identified the approaches to adapt soil and water conservation policies and practices to a changing climate regime are to take account of updating climatic parameters in critical conservation planning tools, investigating sensible estimates of the damage likely to occur, and evaluating benefits of building the risk of damage from severe rainstorms into the conservation planning process through risk-based assessments targeted to key conservation systems and environmental outcomes.

17.4.5 Land Cover Assessment and Monitoring

Land cover assessment and monitoring of its dynamics are essential for sustainable management of natural resources, assessing the vulnerability of ecosystems and planning food security and humanitarian programmes. However, there is still a lack of reliable or comparable baseline data. The Global Land Cover Network (GLCN) led by FAO and UNEP works to harmonize land cover definitions, standardize land cover baseline datasets, facilitate data acquisition and build capacity at the national and regional levels. FAO and the International Institute for Applied Systems Analysis (IIASA) took the initiative of combining the recently collected vast volumes of regional and national updates of soil information with the information already contained within the 1:5,000,000 scale FAO-UNESCO Digital Soil Map of the World, into a new comprehensive Harmonized World Soil Database (HWSD; FAO/IIASA/ISRIC/ISS-CAS/JRC 2008).

The AEZ methodology which utilizes a land resources inventory to assess for specified management conditions and levels of inputs, all feasible agricultural land-use options and to quantify expected production of cropping activities relevant in the specific agro-ecological context can be suggested. The characterization of land resources includes components of climate, soils and landform, which are basic for the supply of water, energy, nutrients and physical support to plants (FAO 2000).

Recent availability of digital global databases of climatic parameters, topography, soil and terrain, and land cover allow for revisions and improvements in calculation procedures.

17.5 Livelihood Adaptation Strategies

In order to increase resilience at all levels and reduce damage and losses from natural disasters, in 2003 the Government of Bangladesh approved the Comprehensive Disaster Management Programme (CDMP) as a key strategy to advance government and agency risk reduction efforts in the country (FAO 2006). The CDMP follows a strategic institutional and programming approach to address risks associated with climate variability and change. CDMP is designed to optimize the reduction of long-term risk and to strengthen the operational capacities for responding to, and recovering from, emergencies and disasters. Efforts to reduce vulnerability to long-term climate change have included livelihood adaptation strategies in the agricultural and allied sectors, particularly for women and poor communities with the lowest capacities to adapt.

FAO assisted the Government of Bangladesh and other key stakeholders in designing and promoting livelihood adaptation strategies in the agricultural sector to help reduce vulnerability to climate change, particularly among women and poor communities. The outputs and deliverables of this effort include (1) survey of climate impacts and local perceptions of climate hazards including assessment and analysis of capacities and coping strategies; (2) increased understanding of the effect of drought on agriculture and allied sectors; (3) analysis of climate analogues and climate change model outputs; (4) documentation of viable adaptation practices and options for livelihood adaptation; and (5) development of extension materials and awareness-raising methods (FAO 2006).

Successful local adaptation to climate variability and change requires multiple pathways with well planned interrelated short- and long-term measures, including:

- Adopting physical adaptive measures – such as excavation, re-excavation of canals, miniponds, irrigation, storage facilities for retaining rain water
- Adjusting existing agricultural practices – such as adjustment of cropping patterns, selection of drought-tolerant crop varieties; better storage of seeds and fodder; dry seedbeds; or adopting alternative, cash crops such as mango (*Mangifera indica* L.) and jujube (*Ziziphus jujuba*)
- Adjusting socio-economic activities – such as livelihood diversification, market facilitation, small-scale cottage industries, integration of traditional knowledge
- Strengthening local institutions – such as self-help programmes, capacity building and awareness raising for local institutions
- Strengthening formal institutional structures – such as local disaster management committees and financial institutions
- Formulating policy to catalyze enhancement of adaptive livelihood opportunities

- Creating awareness and advocacy on climate change and adaptation issues
- Supporting better research – such as on-farm links to new or improved crops including drought-tolerant varieties, and other conducive and adaptive technologies.

17.6 Biofuel

The remarkable increase in the development of the biofuel sector is a response to the virtual panacea of benefits being associated with it. Few other sources of energy can arguably contend to mitigate climate change, enhance national energy security and lead to re-vitalized rural development through the increase of relatively cheap and easily produced agricultural feedstocks. The benefits attributed to biofuels are, however, mirrored by a list of potential threats and pitfalls. Biofuels have been debated in a wide range of social and environmental aspects. Foremost among these high profile issues are food vs. fuel, increased food prices, loss of land access by vulnerable groups, increased demands on water supplies, impacts on forests and biodiversity, and questionable climate change contributions.

The implications of increased biofuel use are that increased land will be converted to the production of biofuel feedstocks. This may occur through several processes, but most likely a patchwork combination of approaches. Portions of farmland under current cultivation of food crops will inevitably be turned over to higher value biofuel crops. Increased demand for arable land will presumably be met as well through the conversion of forested areas, particularly where enforcement and tenure arrangements are weak (FAO 2008b).

Land use planning approaches have been developed that stress participatory approaches for identifying priority areas at district and national level and identifying where investments are most needed under changing climatic conditions. However, any diversion of land from food crops to production of energy crops will influence food prices as both compete for the same inputs. Conflicting demands for arable land among food, fiber, biomaterials and energy production interests cannot be avoided. In India, biofuel plantations are planned for wastelands, which are considered not suited to cultivation of food crops. The categorization of such lands as wastelands has been disputed given the dependence of the rural poor on those lands for grazing and fuelwood collection. It is imperative that their tenure and access rights be strengthened by appropriate policies.

17.7 Conclusions

Improve conservation and efficient use of land, water and bio-diversity resources through available technologies and by ensuring effective reforms in policy are deemed necessary. South Asian governments took significant initiatives to strengthen institutions, improve regulatory systems, and financial and policy reforms (ESCAP 2000).

However, a number of common limitations observed are legal, industrial, policy and NGO communities. Legislation failed to respond to changing paradigms of development and many resource laws and statutes became obsolete. Likewise, regulatory framework is weak and implementation and enforcement are challenging. Environmental policies showed some successful cases such as local level decentralized approaches have larger impact on resource management. The environmental management, the increased involvement of the public sector in environmental decision making, and poverty alleviation are to be further emphasized.

Many countries already are adapting to current climatic events at national, provincial, state, district and local levels in short-, medium- and long-term time frames. However, in the past, many structural, physical and institutional adaptation mechanisms, implemented through conventional top-down approaches, lacked community participation and livelihood focus. Appropriate adaptation strategies also require balancing the need to reduce climate change impact with any constraints in national policy-making processes.

To cope with climate change and increase agricultural productivity through conservation and efficient use of agricultural land and water resources with technical, policy support and environmental considerations, improved land management and land-use planning are possible adaptation methods. Sustainable land management combines technologies, policies and activities aimed at integrating socio-economic principles with environmental concerns to simultaneously maintain or enhance productivity, reduce the level of production risk, protect natural resources and prevent degradation of land and water resources, economically viable and socially acceptable (FAO 1999). Monitoring of climate change may also mitigate the serious implications for the South Asia's agriculture.

It is suggested to improve land-use planning and management through technical and policy support, and increase agricultural production/productivity through conservation and efficient use of agricultural land and water resources with technical and policy support and environmental considerations. Bio-diversity conservation and protection should be promoted through proper planning and policy support.

As the weight of evidence in widespread land degradation in the region has been clearly recognized, immediate integrated actions should be taken to combat both the direct and the indirect causes of land degradation and climate change. Two major proposals are assessment of the severity and extent of the problem and its effects and action to check and reverse land degradation which increase resilience and reduce risks affected by changing climate.

References

ADPC (2007) National Platforms for Disaster Reduction – study on current status of disaster reduction, institutional arrangements and potential for national platforms for disaster reduction in 3 South and South East Asian countries. ADPC, Thailand

ADPC (2008) ADPC website. Country profiles. www.adpc.net

Bai ZG, Dent DL, Olsson L, Schaepman ME (2008) Global assessment of land degradation and improvement 1: identification by remote sensing. Report 2008/01, ISRIC – World Soil Information, Wageningen

Bajracharya SR, Mool PK, Shrestha BR (2007) Impact of climate change on Himalayan glaciers and glacial lakes: Case studies on GLOF and Associated Hazards in Nepal and Bhutan. International Centre for Integrated Mountain Development (ICIMOD). Kathmandu, Nepal.

Dent FJ (1990) Land resources of Asia and the Pacific. Problem soils of Asia and the Pacific. Report of the expert consultation of the Asian Network on problem soils. Bangkok, Thailand. 29 August to 1 September 1989. RAP Report 1990/6. FAO Bangkok

Dobermann A (2007) Nutrient use efficiency – measurement and management. 22 pp. Proceedings of the International Fertilizer Industry Association (IFA) Workshop on Fertilizer Best Management Practices. Brussels, Belgium, March 7–9, 2007

ESCAP (2000) State of the environment in Asia and the Pacific. New York

ESCAP (2006) State of the environment in Asia and the Pacific 2005: economic growth and sustainability. Thailand

FAO (1994) Land degradation in South Asia: its severity causes and effects upon the people. FAO, Rome

FAO (1999) Poverty alleviation and food security in Asia: land resources. RAP Publication 1999/2

FAO (2000) Global agro-ecological zones version1.0 FAO land and water digital media series. FAO, Rome

FAO (2001) Soil carbon sequestration for improved land management. FAO World soil Resources Reports 96. FAO, Rome

FAO (2006) Livelihood adaptation to climate variability and change in drought-prone areas of Bangladesh – developing institutions and options Livelihood adaptation to climate variability and change in drought-prone areas of Bangladesh. FAO, Rome

FAO (2007a) Selected indicators of food and agricultural development in the Asia–Pacific region 1996–2006. RAP Publication 2007/15. FAO, Bangkok

FAO (2007b) Adaptation to climate change in agriculture, forestry and fisheries: perspective framework and priorities. FAO, Rome

FAO (2008a) Climate change and disaster risk management – technical background document from the expert consultation held on 28 to 29 February 2008. RAO, Rome

FAO (2008b) Implications of biofuel development on land use, food security and the environment. Regional forum on biofuel sector development: challenges, opportunities, and the way forward – 23–25 January 2008, Bangkok, Thailand, FAO RAP

FAO/IIASA/ISRIC/ISS-CAS/JRC (2008) Harmonized world soil database (version 1.0). FAO Rome Italy and IIASA, Laxenburg, Austria

Fixen PE (2007) Can we define a global framework within which fertilizer best management practices can be adapted to local conditions? pp. 77–86. In: Fertilizer best management practices: general principles, strategy for their adoption, and voluntary initiatives vs. regulations. IFA International Workshop on Fertilizer Best Management Practices, 7–9 March 2007. Brussels, Belgium

Lal R (2004) Soil carbon sequestration to mitigate climate change. Geoderma 123(1–2):1–22

Lal R, Hobbs PR, Uphoff N, Hansen DO (2004) Sustainable agriculture and the international rice-wheat system. CRC Press, Boca Raton

Nachtergaele F, Petri M (2008) Mapping land use systems at global and regional scales for land degradation assessment analysis. FAO, Rome

Roberts TL (2007) Right product, right rate, right time and right place...the foundation of best management practices for fertilizer. In: Fertilizer Best Management Practices: General Principles, Strategy for their Adoption, and Voluntary Initiatives vs Regulations. IFA International Workshop on Fertilizer Best Management Practices, 7–9 March 2007, Brussels, Belgium, pp 29–32

Sanchez PA (1995) Science in agroforestry. Agroforest Syst 30:5–55

Sanchez PA, Buresh RJ, Leakey RRB (1999) Trees, soils and food security. Philos Trans Roy Soc Lond Ser B 353:949–961

Schroeder P (1994) Carbon storage benefits of agroforestry systems. Agroforest Syst 27:89–97
SDMC (2009) SDMC Newsletter, January 2009, vol. III, no. 1, SAARC
Snyder CS, Bruulsema TW, Jensen TL (2007) Greenhouse gas emissions from cropping systems and the influence of fertilizer management – a literature review. International Plant Nutrition Institute, Norcross, GA
Soil and Water Conservation Society (2003) Conservation implications of climate change: soil erosion and runoff from cropland Ankeny, Iowa
UNEP (1997) Global environmental outlook 1, New York

Chapter 18
Challenges and Opportunities in Composting Organic Waste

Harold M. Keener

Contents

This paper is adapted from Keener, H.M. 2004. Opportunities and challenges in composting organic waste. pp. 27–46. Proceedings International Symposium on Organics Recycling (ISOR), October 5–7, Akita, Japan.

H.M. Keener (✉)
Food, Agricultural, and Biological Engineering, The Ohio State University/OARDC,
Wooster, OH 44691-4096
e-mail: keener.3@osu.edu

R. Lal et al. (eds.), *Climate Change and Food Security in South Asia*,
DOI 10.1007/978-90-481-9516-9_18, © Springer Science+Business Media B.V. 2011

Abstract Minimizing waste generation and recycling have become the focus of governmental agencies in many countries to reduce human impact on the environment. With this focus, composting has received a high ranking in the hierarchy of recycling methods and continues to gain importance throughout the world for the conversion of organic by-products to new resources. Adoption of composting for capturing and recycling organics however is not without environmental and economic issues and requires approaches which manages the process from the cradle (source separation) to the grave (high value markets). This paper addresses the challenges and opportunities in composting organic matter. Its focus is on reducing operational cost and managing odors. Composting systems and principles are presented along with results of optimization studies based on engineering analysis as well as pilot scale and large scale composting studies. Results are from studies on many organics including, municipal solid waste (MSW), biosolids, short paper fiber (SPF), yard trimmings, and animal manures. A list of guidelines for developing and managing efficiently composting systems are presented. . Compost value for growing plants, both as a fertilizer and biocontrol agent are outlined along with other uses in protecting the environment.

Keywords Waste management • Municipal solid waste • Biosolids • Clean air act • Clean water act • Windrow

Abbreviations

Letters

E_a	Activation energy in Arrhenius equation
k	Rate of decomposition of dry matter per unit of compostable dry matter: day^{-1}
k_{vs}	Rate of decomposition based on volatile solid: day^{-1}
m	Dry mass (refers to compost if no substance identified): kg
MiR	Mixing ratio decimal fraction of material of interest to total weight of mix, dry basis: kg kg^{-1}
R	Ideal gas constant in Arrhenius equation
T	Temperature (refers to compost if no substance identified): °C
T*	Optimum temperature: °C
w	Moisture content dry basis: kg kg^{-1}
w	Moisture content *wet basis*: kg kg^{-1}
X	Relationship for k as a function of compost variables X_T, X_w, X_{O_2}, X_φ, $X_{C/N}$.
Y_{O_2}	Mass concentration of oxygen in compost free air space: kg_{O_2} kg_a^{-1}

Greek Letters, Symbols

*	Optimu m or chosen operating point
Φ_w	Moisture loss term aeration factor
∞	Infinity

Letter Subscripts, Superscripts

a	Dry air
abs	Absolute
BL	Broiler liter
c	Dry compost
CAFO	Concentrated animal feeding operations
C/N	Carbon to nitrogen ratio
max	Maximun
0	Initial value at time zero
MiR	Mixing ratios
MSW	Municipal solid waste
PFRP	Process to further reduce pathogens
PM10	Particulate matter <10 μm
RCRA	Resource Conservation & Recovery Act
SPF	Short paper fiber
T	Temperature
USEPA	US Environment Protection Agency
VOC	Volatile organic compound
vs	Volatile solids
w	Water

Conversion Table

g	0.002205 lb (pound)
kg	2.2 lb
Mg	1,000 kg or 2,200 lb
m	3.281 ft (feet) or 1.094 yd (yard)
cm	0.03937 in. (inches) or 0.03281 ft
m^3	35.31 ft^3 or 1.308 yd^3
kg m^{-3}	0.0623 lb ft^{-3} or 1.682 lb yd^{-3}
MJ	948.6 BTU (British Thermal Units)
MJ kg^{-1}	431.2 BTU lb^{-1}
ha	2.471 ac (acres)
Mg ha^{-1}	890 lb/ac
m^3 kg^{-1} day^{-1}	16.05 ft^3 lb^{-1} day^{-1}

18.1 Introduction

Composting is a widely used process throughout the world for the treatment of solid waste (Hoitink et al. 2002). In the USA more than 8,500 composting sites were operating in 1997 (Composting Council 1997) with 15 for mixed municipal solid waste (MSW), 250 for municipal biosolids; 138 for food waste, 3,316 for yard trimmings, and >5,700 for farm residuals. Of the farm composting operations, over 5,000 (Kashmanian and Rynk 1996) were composting animal mortalities. Since then composting sites for all categories have increased except for MSW. For example mortality composting has increased exponentially due to economic and bio-security considerations and in the state of Ohio alone over 3,200 farmers have been added to the list doing mortality composting (Keener et al. 2002b). In addition backyard composting by homeowners has also increased and some industries now use composting as a method for the destruction of toxic by-products. This paper outlines the challenges and opportunities for composting to capture/recycle more of organic waste stream citing examples from the USA experiences. It also presents basic concepts to composting organics safely and efficiently. It does not however, purport to claim that composting is the best choice in all cases for recycling organic waste.

18.2 Challenges and Opportunities for Waste in the USA

The United States Environmental Protection Agency (USEPA) notes, "Nearly everything we do leaves behind some kind of waste. Households create ordinary garbage. Industrial and manufacturing processes create both solid waste and hazardous waste" (USEPA 2008a). USEPA regulates all this waste under the Resource Conservation and Recovery Act (RCRA). The goals of the act are to: protect people from the hazards of waste disposal, conserve energy and natural resources by recycling and recovery, reduce or eliminate waste, and clean up waste which may have been spilled, leaked, or disposed of improperly. To accomplish these objectives, the USEPA has developed a hierarchy for waste management – source reduction/reuse, recycling, *composting*, combustion, and landfills. Composting was selected as a management scheme because it can meet RCRA's objectives by recycling natural resources in an environmentally safe way that are important in the production of agronomic, landscape and nursery crops.

18.2.1 Minimize the Waste Stream

Although reducing the waste stream won't necessarily insure more organics will be composted, it will reduce total societal disposal cost and potentially lead to more monies for investing in recycling programs such as composting. Spiegelman and Sheehan (2004) discuss the future of waste and noted that managing product waste effectively is a responsibility shared by both producers and consumers, but it really begins with product design. These concepts are being advanced in the US, and have already been incorporated into other societies such as Japan and parts of Europe. In this respect, the USEPA received in May, 2004 the prestigious 2004 Shingo Prize for Research, a prize named for the Japanese industrial engineer Shigeo Shingo (USEPA 2008b). The award was for the USEPA's National Waste Minimization Partnership

Program, which promotes reducing the amounts of waste generated and lowering the toxicity and persistence of those wastes that are, of necessity, generated.

18.2.2 Source Separated MSW

The USA with a population of over 299,000,000 generated 228,000,000 Mg of MSW in 2006, or 2.09 kg inhabitant^{-1} day^{-1} (USEPA 2008c). Analysis of the trends over the last 16 years gives an annual growth rate of 1.3%, although from 1990 to 2006 only a minimal increase per inhabitant occurred. Makeup of this waste stream is organics (paper, yard trimmings and food scraps) at 59.3% and non-organics (plastics, metals, rubber, leather, textiles, glass, wood, etc.) at 40.7%. Currently 32.6% is recovered (24.3% recycled, 8.3% composted), 12.5% burned and 55% disposed of in landfills. Kaufman et al. (2004) put the MSW generated at 335,000,000 Mg/year for the USA and discusses in detail the difference between their number and the USEPA.

Compost systems for mixed MSW have had high operational cost and the compost produced has only limited uses (Hoitink et al. 1993b; Spencer 2004). This has lead to the increased emphasis on source-separated organics for composting. Operational cost for controlling odors and odor nuisance complaints at these facilities can also be cited as causing their shutdown. Composting source separated yard trimmings from MSW has proven successful with over 3,200 sites in the USA (Kaufman et al. 2004). In many instances, mulch as well as finished compost is generated from the trimmings. Growth in source separation of yard trimmings can be linked to bans on land filling of yard trimmings by individual states in the 1990s. In states where a ban is not mandated, high tipping fees at landfills can lead to recycling the yard trimmings by composting.

Composting source separated food waste has increased only slightly in the USA because grocers, restaurants, and produce distributors are not willing to participate in source separation (Jamieson et al. 2004). The major issue with source separating food has been the extra labor and expense required to sort. In addition, waste haulers generally prefer co-mingled waste hauled directly to landfills because of its lower cost. A second reason for the difficulty of food waste composting being adopted is the lack of expertise by compost operators in managing high moisture by-products and preventing odor, along with high cost per ton for the finished compost. Development of new concepts which reduce labor and management cost of separation of the MSW is needed. Such items as biodegradable packaging and eating utensils may lead to increased source separation by commercial businesses.

A different approach to centralized composting being promoted for food waste is on-site composting using manufactured small scale composting systems (Deller 2004). One example is the Earth Tub™ that was designed specifically for on-site composting of food-wastes at schools, universities, restaurants, hospitals and supermarkets. It is a fully enclosed composting vessel featuring power mixing, compost aeration, and biofiltration of all process air. Their adoption may be limited, however, because cost of installing and operating (maintenance, labor, energy) may be much greater than cost associated with offsite composting facilities.

18.2.3 Biosolids

The North East Biosolids and Residuals Association (NEBRA 2007) estimated that in 2004 a total of 6,527,300 Mg or 60.9 g inhabitant^{-1} day^{-1} of dry biosolids were generated in the U.S. This is equivalent to 1.22 kg inhabitant^{-1} day^{-1} of biosolids at 5% solids. Beneficial use of these biosolids in 2004 was 3,188,000 Mg or 49% of the waste stream. This is down from the 59% estimated in 1998. Beneficial use was 74% for agriculture, 1% forest land, 3% reclamation, and 22% class A distribution (includes composting and dried pellet fertilizer). This left 51% of the waste stream for disposal by land filling, incineration, and other.

Direct land application has been a lower cost method of handling biosolids in the past, costing ≈1/2 that of composting. For the City of Columbus, Ohio cost in 2004 was $90 per dry Mg for land application and $190 per dry Mg for composting (Hoff 2004). Columbus's compost product, Com-Til, is available for purchase in both 18.2 kg bags or in bulk quantities. In 2004 the traditional Com-Til product's retail bulk pickup price was $19.60 m^{-3}, while the new Com-Til Plus soil conditioner was $13.10 m^{-3}. A challenge to the continued land application of biosolids is public concerns over pathogens, air emissions and water contamination. In addition, in wet climates seasonal wet soils affects timeliness of disposal as well as increasing environmental risk.

Recent improvements in solids separation for biosolids has increased the sludge cake solids content from ~22% to ~30% w/w (Goldstein 2004). This dramatically reduces the amount and cost of amendments required to manage moisture content. However, NH_3 emissions would be expected to increase if C/N decreases due to higher biosolids to amendment ratios. Since biosolids composting facilities are continually challenged by odors, NH_3 emissions and high cost of operation, new ways to lower cost are being tried. One way receiving much emphasis is anaerobic digestion of biosolids to produce methane prior to utilization of the residuals for beneficial use (Schmack BioEenergy 2009).

One approach for small communities to compost biosolids is the example from Lenoir City, TN (Blackburn 2004) The approach is a static pile consisting of 20 cm of a biosolids (9% solids) and green woodchips mixture (1:3 ratio) placed on a bed of woodchips 18–20 cm deep, and then covered with 20–30 cm of clean woodchips. Results on temperature indicated the USEPA's PFRP (Pathogen Free Reduction Process) of 55°C is being met with no turning of bed. In addition, Salmonella tests at interface have met USEPA standards. No cost is given, but for small cities this concept may be economical to produce compost that can be safely land applied. This concept is similar to that used for composting animal mortality (Keener et al. 2000b).

18.2.4 Agricultural Waste

Agricultural waste involves plant residues (such as plant materials from greenhouses, weather damaged forage crops, crop residues, etc.) and animal manures. Due to the nature of the plant residues they can often be composted using low tech approaches with minimal cost. However, animal manures are often high in moisture (fresh dairy and swine manure range from 88% to 98%, caged layer manure ≈75% moisture) and cannot be composted without amendments (Michel et al. 2004) or in the case of caged layer manure drying to <55% moisture (Elwell et al. 1998). Poultry litter and horse manure are solid manures, ranging in moisture from 30% to 55%, depending on the levels of bedding used. All of the animal manures, with the exception of horse manure/bedding, have the potential to emit large quantities of NH_3 because of low carbon to nitrogen ratios (C/N <<25).

Location in the USA does affect approaches to composting animal manures. In the southwest or dryer climates, windrow systems predominate with minimal bedding. Often the manure being composted comes from a screen separator or screw presses (Katers et al. 2003). Jurgens (2004) describes composting at New Era Farms. This 30-year-old operation produces 70,000–90,000 Mg annually, drawing on solid and liquid manures from 10 of the larger dairies within a 60 mi. radius in San Joaquin Valley. The composting is done in windrows and takes 90–120 days. Finished compost is uniform in particle size, stable, dark brown (not black which indicates too hot a process, i.e. >70 C) and is marketed to almost 400 growers farming over 405,000 ha, of which 115,000 ha are biologically influenced fertility and 14,160 ha are certified organic. Charge for stable compost was \$13–\$23 Mg^{-1}. Blending gypsum or lime with the compost may be done, depending on pH of soil to which compost is being applied.

Composting dairy manure in the Midwest is best done under roof or with the option of covering outside windrows with a fleece blanket. Results of studies (Keener et al. 2002c; Michel et al. 2004) showed late fall and winter composting resulted in high moisture contents in the finished compost if no covers were used.

18.2.5 Innovative Uses of Compost

Using compost as a soil amendment has been known for centuries. It improves soil tilth, reduces erosion, provides plant nutrients and helps control plant diseases. Today new uses are being found for compost and composting which help save money, reduce the use of chemical fertilizers and pesticides, and conserves natural resources. A list of these uses are: bioremediation and pollution prevention, erosion control, turf remediation, and landscaping and disease control for plants and animals and animal mortality composting (USEPA 2009b).

18.2.6 Composting Regulations

USEPA compost regulations cover biosolids and materials containing biosolids. These rules, identified as USEPA 503 (Epstein 1997) address pollutants (hazardous organics, heavy metals, pathogens) that can impact the environment. For a detailed description of EPA 503 visit EPA's website (USEPA 2009a). Other regulations on composting are set by the individual states. In Ohio, current regulations for non-biosolids require testing of the finished compost for pathogens, heavy metals, organic chemicals (PCB, Benzene, etc.), pathogens, and foreign matter (plastics, glass, textiles, etc.) before marketing, a goal oriented regulation. Also, testing is required for boron, maturity, pH, salinity, N, C, P and K, although no limits are set. Current rules regarding solid waste composting can be found under Chapter 3745-27 of the Ohio Administrative Code (OAC), specifically Rules 3745-27-40 to 3745-27-47 (Ohio EPA 2009).

One part of USEPA regulations is the Process to Further Reduce Pathogens (PFRP). The PFRP requires (1) maintaining the compost temperature at ≥55°C for 3 days for a within-vessel compost method or the static aerated pile method, or (2) if composting is done in turned windrows the criteria is five turns over 15 days while achieving ≥55°C in the windrow. The goal is minimizing the risk of pathogens in the compost. This PFRP has been adopted as a requirement for compost to be sold to certified organic farmers, even though the compost is made from yard trimmings, food waste and manures. Such regulations focus on the process and not the desired outcome, and by so doing eliminates using low cost composting practices which can still provide pathogen control. Outcome based criteria (such as done in Ohio) on allowable pathogen levels should be the basis of regulations and not a specific processing procedure.

18.2.7 Clean Air Act

The state and federal EPA's are promulgating new rules and regulations under the clean air act. These rules apply directly to waste handling and treatment facilities, including composting. Air regulations emphasize ozone, dust (particulate matter PM10 and PM2.5), CO_2, SO_2 and NO_X (USEPA 2004), although NH_3, CH_4, and volatile organic compounds (VOCs) are also considered from a greenhouse gas effect, or precursor to PM formation. Odors are considered from a nuisance standpoint and are the concern of local and state regulations. In the past, odor has been the primary cause of facilities being shut down. Development of biofilters for VOCs and odor and chemical scrubbers for NH_3 have proven effective in many instances, but the high cost of operation and treatment needs to be improved upon.

US agricultural enterprises are facing new regulations associated with the Clean Air Act. This is evident by new federal legislation for CAFOs (concentrated animal feeding operations) to regulate emissions, as well as stepped up enforcement of provisions in the Clean Air and Water Acts. Recently, Ohio's largest egg producer "agreed to a comprehensive Clean Air Act settlement under which the company would spend $1.6 million to install and test innovative pollution controls to dramatically cut air emissions of particulate matter and NH_3" (Department of Justice 2004). In the San Joaquin Valley an estimated 1,350 farms and dairies now face regulations on smog forming emissions (Anonymous 2004). These challenges are leading to innovative concepts in agriculture, many of which center on composting as a way to create a value added 'by-product' with multiple marketing opportunities. New facility design and management, such as the High Rise™ hog facility coupled to windrow composting (Keener et al. 2001a,b) or manure belt systems for caged layers coupled to in-vessel composting (Keener et al. 2002a), high rise poultry + deep pile composting/storage, and aerated pads with windrow turning for dairy manures are new approaches being developed and adopted. These concepts and others are addressing the odors, flies and pathogens associated with livestock manures, while closing the loop on nutrient cycling by enabling manure nutrients to be cycled back into crop production systems.

18.2.8 Clean Water Act

The state and federal USEPA's are promulgating new rules and regulations under the clean water acts. These rules apply directly to waste handling and treatment facilities, including composting. Currently, in Ohio, all runoff from outdoor composting sites must be contained and reused back into the composting operation or disposed of by irrigating onto land with minimal potential for direct discharge into the "waters of the state".

The challenges and opportunities facing expansion of composting in the USA are varied and will require understanding the design and operation of composting facilities. The remainder of this paper focuses on the current knowledge base associated with the design and management of facilities and ways to minimize fixed and variable cost while controlling odors, pathogens, leachate and dust. In addition it presents information on developing new uses and markets for compost.

18.3 Composting Process

In conventional composting, ingredients are mixed and then placed in a pile to allow bacteria, fungi and other microorganisms to break down the organic materials to a stable mixture called compost. The organisms consume oxygen and release heat, water, and carbon dioxide (CO_2). In some cases NH_3 and other gases are emitted. Generally the mix is turned every 3 or 4 days, but sometimes every day or only weekly or monthly. In some systems air is forced through the compost to control

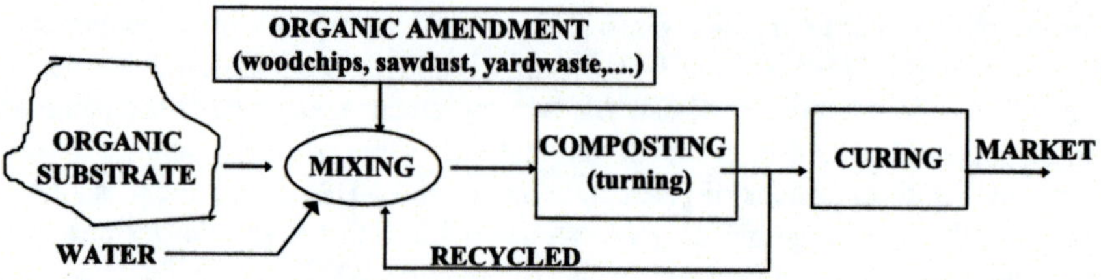

Fig. 18.1 Material flow in conventional composting. Airflow into the composting and curing pile, also a part of the process, is not shown in the figure

temperature and keep the pile supplied with oxygen as composting occurs in a predominantly aerobic environment (Rynk 1992; Haug 1993). When little or no heating is observed, the material is removed, remixed and placed in a curing pile for several months before marketing. Length of time for any one phase of composting is affected by the composition of the parent materials, processes temperature, oxygen levels, C/N ratio, particle size, turning frequency, and many other factors. Length of curing depends on the aforementioned factors and market opportunities (uses) for the compost.

Figure 18.1 is a schematic of the material flow in the process. Volume and mass reductions as high as 75–80% of parent materials have been noted for composting (Michel et al. 2004). A description of the composting process, compost maturity and compost stability are given in numerous references (Haug 1993; Hoitink and Keener 1993; Keener et al. 2000a,d).

18.3.1 Composting Systems

Composting systems can be classified as static piles, aerated static piles, turned windrows and aerated turned windrows. The terms windrow and pile can be used interchangeably in these descriptions. The terms in-vessel and tunnel indicate that the pile or windrow is contained within a structure. Which system represents the best technology depends on the material to be composted and takes into account not only environmental issues (health, safety, public nuisance, etc.) but also the economics. Table 18.1 is an overview of systems which have been used for a particular material. Brief descriptions of the various systems are presented here. It should be noted, *success of any system occurs only when good management practices are followed.*

18.3.1.1 Static Pile

Composting material is placed in a pile (or windrow) and left to compost with minimal turning and without using forced ventilation (Haug 1993; Rynk 1992). This

Table 18.1 Composting concerns for specific substrates and the composting systems used

Material	Odor	Pathogens	Toxic	Technology used
Woody materials				SP,TP
Leaves				SP,TP
Lawn clippings	X			SP,TP,ASP
Food waste	X	X		SP,ASP,TP,ATP[a]
Mortality	X	X		SP,ASP,TP,
Biosolids and manures	X	X	X	SP,ASP,TP,ATP[a]
Municipal waste	X	X	X	SP,ASP,TP,ATP[a]

SP, static pile; ASP, aerated static pile; TP, windrow with turning; ATP, aerated turned windrow (in-vessel, etc.)

[a]System chosen depends on mixing ratios of materials. If a large portion of the material composted is woody then the static pile and turned pile systems are appropriate under good management (good site selection, well-defined operations protocol)

approach is generally used with materials unlikely to generate offensive odors if anaerobic; materials such as leaves and ground yard waste which have high C/N ratios (>50). When escaping odors may be a problem, the pile is capped with a biofilter type material to trap and destroy the odors. This latter approach is used for dead animal composting (Keener et al. 2000b) by capping the pile with 25–50 cm of biofilter. These systems are usually turned once at the end of the high rate phase of composting and once between the stabilizing phase and the curing stage. Screening often proceeds the marketing of material to remove non-organic contaminants and provide a uniform product less than 1 cm in size. The over sized material left from screening, if woody, is often mixed with the incoming material for future composting.

18.3.1.2 Aerated Static Pile

Composting material is placed in piles (or windrows) with ventilation ducting underneath the piles (Haug 1993; Rynk 1992). Past studies on aerated pile systems show they work best using forced ventilation as opposed to suction because of higher pressure drops and accumulation of water in the piping under suction. Even so, many systems ventilate by negative-pressure on the compost and exhaust the resulting gases through a biofilter system (Haug 1993). The rate of aeration required to prevent anaerobic metabolism within the compost varies with the product being composted (Keener et al. 1997a, 2000d, 2002d, 2008).

Pile height is generally limited to 2–1/2 m and less than 30 m length to balance airflow required to control temperature, static pressure drops, and fan power requirements. Spacing of ducts is usually controlled by pile height since duct spacing is approximately equal to the height of the pile. A block approach is often used to maximize use of space. Pipe sizing, both header and aeration duct, and hole placement in the aeration duct are often critical for successful operation of this system

(Rynk 1992). Improper design leads to over ventilation of the pile in one location and under ventilation (with accompanying odor and low decomposition rate) in other places. Matching pile depth to material and moisture of material is critical for the success of this system (Das and Keener 1997). Biosolids mixed with a bulking agent are often composted with this system (Haug 1993; Elwell et al. 1994).

18.3.1.3 Turned Windrow

Composting material is placed in a windrow (pile) and is turned at regular schedules. Pile heights are generally limited to 1.5–2.5 m because of the ability of equipment to handle material and porosity considerations. Equipment ranges from tractors with buckets to payloaders and from pull type turners handling 1.5 (height) × 2.5 m (width) windrows to straddle turners handling 2.5 × 3 m windrows. Length of windrow is generally not critical, but will depend on site location. Windrow systems have been adopted to handle many types of materials such as cattle manure, yard trimmings and MSW.

Compost porosity >30% is important to minimize anaerobic conditions since windrows, as with the static piles, depend on natural convection (chimney effects) to ventilate the pile. Excessive porosity (>70%) however can prevent the windrow from self heating (Michel et al. 2004). Turning the pile can incorporate oxygen, but if the pile lacks porosity, the center of the windrows become anaerobic in a matter of minutes (Hoitink et al. 1993b). Turned windrow can suffer from lack of moisture in dry climates. Many turners designed today introduce water back into the windrow to maintain conditions favorable for high rate composting.

Windrow systems often generate some odors early in the process with highly putrescible material. One method to minimize odor is to keep pile temperature below 58°C as this will minimize volatilization of odorous compounds (Miller 1993). This is accomplished by regulating pile size, moisture content and porosity. Another method is to cap the piles with biofilter type material to trap escaping odors and delay turning the windrows until the temperature of the high rate phase of composting has declined to <58 C, and oxygen levels have recovered in most regions of the compost pile.

18.3.1.4 In-Vessel

This system is an aerated turned windrow with sidewalls. Historically, it has included upright silos, flat rectangular structure with bays, structure with walking floors, etc. (Haug 1993) gives detailed descriptions of in-vessel systems that have been built and used for composting. Operating principles of these systems have been well defined (Haug 1993; Keener et al. 1993). A roofed version of this system is the tunnel system, which completely controls the path of exhaust air leaving the system. In-vessel systems can be used for all types of "prepared (particle size is critical) materials" and have the capabilities of controlling emissions. An approach used to reduce cost with such technology is to use two composting systems – an in-vessel system for 7–21 days and an aerated pile till material is through the high rate phase of composting. In-vessel systems have high fixed and operating cost and should be selected based on materials to be composted and site requirements, rather than an arbitrary criterion of best available technology.

18.3.2 Controllable Factors

While composting does occur naturally, efficient composting requires the control of many factors to avoid nuisance problems such as odors and dust. Keener et al. (2000a, d) identified over 20 controllable factors affecting the cost and performance of the composting system (Table 18.2). Of these factors, nutrient balance, moisture content, particle size, porosity and pH are the most important in formulation of the compost mix. Factors such as oxygen concentration, temperature, and

Table 18.2 Factors affecting performance and operational cost of composting system[a]

Organic amendment	Ambient temperature
Bulking agent	Aeration schedule
Percent recycled compost	Percent recycled air
Nutrient balance, C/N ratio	Stirring frequency
Water content	Moisture control
Particle size	Retention time
Porosity	Curing time
Bulk density	
pH	Pile shape
Oxygen concentration	Pile depth
Compost temperature	Pile volume

[a]Factors presented in boldface are those considered most important in formulation or management of the composting process

Table 18.3 Guidelines for major factors affecting composting

Factor	Reasonable range	Preferred range
Nutrient balance, C/N	25:1–40:1	30:1–40:1
Water content	45–65% w.b.	50–60% w.b.
Particle size	0.8–1.2 cm (1/8–1/2 in.)	Depends on material
Porosity	30–60%	35–55%
Bulk density	<640 kg/m³ (1,100 lb/yd³)	
Temperature	45–68°C (110–155°F)	54–66°C (130–150°F)
Oxygen concentration	>1%	> >5%
pH	6.0–9.0	6.5–8.0

water content are the most important during management of the process. Guidelines for major factors are presented in Table 18.3 (Rynk 1992).

18.3.3 Optimizing Design and Management of Compost Systems

Keener et al. (1993) noted "optimization of a design, whether based on cost of construction and operation, energy use (conservation of resources) or pollution levels (odors, dust, etc.) can be done through field experimentation and mathematical modeling of the composting process. Field experimentation implies collection of basic information for real working systems (pilot or full scale) and evaluation of the results. Modeling the process is done using analytical or numerical models developed on the basis of heat and mass transfer principles and the reaction kinetics of biological systems. These models are used interactively with cost models that describe the system." The following sections present results of both field experimentation and modeling on controllable factors in composting.

18.3.3.1 MiR – Mixing Ratios

Rynk (1992) describes materials encountered in composting as being primary (material of interest), amendment (material to adjust C/N or water content) and bulking agent (material to "open" up the compost mix, giving it porosity so air can move through the pile). Because many materials do not compost rapidly and properly by themselves, efficient composting begins with a proper mix of these materials to control water content, C/N ratio and porosity. Three examples of materials requiring amendments or bulking agents are pine bark (high C/N), short paper fiber (high C/N), and biosolids (low C/N, high water content, low porosity). Some materials serve both as a bulking agent and amendment and in special cases the added material is also a by-product. Examples of co-composting are grass/leaves/brush (Marugg et al. 1993), biosolids/ woodchips/leaves (Elwell et al. 1994), food waste/yard waste/chicken manure (Elwell et al. 1996) and short paper fiber/poultry liter (Eckinci et al. 2002). Mixing ratios or the proportion of the various materials which give the correct levels of C/N and moisture can be calculated using data on chemical composition and density of materials (Fitzpatrick 1993; Keener et al. 2000d). Mixes are usually developed on a volume basis to make it easier for the operator of a facility to manage the process. The website OCAMM (2008) presents computer spreadsheets (Keener et al. 2002d), based on volumes or mass of materials and their measured properties moisture, N, C/N and wet bulk density, used to calculate the mixture's C, N, C/N, moisture, and density.

Sources of data on the chemical and physical properties of waste products are given in Rynk (1992) and Keener et al. (2000a). More recent sources of information on specific products are easily located through computerized literature searches. Since material's properties vary considerably from source to source, each by-product to be composted needs to be analyzed. Recycling compost and overs is sometimes a part of the process (Haug 1993; Elwell et al. 1994). Keener et al. (1993) presented equations for solving the effects of recycling on compost efficiencies and maturities if decomposition rates are known.

18.3.3.2 C/N – Carbon Nitrogen Ratio

Composting is usually successful when the mixture of organic materials contains 25–40 parts of carbon to 1 part of nitrogen. This carbon to nitrogen ratio (C/N ratio) in the compost meets the needs of microorganisms for high rates of decomposition while minimizing the loss of nitrogen as ammonia (Hansen et al. 1993; Michel et al. 2004; Ekinci et al. 2002). Woody materials such as bark, pulp or sawdust are typically enriched with nitrogen as ammonia or urea to avoid nitrogen deficiency. It is best to add nitrogen in organic form, such as in poultry manure, to avoid losses of the added N due to leaching or volatilization. The quantity of nitrogen to be added depends on the C/N ratio, particle size and the biodegradability of the material and typically ranges from 1.0 to 1.5 kg m^{-3} material (Hoitink and Kuter, 1986). Phosphorus, other principal and trace elements are generally available in satisfactory amounts for microorganisms when organic materials are blended to achieve this C/N ratio.

18.3.3.3 C/N and MiR – Maximizing Composting Rate and Capacity of Facility

Regulating C/N ratio is critical to achieving high rates of decomposition. It is adjusted by altering the mixing ratio, MiR, of materials being composted. Rynk (1992) states desired C/N ratio for composting is between 25:1 and 40:1 and is based on the C being readily available to microorganisms. This C/N range favors high growth rates of microorganisms and their efficient processing of degradable organics with limited odor generation. C/N above 40 generally slows decomposition due to limited nitrogen for microbial growth. Ekinci et al. (2002) evaluated the effects of mixing ratio (SPF/[SPF + BL], w/w) of paper mill sludge and broiler litter on k, decomposition rate, and β_o, non-compostable fraction (Fig. 18.2). Maximum decomposition rate, k_{max}, was 0.119 day^{-1} and occurred at C/N = 33.5.

Ekinci et al. (2002) solved for the optimum mixing ratio, MiR, which for materials $\mathbf{x}_1$ and $\mathbf{x}_2$ maximized processing capacity of a composting system for material $\mathbf{x}_1$. Solution was dependent on desired level of stability (maturity ratio, mR*)". In this problem, mixing ratio was defined as

$$\text{MiR} = m_{x1} / [m_{x1} + m_{x2}]$$

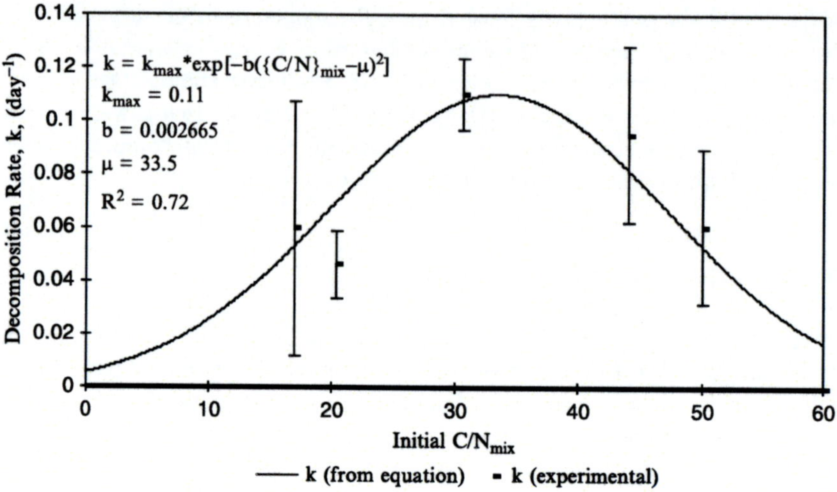

Fig. 18.2 Decomposition rate, k, as a function of C/N for Short Paper Fiber and Broiler Liter. $T_c = 60°C$

Presented was a solution for composting short paper fiber (x_1) with broiler liter (x_2). The procedure outlined can be applied to any compost mix if decomposition rates are known as a function of C/N. Results (Fig. 18.3) showed maximum throughput capacity was occurring at MiR ≅ 0.7 and C/N ≅ 35.

18.3.3.4 Water Content and Moisture Control

Numerous researchers have noted effects of compost moisture content on rates of decomposition. For w_c between 50% and 65% on a wet weight basis, k is not generally affected. When the water content of most materials exceeds 65%, the structural strength of the compost deteriorates below critical threshold levels, oxygen movement is inhibited, and the process tends to become anaerobic (Das and Keener 1997). If the process becomes anaerobic and the compost has a low C/N ratio (e.g. meat wastes), putrefaction occurs. At high C/N values, the materials ferment which result in decreasing pH and inhibition of the process. Both processes produce odors and must be avoided (Miller 1993). Below 50% moisture, bacterial growth becomes a rate limiting factor. The optimum moisture level varies with materials because of their specific water holding properties (Jeris and Regan 1973b). The ideal water content of the compost during the curing stage is 45–50% wet weight basis. This provides an acceptable environment for recolonization by beneficial microorganisms, but minimizes potential anaerobic conditions.

If compost contains a level of inert (ash, plastic, glass,...) > 10%, water content should be at 50–60% for the non-inert fraction (Keener et al. 2000d). Ekinci

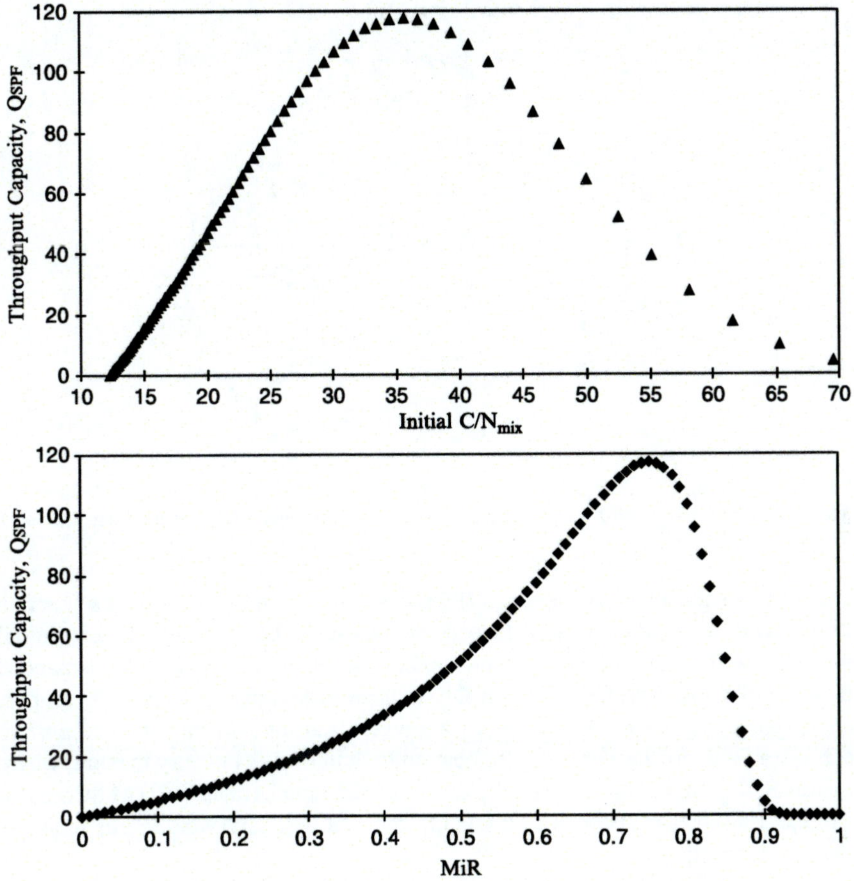

Fig. 18.3 Throughput rate, Q_{SPF} ($kg_{SPF}M_{gc}^{-1}$ day^{-1}), versus MiR and C/N. (mR* = 0.5; T* = 6.34 day)

et al. (2004a) evaluated decomposition as a function of moisture for SPF/BL and found decomposition could be represented as a Gaussian function (Fig. 18.4). The equation incorporating moisture was

$$X_{wc} = e^{-0.5[(w_c - 44.22)/19.87]^2}$$

Because this material had a high ash content of 52%, the optimum moisture level was 44.2%. The ash free moisture was 62% w.b.

Because large quantities of water are evaporated during composting to achieve temperature control, sometimes over two-thirds, optimum water content can only be maintained through rewetting. Time to first remixing event based on moisture

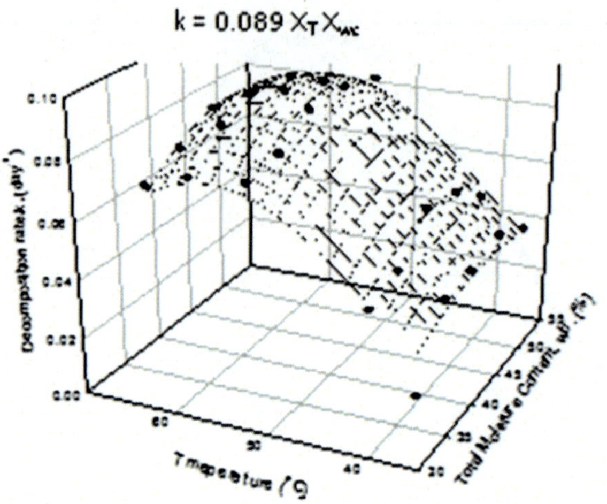

Fig. 18.4 Decomposition rate, k, as a function of (T_c, w_c) for Short Paper Fiber and Broiler Litter Mix. Equations for X_{WC} and X_{Tc} are presented in sections X3.3.4 and X3.3.7, respectively. C/N = 32

control was solved by Keener et al. (1996). Results gave moisture loss as a function of time and decomposition and showed the interaction between ash content, initial moisture, and minimum moisture on remix time. Keener et al. (2000c) expanded those results and used field data on Φ_w, water removed per unit mass of compost decomposed, to solve for maximum allowable starting moisture of compost to achieve a final moisture of $w_{c,\infty}$. Application to field studies on dairy manure, where measured Φ_w was 4.4 kg water per kg dry mass loss (approximately 2/3 of theoretical value), showed that above 81% moisture the compost mix would gain moisture during composting.

18.3.3.5 Particle Size and Distribution

Particle size and distribution are critical for balancing surface area for growth of microorganisms and maintenance of adequate porosity for aeration. As particle size increases, surface area decreases, thus lowering the surface area to mass ratio. Studies on particle size suggest acceptable values to be 0.8–1.2 cm nominal diameters (Rynk 1992). In addition, most of the compost mass in larger particles (interior of particles) does not decompose due to low oxygen levels (Hamelers 1993). These factors are material specific since they relate to porosity of the particle and its chemical composition. Particle size distribution, shape and packing of particles and moisture content control porosity of the compost mix. The percent air-filled pore space of compost piles should be in the range of 35–50%. Optimum recommended particle size for composting varies with materials used, the method of composting used and the intended market for the product.

18.3.3.6 Porosity and Oxygen

Optimum porosity is achieved by balancing materials and their particle sizes, water content of the mix, and pile size (Jeris and Regan 1973b). Greater than 50% porosity generally causes the pile to remain low in temperature because energy lost exceeds heat produced. Building bigger piles can overcome part of the problem if porosity is high. Too little porosity leads to anaerobic conditions and odor generation.

Haug (1993) discussed the effects of oxygen level and free air space, φ, on k. For most composting systems, the rate limiting effects of these parameters are avoided if interstitial oxygen, Y_{02}, is maintained above 5% and φ is greater than 35%. Ekinci et al. (2004b) adapted Haug's model for Y_{02} and φ and incorporated these variables in his simulation models evaluating air recirculation. They found that oxygen would become self-limiting in systems with high recirculation of composting air. For Y_{02} (w/w) above 15% k was reduced less than 4%, but for Y_{02} at 5% k was reduced 25%. The φ did not affect k (<1% drop) if φ exceeded 35% in his model.

Composting in the open air is affected by rainfall and in some regions rainfall saturates compost piles. This results in leachate formation, odors and other problems. In dry regions and in covered facilities, water must be added to avoid process inhibition. As air leaves the pile, CO_2 is also removed which is essential to avoid inhibition of microbial activity. Typically, if enough air enters active compost heaps to control temperatures below 65°C, enough oxygen is supplied as well (Finstein and Miller 1985). Keener et al. (2002d) showed for aerated systems which controlled compost temperatures below 70°C, oxygen level >18% would be maintained in the compost.

18.3.3.7 Temperature and Airflow

The optimum temperature range for composting is 40–65°C (Jeris and Regan 1973a; Kuter et al. 1985; Bach et al. 1987). Temperatures above 55°C kill fecal coliforms, parasites, and plant pathogens (review in Epstein 1997). At temperatures above 57–63°C, microbial activity declines rapidly with activity approaching low values as compost temperatures exceed 72°C (Finstein and Hogan 1993). However, due to gradients in the composting system composting can still be efficient at exhaust temperatures as high as 68–70°C (Keener et al. 1997b).

The decomposition rate k in a biological system is considered to be a function of temperature and has been expressed by the Arrhenius relationship, $e^{-Ea/RTabs}$ (Haug 1993; Weppen 2002) or a linearized form known as "Q_{10}" effect, which states activity doubles for each 10°C increase. Above 60°C a second temperature term is often added to the expression for k (Haug 1993; Weppen 2002) such thatk → 0 as compost temperature exceeds 70°C. Recently, Ekinci et al. (2004a) evaluated decomposition of paper mill sludge mixed with broiler litter (SPF/BL) as a function of composting temperature and used the Gaussian distribution equation (Fig. 18.4) to fit the data over the range of 40–65°C ($R^2 = 0.95$). The developed equation was

$$X_{Tc} = e^{-0.5[(Tc-58.31)/16.72]^2}$$

Based on these results the optimum composting temperature was 58.3°C.

Optimum compost temperatures is achieved by controlling pile size and/or regulating airflow (Finstein and Miller 1985; Haug 1993; Keener et al. 1997a) and allowing heat (≈18 MJ kg^{-1}) generated through microbial activity to leave the pile. Compost mass stored in piles a few meters high by several meters across usually reach 60°C in less than 2 days with minimal aeration. Maximum practical depth ranges from 1.5 to 3.5 m, depending upon the material to be composted. Deep piles (depths >4 m) sometimes lead to spontaneous combustion. Because the temperature of the material in the pile is an indication of how well the material is composting, monitoring temperature is an essential part of running a composting operation. In cold climates, when the ambient temperature drops below –5 C, pre-heating of air may be required for systems employing forced ventilation.

Fundamental equation for design of the aeration system for composting have been present by Keener et al. (1993, 2000a,d, 2002d, 2005) both for continuous and intermittent aeration. Figure 18.5 presents airflow requirements for three compost materials (Keener et al. 1997a): biosolids/woodchips, yard waste, and MSW. Initial airflows were 1.41, 3.48 and 0.76 m^3 $kg_{CO^{-1}}$ day^{-1}, respectively for composting at 55–60°C. Inlet air had a density of 1.18 kg_a m^{-3}. Results suggest airflow requirements can decrease as much as 40-fold from day 0–21. Haug (1993) recommended 3.0–3.8 m^3 $kg_{CO^{-1}}$ $\cdot day^{-1}$ for a biosolids/woodchip mixture and Rynk (1992) 0.47 or 4.7 m^3 $kg_{CO^{-1}}$ $\cdot day^{-1}$ for animal manures (continuous fan operation or temperature control of fan). To minimize fan fixed and operating cost, design should

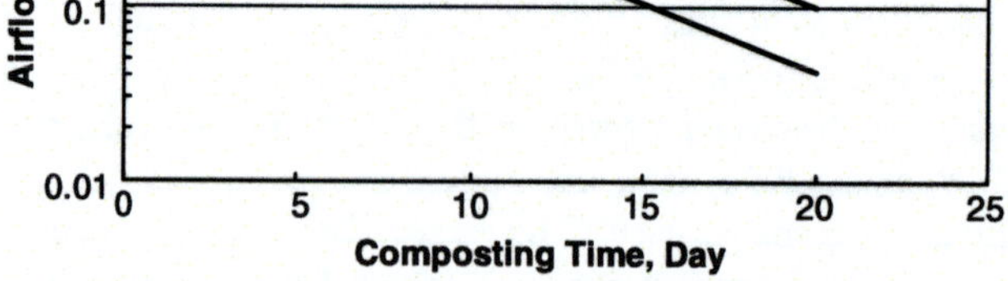

Fig. 18.5 Evaluation of airflow requirements for a yard trimming compost, grass/leaves/brush (G/L/B) with β = 0.51, k = 0.18 day^{-1}, biosolids/woodchip (BS/WC) with β = 0.80, k = 0.176 day^{-1} and municipal solid waste (MSW) with β = 0.393, k = 0.024 day^{-1}. Airflow is m^3 day^{-1} per kg of initial dry compost mass

be based on 3–5°C above optimal temperature for decomposition (Keener et al. 2003). Das and Keener (1997) showed that during forced aeration channeling can occur and lead to inefficient operation, unless remixing occurs to correct the problem.

18.3.3.8 pH

A pH between 6.7 and 9.0 supports good microbial activity during composting. Therefore, pH often is not a factor since most materials are within this pH range. An exception is a material such as cage layer manure. Large quantities of ammonia are lost if the pH is not maintained below 7.5 (Carey 1997). The pH of high C/N ratio materials such as municipal solid wastes which are low in buffer capacity readily reach a pH <6.7 and this leads to inhibition. This problem can almost always be overcome with O_2 levels maintained above 5% within the compost matrix (Haug 1993) since high O_2 concentrations results in respiration and utilization of organic acids which avoids low pH problems. Blending of materials with low and high pH can reduce ammonia losses when it is feasible.

18.3.3.9 Odor Control

Wilbur and Murry (1990) discussed odor control at biosolids composting facilities. Their recommendations on reducing odor were: (1) minimize moisture going into the process, (2) get a good mix, and (3) minimize air to dry the sludge, i.e. operate at the highest temperature possible for moisture and stability control. Haug (1993) discusses use of biofiltration for VOCs and odor control and presents design information. Chapters by Walker, Miller, Dunsen and Williams in Hoitink and Keener (1993a) discuss odor prevention and control.

Experimental studies (Hong et al. 1998; Elwell et al. 2001; Wiles et al. 2001) have explored the effect of continuous and intermittent aeration on NH_3 emissions and odor generation during composting hog and dairy manure. Results of those studies showed NH_3 emission was highly correlated with total airflow, regardless of fan operating mode. Results also suggested odor emissions would be reduced if the airflow were reduced while maintaining aerobic conditions. Results of Elwell et al. (2004) showed preventing anaerobic aging of manure (could be any organics) prior to composting is a major step in preventing the formation of offensive odors.

18.4 Compost Utilization

For a summary of available information on the use of compost for managing hazardous waste streams (as well as other applications) and possible areas for future investigations the reader is refereed to USEPA (2009b).

18.4.1 Determination of Physical and Chemical Properties

Laboratory analysis for physical and chemical properties of materials is necessary to manage the composting process and to provide information for marketing of the compost produced. Moisture (i.e. water content), density, and porosity can be done easily by the composter (Keener et al. 2000d). Other properties such as carbon, nitrogen (total, ammonia, nitrate), pH, nutrients (P, K, Ca,...), heavy metals, volatile solids, and maturity often require the use of commercial laboratories. The US Composting Council (USCC 2009) has developed detailed protocols for the composting industry to verify the physical, chemical, and biological condition of composting feedstocks, material in process and compost products at the point of sale. This protocol is called TMECC, Test Methods for the Examination of Composting and Compost. In addition the USCC has initiated a marketing program called the STA, Seal of Testing Assurance. It is a testing and information disclosure program to inform the consumer the necessary product information on the compost bag (or data sheet) so he can use the compost properly. Both these programs are designed to enhance the marketing of compost in the USA.

18.4.2 Maintaining N, P, K Levels in Compost

Many compost are marketed and priced based on their nutritive value, which is fixed (except for N) by the initial compost mix. This is because the plant nutrients P, K, Ca, Mg, as well as heavy metals do not disappear from the system as the dry matter decomposes (unless leaching occurs). However, N is lost during composting via numerous pathways, with ammonia being the most common. Its loss is directly related to total N content of manure and by C/N ratio. Numerous studies have shown C/N ratios near 40 or above will minimize N loss. However, high nitrogen waste such as caged layer manure and biosolids do not achieve this ratio for any reasonable level of amendment (Hansen et al. 1993). To prevent N losses, unamended caged layer manure can be composted in a enclosed structure with high atmospheric NH_3 to reduced N losses (Keener et al. 2002a). A second method which has been proposed is to use a replaceable biofilter made from curing compost (Tanaka et al. 2003). This system passes the NH_3 laden air from the composting stage to the curing stage for biofiltering and scrubbing out of the ammonia. It is unlikely this method will work for mixes with C/N < 20. A third approach is acid scrubbing and returning the NH_3 salt back to the finished compost (Carey 1997). Still other approaches have used alum (Ekinci et al. 2002) and gypsum (Tubail et al. 2008). Since compost nutrient values can vary within the size ranges of the cured compost (Elwell et al. 1994), screening can also be used to modify NPK levels in products.

18.4.3 Utilizing Compost Effectively in Crop Production

Proper land application of composts for crop production requires careful attention to the source of compost and the time, method and rate of application. Dalzell et al (1987) recommended that mixtures of mineral fertilizers and composts should be such that at least 30% of the N is supplied by each source. However, composts made from biosolids often contain high concentrations of nitrate-N such that these composts require no mineral fertilizer N additions (Chen et al. 1996). The timing of application versus when a crop is grown on the amended soil is important. Immature compost with a high C/N ratio should be applied several weeks ahead of planting to prevent immobilizing soil N and inhibiting crop growth. The method of application, i.e. surface applied or mixed into the soil, is controlled by whether N loss will be excessive from surface application and the tillage program for the crop production system. Lastly, rate of application is important. For crop production a minimum of about 2.5 Mg ha^{-1} is required before benefits of compost application become evident. However, rates 10 times or more above the stated minimum rate can be equally effective if applications are not made annually but only every 3 or 4 years.

18.4.4 Avoiding Toxicity in Compost

Organic acids in immature compost have negatively impacted percentage emergence of seeds (Chanyasak et al. 1983). Low molecular weight fatty acids (acetic, propionic and butyric acids) are most toxic. In practice, producing compost in properly aerated systems so that anaerobic pockets are avoided and preventing anaerobic conditions during curing/storage avoid toxicity caused by organic acids. In regions with high levels of precipitation, stabilized composts should be stored under a roof.

18.4.5 Maintaining Compost Physical Properties

Utilizing compost in container media and soil blending is determined primarily by the compost effects on hydraulic conductivity, water retention and air capacity (Hoitink 1980; Spencer and Benson 1982). Compost addition should maintain air capacities above 25% as air capacity of a potting mix directly affects plant growth and has an impact on root rot severity. For example, observations in nurseries indicate that *Phytophthora* root rots do not occur in media that contain tree barks having air capacities >25% and percolation rates >2.5 cm/min. Since the ratio of compost to soil in land application is small, compost's physical properties have only marginal effects on soil physical properties in land application.

18.4.6 Producing Disease Suppressive Compost

It is recognized that control of root rots with composts can be as effective as that obtained with fungicides (Spring et al. 1980; Ownley and Benson 1991; Hoitink et al. 1997). The ornamental plant industry now relies heavily on compost products for control of diseases caused by soilborne plant pathogens. However, composts must be of consistent quality to be used successfully for biological control of diseases of horticultural crops, particularly if used in container media (Inbar et al. 1993).

Effects of chemical properties of composts on soil borne disease severity are important but often overlooked. Highly saline composts such as those prepared from dairy manure or hog manure (Keener et al. 2001a) enhance Pythium and Phytophthora diseases unless they are applied months ahead of planting to allow for leaching. Compost prepared from municipal sewage sludge has a low carbon to nitrogen ratio. They release considerable amounts of ammonium nitrogen and enhance Fusarium wilt diseases (Quarles and Grossman 1995). Germination and plant growth studies, evaluating the effect of compost age, are a way to maintain quality assurance of the compost.

18.4.7 Recolonizing Compost During Curing with Beneficial Organisms

Most beneficial microorganisms, along with pathogens and weed seeds are killed during the high temperature phase of composting. During curing, mesophilic microorganisms that grow at temperatures <40°C recolonize the compost from the outer low temperature layer into the compost windrow or pile. Therefore, suppression of pathogens and/or disease is largely induced during curing. The reader is referred to Hoitink and Boehm (1999) for more detailed discussion on mechanisms. In order to induce the growth of suppressive organisms, maintain water content during curing at least 40–50%, w/w. Compost pH also affects the potential for beneficial bacteria to colonize composts. A pH <5.0 inhibits bacterial biocontrol agents.

In mature compost, where concentrations of free nutrients are low, plant diseases sclerotia of *R. solani* are killed by the parasite *Trichoderma*, and biological control prevails (Nelson and Hoitink 1983). This reveals that composts must be adequately stabilized to reach that decomposition level where biological control is feasible. Note, however that excessively stabilized or pyrolyzed compost does not support adequate activity of biocontrol agents. Because practical guidelines defining this critical stage of decomposition in terms of biological control is not yet available, industry achieves this end condition by maintaining constant conditions during the entire process and adhering to a given time schedule (Hoitink et al. 1991). In practice, this occurs in composts that have been (1) stabilized far enough to avoid phytotoxicity and (2) colonized by the appropriate specific microflora.

To minimize variability in compost suppressiveness three approaches are suggested: (1) cure the composts for 4 months or more (Kuter et al. 1988); (2) incorporate composts into field soils for several months before planting (Lumsden et al. 1983); and (3) inoculate composts with specific biocontrol agents (Kwok et al. 1987; Hoitink 1990). Composition of the feed stock appears to have an impact on the microflora in composts active in biological control. This factor has to be considered when specific inoculants are introduced into composts on a commercial scale.

18.4.8 Developing Compost for Foliar Disease Control

Papers have been published on the control of plant diseases of above ground plant parts with water extracts, also known as steepages, prepared from composts (Weltzhien 1992; Yohalem et al. 1994; Conforti et al. 2002). Unfortunately, efficacy varies with the compost, batches of steepages produced, crops and the disease under question. Recently it has been shown that composts incorporated into soils can protect the foliage of plants from foliar plant pathogens (Tränkner 1992; Zhang et al. 1998; Hoitink et al. 1997; Khan et al. 2002; Pharand et al. 2002; Krause et al. 2003). It seems possible that composts may be used more widely in the future for control of foliar diseases but currently many uncertainties remain.

18.5 Summary and Conclusions

Opportunities and challenges exist in many countries for recycling more of the MSW and biosolids waste stream and agricultural manures by composting. However for this to happen will require

- Changing the mindset of the public and businesses (source separation) and government policies (use outcome based criteria)
- Keeping composting operations environmentally friendly (manage process to control odors, leachate, emissions; produce compost free of pathogens, organic acids, heavy metals, ….)
- Making composting operation profitable (optimize the process and develop markets for products)

If the opportunity is seized and the challenges met, composting can transform organic wastes into high value environmentally friendly products. Such products (compost) will not only have fertilizer value, but also will promote plant health, reduce erosion and bring about many other beneficial effects.

18.6 Relevance to South Asia

Many organic "waste" streams exist in the cities and rural villages in South Asia. Composting is one way to treat these organic streams and recycle their plant nutrients in an environmentally safe manner back to agricultural production systems.

This recycling can lead to increased food, feed and fiber production, i.e. kg/ha, since soil plant nutrients are often limiting crop yields and farmers are unable to meet the needs with commercial fertilizers due to high prices relative to food prices. However, before these countries can realize the full potential of recycling waste, they need to develop system wide inventories of their waste streams which identify sources of materials, potential collection sites, chemical composition and environmental hazards. From those inventories, studies can then be done (government, universities, private businesses) to evaluate and determine cost effective approaches to *capture*, *treat* and *recycle* those waste.

References

Anonymous (2004) Farms and dairies forced to comply with state air pollution standards. BioCycle 45(7):19–20

Bach PD, Nakasaki K et al (1987) Thermal balance in composting operations. J Ferment Technol 65(2):199–209

Blackburn ML (2004) Biosolids composting for cities under 10,000 population. BioCycle 45(7):24–25

Carey DS (1997) Minimizing nitrogen loss from poultry manure amended with ammonia sulfate. M.S. Thesis. The Ohio State University, Columbus, OH

Chanyasak V, Katayama A et al (1983) Effects of compost maturity on growth of komatsuna *(Brassica rapa* var. *peridis)* in Neubauer's pot. 11. Growth inhibitory factors and assessment of degree of maturity by org.-C/org.-N ratio of water extract. Soil Sci Plant Nutr 29:251–259

Chen L, Dick WA, Streeter JG et al (1996) Ryegrass utilization of nutrients released from composted biosolids and cow manure. Compost Sci Util 4:73–83

Composting Council (1997) A watershed manager's guide to organics. The soil and water connection. The Composting Council, Alexandria, Virginia

Conforti C, Blair M, et al (2002) The effects of compost tea on golf course greens: Presidio Golf Course, San Francisco CA. Proceedings 2002 International Symposium on Composting and Compost Utilization. May 6–8. Columbus, Ohio. Published on CD – Disease Suppression II, May

Dalzell HW, Biddlestone AJ et al (1987) Soil management: compost production and use in tropical and subtropical environments. FAO Soils Bulletin 56, Food and Agriculture Organization of the United Nations (FAO), Rome

Das K, Keener HM (1997) Moisture effect on compaction and permeability in composts. J Envir Eng 123(3):1–7

Deller K (2004) Using on-site systems to compost food residuals. BioCycle 45(7):38–40

Department of Justice (2004) Ohio's largest egg producer agrees to dramatic air pollution reductions from three giant facilities, www.usdoj.gov/opa/pr/2004/February/04_enrd_105.htm

Ekinci K, Keener HM et al (2002) Composting short paper fiber with broiler litter and additives. II. Evaluation and optimization of decomposition rate versus mixing ratio. Compost Sci Util 10(1):16–28

Ekinci K, Keener HM et al (2004a) Modeling composting rate as a function of temperature and initial moisture content. Compost Sci Util 12(4):356–364

Ekinci K, Keener HM et al (2004b) Effects of aeration strategies on the composting process. Part I – experimental studies. Trans ASAE 47(5):1697–1708

Elwell DL, Keener HM et al (1994) Pilot- and full-scale evaluations of leaves as an amendment in sewage sludge composting. Compost Sci Util 2(2):55–74

Elwell DL, Keener HM et al (1996) Controlled, high rate composting mixtures of food wastes, yard wastes and chicken manures. Compost Sci Util 4(1):6–15

Elwell DL, Keener HM et al (1998) Composting unamended chicken manure. Compost Sci Util 6(2):22–35
Elwell DL, Keener HM et al (2001) Odorous emissions and odor control in composting swine manure/sawdust mixes using continuous and intermittent aeration. Trans ASAE 44(5):1307–1316
Elwell DL, Borger DC et al (2004) Changes in concentrations of malodorous compounds during controlled aeration composting. Compost Sci Util 12(2):102–107
Epstein E (1997) The science of composting. Technomic Publishing Company, Inc., Lancaster, PA
Finstein MS, Hogan J (1993) Integration of composting process microbiology, facility structure and decision making. In: Hoitink HAJ, Keener (eds) Science and engineering of composting: design, environmental, microbiological and utilization aspects. Renaissance Publications, Worthington, Ohio, pp 1–23
Finstein MS, Miller FC (1985) Principles of composting leading to maximization of decomposition rate, odour control, and cost effectiveness. In: Gasser JKR (ed) Composting of agricultural and other wastes. Elsevier, London, pp 13–26
Fitzpatrick GE (1993) A program for determining compost blending ratios. Compost Sci Util 1(3):30–33
Goldstein N (2004) Composting facilities apply lessons learned to capital improvements. BioCycle 45(6):26–29
Hamelers HVM (1993) A theoretical model of composting kinetics. In: Hoitink HAJ, Keener HM (eds) Science and engineering of composting: design, environmental, microbiological and utilization aspects. Renaissance Publications, Worthington, OH, pp 36–58
Hansen RC, Keener HM et al (1993) Composting of poultry manure. In: Hoitink HAJ, Keener HM (eds) Science and engineering of composting: design, environmental, microbiological and utilization aspects. Renaissance Publications, Worthington, OH, pp 131–153
Haug RT (1993) The practical handbook of compost engineering. Lewis Publishers, Boca Raton, FL
Hoff J (2004) Personnel communication. 9/1/04 (Residuals Manager, City of Columbus, Department of Public Utilities, Southwesterly Compost Facility, Lockbourne, Ohio. JDHoff@smoc.cmhmetro.net
Hoitink HAJ (1980) Composted bark, a lightweight growth medium with fungicidal properties. Plant Dis 64:142–147
Hoitink HAJ (1990) Production of disease suppressive compost and container media, and microorganism culture for use therein. US Patent 4960348. Feb. 13, 1990
Hoitink HAJ, Boehm MJ (1999) Biocontrol within the context of soil microbial communities: a substrate-dependent phenomenon. Annu Rev Phytopathol 37:427–446
Hoitink HAJ, Keener HM (eds) (1993) Science and engineering of composting: design, environmental, microbiological and utilization aspects. Renaissance Publications, Worthington, OH, [Contact – JG Press Inc., 419 State Ave., Emmaus, PA 18049], p 728
Hoitink HAJ, Kuter GA (1986) Effects of composts in growth media on soilborne pathogens. In: Chen Y, Avnimelech Y (eds) The role of organic matter in modern agriculture. Martinus Nijhoff Publishers, Dordrecht, Netherlands, pp 289–306
Hoitink HAJ, Inbar Y et al (1991) Status of composted-amended potting mixes naturally suppressive to soilborne diseases of floricultural crops. Plant Dis 75:869–873
Hoitink HAJ, Keener HM et al (1993) Key steps to successful composting. BioCycle 4(8):30–33
Hoitink HAJ, Stone AG et al (1997) Suppression of plant diseases by composts. HortScience 32(2):184–187
Hoitink HAJ, Rynk R et al (2002) Role of composting in environmental, soil, and plant health management. Proceedings 2002 International Symposium on Composting and Compost Utilization. May 6–8. Columbus, Ohio. Published on CD – Plenary, May 6
Hong JH, Keener HM et al (1998) Preliminary study of the effect of continuous and intermittent aeration on composting hog manure amended with sawdust. Compost Sci Util 6(3):74–88
Inbar Y, Chen Y et al (1993) Properties for establishing standards for utilization of compost in container media. In: Hoitink HAJ, Keener HM (eds) Science and engineering of composting: design, environmental, microbiological and utilization aspects. Renaissance Publications, Worthington, OH, pp 668–695

Jamieson C, White J et al (2004) Collection and diversion of food residuals in Southwest Florida. BioCycle 45(7):32–37
Jeris JS, Regan RW (1973a) Controlling environmental parameters for optimum composting. I. Experimental procedures and temperature. Compost Sci 14(1):10–15
Jeris JS, Regan RW (1973b) Controlling environmental parameters for optimum composting: II. Moisture, free air space and recycle. Compost Sci 14(2):8–15
Jurgens R (2004) Applying compost in mainstream agriculture. BioCycle 45(7):26–28
Kashmanian RM, Rynk RF (1996) Agricultural composting in the United States: trends and driving forces. J Soil Water Conserv 51(3):194–201
Katers JF, Schulta JM et al (2003) Solids separation of dairy manure. Biosci 44(12):51–54
Kaufman SM, Goldstein N et al (2004) The state of garbage in America. BioCycle 45(1):31–41
Keener HM, Marugg C et al (1993) Optimizing the efficiency of the composting process. In: Hoitink HAJ, Keener HM (eds) Science and engineering of composting: design, environmental, microbiological and utilization aspects. Renaissance Publications, Worthington, OH, pp 59–94
Keener HM, Elwell DL et al (1996) Remix scheduling during composting based on moisture control. Trans ASAE 39(5):1839–1845
Keener HM, Hansen RC et al (1997a) Airflow through compost: design and cost implications. Appl Eng Agric 13(3):377–384
Keener HM, Elwell DL et al (1997b) Specifying design/operation of composting systems using pilot scale data. Appl Eng Agric 13(6):767–772
Keener HM, Dick WA et al (2000a) Composting and beneficial utilization of composted by-product materials. In: Power JF et al (eds) Beneficial uses of agricultural, industrial and municipal by-products. Soil Sci Soc of Amer. Madison, Wisconsin, Ch. 10, pp 315–341
Keener HM, Elwell DL et al (2000b) Procedures and equations for sizing of structures and windrows for composting animal mortalities. Appl Eng Agric 6(6):681–692
Keener HM, Elwell DL et al (2000c) Composting non-separated dairy manure – theoretical limits and practical experience. In: Proceedings Eight Int. Sym. on Animal, Agr. and Food Processing Wastes. Oct. 9–11. Des Moines, Iowa. Amer Soc of Agricultural Engineers. St. Joseph, MI, pp 615–623
Keener HM, Ekinci K et al (2000d) Mathematics of composting-facility design and process control. In: Warman PR, Taylor BR (eds) Proceedings international composting symposium (ICS'99), vol 1. CBA Press, Inc, Nova Scotia, Canada, pp 164–197
Keener HM, Elwell DL et al (2001a) Composting & value-added utilization of manure from a High-RiseTM swine finishing facility. Compost Sci Util 9(4):312–321
Keener HM, Elwell DL et al (2001b) Design and performance of a high-rise hog facility manure drying bed. Appl Eng Agric 17(5):703–709
Keener HM, Elwell DL et al (2002a) NH_3 emissions and N-balances for 1.6 million caged layer facility: manure belt/composting system vs. deep pit operation. Trans ASAE 45(6): 1977–1984
Keener HM, Foster SS et al (2002b) Dealing with dead livestock. Resource 9(8):9–10
Keener HM, Pecchia JA, G Reid et al (2002c) Effects of aeration and covers on NH_3 water and dry matter loss during windrow composting of dairy manure. ASAE Paper 024139. Presented at 2002 ASAE Annual International Meeting/CIGR World Congress. July 29–Aug. 1, Chicago, IL. http://asae.frymulti.com/abstract.asp?aid=9385&t=1
Keener HM, Ekinci K, et al (2002d) Principles of composting process optimization. Proceedings 2002 International Symposium on Composting and Compost Utilization, May 6–8. Columbus, Ohio. Published on CD – Track 1, Process Kinetics And Modeling. May 6
Keener HM, Pecchia JA et al (2003) Optimizing design and operation of dairy manure composting systems – using pilot and full scale kinetic studies, pp 310–324. In: Animal agricultural and food processing waste IX: Proceedings. Amer Soc of Agric Engineers, St. Joseph, MI
Keener HM, Ekinci K et al (2005) Composting process optimization – using on/off control. Compost Sci Util 13(4):288–299
Khan J, Madden LV et al (2002) Systemic resistance induced in cucumber against Phytophthora crown rot and blight by *Trichoderma hamatum 382*. Phytopathology 26:S41

Krause MS, De Ceuster TJJ et al (2003) Isolation and characterization of rhizocbacteria from composts that suppress the severity of bacterial leaf spot of radish. Phytopathology 93:1292–1300

Kuter GA, Hoitink HAJ et al (1985) Effects of aeration and temperature on composting of municipal sludge in a full scale vessel system. J Water Pollut Control Fed 57(4):309–315

Kuter GA, Hoitink HAJ et al (1988) Effects of municipal sludge compost curing time on suppression of Pythium and Rhizoctonia diseases of ornamental plants. Plant Dis 72:751–756

Kwok OCH, Fahy PC et al (1987) Interactions between bacteria and Trichoderma hamatum in suppression of Rhizoctonia damping-off in bark compost media. Phytopathology 77:1206–1212

Lumsden RD, Lewis JA et al (1983) Effect of composted sewage sludge on several soilborne pathogens and diseases. Phytopathology 73:1543–1548

Marugg C, Grebus ME et al (1993) A kinetic model of the yard waste composting process. Compost Sci Util 1(1):38–51

Michel FC Jr, Pecchia JA et al (2004) Mass and nutrient losses during the composting of dairy manure amended with sawdust or straw. Compost Sci Util 12(4):323–334

Miller FC (1993) Minimizing odor generation. In: Hoitink HAJ, Keener HM (eds) Science and engineering of composting: design, environmental, microbiological and utilization aspects. Renaissance Publications, Worthington, OH, pp 219–241

NEBRA (2007) A national biosolids regulation, quality, end use & disposal survey, http://www.nebiosolids.org/pdf/NtlBiosolidsReport-20July07.pdf (9/29/2008)

Nelson EB, Hoitink HAJ (1983) The role of microorganisms in the suppression *Rhizoctonia solani* in container media amended with composted hardwood bark. Phytopathology 73:274–278

OCAMM (2008) Ohio compost operator education course. http://www.oardc.ohio-state.edu/ocamm/COEC.htm (3/9/09)

Ohio EPA (2009) OAC chapter 3745–27: solid & infectious waste regulations, http://www.epa.state.oh.us/dsiwm/pages/3745–27.html (3/6/09)

Ownley BH, Benson BM (1991) Relationship of matric water potential and air-filled porosity of container media to development of Phytophthora root rot of rhododendron. Phytopathology 81:936–941

Pharand B, Carisse O et al (2002) Cytological aspects of compost-mediated induced resistance against Fusarium crown and root rot in tomato. Phytopathology 92:424–438

Quarles W, Grossmam J (1995) Alternatives to methyl bromide in nurseries – disease suppressive media. The IPM Practitioner 17(8):1–13

Rynk R (ed) (1992) On farm composting handbook. NRAES-54. Cooperative extension service, northeast regional agricultural engineering service. Cornell University, Ithaca, NY

Schmack BioEenergy (2009) Our residuals-renewable resources, http://www.schmackbioenergy.com/pages/resources.html (accessed March 2009)

Spencer R (2004) An MSW composting journey. BioCycle 45(8):43–44

Spencer S, Benson DM (1982) Pine bark, hardwood bark compost, and peat amendment effects on development *of phytophthora* spp. and lupine root rot. Phytopathology 72:346–351

Spiegelman H, Sheehan B (2004) The future of waste. BioCycle 45(1):59–62

Spring DE, Ellis MA et al (1980) Suppression of the apple collar rot pathogen in composted hardwood bark. Phytopathology 70:1209–1212

Tanaka A, Yakushido K et al (2003) Adsorption process for odor emission control at a pilot scale dairy manure composting facility. Proceedings of the Third International Conference for Air Pollution from Agricultural Operations, Raleigh, North Carolina, October 12–15, 2003, Published by ASAE, St. Joseph, MI. pp 189–196

Tränkner A (1992) Use of agricultural and municipal organic wastes to develop suppressiveness to plant pathogens. In: Tjamos EC, Papavizas GC et al (eds) Biological control of plant diseases. Plenum, New York, pp 35–42

Trillas I, Avilés M, et al (2002) Evaluation of composted agricultural wastes as a methyl bromide alternative for plant disease control. Principles of Composting Process Optimization.

Proceedings 2002 International Symposium on Composting and Compost Utilization. May 6–8. Columbus, Ohio. Published on CD – Track 2 Disease Suppression
Tubail K, Chen L et al (2008) Gypsum additions reduce ammonia nitrogen losses during composting of dairy manure and biosolids. Compost Sci Util 16(4):285–293
USCC (2009) TMECC, http://www.compostingcouncil.org/programs/tmecc/ (3/9/09)
USEPA (2004) National ambient air quality standards, http://www.epa.gov/ttn/naaqs (9/9/2004)
USEPA (2008a) Waste, http://www.epa.gov/ebtpages/wastes.html (9/29/2008)
USEPA (2008b) Waste minimization, http://www.epa.gov/lean/activity.htm (9/29/2008)
USEPA (2008c) Municipal solid waste, http://www.epa.gov/epaoswer/non-hw/muncpl/facts.htm (9/29/2008)
USEPA (2009a) A plain English guide to the EPA part 503 biosolids rule, http://www.epa.gov/OWM/mtb/biosolids/503pe/index.htm (3/06/09)
USEPA (2009b) Wastes – resource conservation – reduce, reuse, recycle – composting, http://www.epa.gov/epawaste/conserve/rrr/composting/pubs/index.htm (3/06/09)
Weltzhien HC (1992) Biocontrol of foliar fungal diseases with compost extracts. In: Andrews JH, Hirano S (eds) Microbial ecology of leaves. Brock Springer Series in Contemporary Bioscience. BSBN 0387-97579-9. Springer, New York, pp 430–450
Weppen P (2002) Determining compost maturity: evaluation of analytical properties. Compost Sci Util 10(1):6–15
Wilber C, Murray C (1990) Odor source evaluation. BioCycle 31(3):68–72
Wiles MC, Elwell DL et al (2001) Volatile fatty acid emissions during composting of swine waste amended with sawdust as a measure of odor potential. Compost Sci Util 9(1):27–37
Yohalem DS, Harris RF et al (1994) Aqueous extracts of spent mushrooms substrate for foliar disease control. Compost Sci Util 4:67–83
Zhang W, Han DY et al (1998) Compost and compost water-induced systemic acquired resistance in cucumber and arabidopsis. Phytopathology 88:450–455

Chapter 19
Adaptation Strategies in Coping with Climate Change Impacts for Improved Crop Health and Sustainable Food Production

Abul Kalam Samsul Huda, Raj Mehrotra, and Ashish Sharma

Contents

Abstract In this chapter, historical climate information on rainfall, evaporation and temperature is compared with projections of these variables using outputs from the CSIRO Mk3 GCM for the A1B SRES scenario and stochastic downscaling models for two selected locations in Australia. The differences between current climatic parameters and those for 2030 and 2070 provide information on

Regional Symposium on Climate Change, Food Security, Sea Level Rise, and Environment in South Asia, held at University of Dhaka, Bangladesh, 25–30 August 2008, sponsored by Ohio State of University, the World Meteorological Organization (WMO), the Food and Agriculture Organization (FAO), Regional Office for Asia and the Pacific, and the UN Economic and Social Commission for Asia and Pacific (ESCAP).

A.K.S. Huda (✉)
School of Natural Sciences, University of Western Sydney, Hawkesbury, Locked Bag 1797, Penrith South DC, NSW 1797, Australia
e-mail: s.huda@uws.edu.au

R. Mehrotra and A. Sharma
School of Civil and Environmental Engineering, University of New South Wales, Sydney, 2052, Australia
e-mail: raj.mehrotra@unsw.edu.au; a.sharma@unsw.edu.au

R. Lal et al. (eds.), *Climate Change and Food Security in South Asia*,
DOI 10.1007/978-90-481-9516-9_19,

the changes in dependability of crop growing seasons. These analyses could also be used in identifying suitable strategies for adaptation to the projected climate change, and for evaluating alternate management options relating to excess and insufficient water during the growing season. These findings also have implications in resource management for mitigation purposes (e.g. the role of crop models in screening environments for growing of carbon sinks). The issue of the response of pests and diseases, and their interaction with host crops, to climate change is also considered. Information is required on the seasonal and geographic distributions of the occurrence and severity of pests and diseases for future climatic situations. We also report on current research aiming to integrate pathogen, crop and weather information for making tactical crop management decisions.

Even though the study uses Australian data as a case study to test and validate the methodologies presented, it provides a generic framework and research directions which can easily be applied to other parts of the world including South Asia. Implications of research findings to South Asia and future research directions are also discussed.

Keywords Australia • Downscaling • Modelling climate change • Meteorological anomalies • New South Wales • Crop and pests modelling

Abbreviations

APN	Asia Pacific Network
CSIRO	Council of Scientific and Industrial Research
ESCAP	Economic & Social Commission for Asia and Pacific
GCM	Global Circulation Models
ICM	Integrated Crop Management

19.1 Introduction

Various human activities, including the burning of fossil fuels and changes in land vegetative cover and land use patterns, are believed to be responsible for increasing the atmospheric concentrations of greenhouse gases and hence global warming. The warmer climate is expected to accelerate the global water cycle leading to spatial and temporal redistribution of precipitation, faster evaporation (conditions under which evaporation is likely to increase i.e. higher temperature and low humidity), changes in streamflow, and increased frequency of extreme hydrologic anomalies such as floods and droughts (Arora and Boer 2001; Nijssen et al. 2001; Bürger and Chen 2005). Changes in the extremes of floods and droughts could have serious implications on the planning and management of our water resources and are of more concern for the developing countries like Bangladesh where flooding

Fig. 19.1 An example of a common flood situation in Bangladesh (picture supplied by researchers from Bangladesh)

is a common feature (Fig. 19.1). Increased global warming also adds an element of significant uncertainty on how the changes in hydrologic variables will evolve over time and influence crop growth and food production.

19.2 Australia's Variable Climate

The small landmass of Australia in relation to the size of ocean that surrounds it, and its location across tropical, subtropical and temperate climate zones are the major factors causing Australia's climate to be highly variable. The country is well known as the driest inhabited continent, with highly variable rainfall and high rates of evaporation. The year-to-year variations in sea surface temperatures in both the Indian and Pacific Oceans and variations in atmospheric circulation (such as the latitude of the sub-tropical high-pressure belt) exert a significant influence on Australia's climate (Drosdowsky 2002, 2005). Several studies have also noted the larger annual variability of rainfall and streamflow in regions of Australia in comparison to the similar locations in the world (see, for example, McMahon et al. 1992; Nicholls et al. 1997). Considering the fact that for most Australian catchments, average annual streamflow is a small proportion of the overall water balance and therefore, small changes in average rainfall either as a result of natural climate variability at longer time scales, or due to climate change, can result in large changes in streamflow. A study carried out by Sadler et al. (1988) estimates that for a catchment in south-west Western Australia a decrease of 20% in rainfall would result in a 45% decline in streamflow.

To understand the Australia's variable climate, a simple measure is the comparison of the variability of the areal averaged rainfall of different locations or countries

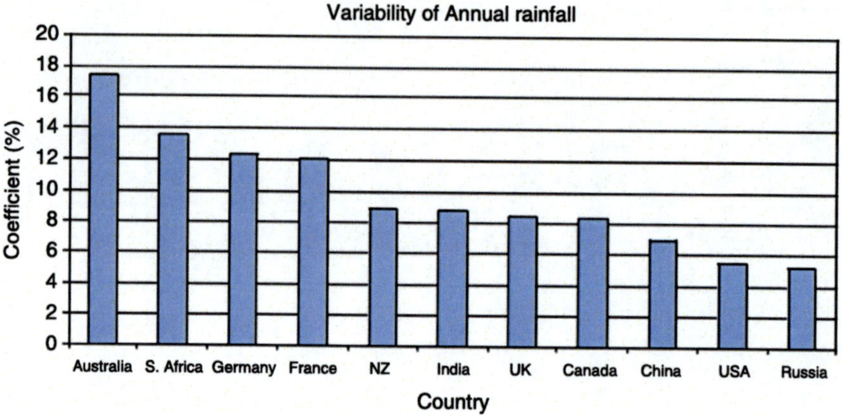

Fig. 19.2 The coefficient of variation of national annual rainfall for Australia and 10 other countries (source: http://www.connectedwater.gov.au/framework/water_availability.html)

(Nicholls and Wong 1990; Nicholls et al. 1997). The coefficient of variation is given by the ratio of standard deviation and the mean. Figure 19.2 provides the coefficient of variation (expressed as percentage) of annual rainfall of a number of representative countries including Australia. From the figure it is evident that Australia's rainfall is highly variable relative to some of the countries that compete with Australia in the world's market for agricultural products.

19.3 Climate Change Scenarios Projection: the Need for Downscaling

General circulation models (GCMs) are among the most advanced tools used to simulate the present climate and to obtain detailed climate information needed for the assessment of the various consequences of future climate changes on ecosystems and societies, both in space and time (e.g. IPCC 2007; Bergstrom et al. 2001; Varis et al. 2004). The GCMs are usually run at coarse grid resolution and provide a reasonably accurate representation of the average planetary climate. However, they are incapable of representing local sub-grid-scale features and dynamics that are often required for impact studies (IPCC 2007; Charles et al. 2004; Vicuna et al. 2007). As a consequence techniques have been developed to transfer the GCM output from the large spatial scales to the local or regional scales by means of downscaling for use in modelling studies at a catchment or regional scale. Statistical downscaling is thus a way to infer local information from coarse scale information by applying empirical or statistical links between large-scale fields and local conditions (Fig. 19.3). Such statistical links may be used both to validate global and regional climate models, and to develop detailed local climate scenarios based upon the output from such climate

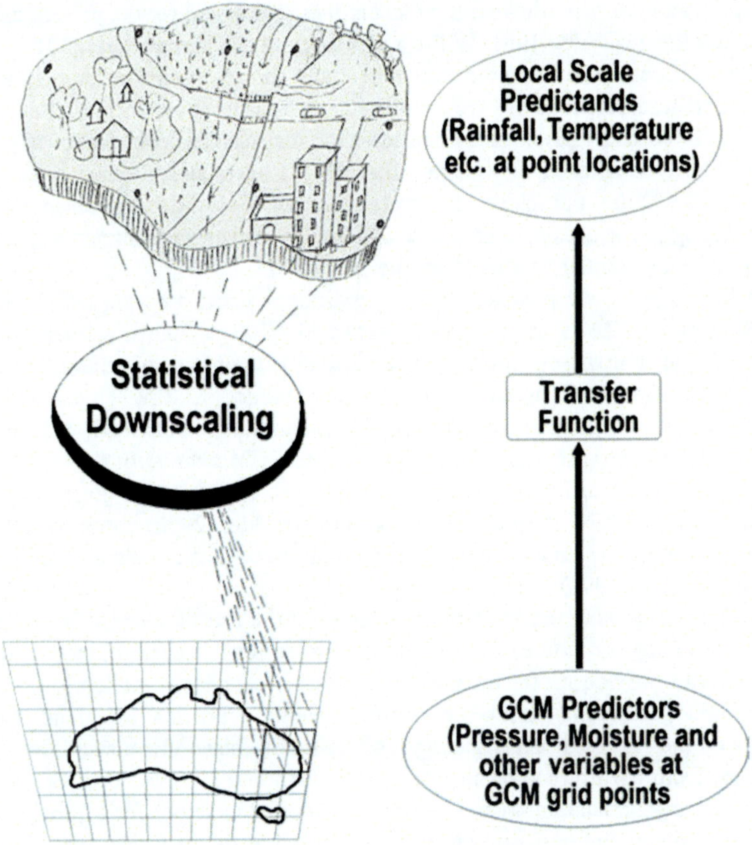

Fig. 19.3 Translating information from GCM scale to point or catchment scale

models. A diverse range of statistical downscaling techniques have been developed over the past few years, with regression-based and weather state-based methods being quite popular. IPCC (2007) and Fowler et al. (2007) provide an excellent review and discussion of various downscaling techniques.

19.4 Issues and Complexities in Downscaling Data for Future Projections

Downscaling focuses on statistical descriptions of the relationships between predictor and local-scale variables, paying little attention to physical linkages between the atmosphere and the surface environment. Some major assumptions include stability of the statistical relationships over time, integrity of the GCM output, and application of these downscaling techniques (calibrated for the present climate) to future climate (Wilby 1997; Wilby and Wigley 1997; Yarnal et al. 2001).

The relationship between the atmosphere and surface variables can be unstable, since short-term relationships are conditional on long-term variations in the climate system. The surface variable under consideration may also be dependent on additional atmospheric variables. It is assumed that GCMs adequately represent the large-scale features of the atmosphere. However, it has been observed that errors in statistical downscaling can be a result of improper simulations of long wave amplitudes, pressure fields and geopotential heights (Yarnal et al. 2001).

In this chapter, we provide climate predictions using a single GCM run (the CSIRO Mk3 GCM is used) for one assumed climate change scenario (SRES A1B). While comparisons have been performed with outputs of other GCMs (The Second Generation Coupled Global Climate Model, CGCM2) and additional scenarios and the general trends in predictor variables confirmed, further work needs to be done to ascertain the stability of the GCM simulations using multiple ensemble runs of different GCMs. This is especially relevant given the added uncertainty in GCM outputs (Phillips and Gleckler 2006) more specifically across the southern oceans, as has been pointed out in many studies in the past (Hennessy et al. 1998).

Whilst recognising the limitations inherent in the use of a single GCM, and in the downscaling procedure, we feel that the projections discussed in the following sections can provide useful insights into possible changes in meteorological and hydrologic parameters under the climate change scenarios examined. Further details on the stochastic downscaling model used in the results that follow, can be obtained from Mehrotra and Sharma (2007). It should be noted that the model is structured to generate rainfall and evaporation at point locations for a future climate, such that spatial and temporal attributes observed in the historical record, can be appropriately represented in the downscaled outputs corresponding to the same time period. It should also be noted that the downscaling model is stochastic, or, it generates multiple realisations, which allow a representation of the uncertainty contained within the downscaling procedure being used.

19.5 Projections for Meteorological Anomalies for the Years 2030 and 2070, for Two Locations in New South Wales, Australia

The SRES A1B emission scenario provides an estimate of the average climate around the nominated years (2030 and 2070) under future greenhouse gas emission scenarios. The 2030 climate is calculated as the average of 2021–2040 data, while the 2070 climate is the average of 2061–2080. Individual years will, of course, show some natural variability about this average.

Two locations in New South Wales (NSW), Australia – Richmond and Goulburn – were chosen for the simulations. Both areas are about 34–35°S latitude, and are within 200 km of each other. However, Goulburn is further from the Australian coast, and is in the rain shadow of coastal mountain ranges. As a result, Goulburn has lower rainfall (674 mm versus 858 mm) than Richmond. Therefore the two areas provide a good comparison from the climate change assessment perspective.

The results of climate change projections have been prepared for the climate changes centered on 2030 and 2070, relative to the baseline period 1960–2002. For convenience, the baseline period is referred to as the current climate. Note that in this analysis daily maximum temperature and pan evaporation data is used. Table 19.1 provides details on the rainfall and pan evaporation anomalies expressed as the percentage changes in relation to the current climate. This table also includes the details on the daily maximum temperature anomalies expressed as the relative change in the temperature with respect to the current climate.

19.5.1 Median Estimate of Changes by 2030

The best estimate of annual mean warming (increase in daily maximum temperature) for Goulburn and Richmond by 2030 is around 0.3°C, with variations across the seasons from –0.1 to 0.7°C. Changes in median estimate of annual wet days suggest no change (–4 to 4%) for Goulburn and about 7% (2–10%) increase for Richmond while for rainfall amount a median increase of about 4% (–4 to 10%) for Goulburn and about 14% (5–22%) for Richmond is expected by 2030. This increase is not evenly spread throughout the year, with large increases in summer and lesser in autumn and winter. The implications of such changes in rainfall distribution on seasonal basis are examined in Section 15.6. The best estimates of pan evaporation for both the stations show no significant changes in 2030.

19.5.2 Median Estimate of Changes by 2070

The best estimate of the mean annual warming by 2070 for both Goulburn and Richmond is 1.4°C with spread of 0.7–2.2°C. Projected median changes in annual wet days show decreases of 12% (–16 to –9%) and –2% (–5 to 1%) for Goulburn and Richmond, respectively. For annual rainfall median estimates indicate a decrease of 13% (–18 to –8%) for Goulburn and an increase of about 4% (–3 to11%) for Richmond in 2070 with large variations across the seasons and realizations. Pan evaporation estimates indicate an annual increase of 20% for Goulburn and about 8% for Richmond with small variations across the seasons (Table 19.1).

The agronomic implications of rainfall and temperature changes must be considered in relation to the changes in the evaporation regime, and this is considered in Section 15.6. Changes in the proportion of wet days may have an effect on the incidence of climate-sensitive diseases such as Downy Mildew. Table 19.1 shows both increases and decreases in the number of seasonal wet days for 2030 and 2070.

Table 19.1 Seasonal and annual wet days total rainfall, pan evaporation and daily maximum temperature anomalies for years 2030–2070 in relation to the current climate for the A1B Scenarios. For rainfall and pan evaporation these are expressed as percentage changes while for daily maximum temperature as relative changes

	In wet days (%)			In rainfall (%)			In pan evaporation (%)			In daily max temperature as relative change (°C)		
Season	5th	Median	95th	5th	Median	95th	5th	Median	95th	5th	Median	95th
Goulburn 2030												
Autumn	11	2	–6	16	–3	–14	6	1	–2	0.5	0.3	0.2
Winter	–2	–7	–12	8	–6	–17	8	2	–2	0.5	0.1	0.3
Spring	2	–3	–10	11	0	–11	8	3	–2	0.4	0.3	0.2
Summer	20	13	6	34	17	4	6	1	–2	0.2	0.0	–0.1
Annual	4	0	–4	10	4	–4	6	1	–2	0.4	0.3	0.2
Goulburn 2070												
Autumn	8	0	–6	18	2	–13	28	20	14	1.3	1.2	1.0
Winter	–10	–16	–20	–4	–17	–26	22	15	10	0.9	0.8	0.7
Spring	–15	–20	–25	–17	–26	–34	27	22	16	1.6	0.5	1.3
Summer	–5	–11	–18	3	–10	–21	25	19	14	2.2	2.0	1.8
Annual	–9	–12	–16	–8	–13	–18	25	20	15	1.4	1.4	1.3
Richmond 2030												
Autumn	11	5	0	26	10	0	5	1	–3	0.5	0.3	0.2
Winter	6	0	–8	27	5	–13	2	–1	–3	0.7	0.6	0.6
Spring	16	10	3	34	17	4	5	2	–2	0.5	0.3	0.2
Summer	15	10	5	32	17	5	4	0	–3	0.2	0.0	–0.1
Annual	10	7	2	22	14	5	3	0	–1	0.4	0.3	0.3
Richmond 2070												
Autumn	13	9	3	33	19	4	13	9	5	1.3	1.1	1.0
Winter	–2	–8	–14	43	19	–3	5	0	–3	1.3	1.2	1.1
Spring	3	–3	–9	5	–5	–17	13	9	5	1.8	1.6	1.5
Summer	1	–6	–11	5	–7	–19	15	11	7	2.1	1.9	1.8
Annual	1	–2	–5	11	4	–3	11	8	6	1.5	1.4	1.4

19.6 Implications of Changes in Rainfall and Evaporation for Crop Growth in 2030

In this section we examine the implications of climate change for cropping in southern NSW through a qualitative interpretation of rainfall (P) and evaporation (E) data, which will exert a dominant influence on the crop water balance.

Rain fed winter cropping is dominant in southern NSW, with some summer cropping if irrigation is available. Wheat is the dominant winter cereal crop, with canola and legumes as the main broadleaf crops. Rotations are widely practiced. Cereals are alternated with broadleaf crops for a number of years, followed by a pasture phase. The major purpose of the rotation strategy is for stubble/residue borne disease and weed control.

Expected changes in annual wet days, rainfall amount and pan evaporation at Richmond and Goulburn in 2030 and 2070 are shown in Table 19.2. There are small reductions (2–7%) in annual wet days and a small increase in annual rainfall in 2030 in relation to the current climate, and this result is expected for areas in southern NSW. For 2070, however, a 13% decrease in annual rainfall for Goulburn is noticed. Similarly, the results for annual pan evaporation show a no appreciable changes in annual pan evaporation in 2030 with appreciable increases of about 20% and 8% at Goulburn and Richmond, respectively. This increase is expected in terms of the increased temperature (Table 19.1) and higher average vapor pressure deficit.

Table 19.2 Median and 5 and 95 percentiles estimated of annual wet days, rainfall totals, pan evaporation and daily maximum temperature for Richmond and Goulburn for the current climate and for years 2030 and 2070 using A1B scenarios. Figures in parentheses represent 5th and 95th percentile limits on the median estimates drawn form the 100 replicates

Time period	Wet days	Rainfall (mm)	Evaporation (mm)	Temperature (°C)
Goulburn				
Observed (1960–2003)	97	674	1,275	19.3
Current climate (1960–2003)	101 (103–98)	704 (737–681)	1,246 (1277–1221)	19.4 (19.4–19.3)
Year 2030 (2021–2040)	101 (105–97)	731 (776–677)	1,263 (1324–1277)	19.7 (19.8–19.6)
Year 2070 (2061–2080)	88 (92–85)	614 (651–578)	1,493 (1559–1430)	20.8 (20.8–20.7)
Richmond				
Observed (1960–2003)	108	858	1,466	23.7
Current climate (1960–2003)	108 (110–105)	766 (801–733)	1,449 (1469–1433)	23.7 (23.7–23.6)
Year 2030 (2021–2040)	115 (118–110)	874 (932–805)	1,453 (1490–1433)	24.0 (24.1–24.0)
Year 2070 (2061–2080)	105 (109–102)	795 (849–744)	1,572 (1615–1534)	25.1 (25.21–25.1)

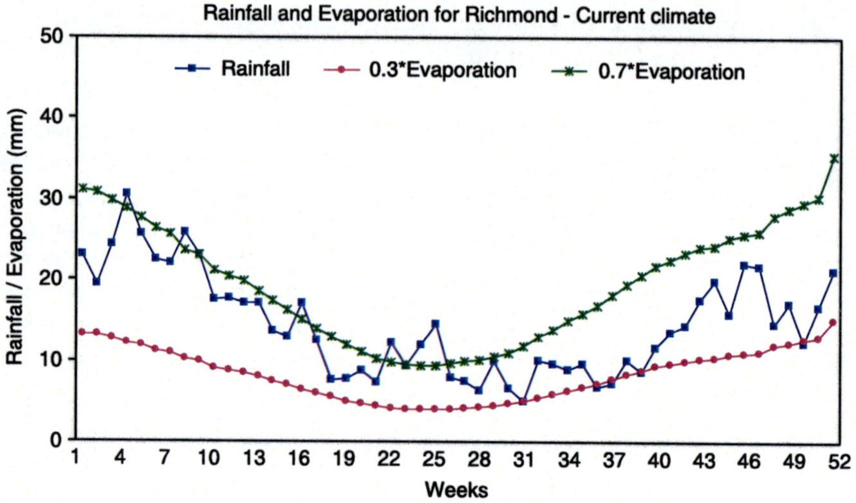

Fig. 19.4 Measured rainfall and evaporation for Richmond, 1960–2004

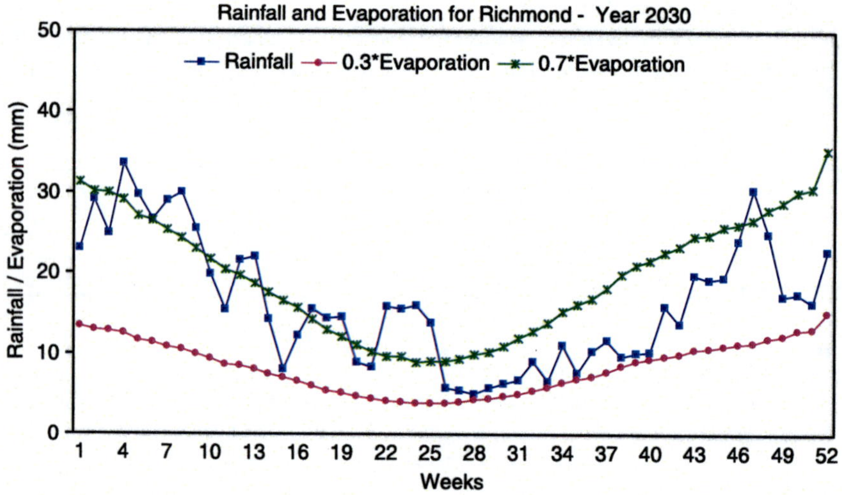

Fig. 19.5 Projections of rainfall and evaporation for Richmond, 2030 (2021–2040)

The rainfall decrease coupled with increased evaporation can have serious implications for cropping in the Goulburn region.

The implications of Climate Change for cropping in 2030 and 2070 may be best examined by analysis of data on rainfall and evaporation on a weekly, monthly and seasonal basis. The rainfall and evaporation regimes for the current climate and for years 2030 and 2070 for Richmond and Goulburn are shown in Figs. 19.4–19.9. For this analysis (0.7*E) is regarded as an estimate of maximum transpiration for more

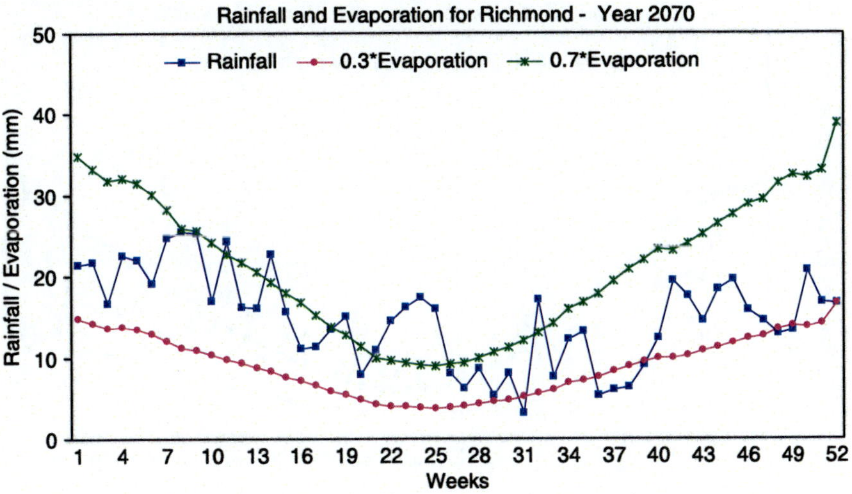

Fig. 19.6 Projections of rainfall and evaporation for Richmond, 2070 (2061–2080)

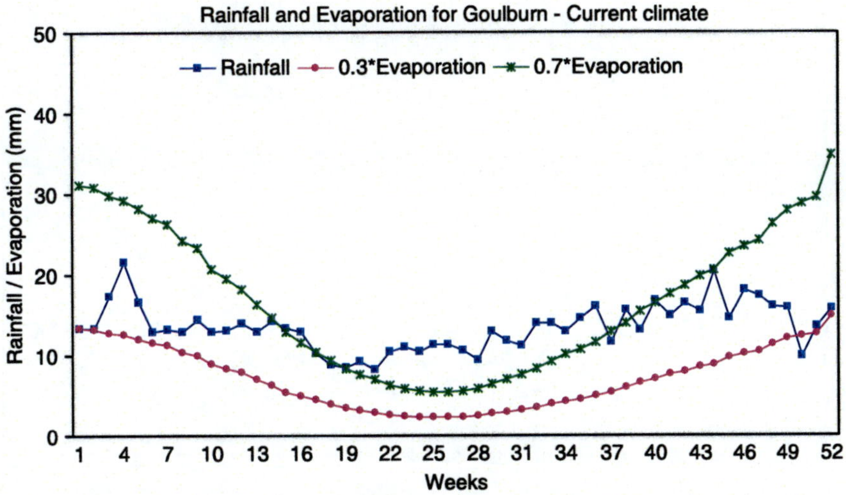

Fig. 19.7 Measured rainfall and evaporation for Goulburn, 1960–2003

developed winter crop, while (0.3*E) is used for a crop earlier in its development. A more quantitative analysis, including estimates of yield for the different scenarios, could be carried out using cropping systems simulation models such as APSIM (Keating et al. 2003). However, for our purposes a simple comparison of rainfall with crop water requirements was regarded as adequate as a method of identifying changes in the soil and crop water balance throughout the year.

Figures 19.4–19.9 indicate that in most cases rainfall is reasonably well distributed throughout the year. However Fig. 19.9 suggests an increasing dominance of summer

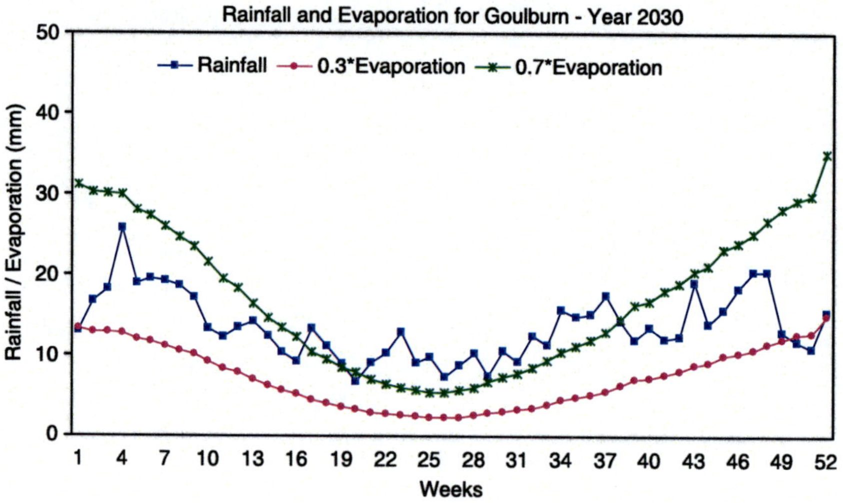

Fig. 19.8 Projections of rainfall and evaporation for Goulburn, 2030 (2021–2040)

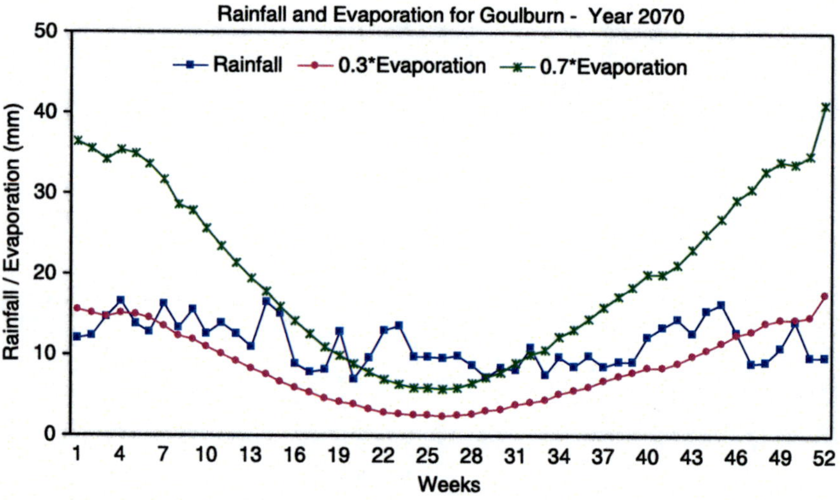

Fig. 19.9 Projections of rainfall and evaporation for Goulburn, 2070 (2061–2080)

rainfall for Goulburn in 2070. This change could have significant implications for the success of winter cropping in this area, and would require changes in cropping strategies.

The performance of a rain fed cropping regime is increased when the majority of rain falls during the cropping season (as is the case for most Monsoon and Mediterranean climates) and when rainfall during the growing season is sufficient to satisfy the requirements of plant transpiration. Based on this argument, the data presented in Figs. 19.4–19.9 has been used to calculate three types of information:

1. Growing Season rainfall and Fallow rainfall: For the purposes of these calculations the growing season is considered to be June to November, while the fallow period is the 6-month period from December to May.
2. (P-0.3 E) as an indicator of the sufficiency of water supply to the crop in the early part of the growing season. This calculation was used for the period June to August.
3. (P-0.7 E) as an indicator of the sufficiency of water supply to the crop in the latter part of the growing season. The period used for this calculation was September to November.

Information on these parameters for the two sites for the current climate and for the 2030 and 2070 climates is shown in Table 19.3. From the table it is evident that almost all of the annual rainfall decrease in future climates occur in the growing season, thus reducing both the efficiency of annual rainfall and the viability of winter cropping. This later result is illustrated by large decreases in the parameter (P-0.7 E) for the spring season in years 2030 and 2070. Changes in annual rainfall and pan evaporation at these reference stations are small in years 2030, however, assume significance in 2070.

The methodologies described above, along with cropping systems simulation modelling, can be used to study the implications of climate change for cropping in any

Table 19.3 Hydrologic parameters for different time periods of a year for the current climate and for 2030 and 2070 scenarios at two sites in New South Wales, Australia

Time period		Annual wet days	Annual rainfall in mm	Fallow rainfall in mm	Growing season rainfall in mm	Rainfall – (0.3 × evaporation) (June–Aug.) (in mm)	Rainfall – (0.7 × evaporation) (Sept.–Nov.) (in mm)
Goulburn							
Current	Median	101	704	344	364	116	−31
	5th percentile	103	738	374	394	129	−23
	95th percentile	98	681	312	341	107	−38
2030	Median	101	731	371	354	105	−37
	5th percentile	105	776	433	401	126	−27
	95th percentile	97	677	330	315	91	−50
2070	Median	88	614	330	284	83	−137
	5th percentile	92	651	377	323	101	−131
	95th percentile	85	578	283	252	71	−141
Richmond							
Current	Median	108	766	456	309	61	−107
	5th percentile	110	801	499	343	78	−98
	95th percentile	105	733	423	280	47	−114
2030	Median	115	874	522	346	67	−81
	5th percentile	118	932	592	405	93	−60
	95th percentile	110	805	469	301	47	−95
2070	Median	105	795	472	323	84	−143
	5th percentile	109	849	530	371	111	−136
	95th percentile	102	744	410	273	59	−154

environment. Information on soil properties including available water-holding capacities along with the presented information would improve the quality of the analysis.

The methodologies can also be used in identifying suitable strategies for sustainable food production. These strategies provide an adaptation in natural or human systems in response to the projected climate change in Australia and South Asia. Some options include:

- Determining the length of the effective growing season.
- Choosing appropriate enterprises to allow adaptation to climate change, including summer cropping and different crops.
- Evaluating alternate management options in addressing issues relating to excess water during water sufficiency periods within the crop growing season, and scheduling supplementary irrigation in dry periods.
- Examining options for mitigation purposes (e.g. the growing of carbon sinks). This includes the role of crop models in screening environments where such long-term mitigation strategies as Carbon Emissions Trading and offsets, including the forest planting as tree carbon sequestration, is viable.

19.7 Crop and Pests Modelling

The potential impacts of Global Warming on pests, and the threat that they pose to our agricultural industries, are not clearly understood. A review of the impacts of climate change on plant biosecurity (Aurambout et al. 2006) indicated a need to document pest and disease responses to climate change and incorporate them into our management and contingency planning. The expected increase in atmospheric CO_2, heavy and unseasonal rains, increased humidity, drought, cyclones and hurricanes and warmer winter temperatures, are likely to affect crops, pests and diseases and pathogen–host interactions. However, the extent to which climate change will affect emergency pests and pathogens, their hosts and the environment is not clearly understood.

For example, it is not known how climate change will affect the biology and distribution of known plant hosts. Nor do we understand the effects of climate change on our currently identified key pests and pathogens. While future spatial distribution can be predicted under climate change scenarios using models and examining historical trends, there appears to be limited knowledge of how climate change will impact on the biology of the pest or pathogen and how the interaction with their host responds to the climatic factors listed above.

19.8 Implications of the Information to South Asia

The assessment of potential impacts of climate change on agriculture in South Asia is crucial because of agriculture's role in providing food and fiber to South Asia's human population. A clear understanding of the relationship between climatic variability, rainfall, temperature, soil moisture, crop management, agricultural productivity and pest management is crucial in assessing the impacts of climatic variability and change on ultimate crop production, the identification of adaptation strategies and appropriate management practices, and the formulation of mitigating measures to minimize the negative effects of climatic variability on agricultural productivity. In the future, food security will be at the top of the agenda in Asian countries because of growing population. Greatly enhanced efforts to understand the relationship between key climate elements and agriculture should provide a sound basis for meeting the challenges of optimizing the benefits of changing climatic resources. While the examples discussed in this paper are related to Australian locations, the principles developed are of direct relevance to other parts of the world including South Asia and offer many useful suggestions and research directions including the following:

- Extending the research to South Asia including output of multiple GCM and climate change scenarios to encompass the possible spectrum of climate change projections
- Forming a network of able regional researchers willing to collaborate in research and application activities in this area including weather-crop-disease modelling and advisories
- Strengthening regional collaboration for proactive crop management
- Initiation of similar sets of experiments for testing predictions of selected disease epidemic and consequences on growth and yield of selected crops, to include new efforts to improve model code and structure
- Devising a system for validation of existing and possibly future results using local index crops
- Collection of data on crop growth and disease occurrence from a set of pilot projects (field trials) in South Asian countries
- Elucidation of the relationship between data obtained from meteorological stations and microclimatic conditions in nearby crop fields, and comparing with future climate change projections
- Involvement of stakeholders and national policy makers for drawing contingency plans for the actions they will need to take to mitigate such impacts

19.9 Research and Development Needs for South Asia

To respond to future climates, changes to industry practices and government policies may be required. A wide range of stakeholders therefore need to be involved in development of policies and strategies for managing crops, pests and diseases in a changing climate.

As indicated in the previous section, the information needs to be developed at two levels:

1. At a strategic level, where information on both the seasonal and geographic distributions of the occurrence and severity of pests and diseases is required for future climatic situations as projected by various GCMs and scenarios.

These changes need to be modelled to make strategic decisions about crop management. Such information can be used to target research and extension efforts.

2. At a tactical level, where information on the biology of the pest or pathogen, and on the interaction with host plants, is used to make tactical decisions on control options such as spraying. As discussed earlier, such information is not yet available for future climates. However, the potential of using dynamic crop simulation models in making reliable agronomic and policy decisions is highlighted as a means of reducing ecological and economic risks.

Experience in the use of dynamic crop simulation models in the current climate has been gained from the Asia–Pacific Network for Global Change Research (APN) funded project on "climate and crop disease risk management", led by the University of Western Sydney. As illustrated in Fig. 19.10, the aim of the project was to integrate a range of information including historical climate data, medium-range weather forecasts, crop growth and plant disease models with economic analysis to produce a crop disease risk model and appropriate management tools. The project has demonstrated the value of climate information in developing integrated pest management strategies for coping with climate change (Coughlan et al. 2009; Huda et al. 2007).

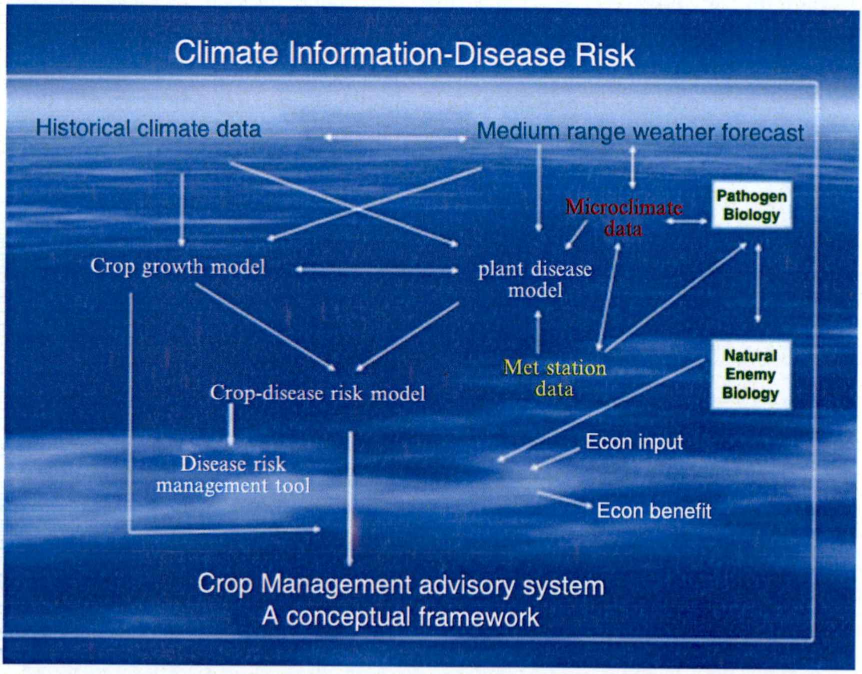

Fig. 19.10 Integrating climate and crop disease. (The figure originated from collaboration *led by University of Western Sydney* with the Central Research Institute for Dryland Agriculture (CRIDA), and the International Crops Research Institute for the Semi-Arid Tropics (ICRISAT), Hyderabad, India for climate and crop disease risk management research)

Future programs should build on the previous research including the APN funded project's (Coughlan et al. 2009) success in disease management and broaden its sphere of activity to Integrated Crop Management (ICM), including decisions such as fertilizer application, planting, and other agronomic decisions

19.10 Conclusions

The Intergovernmental Panel on Climate Change (IPCC) fourth assessment report (IPCC 2007) concludes that it is very likely that changes in the global climate system will continue well into the future. These changes are expected to have significant implications for our environment, affecting water resources, rainfall, agriculture, economy, health, biodiversity and many other factors. Sustainable management of water is probably the biggest challenge we face in this era of increasing demands (through population growth and better living standards) and climatic uncertainty and change.

This chapter illustrates the use and value of GCM-based scenarios in examining the likely future impact of climate change on water availability and viability of cropping options. The simple method of data analysis used here suggests options for strategies for adaptation to climate change from the time distribution of rainfall and evaporation. Although dynamic cropping systems simulation models provide quantitative data on projected crop yields, they require additional inputs such as available water-holding capacity of the soil and drained upper limit to allow calculation of the soil water balance.

While the examples discussed in this paper are related to two selected Australian locations, the principles developed will be of direct relevance to other regions. The methodology presented can be tested and used for developing mitigation and adaptation strategies in both South Asia and Australia.

The chapter has also demonstrated the value of climate information in developing integrated pest management strategies for coping with the predicted changes in the climate system as a result of global warming.

Acknowledgements The authors would like to express their sincere thanks to Dr. Kep Coughlan and Dr. Michael Brennan for their assistance in data analysis, and to the Asia–Pacific Network for Global Change Research (APN) for their support in the project on "Climate and crop disease risk management: an international initiative in the Asia–Pacific region." Thanks are due to the University of Western Sydney and the University of New South Wales.

References

Arora VK, Boer GJ (2001) Effects of simulated climate change on the hydrology of major river basins. J Geophys Res 106:3335–3348

Aurambout JP, Constable F, Luck J, Sposito V (2006) In: Tignor M, Miller HL (eds) The impacts of climate change on plant biosecurity, Victorian Government Department of Primary Industries, Melbourne, Victoria, Averyt. Cambridge University Press, Cambridge, p 996

Bergstrom S, Carlsson B, Gardelin M, Lindstrom G, Pettersson A, Rummukainen M (2001) Climate change impacts on runoff in Sweden – assessments by global climate models, dynamical downscaling and hydrological modeling. Clim Res 16:101–112

Bürger G, Chen Y (2005) Regression-based downscaling of spatial variability for hydrologic applications. J Hydrol 311:299–317

Charles SP, Bates BC et al (2004) Statistical downscaling of daily precipitation from observed and modeled atmospheric fields. Hydrol Process 18:1373–1394

Coughlan KJ, Huda, AKS, et al (eds) (2009) Climate and crop disease risk management: an international initiative in the Asia–Pacific Region. Proceedings of the Project Review and Planning Workshop, 11–14 February 2008, Bangladesh Science Foundation, Dhaka, Bangladesh, p 25 (ISBN 978-984-33 0096-6)

Drosdowsky W (2002) SST phases and Australian rainfall. Aust Meteorol Mag 51:1–12

Drosdowsky W (2005) The latitude of the sub-tropical ridge over eastern Australia: the L index revisited. Int J Climatol 25:1291–99

Fowler HJ, Blenkinsop S, Tebaldi C (2007) Linking climate change modeling to impacts studies: recent advances in downscaling techniques for hydrological modeling. Int J Climatol 27(12):1547–1578

Hennessy KJ, Whetton PH et al (1998) Fine resolution climate change scenarios for new south wales: annual report 1997–98. CSIRO Atmospheric Research, Aspendale, Victoria, Australia, p 48

Huda AKS, Hind-Lanoiselet T et al (2007) Examples of coping strategies with agrometeorological risks and uncertainties for integrated pest management. In: Sivakumar M (ed) Managing weather and climate risks in agriculture. Springer, New York

IPCC (2007) In: Solomon SD, Qin M (eds) Climate Change 2007: the physical science basis. Contribution of Working Group I to the Fourth Assessment Report of the Intergovernmental Panel on Climate Change. Cambridge University Press, Cambridge, p 996

Keating BA, Carberry PS et al (2003) The Agricultural Production Systems Simulator (APSIM): its history and current capability. Eur J Agron 18:267–288

McMahon TA, Finlayson BL et al (1992) Global runoff – continental comparisons of annual flows and peak discharges. Catena Paperback, Cremlingen-Destedt

Mehrotra R, Sharma A (2007) A multisite stochastic downscaling model of daily rainfall occurrences with long term persistence. MODSIM 2007 Congress 10–13 December, 2007. Christchurch, New Zealand

Nicholls N, Wong KK (1990) Dependence of rainfall variability on mean rainfall, latitude, and the southern oscillation. J Climate 3:163–170

Nicholls N, Drosdowsky W et al (1997) Australian rainfall variability and change. Weather 52:66–71

Nijssen B, O'Donnell GM et al (2001) Hydrologic sensitivity of global rivers to climate change. Climate Change 50:143–175

Phillips TJ, Gleckler PJ (2006) Evaluation of continental precipitation in 20th century climate simulations: the utility of multimodel statistics. Water Resour Res 42:W03202. doi:10.1029/2005WR004313

Sadler BS, Mauger GW et al (1988) The water resource implications of a drying climate in south-west Western Australia. In: Pearman GI (ed) Greenhouse: planning for climate change. CSIRO, Melbourne, pp 296–311

Varis O, Kajander T et al (2004) Climate and water: from climate models to water resources management and vice versa. Clim Change 66:321–344

Vicuna S, Murer PE et al (2007) The sensitivity of California water resources to climate change scenarios. J A Water Resour Assoc 43:482–499

Wilby RL (1997) Nonstationarity in daily precipitation series: implications for GCM downscaling using atmospheric circulation indices. Int J Climatol 17:439–454

Wilby RL, Wigley TML (1997) Downscaling general circulation model output: a review of methods and limitations. Progress in Phys Geography 21:530–548

Yarnal B, Comrie AC et al (2001) Developments and prospects in synoptic climatology. Int J Climatol 2:1923–1950

Chapter 20
Options for Forest Management for Coping with Climate Change in South Asia

N.H. Ravindranath, Indu K. Murthy, and Shilpa Swarnim

Contents

Abstract Projected climate change will adversely impact distribution and diversity of forests in South Asia. It may also lead to increase in vulnerability to pests and pathogens including risks of forest dieback. While there may be short-term increase in biomass production because of the CO_2 fertilization effect, eventually climate change will reduce NPP. Thus, there is a strong need for identification of adaptation strategies to minimize risks of forest fragmentation. Research priorities include development of dynamic vegetation models, and initiation of long-term experiments. Identification of appropriate policies is essential to minimize the adverse impacts of climate change on forest production.

Keywords Ecosystem service • Forest biodiversity • Dynamic vegetation models • Ecological research • Mitigation potential policies

---

N.H. Ravindranath (✉), I.K. Murthy, and S. Swarnim
Centre for Sustainable Technologies, Indian Institute of Science, Bangalore 560 012, India
e-mail: ravi@ces.iisc.ernet.in; indu@ces.iisc.ernet.in; swarnimshilpa@gmail.com

R. Lal et al. (eds.), *Climate Change and Food Security in South Asia*,
DOI 10.1007/978-90-481-9516-9_20, © Springer Science+Business Media B.V. 2011

Abbreviations

CCCM	Canadian Climate Change Model
GCM	General Circulation Model
GFDM	Geophysical Fluid Dynamic Model
HadRM	Hadley Center Regional Model
Mha	Million hectares
MMD	Multi-model data
RCM	Regional Climate Model

20.1 Introduction

Forests account for nearly 20% of the geographic area of the south Asian region. Forests in south Asia are also known for their rich biodiversity with many biodiversity 'hotspots' (http://chm.aseanbiodiversity.org/featured-articles/47-latest-features/78-revisiting-aseans-biodiversity-hotspots-). Forests provide a number of environmental services as well as livelihoods to a large number of forest dependent and rural communities. Forests are critical for protection from floods, cyclones, etc. in particular mangroves in south Asia are critically important. According to the Intergovernmental Panel on Climate Change (IPCC) and several research studies in the region as well as the reports of the National Communication, climate change is projected to have adverse impacts on forest ecosystems and biodiversity. This could have adverse implications for the livelihoods of the forest dependent communities. The forests in south Asia are also subjected to human pressure leading to degradation of forest biodiversity and biomass. The disturbed or fragmented forests are more vulnerable to climate change (Laurance and Peres, 2006). Some of the impacts of climate change are likely to be irreversible, e.g., loss of biodiversity and even forest biomes. Thus, there is a need for development and implementation of adaptation practices and policies to reduce the vulnerability of forest ecosystems and forest dependent communities. In this chapter, an attempt is made to present the state of forests in the south Asian region, review the studies on the impacts of climate change, and explore potential adaptation practices and policies in the context of the south Asian region.

20.2 Status of Forests in South Asia

20.2.1 Area Under Forest in South Asia

The total area under forest in south Asia according to FAO (2009) is 79 Mha (Table 20.1), accounting for 19% of the total geographic area. India dominates by accounting for 67.7 Mha (22.8% of land area), according to the State of Forest Report (FSI 2005) followed by Nepal 3.6 Mha (25% of land area) and Bhutan

Table 20.1 Area under forests in South Asia (FAO 2009)

	Extent of forests 2005	Annual change rate		1990–2000		2000–2005	
Country	Forest area (1,000 ha)	% of land area	Area per 1,000 people (ha)	1,000 (ha)	%	1,000 (ha)	%
Bangladesh	871	6.7	6.0	6	0.0	–2	–3
Bhutan	3,195	68	4,931	11	0.3	11	0.3
India	67,701	22.8	59	362	0.6	29	0.0
Maldives	1	3.0	3	0	0.0	0	0.0
Nepal	3,636	25.4	132	–92	–2.1	–53	–1.4
Pakistan	1,902	2.5	12	–41	–1.8	–43	–2.1
Sri Lanka	1,933	29.9	101	–27	–1.2	–30	–1.5
Total	79.239	19.252	213	0.27	–88	–0.11	

3.2 Mha (68% of land area). Pakistan and Bangladesh have the lowest area under forest with only 2.5% and 6.7% of the land area under forests, respectively.

When the area per 1,000 population is considered, among the major countries, Sri Lanka (101 ha/1,000 people) is followed by India (59 ha/1,000 people), Pakistan (12 ha/1,000 people) and Bangladesh (6 ha/1,000 people). This indicates that the area under forests in the south Asian region per unit of population is very low among all the major countries except Bhutan (Table 20.1).

20.2.2 Changes in Area Under Forest

According to FAO 2009, the area under forests for south Asia as a whole has declined marginally in other words, the area under forests has stabilized during the period 2000–2005. However, during the period 1990–2000, forest area increased marginally. Thus on the whole, during the period 1990–2005, the area under forests in south Asia as a whole has stabilized. Among the south Asian countries, area under forests has increased significantly in India during the period 1987–2003 (Fig. 20.1).

In south Asia, India has been implementing a large afforestation programme under which annually about 1.3 Mha is afforested over a period of 1980–2005 (Ravindranath et al. 2008). Similarly afforestation programmes have been implemented in other countries of the region. The afforestation programme largely caters to meeting the biomass requirements, particularly of the rural communities.

20.2.3 Status of Forests

South Asia is a highly populated region of the world with a total population of over 1.5 billion and a high population density of 367 persons/km^2. Rural population dominates the south Asian region accounting for 71% of the total population,

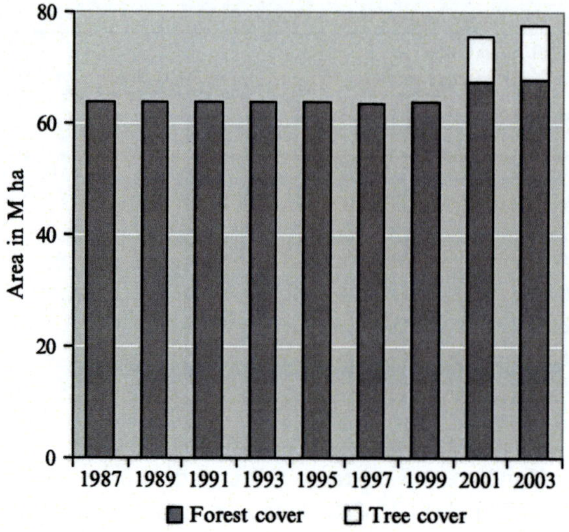

Fig. 20.1 Forest cover trends in India (Chaturvedi et al. 2008)

with many countries having over 80% of the population in rural areas. The rural communities in the south Asian region depend on fuelwood as the dominant source of cooking fuel. The fuelwood consumption in the south Asian region is estimated to be around 382 million m^3. Though it is not possible to say exactly what percent of this fuelwood comes from forests, it is likely that significant part of fuelwood consumed is extracted from the forests, leading to degradation. The degradation of the forests in the south Asian region can be observed from the fact that the growing stock in the region is 79 m^3/ha compared to 102 m^3/ha in Africa and 155 m^3/ha in south America (FAO 2009). Further according to FAO 2007, the growing stock in the region has declined from 110 in 1990 to 88 m^3/ha in 2005 and further to 79 m^3/ha in 2007 (FAO 2009). Further the biodiversity in forests is declining in the region and at the same time, the problem with forest fire is increasing (FAO 2007). Thus, it is possible to conclude that the forests in south Asian region are subjected to degradation, leading to decline in growing stock as well as biodiversity.

20.3 Climate Change Projections for South Asia

Climate change is projected to have significant impacts on forest ecosystems. The climate projections for south Asia according to IPCC 2007b are summarized below.

20.3.1 Temperature Projections for South Asia

For the A1B scenario, the multi-model data set (MMD)-A1B models show a median increase of 3.3°C in annual mean temperature by the end of the twenty-first century.

The median warming varies seasonally from 2.7°C in June–July–August to 3.6°C in December–January–February, and is likely to increase northward in the area, particularly in winter, and from sea to land. The tendency of the warming to be more pronounced in winter is also a conspicuous feature of the observed temperature trends over India. Downscaled projections using the Hadley Centre Regional Model (HadRM2) indicate future increases in extreme daily maximum and minimum temperatures throughout south Asia due to the increase in greenhouse gas concentrations. This projected increase is of the order of 2–4°C in the mid-twenty-first century under the IPCC Scenario IS92a in both minimum and maximum temperatures (Krishna Kumar et al. 2003). Results from a more recent Regional Climate Model (RCM), PRECIS, indicate that the night temperatures increase faster than the day temperatures, with the implication that cold extremes are very likely to be less severe in the future.

20.3.2 Precipitation Projections for South Asia

Most of the MMD-A1B models project a decrease in precipitation in December–January–February (the dry season), and an increase during the rest of the year. The median change is 11% by the end of the twenty-first century, and seasonally is –5% in December–January–February and 11% in June–July–August, with a large inter-model spread. In a study with four General Circulation Models (GCMs), Douville et al. (2000) find a significant spread in the summer monsoon precipitation anomalies despite a general weakening of the monsoon circulation. They conclude that the changes in atmospheric water content, precipitation and land surface hydrology under greenhouse forcing could be more important than the increase in the land-sea thermal gradient for the future evolution of monsoon precipitation.

Time-slice experiments with ECHAM4 indicate a general increase in the intensity of heavy rainfall events in the future, with large increases over the Arabian Sea and the tropical Indian Ocean, in northern Pakistan and northwest India, as well as in northeast India, Bangladesh and Myanmar (May 2004). The HadRM2 RCM shows an overall decrease by up to 15 days in the annual number of rainy days over a large part of south Asia, under the IS92a scenario in the 2050s, but with an increase in the precipitation intensity as well as extreme precipitation (Krishna Kumar et al. 2003). Simulations with the PRECIS RCM also project substantial increases in extreme precipitation over a large area, particularly over the west coast of India and west central India (Rupa Kumar et al. 2006). Based on regional HadRM2 simulations, Unnikrishnan et al. (2006) report increases in the frequency as well as intensities of tropical cyclones in the 2050s under the IS92a scenario in the Bay of Bengal, which will cause more heavy precipitation in the surrounding coastal regions of south Asia, during both southwest and northeast monsoon seasons.

Thus climate is projected to change significantly in the south Asian region, potentially leading to impacts on forest ecosystems, which is presented in the following section.

20.4 Impact of Climate Change on Forests

20.4.1 IPCC Conclusions

IPCC 2007a has presented an assessment of the impacts of climate change on forest ecosystems. A brief summary of the overall impacts of climate change are presented:

- Populations of threatened species are expected to be at greater risk
  - Species that are currently classified as "critically endangered" will become extinct
  - 1/3 to 2/3 species at risk of extinction
  - Loss of biodiversity
- Species composition and dominance will be altered, resulting in ecosystem changes
- Shifts in forest types boundary
  - Altitude and latitude
- Forest die-back/mortality
  - Climate will change faster than capacity of plants to migrate
- Increase and later decrease in biomass productivity

IPCC 2007a projects changes in the structure, production and function of forests. Further, it concludes that the tropical forests are likely to be subjected to significant impacts with respect to changes in the geographical spread of forests, changes in biodiversity largely leading to decline of threatened species and finally, positive implications for biomass productivity in the short-term followed by negative impacts in the long-term with increased warming.

20.4.2 Impacts of Climate Change on Forests in South Asian Countries

The implications of projected climate change on forest ecosystems in different south Asian countries from published literature and the reports of the National Communications (www.unfcccc.int) are presented in the following sections.

India A study by Ravindranath et al. (2006) has assessed the impacts of climate change on forest ecosystems in India. This assessment is based on climate projections of Regional Climate Model of the Hadley Centre (HadRM3) using the A2 (740 ppm CO_2) and B2 (575 ppm CO_2) scenarios of Special Report on Emissions Scenarios and the BIOME4 vegetation response model.

When the impact of projected climate on the forested grids is considered, the dominant Miscellaneous forest type (where no species dominates) distributed across different parts of India, occurring in different rainfall and temperature zones and

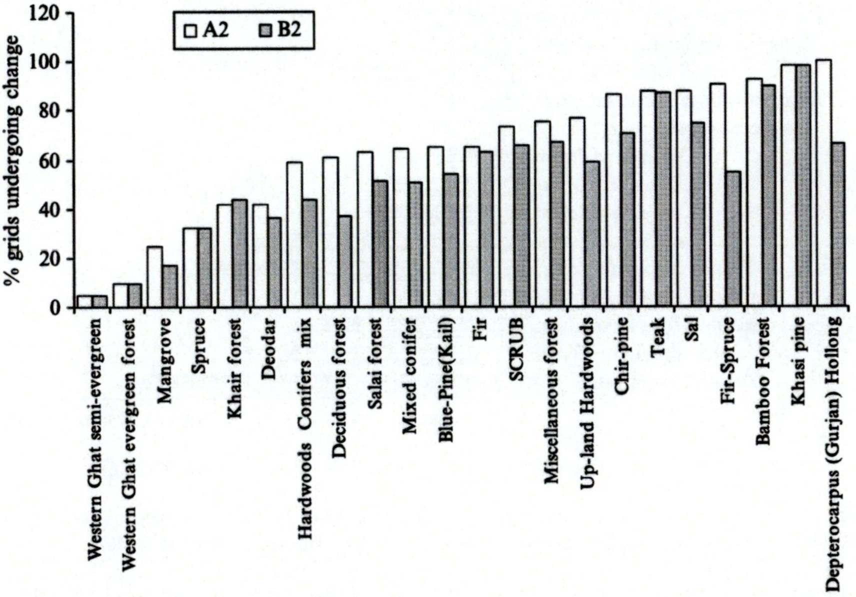

Fig. 20.2 Percentage of grids under different forest types undergoing change in A2 and B2 GHG scenarios (Ravindranath et al. 2006)

dispersed in fragments of varying sizes, is projected to undergo large-scale changes with 75% of the grids in A2 and 67% of the grids likely to be subjected to change of forest type (Fig. 20.2). The economically important forest types, such as Tectona grandis (Teak), Shorea robusta (Sal), Bambusa spp. (Bamboo), Upland Hardwoods and Pinus spp. (Pine), are projected to undergo change. When Pine, Teak, Sal and Bamboo are considered, over 75% of the grids are projected to undergo change in A2 and B2 scenarios. The forest types, which are likely to undergo minimal or no change under both A2 and B2 scenarios are Western Ghats Evergreen, Semi-evergreen and Mangrove forest types.

The overall impact of climate change could be assessed by considering the percentage of grids or area that show a change in the forest types under climate change scenarios. This change in the vegetation or forest type may be taken as an indicator of the vulnerability of the forest ecosystems to the projected climate change. Analysis for the 35,190 forested grids shows that 77% of the grids under A2 and 68% under B2 scenario are likely to undergo vegetation change. This indicates that well over half of the area under forests in India is vulnerable to the projected climate change under both A2 as well as the moderate B2 GHG scenarios. Similar trends were observed using IS92a scenario based HadRM2 climate outputs and BIOME3 model, which reported over 70% of all the grids projected to undergo change in vegetation types and potential vegetation (Ravindranath et al. 2003).

A review of studies by (IPCC 2001) and Gitay et al. (2002) has shown that forest biodiversity or the species assemblage is projected to undergo changes due to the projected climate change. Biodiversity is likely to be impacted under the projected

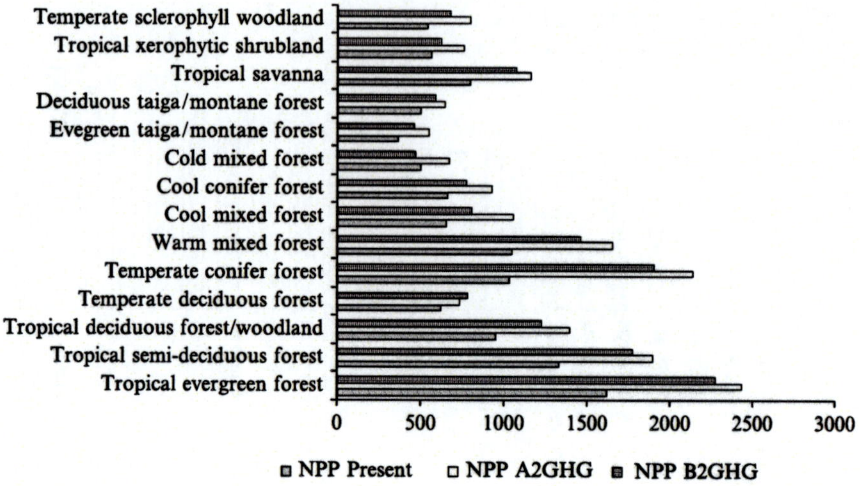

Fig. 20.3 Climate impacts on NPP; % forest biome-RCM grids subjected to change in NPP under GHG scenario over the current scenario under B2 Scenario (Ravindranath et al. 2006)

climate scenarios due to the changes or shifts in forest or vegetation types (in 57–60% of forested grids), forest dieback during the transient phase, and different species responding differently to climate change (IPCC 2001) even when there is no change in forest type. Climate change will be an additional pressure and will exacerbate the declines in biodiversity resulting from socio-economic pressures (Gitay et al. 2002).

Impact of Climate Change on Net Primary Productivity The impact of climate change on net primary productivity (NPP in g C/m^2 per year) was estimated under the Current and GHG scenarios. The mean value of NPP is estimated to be 835 g C/m^2 per year under Current climate scenario for the forested grids. Under the A2 GHG scenario, a doubling of NPP is predicted, while the moderate B2 GHG scenario projects an increase of about 73% for the forested grids. NPP is projected to increase in all the forested grids mainly due to the CO_2 fertilization effect on forest ecosystems. The impact of climate change on NPP varies according to vegetation types (Fig. 20.3).

It can be observed from Fig. 20.3 that among the dominant vegetation types (Tropical Xerophytic Shrubland, Tropical Deciduous Forest, Warm Mixed Forest and Tropical Semi-deciduous Forest), the NPP increases by 1.35–1.57 times under the GHG scenarios (A2 and B2) over the Current scenario NPP. The NPP under Tropical Evergreen Forest increases by 1.5 times under the GHG scenarios. The rate of increase on NPP was lower for Cool Conifer Forest, Cold Mixed Forest and Temperate Deciduous Forest. Generally the rate of increase is higher for warmer vegetation types.

Pakistan The possible impacts of climate change on the forestry sector assessed include changes in forest area, productivity and species composition as reported in the Pakistan's Initial National Communication on Climate Change (unfccc.int/resource/docs/natc/paknc1.pdf). Climate impacts are projected under two scenarios of CO_2; 360 and 550 ppm, temperature increase of +0.9°C and +1.8°C in 2020 and 2050 and rainfall increase by 3% and 6% by 2020 and 2050, respectively.

Changes in the Location of Optimal Growing Areas In general, a shift in the location of different biomes is likely under the change in precipitation scenarios. Cold and temperate conifers will show a northward shift, pushing against the cold conifer/mixed woodland, which in turn encroach upon the southern and lower edges of the alpine tundra. Similarly, the northern boundaries of warm conifer/mixed forest will also move north, pushing against the southern boundaries of the temperate conifer/mixed forest. This northwards shift of coniferous biomes will increase their size at the cost of the extent of the alpine tundra. A change in species composition may also occur, as those species that are hardier and have a wider distribution are likely to shift to other biomes in the north and south. Due to less severity and frequency in the incidence of frost and rise in temperature because of climate change, the frost tender species, which are at present confined to the southern biomes, will start moving northwards.

The size of biomes with considerable economic value such as the cold conifer/mixed woodland, cold conifer/mixed forest, temperate conifer/mixed woodland, etc., increase in general, while the size of biomes of degraded vegetation/scrub such as the alpine tundra, xerophytic wood/scrub, grassland/arid woodland and desert decrease. Biome size showed a consistent relation of increase or decrease with rising temperature, but remained independent of changes in precipitation.

Changes in Productivity The increase in temperature scenario tended to increase NPP in all biomes in the year 2020 and 2050 due to CO_2 fertilization effect. Under increase or decrease in rainfall scenarios, the NPP of all biomes did not show much increase in the years 2020 and 2050.

Changes in Carbon Stored The NPP increase of different biomes, over the base year of 1990 is estimated as 12% in the year 2020 and 19% in the year 2040–2050 under the climate change scenarios. Therefore, the carbon stored in the dry woody forest biomass is expected to increase from 111.75 million tons to 125.16 million tons in the year 2020 and to 132.98 million tons in the year 2050.

Changes in Nutrient Retention and Litter Decay Rate Under the prediction of increase in NPP and forest biomass as a result of climate change, there is likely to be a drain of soil nutrients, which are normally found in woody and non-woody forest biomass. In forests that are used for commercial harvesting, there is no immediate risk of nutrient depletion under climate change scenarios. In coniferous biomes, the soils are shallow but the nutrient balance shall remain undisturbed due to long rotation, low yield and selective cutting, as well as the increased activity of nutrient rebuilding process under climate change.

Changes in Pests and Weeds High temperature and increased precipitation reduces the dormant period for insects and increases the length of active period. Moreover, longer summers with early onset of growth may lead to more development of weeds in spring, which in turn can aggravate the spread of forest pests and pathogens. This may result in greater damage to forest vegetation due to the increase in prevalence of defoliating, sap sucking and stem boring insects.

A study by Siddiqui et al. (1999) using BIOME3 model and a 0.3°C rise in temperature per decade and a CO_2 concentration change from 350 to 425 ppm showed changes in area under different biome types with three biomes showing decline in area and five biomes showing an increase in area. Further, the study projects an increase in NPP for all the biomes.

Nepal Forests are the most important natural resources after water in Nepal. Majority of people use forest products as firewood, food, fodder, timber and medicines. Extensive utilization of and increasing demands for forest products have led to its dwindling both in area and quality. Further, Global Warming may cause forest damage through migration towards the polar region, changes in their composition, extinction of species etc. The consequence of this situation could affect directly not only the environment of Nepal but also lives of majority of the people.

To assess potential impacts on biodiversity, climate data for $1 \times CO_2$ and $2 \times CO_2$ were obtained from Climate Change scenarios developed using Canadian Climate Change Model (CCCM) and Geophysical Fluid Dynamics Model (GFD3) models. Increase of 2°C of temperature and 20% of rainfall to the observed data of 80 stations under the existing $1 \times CO_2$ level were also used for projecting scenarios of forest types. As one of the final outcomes, Holdridge vegetation map was prepared from data of 80 meteorological stations using Holdridge model system. The vegetation change due to double Carbon dioxide ($2 \times CO_2$) was also mapped using this Holdridge system, and variations in vegetation patterns were recorded.

Out of the 39 forest type zones categorized by Holdridge model, Nepal is depicted having 15 types (excluding snow area) under the existing $1 \times CO_2$ condition. These forests are tropical moist, tropical dry, tropical wet, subtropical moist, subtropical wet, subtropical dry, warm temperate rain, warm temperate moist, warm temperate wet, warm temperate dry, cool temperate moist, cool temperate wet, cool temperate dry, cool temperate steppe and boreal dry bush. Under $2 \times CO_2$ condition, there will remain only 12 types of vegetation: tropical dry, tropical moist, tropical rain, subtropical dry, subtropical moist, subtropical wet, subtropical rain, warm temperate dry, warm moist, warm wet, warm thorn steppe and desert bush. But under the incremental scenario (2°C temperature rise and 20% increase in precipitation) projected by the model, vegetation pattern in Nepal would be different than the existing ones. Out of eighty, 38 station locations will have change in vegetation pattern; tropical and subtropical moist forest will extend in area, subtropical and warm temperate dry forest will change to warm temperate moist forest in Jumla, cool temperate, steppe and thorn steppe will emerge in warm temperate and desert bush in existing cool temperate regions of Mustang, and forest cool temperate moist forest of this area will convert into warm temperate moist forest.

This study indicates that the tropical wet forest and warm temperate rain forest will disappear, and cool temperate vegetation will turn to warm temperate vegetation. Existing classification reveals that there does not exist rain forest in tropical and subtropical regions in the country, but they are expected to emerge under $2 \times CO_2$ condition. Warming effect will be significant in the sub-alpine and alpine regions of Nepal. The vegetation regime in general could be pushed up as much as 500 m in elevation.

Bangladesh Bangladesh has a diverse range of forest ecosystems, including savannah, bamboo, freshwater swamp forests and mangroves. The Sundarbans of Bangladesh, a world heritage site, is the single largest mangrove area in the world, comprising an area of 577 ha, and housing one of the richest natural gene pools. A total of 425 species have been identified there, the most notable of which is the Bengal tiger, which is endemic to the area. Forests contribute to a great extent to the ecological and economic stability of Bangladesh. The area under forest cover is shrinking rapidly due to over exploitation of forest resources and gradual conversion of forestlands to other uses. In Neilson et al. (1998), the IPCC reports that under some scenarios of climate change for late in the twenty-first century, Bangladesh would become a savanna/woodland, whereas under other scenarios, conditions would become wet enough to support tropical broad-leaf forests. The modeling results are inconsistent on whether the density of vegetation will increase or decrease. Studies on vegetation changes in Bangladesh under scenarios corresponding to the first half of the next century are not available. It is, however, reported that climate change induced additional flooding would adversely affect the Artocarpus species, especially *Artocarpus heterophyllus*. Similarly, other flood vulnerable species including *Azadirachta indica, Cajanus cajan, Leucaena leucocephala* would be affected. It is also feared that the Sal forest ecosystems in the Madhupur and the Barind tract areas would face additional moisture stress. Further water stress due to increased groundwater demand for irrigation would have compounding effects on the regeneration process of the species in those areas.

Climate change will have a detrimental impact on all of the forest ecosystems in Bangladesh, and the Sundarbans are likely to be the worst affected (Rahman and Alam 2003). The changes in temperature and water resources with climate change will result in direct pressure on many climate-sensitive species, and thereby increased erosion and deterioration of soil quality in may upland forested areas. Increased rainfall intensity will cause enhanced erosion upstream and thereby sedimentation. Saline intrusion is already a major problem in the Sundarbans, however it should be noted that climate change will also cause an increase in freshwater flows from the major distributaries with increased precipitation, and the extent to which this may offset salinity intrusion is uncertain. The Sundarbans also offer subsistence to around 3.5 million inhabitants who live within and around the forest boundary. The inundation and intruding salinity are interrupting traditional practices in the Sundarbans.

Sri Lanka The National Communications Report of Sri Lanka to UNFCCC (unfccc.int/resource/docs/natc/srinc1.pdf) provides limited information on the projected impacts of climate change on forest ecosystems, indicating the limitations of scientific assessments. Drought could lead to increased fire hazard in forests. Furthermore, under extreme drought conditions, small plants below 2 years of age could die due to water stress. Coastal wetlands are generally found at elevations just above mean sea level and below the highest tide. These wetlands account for a significant proportion of land less than 1m above sea level. With the rise in sea level marshes have generally kept pace by migrating inland and this has helped the prevention of wetland loss. However, if marshes are unable to keep pace with sea level rises, it would lead to a net loss of wetlands. According to IPCC (1997), in Sri Lanka, a significant increase in dry forest and a decrease in wet forest could occur. Projected increases in evapotranspiration and rainfall variability are likely to have a negative impact on the viability of freshwater wetlands, resulting in shrinkage and desiccation.

A study by Somaratne and Dhanapala (1996) has evaluated the potential impact of climate change on forest distribution in Sri Lanka. The Holdridge Life Zone Classification was used along with current climate and climate change scenarios derived from two general circulation models, the Geophysical Fluid Dynamics Laboratory model and the Canadian Climate Centre Model, at a 0.5° × 0.5° resolution. Current and future distributions of life zones were mapped with a Geographic Information System. These maps were then used to calculate the extent of the impact areas for the climate change scenarios. The current distribution pattern of forest vegetation includes tropical very dry forest (6%), tropical dry forest (56%), and tropical wet forest (38%). Results obtained using the Geophysical Fluid Dynamics Laboratory model show an increase in tropical dry forest (8%) and decrease in tropical wet forest (2%). The Canadian Climate Centre Model scenario predicted an increase in tropical very dry forest (5%) and tropical dry forest (7%), and a decrease in tropical wet forest (11%). Both models predicted a northward shift of tropical wet forest into areas currently occupied by tropical dry forest.

Bhutan Bhutan is part of one of the 10 global biodiversity "hotspots". Today its rich biodiversity resources still make a large contribution to the economy. Unfortunately, human activities are increasingly threatening all the existing ecosystems. In addition, several species are already endangered by climate change and extreme events. In the recent years, frequent landslides and prolonged dry periods and unprecedented heavy monsoon rain affecting agriculture and biodiversity are visible because of climate change. Given the lack of climate change scenarios for Bhutan, the National Communications Report analyses the vulnerability and potential impacts on the basis of available information and the current state of knowledge under the IPCC and not based on Bhutan-specific model-based analysis. The combination of climate change with the pressures of deforestation, land use changes, habitat degradation and fragmentation presents a significant threat to biodiversity. Climate change can affect biodiversity either directly, by changing the physiological responses of species, or indirectly, by changing the relationships between species. For example, a change in the insect population could influence the evolution of plant biodiversity and vice versa. Ironically, the projected climate change, with some increase in rainfall and temperature, is favorable for the richness of diversity; this is because a warmer world has greater potential for plant productivity, together with a movement of many species towards higher latitudes. Some species, however, will die out or become extinct as temperatures exceed their tolerance limits. Sufficient water supply will mean enough water for plants and animals for their survival and production. An increase in rainfall, on the other hand, also will enhance soil erosion and will affect vegetation and other biodiversity.

20.5 Adaptation to Climate Change in Forest Sector

According to IPCC 2007a as well as south Asia country-specific studies, climate change could cause irreversible damage to unique forest ecosystems and biodiversity, rendering several species extinct, locally and globally. Forest ecosystems require the longest response time to adapt, say through migration and regrowth. Further, a long gestation period is involved in developing and implementing adaptation strategies in the forest sector (Leemans and Eickhout 2004). Thus there is a need to develop and implement adaptation strategies. Adaptation is adjustment in natural or human systems in response to actual or expected climatic stimuli and their impacts on natural and socio-economic systems. Several 'no regret' or 'win–win' policies and forest management practices could be considered to adapt to the impacts of climate change. As a first step to adaptation, there is a need to conduct model-based projection of impacts of climate change at regional level using global dynamic vegetation models to identify vulnerable regions, forest types and species. This would enable planning adaptation practices and strategies. Adaptation practices are likely to vary for different forest types and regions, depending on the current status of the forests, knowledge of the projected impacts and access to information on suitable silvicultural practices and forest management strategies.

Adaptation practices: Some examples of the 'win–win' practices are as follows:

- Anticipatory planting of species
  - Along latitude and altitude
  - Promote assisted natural regeneration
- Promote mixed species forestry
  - Species adapted to different temperature tolerance regimes
- Develop and implement fire protection and management practices
- Adopt thinning, sanitation and other silvicultural practices
- Promote in situ and ex situ conservation of genetic diversity
- Develop drought and pest resistance in commercial tree species
- Develop and adopt sustainable forest management practices
- Expand Protected Areas and link them wherever possible to promote migration
- Conserve forests and reduce forest fragmentation
- Adoption of energy efficient fuelwood cooking devices to reduce pressure on forests

Adaptation Strategies and Policies Forest planning and development programs may have to be altered to address the likely impacts of climate change and appropriately adopt various policy and management practices to minimize the adverse impacts and vulnerability. Examples of policies and strategies include:

- Incorporate adaptation practices in forest planning both for the short and long term
  - Promote forest conservation since biodiversity rich forest are less vulnerable due to varying temperature tolerance
  - Halt forest fragmentation to promote migration of species
  - Link protected areas and create corridors to promote migration
- Promote community forest management to create long-term stake for communities in conserving forests and biodiversity
- Build capacity to develop and implement adaptation strategies

20.6 Conclusions and Future Direction

Forest account for nearly 20% of the geographic area of south Asia and includes many 'hotspots' of biodiversity. There is a large dependence of communities on the forests and biodiversity for their livelihoods. Any adverse impacts of climate change could have implications for biodiversity, livelihoods of dependent communities and many other ecosystem services. Very limited studies have attempted to assess the potential impacts of climate change on forest ecosystems, biodiversity and biomass production using vegetation models at the regional level. The existing studies, with all the limitations, conclusively show that the projected climate change is likely to have adverse implications for the distribution of different forest types and biodiversity. Studies state that forest types will undergo change or shifts in the forest boundaries, leading to forest die back in the transition period. Forest biomass production may increase initially with warming and rise in CO_2 concentration in the short-term but may experience decline in the long-term. All the studies and the reports of the National Communications of the region highlight the need to develop and implement adaptation strategies. This may involve implementing adaptation practices, incorporating climate change adaptation in forest planning as well as implementing policies to prevent forest fragmentation, conservation of forests and expansion and linking of protected areas. Planning and implementation of adaptation strategies requires reliable data and information on the projected climate change and impacts at regional level for different forest types. However, currently there are uncertainties with respect to climate projections at the regional level as well as the projections of impacts of climate change at the regional level. Some of the limitations in development and implementation of adaptation strategies are:

- Absence of reliable regional climate projections
- Limitations of dynamic vegetation models specific to tropical forests
- Limitations of data on vegetation, soil, water, etc., parameters for modeling
- Absence of models incorporating mitigation and adaptation components
- Absence of long-term forest vegetation monitoring studies
- Absence of long-term adaptation studies

There is a need for long term collaborative research into development of dynamic vegetation models, generating parameters required for developing country-specific forest types (or Plant Functional Types), assessment of climate impacts at regional level and development of adaptation practices and policies. This requires networking of different research institutions in the region as well as collaboration with international organizations such as FAO, UNEP, CIFOR and ICRAF. Some of the broad research needs in the climate change impact and adaptation area include the following:

- Ecological research on plant and animal species and communities in relation to climate variability and change
- Dynamic vegetation modeling of climate change impacts on forest ecosystems, biodiversity and adaptation

- Impact of climate change on mitigation potential, carbon sinks and adaptation
- Development of adaptation practices, strategies and policies

Acknowledgments We thank the Ministry of Environment and Forests of Government of India for the support to Indian Institute of Science under the GEF-UNDP National Communications Project. We also thank Norwegian Embassy for the support for conducting research on climate change at the Centre for Sustainable Technologies.

References

Chaturvedi RK, Tiwari R, Ravindranath NH (2008) Climate change and forests in India. Int Forest Rev 10:216–222

Douville H, Royer JF, Stephenson DB, Tyteca S, Polcher J, Cox P, Gedney N, Valdes P (2000) Impact of CO_2 doubling on the Asian summer monsoon: robust versus model dependent responses. J Meteorol Soc Jap 78:1–19

FAO (2007) State of the World's Forests 2007. Food and Agriculture Organization of the United Nations, Rome

FAO (2009) State of the World's Forests 2009. Food and Agriculture Organization of the United Nations, Rome

FSI (2005) State of Forest Report 2005, Forest Survey of India (FSI) Ministry of Environment and Forest. Dehra Dun, India

Gitay H, Suarez A, Watson RT, Dokken DJ (2002) Climate change and biodiversity. Intergovernmental Panel on Climate Change

IPCC (1997) The regional impacts of climate change: an assessment of vulnerability the regional impacts of climate change: an assessment of vulnerability. A special Report of IPCC Working Group II

IPCC (2001) Climate Change 2001: Synthesis Report 2001a. Intergovernmental Panel on Climate Change, Geneva, Switzerland

IPCC (2007a) Summary for policymakers, in climate change 2007: impacts, adaptation and vulnerability. In: Perry ML (ed) Contribution of Working Group II to the Fourth Assessment Report of the Intergovernmental Panel on Climate Change. Cambridge University Press, Cambridge, p 1000

IPCC (2007b) Summary for policymakers. In: Climate Change 2007: The Physical Science Basis. In: Solomon S, Qin D, Manning M, Chen Z, Marquis M, Avery KB, Tignor M, Miller HL (eds) Contribution of Working Group I to the Fourth Assessment Report of the Intergovernmental Panel on Climate Change. Cambridge University Press, Cambridge

Krishna Kumar K et al (2003) Future scenarios of extreme rainfall and temperature over India, Proceedings of the workshop on scenarios and future emissions, Indian Institute of Management (IIM), Ahmedabad, July 22, 2003. NATCOM Project Management Cell, Ministry of Environment and Forests. Government of India, New Delhi, pp 56–68

Laurance WF, Peres CA (2006) Emerging threats to tropical forests. The University of Chicago Press, Chicago

Leemans R, Eickhout B (2004) Another reason for concern: regional and global impacts on ecosystems for different levels of climate change. Glob Environ Change 14:219–228

Neilson RP, Prentice IC, Smith B, Kittel T, Viner D (1998) Simulated changes in vegetation distribution under global warming. In: Watson RT, Zinyowera MC, Moss RH, Dokken DJ (eds) Regional impacts of climatic change: an assessment of vulnerability. A special report of the Intergovernmental Panel on Climate Change (IPCC) Working Group II. Cambridge University Press, Cambridge, UK, pp 439–456

Rahman A, Alam M (2003) Mainstreaming adaptation to climate change in least developed countries (LDCs): Bangladesh Country Case Study. IIED Working Paper

Ravindranath NH, Joshi NV, Sukumar R, Murthy IK, Suresh HS (2003) Vulnerability and adaptation to climate change in the forest sector. In: Shukla PR, Sharma SK, Ravindranath NH, Garg A, Bhattacharya S (eds) Climate change and India: vulnerability assessment and adaptation. Universities Press, Hyderabad

Ravindranath NH, Joshi NV, Sukumar R, Saxena A (2006) Impact of climate change on forests in India. Curr Sci 90:354–361

Ravindranath NH, Chaturvedi RK, Murthy IK (2008) Forest conservation, afforestation and reforestation in India: implications for forest carbon stocks. Curr Sci 95:216–222

Rupa Kumar K, Sahai AK, Krishna Kumar K, Patwardhan SK, Mishra PK, Revadekar JV, Kamala K, Pant GB (2006) High-resolution climate change scenarios for India for the 21st century. Curr Sci 90:334–345

Siddiqui KM, Mohammad I, Ayaz M (1999) Forest ecosystem climate change impact assessment and adaptation strategies for Pakistan. Climate Res 12:195–203

Somaratne S, Dhanapala AH (1996) Potential impact of global climate change on forest distribution in Sri Lanka. Water Air Soil Pollut 92:129–135

Unnikrishnan AS, Rupa Kumar K, Fernandes SE, Michael GS, Patwardhan SK (2006) Sea level changes along the Indian coast: observations and projections. Curr Sci 90:362–368

Chapter 21
Options on Fisheries and Aquaculture for Coping with Climate Change in South Asia

Elayaperumal Vivekanandan

Contents

Abstract Fisheries and aquaculture have very important roles for food supply, food security and income generation in South Asia. About 7.5 million people work directly in the sector in this region, producing around 8.5 million tonnes annually. Due to several reasons, production from fisheries is stagnant in the last ten years, and aquaculture is not expanding as anticipated. Climate change is projected to exacerbate this situation. The potential outcome for fisheries may be decrease in production and value of coastal and inland fisheries, and decline in the economic returns from fishing operations. The potential outcome for aquaculture may be higher capital, operating and insurance costs, loss of fish stocks, damage to facilities, conflict with other water users, reduced production capacity and increased per unit production costs.

E. Vivekanandan (✉)
Central Marine Fisheries Research Institute, Cochin 682 018, India
e-mail: evivekanandan@hotmail.com

R. Lal et al. (eds.), *Climate Change and Food Security in South Asia*,
DOI 10.1007/978-90-481-9516-9_21, © Springer Science+Business Media B.V. 2011

Despite the uncertainties and potential negative impacts of climate change on fisheries and aquaculture, there are opportunities to reduce the vulnerability to climate-related impacts. The following measures could contribute to coping with climate change: (i) evaluating the adaptive capacity of important fish groups; (ii) identifying adaptive fishing and post-harvest practices to sustain fish production and quality; (iii) supporting energy efficient fishing craft and gear; (iv) identifying new land use system for aquaculture; (v) identifying new candidate species and developing hatchery and grow-out technologies; (vi) cultivating aquatic algae, which have positive response to climate change for food and pharmaceutical purposes and for production of biodiesel; (vii) investigating the potential fish diseases in the natural and aquaculture systems; (viii) increasing climate literacy among the fishing and farming communities; (ix) establishing Weather Watch Groups; and (x) evolving decision support systems for fisheries and aquaculture in the region. It is also important to recognize the synergies between adaptive and mitigation options related to climate change and non-climatic factors such as responsible fisheries, and ecofriendly aquaculture.

Keywords Fish production • Marine fish • Phenological changes and climate change • Corals • Iron fertilization • Sustainable fishing

Abbreviations

APR	Annual production rate
CMFRI	Central Marine Fisheries Research Institute
HABS	Harmful algal blooms
ICAR	Indian Council of Agricultural Research
Mt	Million tonnes
NDI	National Dependency Index

21.1 Introduction

South Asia includes some of the most productive fishing waters in the world. With more than 2,500 species of finfish alone, it is one of the regions of richest biodiversity. The sector is one of the important revenue-earning and employment-opportunity sectors, contributing significantly to the economy of the region. The annual fish production during 2000–2007 was about 8.5 million tonnes (Table 21.1). The region contributes about 6.5% to the overall fish production in the world. India contributes 70.2% to the region's total followed by Bangladesh (17.4%). Productions from countries bordering Arabian Sea (west coast of India and Pakistan) contribute 56.3%, countries bordering Bay of Bengal (east coast of India and Bangladesh) 38.9%, and Indian Ocean islands (Sri Lanka and Maldives) 4.8%. Production comes from three subsectors, viz., marine capture fisheries (50.3% to the region's total production), inland capture fisheries (11.5%), and inland and brackishwater

Table 21.1 Profile of fisheries and aquaculture in South Asia during 2000–2007 (Source: FAO 2004; CMFRI 2007; www.earthtrends.wri.org)

	Annual production (in thousand tonnes)							
	Marine	Inland			Fishermen and farmers	Export (million	GDP	
Country	Capture	Capture	Culture	Total	(in thousands)	US $)	Million US $	%
Bangladesh	620.0		857	1,477.0	Fulltime & Part-time: 4,360	420	1,181	4.7
Bhutan	–			0.3				
India	2,810	800	2,352	5,962	Fulltime : 930 Part-time: 1,070	1,900	4,770	1.1
Maldives	158.6	–	–	158.6	Fulltime : 22 Part-time: 5	56	127	11.0
Nepal	–	11.3	15.0	26.3	Fulltime & Part-time : 266		21	1.5
Pakistan	436.0	166.5	12.5	615.0	Fulltime : 287 Part-time: 184	126	492	1.5
Sri Lanka	253.1	NA	6	259.1	Fulltime : 250 Part-time : 100	94	207	2.0
Total	4,277.7	977.8	3,242.5	8,498.0	7,474	2,596	6,798	

aquaculture (38.2%). Mariculture is in its infancy, but is poised for growth. About 7.5 million people are employed either fulltime or part-time in fisheries and aquaculture. Fish trade has expanded significantly in the region over the last decade, with annual export reaching 2,596 million US$. The Nutritional Dependency Index (NDI is calculated from per caput consumption of fish to total animal protein intake, and scaled to 100) is very high for Maldives (100), Sri Lanka (62) and Bangladesh (58), indicating the high level of dependency on fish as food among the populations (Allison et al. 2004).

In recent years, the sector is facing serious challenges. Production from capture fisheries is stagnant for the last ten years as there is overexploitation, depletion of coastal fish stocks and competition among stakeholders in sharing the renewable, but limited resources. Aquaculture is not expanding to new species and areas due to legal, social and trade issues, and challenges posed by disease problems. Climate change exacerbates this situation. The region is recognized as being very vulnerable to climate change and sea level. Modeling studies show that climate will have the greatest economic impact on the fisheries sector in Asian countries (Brander 2007). The majority of South Asia's 7.5 million fishermen and fish farmers live in areas vulnerable to sea level rise. Bangladesh, for instance, is an extremely flat and low lying deltaic country. The east coast of India is prone to annual events of storms and cyclones. It is being increasingly realized that all the subsectors of fisheries and aquaculture will be impacted by climate change. Analyzing the impact of climate change on production systems, the Intergovernmental Panel on Climate Change (IPCC) ranked that small rivers and lakes in areas of higher temperatures and less rain will be affected the maximum followed by coastal waters, large rivers and lakes, estuaries and high seas, in that order (IPCC 1995).

In spite of the realization that fisheries and aquaculture have very important roles for food supply, food security and income generation in South Asia, research on the impacts, vulnerability and adaptation of this sector to climate change is limited. In the absence of peer-reviewed research papers from the region, grey literature in the form of reports and articles prepared by different organizations were reviewed for preparation of this paper (IPCC 1995, 2007; Allison et al. 2004; Adam 2006; World Fish Center 2007; FAO 2007a, 2008).

21.2 Historic Trends and Future Projections in Fish Production and Aquaculture

In 1995, South Asia produced 1.96 million tonnes (m t) of aquatic animals and plants valued at 279,000 million US $ through aquaculture, contributing 7.1% by weight and 6.6% by value to global aquaculture production. Production from aquaculture increased by 206.3% between 1984 and 1995 at an annual production rate (APR) of 10.7, with India being the largest producer (1.61 m t in 1995) within the sub-region. The share of aquaculture in total fisheries production in South Asia increased significantly from 15.2% in 1984 to 27.8% in 1995. In contrast, total capture fisheries production in South Asia increased by only 42.0% from 3.59 m t in 1984 to 5.10 m t in 1995. Thus the production from capture fisheries as well as aquaculture increased from 4.49 m t in 1984 to 7.06 m t in 1995 to 8.50 m t during 2000–2007.

The projected total production in South Asia for the year 2020 is 14.5 m t (India: 10.0 m t), and the share from aquaculture is expected to range from 60% (India) to 50% (other South Asian countries). The projected APR is 1.0% and 8.5% for capture fisheries and aquaculture, respectively for India; and 0.8% and 3.5% for capture fisheries and aquaculture, respectively for the other South Asian countries.

21.3 Impact of Climate Change on Marine Fish

Fish is the largest living resource exploited from the wild. The most important characteristic of capture fisheries in South Asia is that the resource is a common property, the access to which is free and open. Second, property rights to fisheries are difficult to establish, leading to intrasectoral conflicts. In addition to these, other anthropogenic interventions such as discharge of untreated domestic and industrial effluents into the sea, bioaccumulation and biomagnification of toxic chemicals and trace metals in fish tissues, ballast water discharge, coral mining and habitat destruction cause considerable damage to fish populations. For India, the catches have been either very close to or have exceeded the potential yield in the late 1990s for several fish stocks due to overfishing (Vivekanandan 2001). Climate change is a dispensatory factor on fish populations.

Most fish species have a fairly narrow range of optimum temperatures related to their basic metabolism and availability of food organisms. Being poikilotherms, even a difference of 1°C or 0.1 unit pH in seawater may affect their distribution and life processes. The more mobile species should be able to adjust their ranges over time, but less mobile and sedentary species may not. Depending on the species, the area it occupies may expand, shrink or be relocated. This will induce increases, decreases and shifts in the distribution of marine fish, with some areas benefiting while others lose. From the recent investigations carried out by the Indian Council of Agricultural Research (ICAR) and Central Marine Fisheries Research Institute (CMFRI), the following responses to climate change by different marine fish species are discernible in the Indian seas: (i) Extension of distributional boundary (Vivekanandan et al. 2009a); (ii) Shift in latitudinal distribution; (iii) Shift/extension of depth of occurrence (CMFRI 2008); and (iv) phenological changes (Vivekanandan and Rajagopalan 2009). Some evidences of the responses are given below:

21.3.1 Extension of Distributional Boundary

The oil sardine *Sardinella longiceps* and the Indian mackerel *Rastrelliger kanagurta* are tropical coastal and small pelagic fish, forming massive fisheries in India (catch during 2007: 0.7 million tonnes valued at about 150 million US$). They are governed by the vagaries of ocean climatic conditions, and have high population doubling time of 15–24 months. They are cheap source of protein, and form a staple, sustenance and nutritional food for millions of coastal people. They were known for their restricted distribution between latitude 8°N and 14°N and longitude 75°E and 77°E (Malabar upwelling zone along the southwest coast of India) where the annual average sea surface temperature ranges from 27°C to 29°C. Until 1985, almost the entire catch was from the Malabar upwelling zone and the catch was either very low or there was no catch from latitudes north of 14°N. In the last two decades, however, the catches from latitude 14°N to 20°N are increasing, reaching 150,000 t (21% of all-India catch during 2007). A positive correlation was found between the catches and sea surface temperature (SST). The surface waters of the Indian seas are warming by 0.04°C per decade, and the warmer tongue (27–28.5°C) of the surface waters expanded to latitudes north of 14°N, enabling the oil sardine and Indian mackerel to extend their distributional boundary to northern latitudes. It is also found that the catches from the Malabar upwelling zone has not decreased indicating the distributional "extension" and not distributional "shift". Considering the catch as a surrogate of distribution and abundance, it is also found that the two most dominant fish are able to find temperature to their preference especially in the northern latitudes in recent years, thereby establishing fisheries in the extended coastal areas. Assuming further extension of warmer SST tongue in the future, it is expected that the distribution may extend further north of latitude 20°N. However, if the SST in the Malabar upwelling zone increases beyond the physiological optimum of the fish, it is possible that the populations may be driven away from the southern latitudes.

21.3.2 Shift in Latitudinal Distribution and Abundance

Catfish are one of the major resources along the southwest and southeast coasts of India (latitude: 8°N–14°N). During 1970–2007, the catches from these coasts decreased from 35,000 to 7,800 t. On the other hand, the catches from the northwest and northeast coasts (latitude: 15°N–22°N) increased from 16,000 to 42,500 t during the same period. There was a strong negative correlation between catfish catch and SST along the two southern coasts whereas the correlation between catch and SST was positive along the northern coasts. As the average seawater temperature in the southern latitudes exceeded 29°C in the last one decade, it appears that the catfish have shifted their distribution to the northern latitudes where the seawater temperature is between 27°C and 28.5°C. This is a response different from that shown by the oil sardine and Indian mackerel to increase in seawater temperature.

21.3.3 Shift/Extension of Depth of Occurrence

The Indian mackerel, *Rastrelliger kanagurta*, in addition to extension of its northern boundary, is found to descend to deeper waters in the last two decades.

The fish normally occupies surface and subsurface waters. During 1985–1989, only 2% of mackerel catch was from bottom trawlers, and the rest of the catch was contributed by pelagic gear such as drift gillnet. During 2003–2007, it is estimated that 15% of mackerel catch is contributed by bottom trawlers along the Indian coast. The Indian trawlers operate at a depth ranging from 20 to 80 m by employing high opening trawlnets. In the last 25 years, the specifications of trawlnet such as mouth opening, headrope length, otterboard and mesh size have not been modified, and hence the increase in the contribution of trawlers to the mackercl catch is not gear-related. As the subsurface waters are also warming up, it appears that the mackerel, being a tropical fish, has extended its veritical boundary to deeper waters.

21.3.4 Phenological Changes

Fish have strong temperature preferences to spawning. The process of spawning is known to be triggered by pivotal temperatures. The annually recurring life cycle events such as timing of spawning can provide particularly important indicators of climate change. Though sparsely investigated, phenological changes such as seasonal shift in spawning season are now evident in the Indian seas.

The threadfin breams *Nemipterus japonicus* and *N. mesoprion* are distributed along the entire Indian coast at depths ranging from 10 to 100 m. They are short-lived (longevity: about 3 years), fast growing, highly fecund (annual egg production around 0.2 million per adult female) and medium-sized fishes (maximum length: 35 cm). Data on the number of female spawners collected every month off Chennai (southeast coast of India) from 1981 to 2004 indicated wide monthly fluctuations. However, a trend in the shifting of spawning season from warmer (April–September) to cooler months (October–March) was discernible. Whereas 35.3% of the spawners of *N. japonicus* occurred during warm months during 1981–1985, the number of spawners gradually reduced and only 5.0% of the spawners occurred during the same season during 2000–2004. During 1981–1985, it was observed that 64.7% of the spawners occurred during October–March, whereas as high as 95.0% of the spawners occurred during the same season in 2000–2004. A similar trend was observed in *N. mesoprion* too. The percent occurrence of spawners of the two species linearly decreased with increasing temperature during April–September, but increased with increasing temperature during October–March over the time scale. It appears that SST between 28°C and 29°C may the optimum and when the SST exceeds 29°C, the fish are adapted to shift the spawning activity to seasons when the temperature is around the preferred optima.

Currently, it is difficult to find out how much of catch fluctuation is due to changes in fish distribution and phenology. However, these changes may have impact on nature and value of fisheries (Perry et al. 2005). If small-sized, low value fish species with rapid turnover of generations are able to cope up with changing climate, they may replace large-sized high value species, which are already showing declining trend due to fishing and other non-climatic factors (Vivekanandan et al. 2005).

Phytoplankton are the basis for the productivity of the oceans and are critically important to the flow of resources. Plankton are particularly sensitive to environmental fluctuations, and therefore those fish, which are directly dependent on plankton for food, will be strongly influenced by climate change (Briones et al. 2006). The larvae of several marine fish have wider dispersal ranges (aided by currents) than terrestrial organisms. Major changes to ocean circulation will cause changes in dispersal pattern of larval fish, particularly the pelagics. This will change the food webs as well fish catch. Such distributional changes would lead to novel mixes of organisms in a region, leaving species to adjust to new prey, predators, parasites, diseases and competitors (Kennedy et al. 2002), and result in considerable changes in ecosystem structure and function.

21.4 Vulnerability of Corals

Coral reefs are the most diverse marine habitat, which support an estimated one million species globally. They are highly sensitive to climatic influences and are among the most sensitive of all ecosystems to temperature changes, exhibiting the phenomenon known as coral bleaching when stressed by higher than normal sea temperatures. Reef-building corals are highly dependent on a symbiotic relationship with microscopic algae (type of dinoflagellate known as zooxanthellae), which live within the coral tissues. The corals are dependent on the algae for nutrition and coloration. Bleaching results from the ejection of zooxanthellae by the coral polyps and/or by the loss of chlorophyll by the zooxanthellae themselves. Corals usually recover from bleaching, but die in extreme cases.

In the Indian seas, coral reefs are found in Gulf of Mannar, Gulf of Kachchh, Palk Bay, Andaman Seas and Lakshadweep Seas. Indian coral reefs have experienced 29 widespread bleaching events since 1989 and intense bleaching occurred in 1998 when the SST was higher than the usual summer maxima. By using the relationship between past temperatures and bleaching events and the predicted SST for another 100 years, Vivekanandan et al (2009b) projected the vulnerability of corals in the Indian Seas. The outcome of this analysis suggests that if the projected increase in seawater temperature follows the trajectory suggested by the HadCM3 for an SRES A2 scenario, reefs should soon start to decline in terms of coral cover and appearance. The number of decadal low bleaching events will remain between 0 and 3 during 2000–2099, but the number of catastrophic events will increase from 0 during 2000–2009 to 10 during 2000–2099.

Given the implication that reefs will not be able to sustain catastrophic events more than three times a decade, reef building corals are likely to disappear as dominant organisms on coral reefs between 2020 and 2040 and the reefs are likely to become remnant between 2030 and 2040 in the Lakshadweep region and between 2050 and 2060 in other regions in the Indian seas. These projections on coral reef vulnerability have taken into consideration only the warming of seawater. Other factors such as increasing acidity of seawater would slow down formation of exoskeleton of the reefs, and if acidification continues as it is now, all the coral reefs would be dead within 50 years. Given their central importance in the marine ecosystem, the loss of coral reefs is likely to have several ramifications.

21.5 Status of Aquaculture in South Asia

India is world's second largest aquaculture producer and contributes 70.3% by weight and 65.8% by value to production from South Asia. The contributions (by weight) of other major fish-producing countries in the region are: Bangladesh 26.5%, and Nepal 0.4%.

Aquaculture systems in South Asia range from small low intensity operations such as small family-owned ponds that require little in the way of inputs to large-scale high intensity systems such as shrimp. However, majority are extensive and traditional freshwater and brackishwater farming systems. The South Asian aquaculture production is dominated by freshwater fish (>90%); the remainder is nearly all from brackishwater; and production from marine environment is insignificant (Handisyde et al. 2005). Production originating from stocked lakes (man-made and natural), floodplains and reservoirs also contributes significantly. Countries are looking into integration of aquaculture with agriculture, rice-cum-fish culture and other forms of integrated aquaculture. Freshwater carp culture is expanding at a considerable rate over the past decade. Over 75 species have been identified as potential candidates for freshwater aquaculture. However, only a few species (Indian major carps, Chinese carps, and a very few indigenous species) are currently cultured commercially. High value crustacean, the scampi, *Macrobrachium rosenbergii* is also cultured. Organized farming of other species has not developed on a commercial scale. Bangladesh, with substantial annual flooding, continues to increase fish production through floodplain stocking. However, lack of adequate seed, disease prevention and control programs constrain the future development of this sector.

It is likely that land and water may become limiting factors to the growth of freshwater aquaculture in most countries in the region, especially in some areas of India, where freshwater availability may become a severe limiting factor. The national governments should give priority for allocation of water and land to aquaculture. Aquaculture diversification and intensification are necessary and there is also a need for technological advancement for commercial-scale farming of new species.

Brackishwater aquaculture is focused almost exclusively on the penaeid shrimp, *Penaeus monodon*. Shrimps are produced mainly for export markets and the main producers are India, Bangladesh and Sri Lanka. Environmental costs, social equity and sustainability have become important issues in shrimp culture development in these countries. Besides the Whitespot Syndrome disease outbreaks, which have devastated the shrimp aquaculture sector in the last ten years, issues of shrimp farm effluent management have to be adequately addressed. In recent years, culture of the seabass *Lates calcarifer* is gaining popularity in India.

Development of mariculture has commenced in a modest scale in India and Sri Lanka. In India, mussel and ornamental fish farming are becoming popular.

In Sri Lanka, marine ornamental fish farming is being practiced on a commercial scale. Finfish culture in open cages is in infancy and at present does not contribute significantly to production. Global demand for mariculture products is creating interest in developing appropriate technology and trained personnel. Effective integrated coastal area management needs to be adopted to avoid environmental issues of open sea and coastal aquaculture development, and to control impact from other non-aquaculture activities.

Overall, the prospects for aquaculture development in South Asia are good, as there is scope for growth of commercial and subsistence farming. The countries have optimistic projections for the expansion of aquaculture. Projections by the Indian Council of Agricultural Research indicate that only 20% of available coastal land for shrimp farming is currently being used in India. India and Bangladesh have projected an annual growth rate of 8.5% and 3.5% respectively for another 10 years (FAO 2004). However, future aquaculture development in South Asia calls for developing hatchery technology and locally made artificial feed. If large-scale aquaculture programs are to be successful in the long run, the socioeconomic issues such as land ownership, leasing procedures, common property rights, etc. need to be resolved.

21.5.1 Impact of Climate Change on Aquaculture

The possible sensitivities of climate change on aquaculture is difficult to assess due to several uncertainties such as extent and rate of predicted changes and the biological and physical impacts. There are only a limited number of publications on the possible impacts of climatic factors on aquaculture in South Asia (Handisyde et al. 2005). Climate change issue has not been considered a priority by the aquaculture sector so far in spite of the understanding that increase in fish production in the future has to come from aquaculture.

Small- scale aquaculture farmers represent a large group in the sector. They are many in number and each manages small areas of land and water resources. However, the total area availed by them is quite large, which are ecologically sensitive and highly vulnerable to climate change impacts. The resource – poor small farmers will be particularly vulnerable because of their dependence on biological and economic resources, and their low ability to adapt.

The potential impacts of climate change on aquaculture will be either direct or indirect. The predicted direct impacts are (i) changes in water availability, temperature and salinity; (ii) changes in the season and quantity of seed availability; (iii) damage by extreme climatic events such as storms, floods and droughts; and (iv) adaptability of the candidate species to the changes (Table 21.2). The indirect impacts are (i) availability of fishmeal and its cost; and (ii) increased frequency of disease outbreaks. The impact on aquaculture and the related livelihoods will be linked to the type, scale and intensity of the aquaculture production system and the location and environment in which it is being carried out.

Table 21.2 Impact of climate change on aquaculture systems

Causes	Impact	Adaptation
Higher temperature	– Changes in metabolism of candidate species – Decrease in dissolved oxygen in ponds – Increased incidents of fish diseases	– Identify and grow thermal resistant species, and genetically modified species
Salinization of inland water due to sea level rise	– Candidate species may not tolerate higher salinity – Reduction in growth	– Identify and grow salt resistant species
Storm surges	– Physical damage to farm structures – Loss of stock – Abrupt loss of capital	– Proper planning in selection of aquaculture sites – Adequate fiscal back-up such as insurance
Floods	– Physical damage to structures – Loss of stock – Abrupt loss of capital – Reduction in salinity – Fish disease occurrence	– Proper planning in selection of aquaculture sites – Adequate fiscal back-up such as insurance
Drought	– Great risk of water availability – Erosion of water quality leading to diseases – Reduction in production	– Proper planning in selection of aquaculture sites – Adequate fiscal back-up such as insurance

21.6 Options for Fisheries and Aquaculture Sector for Adaptation

21.6.1 Tackling Overfishing

Options for adaptation are limited, but they do exist. The impact of climate change depends on the magnitude of change, and on the sensitivity of particular species or ecosystems (Brander 2008). Fish populations are facing the familiar problems of overfishing, pollution and habitat degradation. Food and Agriculture Organization has estimated that about 25% of all fish stocks are overexploited and 50% are fully exploited (FAO 2007b). About 1.2% of global oil consumption is used in fisheries, and it is found that fish catching is the main contributor to global warming in the fish production chain (Thrane 2006). Reduction of fishing effort will benefit in relation to adaptation of fish stocks and marine ecosystems to climate impacts; and mitigation by reducing greenhouse gas emissions. Hence, some of the most effective actions which we can be taken to tackle climate impacts are to deal with the old familiar problems such as overfishing (Brander 2008) and adapt Code of Conduct for Responsible Fisheries (FAO 2007b).

21.6.2 Plankton Restoration by Iron Fertilization

Global phytoplankton production is reported to have declined by 6 to 9% since 1980 (Gregg et al. 2003). It is suggested that a massive international plankton restoration program by iron fertilization could regenerate carbon sequestering capacity of the world oceans. Iron fertilization is the physical distribution of microscopic particles of the vital micronutrient, viz., iron in the upper oceans where the growth and/or reproduction of phytoplankton is limited by the amount of iron in the seawater. Being a vital micronutrient, it encourages growth of phytoplankton blooms, increases energy flow in marine food chain, and sequesters CO_2 from the atmosphere. Each kilogram of iron can fix 83 t of CO_2 and generate 100 t of phytoplankton (Sunda and Huntsman 1995). Advocates of iron fertilization say that, using this technique, the ocean plankton populations can be restored, reduce the climate change problem, revive major fisheries and other endangered marine organisms. Whereas iron fertilization is considered as one of the effective mechanisms to slow global warming, a number of criticisms has also arisen (Table 21.3). The criticisms are concerned with some serious consequences as given below: (i) The plankton bloom is short-lived. (ii) Possible side effects on the ecosystems are not known. (iii) Carbon sequestration is very little as most of the plankton is eaten and does not settle to ocean bottom. (iv) The bloom induces anoxic condition in the sea bottom, and may cause red tides and other harmful algal blooms (HABs). It also should be recognized that in the iron fertilization program, what holds true for one region may not be true in another (Watson 1997). Hence, iron fertilization programs need to be viewed with caution.

Table 21.3 Arguments for and against iron fertilization

For	Against
Natural blooms occur without ill effects.	Possible side effects not known; the ecosystem may change.
CO_2-induced surface warming is already shifting the distribution of marine organisms on a massive scale.	
Carbon that sinks below marine thermocline is effectively removed from the atmosphere for thousands of years.	Carbon is not removed from ocean system unless it settles to ocean floor where it is sequestrated.
Some ocean trials report remarkable results of CO_2: iron fixation of 300,000 to 1.	Sequestration may be very little, with most of the plankton being eaten rather than deposited to ocean floor.
Red tides and HABs are coastal phenomena. Iron fertilization is done in deep oceans where iron deficiency is the problem.	Plankton blooms may cause red tides and other harmful algal blooms (HABs).
Many extremely massive natural blooms have been studied and no deep-water anoxia has been reported.	Organic bloom sinks, and may induce anoxic condition in the sea bottom.
Even medium-scale fertilization produces ecosystem-level blooms.	Large-scale iron fertilization experiments have not been attempted so far to make conclusions.

21.6.3 Cultivation of Sea Plants

Sea plants are excellent carbon sequestration agents and many of them sequester at a rate better than their terrestrial counterparts (Zon 2005). The emission of CO_2 through their respiratory activity of sea plants is very low as they are capable of reutilizing the respiratory release of CO_2 within their cellular interspace for subsequent photosynthesis (Kanwisher 1966). CO_2 sequestration by the common sea plants such as the red algae *Gracilaria corticata* and *G. edulis*, brown alga *Sargassum polycystum* and the green alga *Ulva lactuca* has been quantified in laboratory studies in India by Kaladharan et al (2009). The carbon sequestering efficiency was found unaltered even in higher levels of dissolved CO_2. Green algae were found to have better CO_2 sequestering ability than the red and brown algae. It is estimated that the standing crop of sea plants in the Indian waters is 260,876 tonnes, comprising 14% agar and carrageenan yielding red algae (Rhodophyceae), 16% algin yielding brown algae (Phaeophycea) and 70% green algae (Chlorophyceae). Estimates indicate that the standing crop in the Indian Seas is capable of utilizing 9,052 tonnes of CO_2 per day. Mass cultivation of these plants will help reducing the CO_2 concentration from the seawater. Coastal communities along the Palk Bay (southeast coast of India) have taken up commercial cultivation of *Kappaphycus alvarezii* for the last three years.

There are two methods for cultivation of sea plants, one by vegetative propagation and the other by reproductive method. In the vegetative propagation method, the fragments of the plant are inserted in the twist of coir ropes and the fragments tied to nylon twine are allowed to grow in the coir mat, which are fixed off bottom in the coastal waters or in floating raft/cages (Kaliaperumal 2005). In 60 days, *G. edulis* shows upto 30-times increase in yield. Similar technique is being followed for commercial cultivation of another carrageenan and high yielding plant, *Kappaphycus alvarezii* (Eswaran et al. 2005). The bays, creeks and lagoons in the open shore in the South Asian region are suitable for cultivation of sea plants.

There is increasing international demand for products from sea plants. These plants can be used as human food, cattle food, fertilizer, and are rich source of agar and algin, which form the basis of confectionary and pharmaceutical industries. Mass cultivation will effectively sequester carbon, augment supply of raw materials to the food and pharmaceutical industries and provide employment to the coastal population.

21.6.4 Cultivation of Halophytes

In coastal areas and mudflats near the sea, where the salinity does not allow farming of the usual food crops, plants that grow and flourish those conditions are advocated. One such plant is the sea asparagus, *Salicornia*. It is a succulent, bushy plant found in the salty terrains along the east coast of India, Bangladesh and Sri Lanka. The variety, SOS-10, grows well in desert sands irrigated with seawater. The tender stems and tips of the plant are edible by humans. The seeds yield edible oil (rich in polyunsaturates), which is similar to safflower oil in fatty acid composition and is usable as biodiesel. The plant grows well with maximum yields in hot climates if the seeds are sown in cool season so as to reach maturity during the hot months (www.hindu.com/seta/2003/09/05.htm).

Salicornia is a better photosynthesizer than several food grains. It uses C-4 pathway, converting the captured CO_2 first into compound containing four carbons (oxaloacetate) using the enzyme PEPCase. Thus *Salicornia* sequesters CO_2 better, grows in saline water, and gives edible oil and biodiesel. With the rising seas inundating the low-lying coastal areas, growing *Salicornia* that is available in South Asia may be an opportunity. A 2,000 ha farm would yield biomass of 30,000 and 2,500 t of seeds. It suits small, labor-intensive farms as well as highly mechanized farms. The Seawater Foundation, Arizona is advocating diversion of flow of inundating seawater inland through ocean canals in the northern Mexican State of Sonora and nourish commercial fish and shrimp aquaculture operations, mangrove forests and crops such as *Salicornia* that produce food and fuel (www. seawater foundation.org).

21.6.5 Artificial Reefs for Coast Protection

With the rising sea level, erosion of coastal areas is a cause for major concern along the South Asian coasts. Coastal areas along Gujarat and Kerala (west coast of India), Tamil Nadu and Orissa (east coast of India), Sunderban (West Bengal in northeast coast of India), and Bangladesh are facing serious threats of sea erosion, risking the lives and properties of millions of people. Along the 600-km coastline of Kerala (southwest coast of India), about 200 fishing villages are within 100 m from the high-tide line (CMFRI 2008). The government has constructed coast protection structures (sea walls, boulders, groynes) for 350 km. However, the construction appears to be a temporary solution as the sea erodes at the two ends of the construction and the coastal villages get inundated.

One of the structures advocated as an opportunity for multipurpose use including coast protection is the Artificial Surfing Reef, which are constructed in the coastal waters. The reefs are constructed using sand-filled geotextile bags or containers, or rocks (granite). The structures have slopes, which increase the reef's volume. The back gradient of the reef is as steep as possible (for example, 1:1) whereas the face gradient is gentle (for example, 1:20). The opportunities of the Artificial Surfing Reef are projected as follows:

i. The upper part of the slope of the reef affects wave break. The coast protection function is largely derived from the widening of the beach in the lee of the reef due to sheltering and wave rotation caused by the reef. Although the shoreline oscillates back and forth due to storms and swell conditions, the shoreline remains further seaward than adjacent coast unprotected by the reef (Black 2000).
ii. The reef provides habitat for colonization and occupation by many marine fauna that would otherwise be unlikely to persist at that location due to high-energy hydrodynamic conditions and sand-dominated substrate. It has the potential to increase local biodiversity and may contribute to biological productivity and fisheries at a regional scale (Edwards 2004).

iii. The reef can be used for promoting surfing, which is not popular in South Asia. The reefs can deliver tourism and amenity facilities with a range of socio-economic benefits (Arup 2001).

Some sites of Artificial Surfing Reefs construction are at Boscombe (Bournemouth), Borth (Ceredigion), Newquay (Cornwall) and Opunake (New Zealand).

21.6.6 Biofuel from Algae

Biodiesel is a clean burning alternative fuel for diesel engines. It is produced from renewable resources, biodegradable, non-toxic, and as a blended product, can be used in most diesel engines without modifications. Biofuels come from agricultural crops such as soy, jatropha and jajoba. Considering the current debate about agricultural land being used to produce biofuels, algae may be a potential alternative fuel source. Algae can be grown easily in ponds or tanks constructed in poor quality land. The first stage should be identification and development of suitable algal strains to achieve stable, continuous and high yield production. The process will also produce byproduct for animal feed stock. Products from *Spirulina* can be used for human consumption.

Algae require large volumes of CO_2 to grow, which means less CO_2 is released into the atmosphere. It may be a good option to locate the farms near major industries and power stations. Companies that produce CO_2 will eventually be able to claim carbon credits as the CO_2 can be stored or captured and released into algae farms (www.greencarcongress.com). The production of biodiesel complements ethanol as an alternative renewable fuel.

21.7 Strategies for Evolving Adaptive Mechanisms

In the context of climate change, the primary challenge to the fisheries and aquaculture sector will be to ensure food supply, enhance nutritional security, improve livelihood and economic output, and ensure ecosystem safety. These objectives call for identifying and addressing the concerns arising out of climate change; evolve adaptive mechanisms and implement action across all stakeholders at national, regional and international levels (Table 21.4). In response to shifting fish population and species, the fishing sector may have to respond with the right types of craft and gear combinations, on-board processing equipments etc. Governments should consider establishing Weather Watch Groups and decision support systems on a regional basis. Allocating research funds to analyze the impacts and establishing institutional mechanisms to enable the sector are also important. The relevance of active regional and international participation and collaboration to exchange information and ideas is being felt now as never before.

For the fisheries and aquaculture sector, climate change notwithstanding, there are several issues to be addressed. Strategies to promote sustainability and improve

Table 21.4 Options for coping with climate change in fisheries and aquaculture (Modified after Allison et al. 2004; Handisyde et al. 2005; FAO 2008)

Concerns	Adaptive mechanisms
Uncertainties in fish availability and supply	i. Adapt code of conduct for responsible fisheries; ii. Develop knowledge-base for climate change impact on fisheries and aquaculture; iii. Predict medium and long term probabilistic production; iv. Predict specific fisheries and aquaculture systems; v. Assess the adaptation capacity, resilience and vulnerability of marine and freshwater production systems; vi. Adjust fishing fleet and infrastructure capacity; vii. Consider the synergistic interactions between climate change and other factors such as fishing, water availability, energy and agriculture;
New challenges for risk assessment	i. Consider increasing frequency of extreme weather events; ii. Consider past management practices to evolve robust adaptation systems; iii. Identify and address the vulnerability of specific communities; consider gender and equity issues
Complexities of climate change interactions into governance of frameworks to meet food security objectives	i. Recognition of climate-related processes, and their interaction with others; ii. Action plans at national level based on (a) code of conduct for responsible fisheries; (b) Integrated ecosystem-based fisheries and aquaculture management plans, (c) framework for expansion of aquaculture; (d) linkage among cross-sectoral policy frameworks such as insurance, agriculture, rural development and trade; iii. Action plans at regional level by (a) strengthening regional organizations and place climate change agenda as a priority; (b) addressing transboundary recourse use; (c) evolving common platforms and sharing the best practices; iv. Action plan at international level by (a) linking with mitigation activities; (b) enhancing co-operation and partnerships; (c) applying international fishery agreements
Fisheries and aquaculture may be more vulnerable in conflicts with other sectors	i. Action plans should involve not only fisheries institutions/ departments, but also those for national development planning and finance; ii. Sharing and exchange of information with other sectors; iii. Existing management plans for fisheries and aquaculture need to be reviewed by considering climate change.
Financing climate change adaptation and mitigation measures	i. Fishermen, fish farmers, processors, traders and exporters should increase self protection through financial mechanisms; ii. Improving equity and economic access such as microcredit should be linked to adaptation responses; iii. Investment on infrastructure, such as construction of fishing harbor, should consider climate change; iv. Financial allocation in national budget for risk reduction and prevention practices such as early warning systems and disaster recovery programmes and for relocation of villages from low lying areas; v. Fiscal insensitive for reducing the sector's carbon footprint and other mitigation and adaptation options; vi. Full potential of existing financial mechanisms has to adapt and mitigate the issue of climate change.

the supplies should be in place before the threat of climate change assumes greater proportion. While the fisheries sector cannot do much to mitigate climate change, it could contribute to reduce the impact by following effective adaptation measures.

References

Adam MS (2006) Vulnerability and Adaptation Assessment of the Fisheries Sector in the Maldives – NAPA Project. Report for Ministry of Environment Male, Republic of Maldives, 32

Allison EH, Adger WN, Badjeck MC et al (2004) Effects of climate change on the sustainability of capture and enhancement fisheries important to the poor: analysis of the vulnerability and adaptability of fisherfolk living in poverty. Fisheries Management Science Programme, DFID, UK, Project Summer Report, p 21

Arup O (2001) Assessment of the potential contribution of marinas and watersports to increasing prosperity in Cornwall. Ove Arup and Partners, Bristol, 23

Black KP (2000) Artificial surfing reefs for erosion control and amenity: theory and application. J Coast Res 2000:1–7

Brander KM (2007) Global fish production and climate change. Proc Natl Acad Sci USA 104:19,709–19,714

Brander KM (2008) Tackling the old familiar problems of pollution, habitat alteration and overfishing will help with adapting to climate change. Marine Pollut Bullet 56:1957–1958

Briones R, Garces L, Ahmed M (2006) Climate change and small pelagic fisheries in developing Asia. Edward Elgar Publications, pp 21

CMFRI (2007) Annual Report 2006–07. Central Marine Fisheries Research Institute, Cochin, India, 126

CMFRI (2008) Research highlights 2007–2008. Central Marine Fisheries Research Institute, Cochin, India, 36

Edwards R (2004) A brief description of the biological assemblages associated with Narrowneck artificial reef and nonwoven geotextile substratum. A report prepared for Soil Filters. Australia by the National Marine Science Centre, NSW, Australia, 34

Eswaran K, Ghosh PK, Jha B (2005) Seaweed cultivation by CSMCRI; retrospect, prospect and future strategies. Proceedings of Ocean Life Food & Medicine Expo 2004. Aquaculture Foundation of India, Chennai, pp 394–397

FAO (2004) State of world fisheries and aquaculture. Food and Agriculture Organization of the United Nations, Rome, p 135

FAO (2007a) Building adaptive capacity to climate change. Policies to sustain livelihoods and fisheries. Food and Agriculture Organization, Policy Brief 8, pp 16

FAO (2007b) State of World Fisheries and Aquaculture. Food and Agriculture Organization of the United Nations, Rome, 153

FAO (2008) Summary proceedings of workshop on climate change and fisheries and aquaculture: "options for decision makers". Food and Agriculture Organization, Rome, 6

Gregg WW, Conbright ME, Ginoux P, O'Reilly JE, Casey NW (2003) Ocean primary production and climate: global decadal changes. Geophys Res Lett 30:1809

Handisyde NT, Ross LG, Badjeck MC, Allison EH (2005) The effects of climate change on world aquaculture: a global perspective. Department for International Development, UK, p 151

IPCC (1995) Impacts, Adaptations and Mitigation of Climate Change. Scientific and Technical Analyses. Working Group II, Second Assessment Report, Intergovernmental Panel on Climate Change, p 878

IPCC (2007) Impacts, adaptation and vulnerability summary for policy makers. Working Group II, Fourth Assessment Report, Intergovernmental Panel on Climate Change, p 16

Kaladharan P, Veena S, Vivekanandan E (2009) Carbon sequestration by a few marine algae: observation and projection. J Marine Biol Assoc India 51:107–110
Kaliaperumal N (2005) Prospects of seaweed farming in India. Proceedings of Ocean Life Food & Medicine Expo 2004. Aquaculture Foundation of India, Chennai, pp 384–393
Kanwisher JW (1966) Photosynthesis and respiration in some seaweeds. In: Barnes H, Allen G (eds) Some contemporary studies in marine science. Unwin Publishers, London, pp 407–420
Kennedy VS, Twilley RR, Kleypas JA et al (2002) Coastal and marine ecosystems & global climate change. Potential effects on U.S. resources, Pew Center on Global Climate Change, Arlington, USA, 52
Perry AL, Low PJ, Ellis JR, Reynolds JD (2005) Climate change and distribution shifts in marine fishes. Science 308:1912–1915
Sunda WG, Huntsman SA (1995) Iron uptake and growth limitation in oceanic and coastal phytoplankton. Marine Chem 50:189–206
Thrane M (2006) LCA of Danish fish products – new methods and insights. Int J Life Cycle Assess 11:66–74
Vivekanandan E (2001) Sustainable coastal fisheries for nutritional security. In: Pandian TJ (ed) Sustainable Indian Fisheries. National Academy of Sciences, New Delhi, pp 19–42
Vivekanandan E, Rajagopalan M (2009) Impact of rise in seawater temperature on the spawning of threadfin breams. In: Aggarwal PK (ed) Impact, adaptation and vulnerability of Indian agriculture to climate change. Indian Council of Agricultural Research, New Delhi, pp 93–96
Vivekanandan E, Srinath M, Kuriakose S (2005) Fishing the food web along the Indian coast. Fish Res 72:241–252
Vivekanandan E, Rajagopalan M, Pillai NGK (2009a) Recent trends in sea surface temperature and its impact on oil sardine. In: Aggarwal PK (ed) Impact, adaptation and vulnerability of Indian agriculture to climate change. Indian Council of Agricultural Research, New Delhi, pp 89–92
Vivekanandan E, Hussain Ali M, Rajagopalan M (2009b) Vulnerability of corals to seawater warming. In: Aggarwal PK (ed) Impact, adaptation and vulnerability of Indian agriculture to climate change. Indian Council of Agricultural Research, New Delhi, pp 97–100
Watson AJ (1997) Volcanic iron, CO_2, ocean productivity and climate. Nature 385:587–588
World Fish Center (2007) Fisheries and aquaculture can provide solutions to cope with climate change. WorldFish Center, Issues Brief, Penang1701, p 4
Zon D (2005) Effects of elevated atmospheric CO_2 on growth, photosynthesis and nitrogen metabolism in the brown seaweed, *Hizikia fusiforme* (Sargassaceae, Phaeophyta). Aquaculture 250:726–735

Chapter 22
Economic Impacts of Climate Change on Global Food Supply and Demand

Jun Furuya, Shintaro Kobayashi, and Seth D. Meyer

Contents

Abstract Global warming caused by a concentration of carbon dioxide in the atmosphere will be a major issue in world food markets over the next century. Agricultural production will be affected by climatic changes, such as rising temperature or droughts, mainly through changes in crop yields. When considering the relationships between food producers and consumers through trade, it is likely that climate change, such as global warming, may cause significant changes in agricultural markets even in the mid-term. Studies focusing on the impact of climatic change on agricultural product markets are very few. This research examines possible results of climatic change focusing on global warming and its impacts on world agricultural product markets, by using the JIRCAS (IFPSIM) world food model which considers aspects of climate changes. Yield functions of crops including temperature and rainfall as independent variables are estimated, and

J. Furuya (✉) and S. Kobayashi
Development Research Division, Japan International Research Center for Agricultural Sciences (JIRCAS), 1-1 Ohwashi, Tsukuba, Ibaraki 305–8686, Japan
e-mail: furuya@affrc.go.jp; shinkoba@affrc.go.jp

S.D. Meyer
Market/Policy Section, Food and Agricultural Policy Research Institute at University of Missouri-Columbia, 101 Park DeVille Dr., Suite E, Columbia, MO 65203
e-mail: MeyerSe@missouri.edu

R. Lal et al. (eds.), *Climate Change and Food Security in South Asia*,
DOI 10.1007/978-90-481-9516-9_22, © Springer Science+Business Media B.V. 2011

replaced the original functions in the IFPSIM model. The basic world food model is extended to a stochastic model considering correlations among countries for temperature and rainfall. The term of the outlook is 25 years, which is considered a mid-term projection. The Hadley Centre provides grid data of the A2 scenario from IPCC (HadCM3-A2) for expectations of future climate behavior. The simulation results of temperature and rainfall are used for estimating the baseline of food supply and demand 2005–2030. Simulation results show that crop production in some countries or regions will be affected greatly by rising temperatures with increased fluctuation. Crop production by the U.S., the European Union and South Asian countries could suffer severe damage from global warming. The results of simulation using the world food model show that the changes in production resulting from variations of temperatures are quite different for each crop in each country or region. However, world total production for most crops other than soybeans is not severely affected.

Keywords World food model • Stochastic model • HadCM3 Scenario A2 • Global warming

Abbreviations

DDC	Data Distribution Center
GHCN	Global Historical Climatology Network
HadCM3	Hadley Center Unified Model 3
IFPSIM	International Food and Agric. Policy Simulation Model
JIRCAS	Japan International Research Center for Agricultural Science
MT	Metric ton
PPP	Purchasing power parity

22.1 Introduction

Global warming caused by concentration of carbon dioxide in the atmosphere will be a major issue in world food markets over the next century. Agricultural production will be affected by climatic changes, such as rising temperature or droughts, mainly through changes in crop yields. The Intergovernmental Panel on Climatic Change (IPCC) claimed that most studies indicate projected warming will negatively affect crop yields in the long term. Considering relationships between food producers and consumers through trade, it is likely that climate change, such as global warming, may cause significant changes in agricultural markets even in the mid-term.

Twenty percent of the world population is concentrated in the South Asia region where the rate of poverty is very high. The shares of person whose Purchasing Power Parity (PPP) is below $2 in India and Bangladesh were 75.6 and 80.3 respectively in 2005. Food security for the people in the region is quite important and stable agricultural production is necessary factor in addressing this issue. The production share of all crops in the region relative to world production was 11.1% in 1960s and it increased to 14.1% in 2000s. The share of rice is highest for all crops, and the production share of rice in the region has been stable at about 30% over the last 50 years. This stability in cropping patterns in this region could be threatened by increasing variation in temperature and rainfall. The coefficient of variation (CV) of rainfall in the flowering season in the wheat cropping region in India increased from 27.4% in 1960s to 35.5 in 1990s. Furthermore, the CV of temperature in the flowering scason in the rice cropping region in India increased from 0.45% in 1960s to 1.12% in 1990s. The data suggests climate conditions have become more variable in the region.

This paper examines possible impacts of climatic change focusing on global warming and its impacts on world agricultural product markets, by using a stochastic version of the world food model developed by JIRCAS. The basic world food model is developed by Oga and Yanagishima (1996) and it has been extended to a stochastic model which considers variation and correlations among countries for temperature and rainfall. The term of the outlook is 25 years, which is considered a mid-term projection in this context.

22.2 Model

22.2.1 World Food Model

The JIRCAS world food model, which is named the International Food and Agricultural Policy Simulation Model (IFPSIM), consists of yield, area, demand, export, import, stock and price linkage functions from fourteen commodities across thirty two countries or regions. Among the commodities, wheat, maize, other coarse grains, rice and soybeans are the crops covered, with other coarse grains the aggregate of barley, rye, oats, millet and sorghum. Furuya and Koyama (2005) estimated yield functions of crops including temperature and rainfall as variables, and replaced the original yield functions with the newly estimated functions in the IFPSIM model. The estimated yield function is as follows:

$$\ln YH_t = a + b_1 T + b_2 \ln TMP_t + b_3 \ln PRC_t \tag{22.1}$$

where *YH* is yield, *T* is time trend; *TMP* and *PRC* are temperature and rainfall in the flowering or silking season. If the climate data are non-stationary, the following difference function is estimated:

$$\mathrm{dln}YH_t = a + b_2 \mathrm{dln}TMP_t + b_3 \mathrm{dln}PRC_t \tag{22.2}$$

where $\mathrm{dln}YH_t = \ln YH_t - \ln YH_{t-1}$, $\mathrm{dln}TMP_t = \ln TMP_t - \ln TMP_{t-1}$, $\mathrm{dln}PRC_t = \ln PRC_t - \ln PRC_{t-1}$. In this case, the parameter a in the function (22.2) is equivalent to the parameter b_1 in the function (22.1).

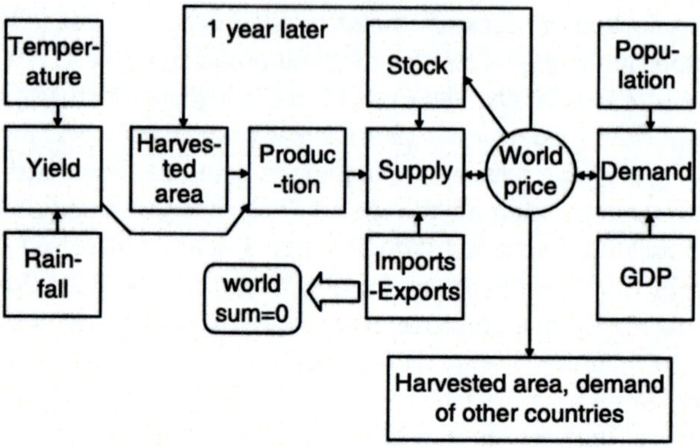

Fig 22.1 Flowchart of a leader country of crop sector in the world food model

Figure 22.1 shows the flowchart of a leader country for a commodity in the crop sector of the world food model. In this model, yield, area, production, imports, exports, stock and demand are endogenous variables. Population, gross domestic products (GDP), temperature and rainfall are exogenous variables.

22.2.2 Partial Stochastic Model

The temperature and rainfall variables entering into the yield functions are exogenous to the supply and demand model. To evaluate the effect of changes in temperature and rainfall during flowering or silking seasons on the world food market, these climate variables must be endogenized in a model which will then recursively feed the supply and demand model. The following simple linear temperature and rainfall models are estimated:

$$TMP_{ijt} = a^T + b^T T \tag{22.3}$$

$$PRC_{ijt} = a^P + b^P T \tag{22.4}$$

where i is the country number and j is the crop number.

Equation errors are retained when comparing the estimates to the actual data. The distribution of the empirical errors and correlation of those errors among countries for the two climate variables are retained and employed to construct a set of random temperature and rainfall variables consistent with history. With the use of the historical error correlation matrix for countries and random draws on a normal distribution, correlated uniform deviates for each country are created through the standard normal cumulative distribution. These random numbers are transformed into draws on the empirical error distributions which maintain the variables historical correlated relationship and distributions. This process creates 150 sets of error draws, with behavior consistent with history, which are then inserted into the temperature and rainfall forecasting model with the same functional forms as function (22.3) and (22.4). Data sets are forecasted based on the A2 scenario and used to create 150 simulated future temperature and rainfall paths. The procedure for creating correlated random climate variables is based on the program of Richardson et al. (2000). Furuya and Meyer (2008) applied this multi-variate stochastic simulation system to evaluate of impacts of water cycle changes on the supply and demand of ricc in Cambodia. The system is shown in Fig. 22.2 below.

22.3 Data

The temperature and rainfall data for estimating functions (22.1) or (22.2) are the monthly average numbers reported by the Global Historical Climatology Network (GHCN). The climatic variables selected are monthly data for the flowering or silking season of each crop, as indicated in the cropping calendar of the United States Department of Agriculture (USDA) (1994). Temperature and rainfall in large countries, such as the U.S., vary greatly across regions. Large countries are divided into regions based on the cropping map of the (USDA 1994). The yield function for the "other Africa" region is not estimated due to insufficient climate data. The basic estimation period is from 1961 to 2000.

The temperature and rainfall data for estimating functions (22.3) and (22.4) are the average number reported by the Data Distribution Centre (DDC) of the IPCC. The data is based on that of the GHCN and modified to 0.5° grid data. The results of simulation by the Hadley Centre are reported by the grid data, then, the DDC data is used for estimating these functions. Table 22.1 shows the results of a correlation matrix of percentage errors for major wheat producing countries.

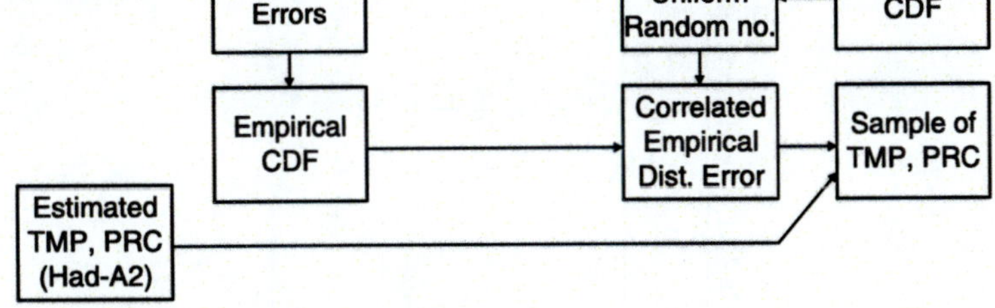

Fig. 22.2 Flowchart for creating random climatic data

Table 22.1 Correlation matrix of % error of major production countries of wheat

		Temperature				Rainfall				
	Country	EU	Ex-USSR	India	China	USA	EU	Ex-USSR	India	China
Temperature	USA	–0.257	0.228	0.268	–0.031	–0.460	0.022	0.187	–0.063	0.110
	EU		0.014	0.216	0.286	–0.045	–0.412	–0.447	0.216	0.166
	Ex-USSR			0.058	–0.122	0.152	0.009	–0.522	0.100	0.169
	India				–0.035	–0.246	0.125	0.021	–0.115	0.052
	China					–0.150	0.148	–0.053	–0.134	0.101
Rainfall	USA						–0.017	–0.138	0.020	–0.138
	EU							0.212	–0.394	–0.047
	Ex-USSR								–0.362	–0.092
	India									–0.050

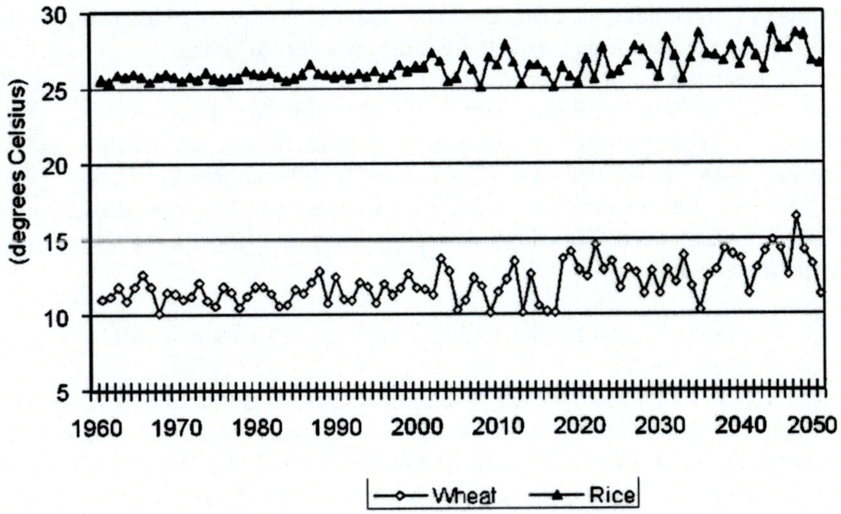

Fig. 22.3 Temperature of flowering season in India (Source: 1961–2000: IPCC Data Distribution Center [DDC], 2001–2050: Hadley Center HadCM3 A2 scenario output)

The Cholesky decompositions of these matrices are used for creating correlated standard normal random numbers.

The Hadley Centre provides grid data for the A2 scenario from IPCC (HadCM3-A2). The A2 scenario assumes that each country holds its own culture and trade, labor movement, and that technology transfer is restricted. Therefore, per capita GDP grows slowly and the annual average per capita income is $7,200 in 2050, while the world population reaches 15 billion people. One of the simulation results is used for estimating the baseline of temperature and rainfall from 2001 to 2050 using functions (22.3) and (22.4). The correlated random errors are added to the baseline and 150 sets of climate data following the A2 scenario are obtained. Figure 22.3 shows the temperature of the flowering season in India for actual historical data of DDC and forecasted data of the HadCM3-A2 scenario.

22.4 Simulation of the World Food Model

The estimated parameters of rainfall and temperature by Furuya and Koyama (2005) are used in the stochastic version of the IFPSIM. The model is written in the FORTRAN programming language and functions are included in the program for ease of development, while these functions are separate files in the original model. Parameters and data for rainfall and temperature are added to the database model and yield functions are modified as discussed later. If the estimated parameters are not significant at the 10% level, these parameters are set equal to zero. The model covers 14 commodities, including livestock products. The base year of the simulation is 1998 and the projection period is from the base year to 2030.

The assumptions for the simulation are as follows; (1) the cropping calendar is fixed, (2) the cropping region is fixed, (3) the climatic variables directly affect yields, (4) the temperature is measured in degrees Celsius for all countries and regions follows the data of HadCM3-A2, and (5) all parameters are fixed.

The yield functions of the simulation model for the U.S. and the European Union for wheat, maize and other coarse grains, rice, and soybeans for all countries, is specified as follows:

$$\ln YH_t = a + 0.1\ln\left(PI_{t-1} / PI_{t-2}\right) + b_1 T + b_2 \ln TMP + b_3 \ln PRC \qquad (22.5)$$

where a is the calibrated intercept of these functions, PI is the subsidized producer price, b_1 is the parameter of the time trend, i.e., the annual increase in yield, b_2 is the parameter of temperature, and b_3 is the parameter of rainfall. The yield function of the simulation model of other countries for these crops is specified as function (22.1).

To investigate impacts of a higher probability of extreme weather on food markets, a scenario which simulates the case of increasing the temperature and rainfall error band by 10% is conducted.

22.5 Results

First, simulation results of crop production in South Asian countries, mainly India are investigated. Figure 22.4 through 22.9 show production of major crops in India and wheat in Pakistan. Averages, 90th and 10th percentiles of the variable distributions

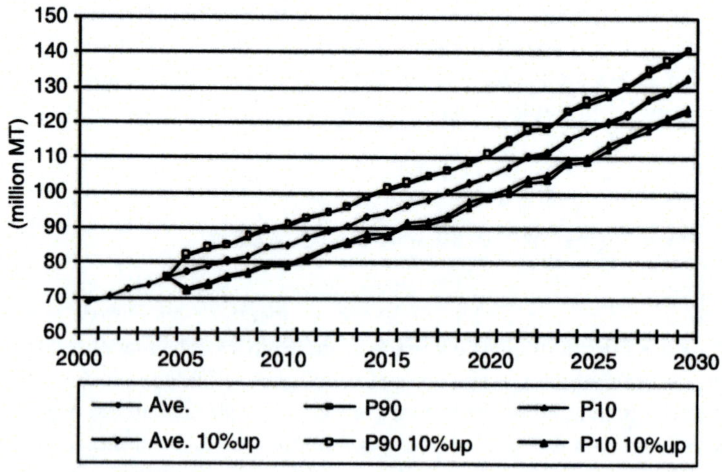

Fig. 22.4 Production of wheat in India

for each year are shown for the baseline, in which temperatures and rainfalls are numbers are taken from the A2 Scenario of HadCM3, as well as the case of increasing temperatures and rainfalls error bands by 10% are shown in these figures. Figure 22.4 shows that the average production of wheat in India will increase about 48 million metric tons (MT) between 2010 and 2030 due to increased yields and a expansion of planted area. The differences between 90th and 10th percentiles are 10.9 million MT in 2010 and 16.7 million MT in 2030, which is equivalent to about 14% of domestic production and about 4% of world production. If the random error of climate variables expands by 10%, the difference between the two percentiles will increase to 12.0 million MT in 2010 and 18.3 million MT in 2030. The distribution of the impact is negatively skewed, i.e., the lower band is thicker. This suggests that if there is an extreme weather event such as a heat wave, the probability of decreasing production of wheat will rise in India.

Figure 22.5 shows that the average production of wheat in Pakistan will increase about 15 million MT during the 20 years. Differences in the two percentiles bounds are 2.5 million MT for baseline and 2.9 million MT for 10% expanding random error case. The distribution of the impact is positively skewed, i.e., the upper band is thicker, and the direction of the impact is opposite of the Indian case. The elasticity of yield for rainfall in Pakistan is –0.0413, while that in India is 0.0504 in our model, thus, rainfall which is well above normal will decrease production of wheat in Pakistan. Higher temperature leads to less rainfall in many countries as shown in Table 22.1, then, expanding random errors for climate variables will increase the probability of substantially above normal production of wheat in Pakistan in a given year.

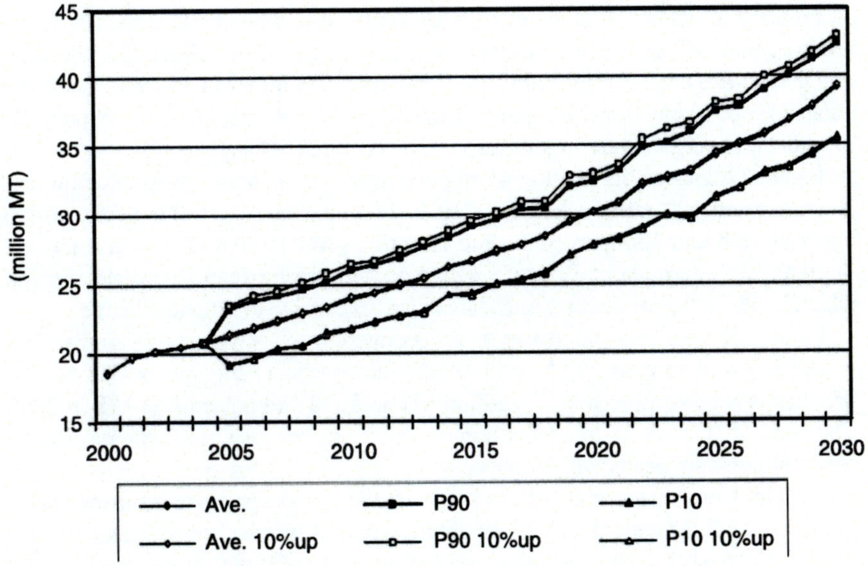

Fig. 22.5 Production of wheat in Pakistan

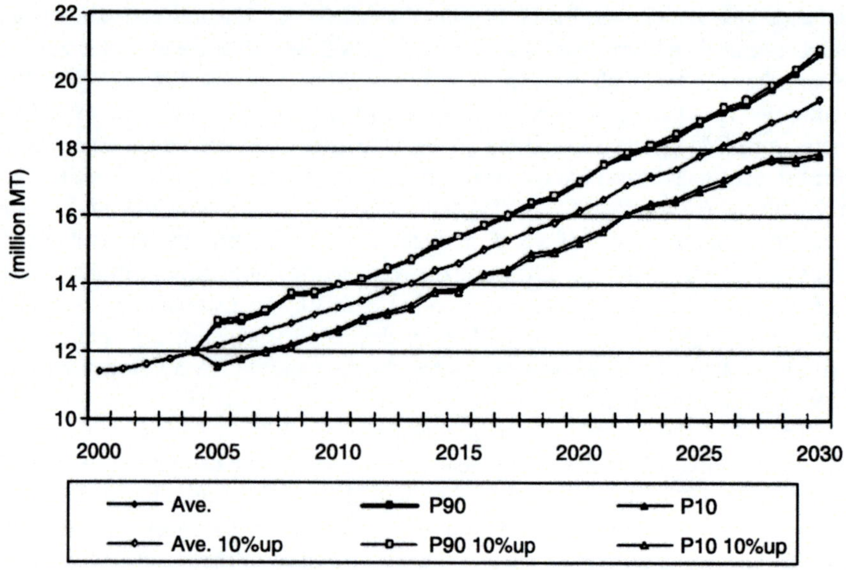

Fig. 22.6 Production of maize in India

Figure 22.6 shows that the average production of maize in India will increase six million MT during the 20 year projection period. The widths between two percentile bounds are 1.3 million MT in 2010 and 3.0 million MT in 2030 and those are almost same in the case of 10% expanding variances of climate variables. Production of maize in India is not as affected by extreme weathers.

Figure 22.7 shows production of other coarse grains which is primarily millet and sorghum will decrease first and then increase after 2010. The widths between two percentiles are 5.3 million MT in 2010 and 8.5 million MT in 2030. These are about 35% of productions and greater than the world average, i.e., 16%. Production of millet and sorghum are highly responsive to climate changes.

Figure 22.8 shows fluctuation of production of rice in India. The production will increase from 109 million MT in 2010 to 167 million MT in 2030. The widths between 90th and 10th percentiles are 20 million MT in 2010 and 28 million MT in 2030. These are about 17% of production and greater than the world average, which is 5%. Rice production in India is also responsive to climate change.

Figure 22.9 shows fluctuations of production of soybeans in India. The production will increase 12.8 million MT during the twenty years. The widths between two percentiles are 1.1 million MT in 2010 and 6.3 million MT in 2030. The band of fluctuation of production will increase because the fluctuation in the world price of soybeans will increase.

Second, simulation results of production at the world aggregate are investigated. Figures 22.10 through 22.13 show fluctuations of production of major crops. Figure 22.10 shows that average production of wheat will increase from 701 million MT in 2010 to 983 million MT in 2030. On the other hand, the differences of 90th and 10th percentiles are 29.3 million MT in 2010 and 46.0 million MT in

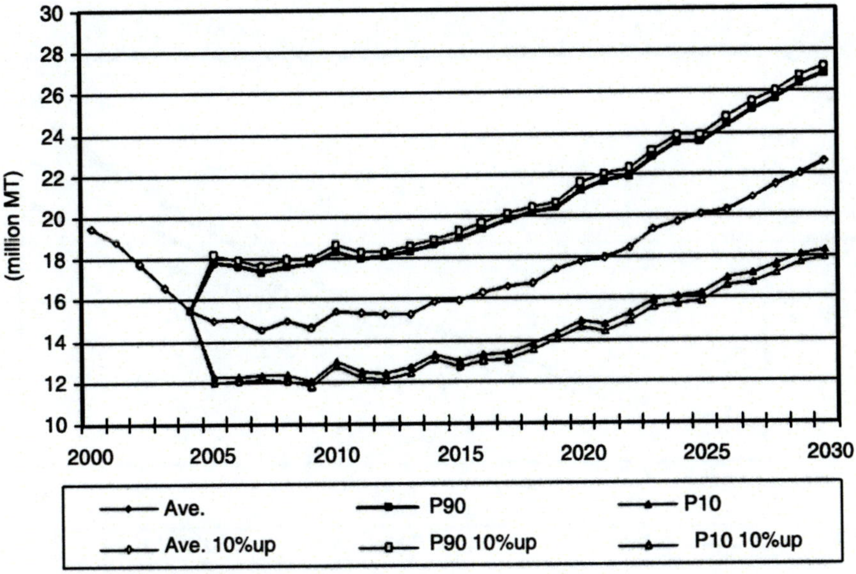

Fig. 22.7 Production of coarse grains in India

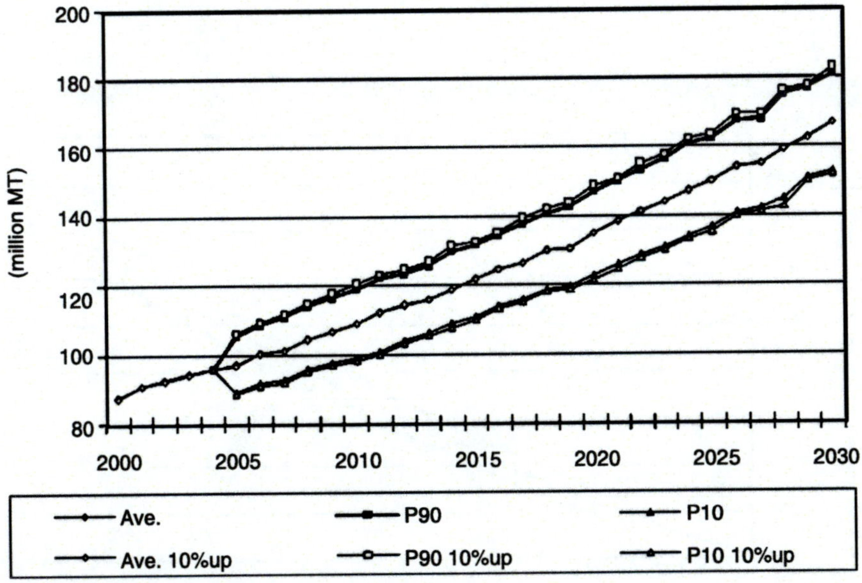

Fig. 22.8 Production of rice in India

2030. The variance of production that is attributable to included climate change variables is smaller than other crops.

Figure 22.11 shows that the average production of maize will increase 294 million MT during the twenty years. Maize is one of the main inputs for bio-ethanol production,

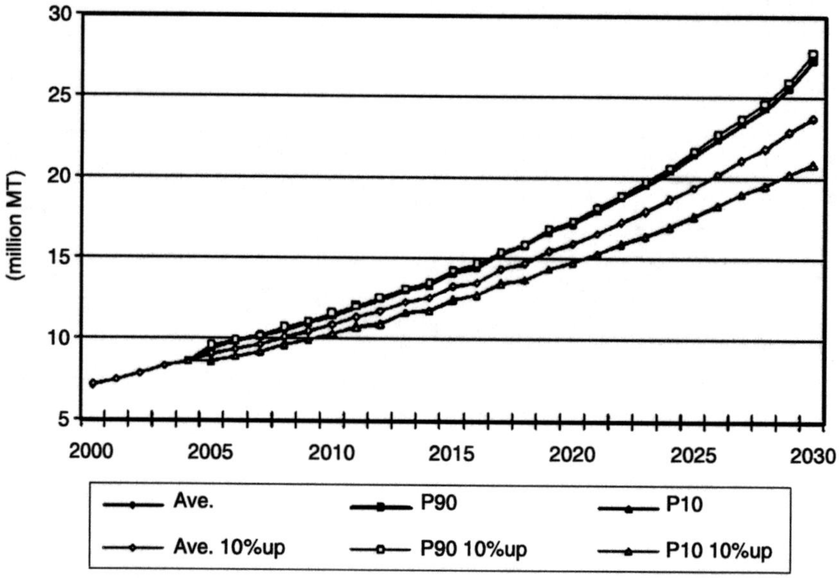

Fig. 22.9 Production of soybeans in India

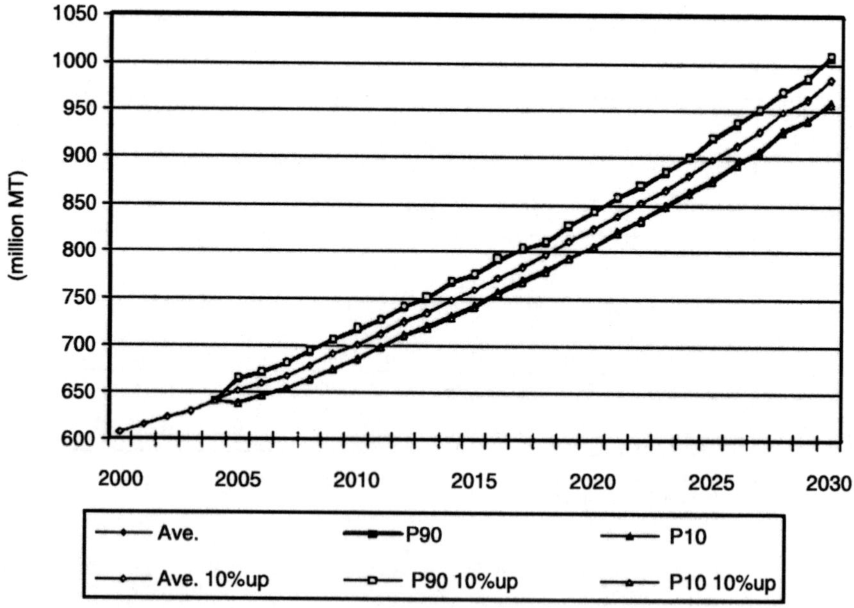

Fig. 22.10 Production of wheat in the world

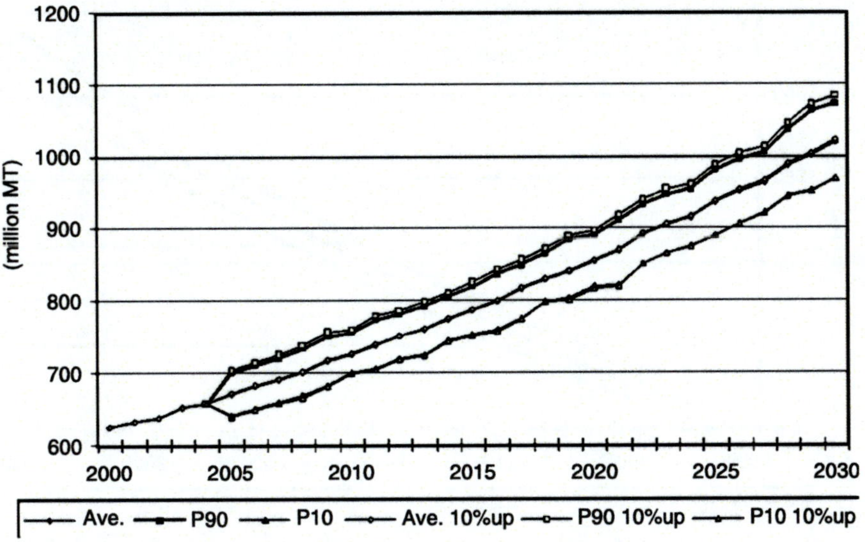

Fig. 22.11 Production of maize in the world

and the outlook will be depend on petroleum price, however, the model, IFPSIM, does not currently follow changes in the fuel market. Model development consider the bio-fuel market is planned. The widths between two percentiles are 54.6 million MT in 2010 and 103.8 million MT in 2030. The band of fluctuations is wider than that of wheat.

Figure 22.12 shows that the average world production of rice will increase from 459 million MT in 2010 to 622 million MT in 2030. The widths between two percentiles are 20.7 million MT in 2030 and 27.7 million MT in 2030. The band of fluctuations of production is narrower than other crops. If temperature increases, yield of rice in East Asian countries will increase due to decreasing probability damage from cool temperatures. Climate change will not affect to total production, however, as production in some countries, such as India, will take damage by higher temperatures.

Figure 22.13 shows that the average production of soybeans will increase from 185 million MT in 2010 to 307 million MT in 2030. The widths between 90th and 10th percentiles will also increase from 29 million MT in 2010 to 121 million MT in 2030. Such a rapid increase of variance of production of soybeans is due to the strong negative value of the parameter of temperature in the yield function in the U.S.A. which is the world price leader in the model. Furthermore, these yield functions respond to price in the model. It indicates that the world production of soybeans is the crop most at risk from climactic change.

Third, variations in farm price of five crops in the U.S.A. i.e., the world price, are investigated. Table 22.2 shows the average (AVE), standard deviation (SD), coefficient of variation (CV), which is SD over AVE, for the baseline and those for the case that the random error of the simulation increases 10% in 2010, 2020, and

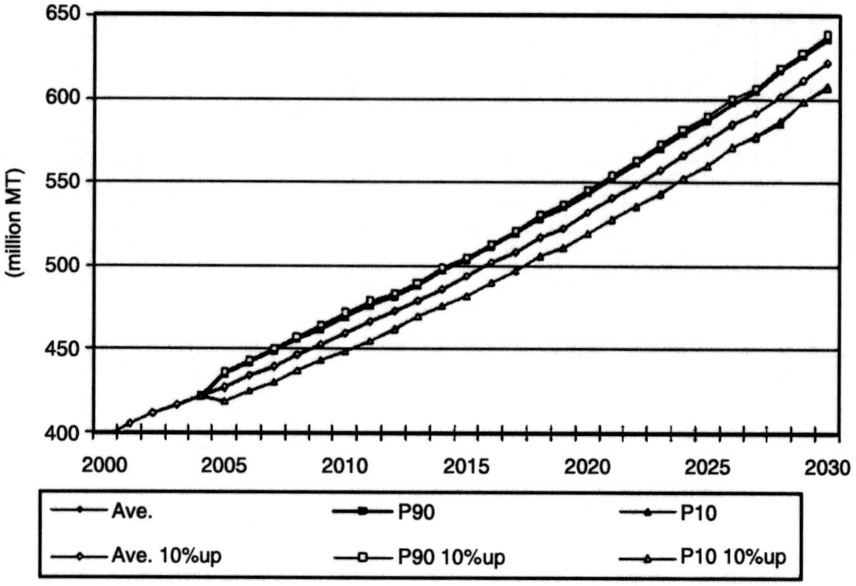

Fig. 22.12 Production of rice in the world

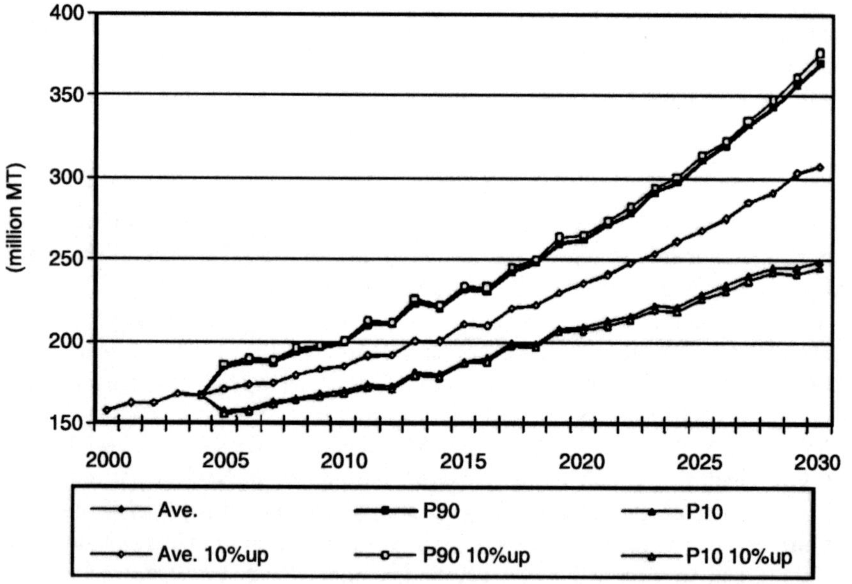

Fig. 22.13 Production of soybeans in the world

2030. Variation of farm prices will increase for all crops for all countries. Soybeans, in particular, will increase from 16.76% in 2010 to 40.75% in 2030, while those of other crops increase 1–3%. The evolutional plant breeding of soybeans in the last two decades might lead the higher production response to price changes in the

Table 22.2 Changes in producer prices in the USA

			Wheat			Maize			Coarse grains		
	Item	Unit	2010	2020	2030	2010	2020	2030	2010	2020	2030
Base-line	Ave.	$/MT	140.7	148.8	159.2	96.0	103.5	109.7	94.6	99.6	105.8
	S. D.	$/MT	5.2	6.0	7.9	7.2	9.2	12.5	3.0	3.4	4.5
	C. V.	%	3.69	4.06	4.98	7.50	8.92	11.38	3.19	3.43	4.21
10% dev. increase	Ave.	$/MT	140.6	148.8	159.2	95.9	103.6	110.1	94.4	99.5	105.8
	S. D.	$/MT	5.7	6.7	8.8	7.9	10.2	13.7	3.3	3.7	4.9
	C. V.	%	4.07	4.49	5.53	8.27	9.84	12.47	3.50	3.76	4.59
Diff.	C. V.	%	0.38	0.43	0.55	0.77	0.92	1.09	0.31	0.33	0.38

			Rice			Soybeans		
	Item	Unit	2010	2020	2030	2010	2020	2030
Base-line	Ave.	$/MT	148.6	155.2	162.9	215.3	228.7	264.6
	S. D.	$/MT	1.9	2.7	2.7	36.1	54.4	107.8
	C. V.	%	1.30	1.71	1.63	16.76	23.78	40.75
10% dev. increase	Ave.	$/MT	148.5	155.2	162.8	216.2	229.7	267.9
	S. D.	$/MT	2.1	2.6	2.7	39.9	59.7	116.4
	C. V.	%	1.42	1.65	1.67	18.47	26.01	43.46
Diff.	C. V.	%	0.12	–0.06	0.04	1.71	2.23	2.71

Ave., average: S.D., standard deviation; C.V., coefficient of variation; Baseline, temperature and rainfall are numbers of HadCM3 A2 scenario; 10% dev., case of the variation of temperature and rainfall increases 10% baseline above; Diff. difference in C.V. between baseline and 10% dev

Table 22.3 Changes in producer prices in India

	Item	Unit	Wheat			Maize			Coarse grains		
			2010	2020	2030	2010	2020	2030	2010	2020	2030
Base-line	Ave.	R/MT	5,388	5,685	6,060	4,295	4,567	4,793	2,457	2,640	2,865
	S. D.	R/MT	188	220	288	261	335	453	110	124	162
	C. V.	%	3.50	3.86	4.75	6.08	7.33	9.45	4.46	4.70	5.65
10% dev. increase	Ave.	R/MT	5,385	5,684	6,062	4,292	4,572	4,806	2,451	2,636	2,865
	S. D.	R/MT	208	243	320	288	370	498	120	136	177
	C. V.	%	3.85	4.27	5.28	6.71	8.10	10.37	4.89	5.15	6.16
Diff.	C. V.	%	0.35	0.41	0.53	0.63	0.77	0.92	0.43	0.45	0.51

	Item	Unit	Rice			Soybeans		
			2010	2020	2030	2010	2020	2030
Base-line	Ave.	$/MT	6,952	6,939	6,879	10,149	10,633	11,937
	S. D.	$/MT	727	743	684	1,311	1,974	3,915
	C. V.	%	10.46	10.70	9.94	12.92	18.57	32.8
10% dev. increase	Ave.	$/MT	6,940	6,945	6,894	10,181	10,671	12,058
	S. D.	$/MT	706	759	695	1,450	2,169	4,228
	C. V.	%	10.17	10.93	10.08	14.25	20.33	35.07
Diff.	C. V.	%	–0.29	0.23	0.14	1.33	1.76	2.27

R, India rupee

simulation. If price variations for wheat and other coarse grains are smaller than those of other crops, then risks related to climate changes of farmers cultivating these crops may be relatively small.

The CV's of rice production are smaller than those of other crops except in India. The ratio of exports to production of rice in the world is only 6.7% in the base year, i.e., 1998, while those of wheat and soybeans are over 20%. Even though the variation of production of rice is smaller than those of other crops, farmers cultivating rice may face latent price risks of climate changes due to the relatively small world trade market.

If the variance of temperature and rainfall increase, the CV of production for all crops will increase with the exception of rice. As mentioned above, some parameters of yield functions of rice of temperature are positive, then, the impacts of extreme weather is ambiguous for the world total.

Finally, variations of farm price of five crops in India are investigated (Table 22.3). The characteristics of the variance of farm prices in India are almost same as those in the world total. However, the impacts of an increase in variance of climate variables for other coarse grains are much greater than that of the world total. This suggests that farmers who produce millet and sorghum face higher risk to climate change in India.

22.6 Conclusions

Simulation results show that crop production in some countries or regions will be affected greatly by rising temperatures with increased fluctuation. Crop production by the U.S., the European Union and South Asian countries could suffer severe damage from global warming. The results of simulation using the world food model show that the changes in production resulting from variations of temperatures are quite different for each crop in each country or region. However, world total production for most crops other than soybeans is not severely affected. The production of soybeans will be affected because yields are highly responsive to prices in the U.S.

These results are based on the mid-term simulation where available cropping regions, and the parameters of the supply and demand model, are fixed. To obtain more accurate simulation results, it is very likely that a different long-term supply and demand model is required. Using forecasting climate data from the A2 scenario by the Hadley Centre, the drastic variations in crop production for some countries is remarkable. The countries which suffer higher price risk by temperature variations may need to consider changes in cropping patterns and practices.

Acknowledgement I would like to thank Dr. M. Nishimori of National Institute for Agro-Environmental Sciences for providing climate forecasting data for the A2 Scenario of HadCM3 and actual data of DDC. He calculated the average of these climate variables in each country and regions for flowering and silking seasons based on the cropping map of USDA (1994). This research is conducted by the project S4 of the Global Environment Research Fund of the Ministry of the Environment of Japan.

References

Furuya J, Koyama O (2005) Impacts of climatic change in world agricultural product markets: estimation of macro yield functions. Jpn Agric Res Q 39:121–134

Furuya J, Meyer SD (2008) Impacts of water cycle changes on the rice market in Cambodia: stochastic supply and demand model analysis. Paddy Water Environ 6:139–151

Oga K, Yanagishima K (1996) International food and agricultural policy simulation model (user's guide). JIRCAS Working Report no. 1

Richardson JW, Klose SL, Gray AW (2000) An applied procedure for estimating and simulating multivariate empirical (MVE) probability distributions in farm-level risk assessment and policy analysis. J Agric Appl Econ 32:299–315

United States Department of Agriculture (USDA) (1994) Major world crop areas and climatic profiles. Agricultural Handbook, vol 664, Washington, DC

Part VII
Climate Change and Agriculture in Bangladesh

Chapter 23
Effect of Climatic Changes on the Incidence of Disease of Winter Pulses

M.A. Bakr, M.H. Rashid, M.S. Hossain, and A.U. Ahmed

Contents

Abstract The accelerated population growth of Bangladesh has triggered the need for increased food production. To meet the demand of food for increasing population Bangladesh has no other alternative to increase its crop production per unit area, which is a great challenge. Bangladesh has acute shortage of food legumes. The major food legumes generally called pulses are grown during winter months out of which lentil and chickpea are most important. But these crops suffer from disease causing remarkable yield reduction. In the epidemiological outbreak of the disease the climatic factors like temperature, relative humidity rainfall etc. have direct contribution. The best instance of which are the occurrence of *stemphylium* blight diseases of lentil and Botrytis Gray Mold of chickpea.

M.A. Bakr (✉)
Pulses and Oilseed Project, BARI, Gazipur, Bangladesh
e-mail: pd.oprc@bari.gov.bd

M.H. Rashid
SSO, Pulses Research Center, BARI, Ishurdi, Pabna, Bangladesh

M.S. Hossain
PSO, Oilseed Research Center, BARI, Gazipur, Bangladesh

A.U. Ahmed
SSO, Plant Pathology Division, BARI, Gazipur, Bangladesh

R. Lal et al. (eds.), *Climate Change and Food Security in South Asia*,
DOI 10.1007/978-90-481-9516-9_23, © Springer Science+Business Media B.V. 2011

However there are some other diseases also occurring on the winter pulses, which was not present in the crop even before twenty years. There is a gradual change in climatic factors on global basis particularly the trend of increasing ambient temperature seems to have contributed in wide spread out break of the damaging diseases due to which some minor disease is becoming factor of major damages. Rainfall is another important factor responsible for causations crop diseases. Increase or decrease and erratic distributions of rainfall seem to have influenced the severity and occurrence of chickpea and lentil diseases. Change in relative humidity is also responsible for increasing disease severity resulting higher magnitude of crop loss. The relative humidity is again the function of other climatic factors like temperature and rainfall. On careful analysis of the long ranges and gradual change of climatic factors like temperature, relative humidity, rainfall seems to influence the occurrence and severity of crop diseases particularly of winter pulses like lentil and chickpea. To mitigate the damaging affect of winter pulses appropriate variety should be developed having potential for withstanding long range of climatic variations and weather factors.

Keywords Food legumes • Pulses • Epidemiology • Plant diseases • Yellow mosaic • Crop losses

Abbreviations

BGM	Botrytis Gray Mould
STB	Stemphylium blight

23.1 Introduction

Food legumes generally called pulses are cultivated in cool season as well as in summer in Bangladesh. Winter pulses are called rabi pulses and summer pulses are called Kharif pulses. Lentil (*Lens culinaris*), Chickpea (*Cicer arietinum*), Grasspea (*Lathyrus sativus*) and field pea (*Pisum sativum*) are grow during rabi season while Mungbean (*Vigna radiata*), Blackgram (*Vigna mungo*) and Felon (*Vigna ungueculata*) are grown during kharif season. Among the winter pulses lentil and chickpea are most important ones upon which lot of research works have been done. Although both the crops are originated in temperate regions but they have been adapted to center regions of sub-tropics and tropics (deCandolle A 1883; Vavilov 1926). But it cannot tolerate extreme cold or hot climate and are vulnerable to terminal heat stress and other biotic and abiotic stresses. The gradual changes in climatic factors on global basis, particularly the trend of increasing ambient temperature mediated through the greenhouse effect seem to have contributed in wide spread outbreak of the damaging diseases due to which minor disease is becoming factor of major damages. Rainfall and relative humidity are other important factors responsible for causation of crop diseases particularly of winter pulses like lentil and chickpea. In this paper the long-range effect of climatic factors on occurrence and severity of major diseases of chickpea and lentil and magnitude of yield loss in relations to change of climatic factors are discussed.

23.2 Occurrence of Diseases on Lentil and Chickpea

Record of crop diseases in Bangladesh was published by Jalaluddin Talukder in 1974 for the first time (Talukder 1974) in which he mentioned two diseases of lentil and three diseases of chickpea. A second list of diseases of four pulses, namely chickpea, lentil, Blackgram and Khesari diseases, was reported in 1977 (Anonymous 1977). Later on, a complete checklist of eight pulses disease was published in 1994 (Bakr 1994). He reported 16 diseases of lentil and 13 diseases of chickpea. The updated version of the list of pulse diseases was however published during 2007 (Bakr and Rashid 2007). Botrytis gray mold is a new disease of lentil reported by Rashid et al. 2008. Among the listed diseases, 17 were on lentil and 17 were on chickpea; however, good number of diseases of lentils and chickpea presented in Table 23.1 did not occur during the 1970s. Some occurred even later. Among the diseases listed in Tables 23.1 and 23.2, some occurred as a minor disease which in later decades became causes of major crop damage, some occurred as more damaging which later on became a disease of minor importance. Some of the diseases although initially occurred as cause of less damage of the crop which over time turned into a disease of major concern but during next decade severity came down making it cause of minor damage to the crop. These ups and downs in manifestation of crop diseases sometimes have been attributed in response to the effects of environmental changes.

In the published list of lentil and chickpea diseases, there are few which could not be recorded in early observations. The reasons for this may be that due to favorable climatic conditions the new diseases have occurred, while there may have probability of escaping the exploration due to which it was not recorded earlier. However, the severity of some diseases has increased many folds than it was recorded 15–20 years back. Botrytis Gray Mold (*Botrytis cineria*) of chickpea and *Stemphylium* blight of lentil are two such diseases, severity as well as incidence of which have increased remarkably. It has been reported that yield can be reduced to 44% by foot rot, 34% by rust, 25–95% by wilt and 88% by stemphylium blight and their distribution is almost uniform throughout the country. This was considered to be due to changes in climatic factors.

Among the recorded diseases of lentil *Stemphylium* blight has caused a havoc on the lentil growers as well as lentil researchers in the country. The disease was first reported in 1985 by Bakr and Zahid (1986). The disease starts appearing from first week of January at a cooler temperature of maximum 25.4°C and minimum of 11.0°C. The disease manifested with first appearance as small pin-headed white

Table 23.1 Status of different diseases of lentil and chickpea in Bangladesh

Crop/disease	Status during different decades				
	1970s	1980s	1990s	2000s	Present
Lentil					
Foot rot *Sclerotium rolfsii* *Fusarium lentis*	Not recorded	Recorded major	Major	Major	Major
Wilt *Fusarium lentis*	Not recorded	Recorded major	Major	Major	Minor
Botrytis gray mold *Botrytis ceneria*	Not recorded	Not recorded	Not recorded	Not recorded	Minor
Bushy stunt Mycoplasma	Not recorded	Recorded major	Major	Minor	Minor
Cercospora leaf spot *Cercospora cruenta*	Not recorded	Recorded Minor	Minor	Minor	Not seen
Downy mildew *Peronospora* sp.	Recorded minor	Minor	Minor	Not seen	Not seen
Leaf roll *PLRV*	Recorded	Minor	Minor	Minor	Minor
Leaf rot *Choanephora* sp.	Recorded	Minor	Minor	Minor	Not seen
Leaf spot *Alternaria* sp.	Recorded	Major	Major	Minor	Minor
Powdery mildew *Oidium* sp.	Recorded minor	Minor	Minor	Minor	Not seen
Rust *Uromyces fabae*	Recorded	Major	Major	Major	Major
Stemphylium blight *S.sauiniformis*	Not recorded	Recorded minor	Major	Major	Major
Chickpea					
Aschochyta blight *A. rabiei*	Recorded major	Minor	Not seen	Not seen	Not seen
Botrytis grey mold *Botrytis ceneria*	Not recorded	Recorded minor	Major	Major	Major
Chickpea stunt *PLRV*	Not recorded	Recorded minor	Major	Minor	Minor
Collar rot *S. rolfsii*	Not recorded	Recorded minor	Major	Major	Major
Dry rot *Rizoctonia bataticola*	Not recorded	Recorded minor	Major	Major	Major
Foot rot *Fusarium oxysporum*	Recorded major	Major	Major	Major	Major
Leaf spot/blight *Alternaria* sp.	Recorded major	Major	Minor	Minor	Minor
Powdery mildew *Oidium* sp.	Recorded minor	Major	Minor	Minor	Minor
Rust *Uromyces* fabae	Recorded minor	Major	Minor	Minor	Minor
Stem rot *Sclerotinia sclerotiorum*	Not recorded	Recorded minor	Minor	Minor	Minor
Wilt *F. oxysporum*	Recorded major	Major	Major	Major	Major

Table 23.2 Factors influencing progress of stemphylium blight

Factors	Genotypes		
	L-81124	L-5	L-84143
Temperature inside canopy (°C)			
Maximum	20–21	21–22	21–22
Minimum	5.0–6.0	4.0–6.0	4.0–6.0
RH inside canopy (%)			
Maximum	92–96	90–95	90–96
Minimum	42–46	40–48	40–44
Period of maximum infection (January)	2–26	7–2	9–22
Time required to infect entire leaflet area (h)	36	36	36
Leaflet infection (%)	73–84	30–38	37–46
Twig infection (%)	65–82	22–33	26–37
Date of initiation of disease (January)	2–9	7–13	7–13

spots on leaf-lets which enlarged rapidly covering the entire leaf surface within a few days. The foliage and twigs gradually turned dull yellow color giving a blighted appearance of the affected crop. The infected leaves shed severely leaving only the terminal leaves on the twigs, the twig bended down dry up and gradually turned ashy white in color. These are the common stages of development of disease symptoms in stemphylium blight of lentil.

Study on epidemiology of stemphylium blight revealed that the disease generally initiate during first week to second week of January. There were, however, slight differences depending on the genotype while maximum infection period was recorded 2–26 January. The pathogen infected entire leaflet surface within 36 h with day 22–30°C and night temperature range of 4–6°C and the relative humidity (RH) inside canopy 90–96%, maximum to 40–48% minimum. 7.7 h or less of daily sunshine, cloudy or foggy weather is suitable for disease development (Bakr and Ahmed 1993).

Out of the total of 17 diseases recorded in chickpea, only four diseases, namely Botrytis gray mold (*Botrytis cinerea* Pers. ex. Fr.), wilt (*Fusarium oxysporum* f. sp. *ciceri,* foot rot (*Fusarium oxysporum* & *Sclerotium rolfsii*) and collar rot (*Sclerotium rolfsi*), have been considered to be of major importance. BGM, collar rot, blight and wilt may cause up to 100, 84, 47 and 62% yield losses, respectively. All the diseases are wide-spread over the chickpea growing zones, except that BGM infestation is less in the Barind tract due to dry weather.

Botrytis Gray Mold (BGM) was first reported in 1981 (Ahmed et al. 1981) but its recurrence after 1985 drastically reduced the chickpea production in Bangladesh. Eight varieties have been released by BARI but they were not able to enhance chickpea production in the country because of BGM problem. The disease becomes serious following frequent winter rains that result in excessive vegetative growth and high humidity (95% or above), which favor its infection, epidemic and severity. Day temperature of 25°C and dense canopy favor BGM development. The disease is seed, soil and air borne. In recent years, this disease has become a great threat to chickpea cultivation in Bangladesh.

23.3 Effect of Climatic Factors on Occurrence of Lentil and Chickpea Diseases

The initial flower dropping has been attributed to infection of botrytis gray mold. Lentil and chickpea require a cold climate. It is sown as a winter season crop. It requires cold temperature during its vegetative growth and warm temperature at the time of maturity. The optimum temperature for growth is 18–30°C. High humidity and excessive rainfall during the season encourages vegetative growth, which prevents good yield and can reduce seed quality. Excessive drought and/or high temperatures during the flowering and pod-fill period also reduce yields. Thus, lentil often experiences considerable drought stress during reproductive development, which affects yield.

The weather parameters viz. maximum and minimum temperature, relative humidity, rainfall and sunshine hours have significant effect on disease development. The rainfall (mm) and sunshine have shown marked influences on disease severity and crop yield. Higher relative humidity, the rainy days and foggy (zero sunshine hours or less sunshine) weather had also significant effect on disease development. Low winter temperatures encountered during December to February particularly in the northern and western districts of Bangladesh can minimize vegetative growth, abort flowers, and thus delay pod setting and filling in chickpea and lentil. This leads to extension of the growing season into March and April, when terminal heat stress is a serious constraint for realizing the yield potential in these legumes. Bakr et al. (2004) reported that widespread and severe infection of BGM usually resulted from frequent and heavy rainfall.

23.4 Impact of Change of Temperature Rainfall and Humidity on Production of Lentil and Chickpea

Systematic research on this aspect has not yet been done in Bangladesh. So enough data has not yet been generated by the researchers in this field. But the observation during the working period in this field as well as the reported data on incidence and severity of major diseases of winter pulses have been simulated to explain the effect of climatic factors on causation of the diseases.

Most plant and plant pathogens grow best within a temperature range 15–30°C. Plants are generally infected faster and to a greater extent when the temperature becomes higher and lower than the maximum and minimum limits for growth. Duration of winter greatly influences the growth of crops and pathogens. A simple explanation of weather factor influencing development of Stemphylium blight can be made from the data presented in Table 23.2.

During the three seasons (1988–1989, 1989–1990 & 1990–1991), the disease initiated during 2–13 January on the three genotypes having some difference among the genotypes, while leaflet infection ranges from 30 to 84 among the genotypes.

The maximum leaflet infection period was 2–26 January in L-81124 and 7–22 in other genotypes. The pathogen infected entire leaflet surface within 36 h. During the period of infection peak, the maximum temperature inside the canopy ranged from 20°C to 22°C and the minimum was 4°C to 6°C. Maximum RH inside canopy was recorded to range from 90% to 96% and minimum 40% to 48%.

23.4.1 *Effect of Temperature on Development of Winter Pulse Diseases*

It is clear from the recorded data that the genotype L-81124 was more prone to infection of stemphylum blight. The infection started at least 5 days earlier and infection period was also more (24 days) as compared to 15 and 13 days in L-5 (BARI Masur-1) and L-84143 respectively. The leaflet and twigs infections were appreciably higher in L-81124. Progress of infection was also rapid. So, it may be perceived that the genotype L-81124 was more vulnerable to the disease infectors than other two genotypes used in the study.

Considering the weather factor and disease development relationship, we can analyze the long-range effect of climatic factors on development of diseases and their impact on crop yields. Analysis of long-term data showed that Stemphylium blight disease incidence of lentil was gradually increased 60–100% with the increase of maximum and minimum temperature of 22.2°C in 1985 and 24.3°C in 2005 (Table 23.3). Similarly Stemphylium blight disease severity was gradually increased 3.0–4.5 scoring scale with the increase of maximum temperature at 22.2°C and 24.3°C and minimum at 9.2°C–11.8°C in 1985 and 2005, respectively. Extent of damage due to stemphylium blight was 40% in 1985, while the highest was 88% in 1995. The concern is that global warming may have adverse effects on plants, and may affect crop production. Through making observation of the climatic factors during a long period, some inferences can be made on development of crop diseases and their subsequent effect on crop damages reducing the yield.

We can see from the Table 23.2 that during a period 20 years there is an increase of 2.1°C and 2.6°C in the maximum and minimum temperatures have subsequently

Table 23.3 Temperature rage in January and its effect on incidence and severity of stemphylium blight disease of lentil

Year	Maximum temp. (°C)	Minimum temp. (°C)	Disease incidence (%)	Disease severity (0–5)	Extent of crop damage (%)
1985	22.2	9.2	60	3.0	40
1990	23.0	9.5	80	3.5	60
1995	23.4	9.8	95	4.5	88
2000	24.1	11.4	100	4.0	70
2005	24.3	11.8	100	4.5	65

increased the incidence of the diseases from 60% to 100% and raising the severity from 3.0 to 4.5. The incidence and severity of the diseases have increasingly affected crops in the last 20 years.

23.4.2 *Effect of Rainfall and Relative Humidity*

Moisture is indispensable for the germination of fungal spores and penetration of host tissues, and for the activation of bacterial and nematode pathogens before they can infect a plant. Rainfall and irrigation is critical for the distribution and spread of pathogens from plant to plant, and field to field. Access to abundant moisture increases the succulence of host plants and thus increase their susceptibility to certain pathogens, which in turn influence the extent and severity of the diseases. Bakr et al. (2004) reported that widespread and severe infestation of BGM resulted usually from frequent rainfalls. (Table 23.4)

When analyzing the 20 years of data (Table 23.3), it clearly showed that there was a gradual increase in the rainfall during maximum growth period of lentil and chickpea i.e. January, February and March. Similarly the relative humidity during these months were also higher in both maximum and minimum ranges. In maximum range it was always higher than 90% which was favorable for initiation and development of stemphylium blight and botrytis gray mould disease (BGM). There was a gradual increase in disease incidence as well as severity except during 2005 the incidence of BGM was lower than 95%, as compared to earlier years. But the extent of damage in chickpea due to BGM was higher than that of recorded in 1985. There was a gradual increase of rainfall during last 20 years at least during the cropping months of chickpea and lentil which might have been the cause of maintaining humidity above 90% during disease initiation time. This resulted higher incidence and severity of the most damaging disease of chickpea botrytis gray mold and lentil stemphylium blight. The reason for decrease of crop damage during 2005 onwards was that due to greater use of stemphylium blight tolerant variety lentil which is Barimasur-4 and Barichhola-5.

Table 23.4 Long range effect of rainfall and relative humidity on development and magnitude of damage due stemphylium blight of lentil and Botrytis Gray Mold of chickpea

Year	Rainfall during January (mm)	Relative humidity (%)	Disease incidence (%)		Disease severity		Extent of crop damage (%)	
			Lentil STB	Chickpea BGM	Lentil STB (0–5)	Chickpea BGM (1–9)	Lentil STB	Chickpea BGM
1985	2.0	42–96	60	95	3	7	60	65
1990	0.0	48–97	80	98	3.5	8	70	70
1995	8.0	54–95	95	100	4.5	9	80	80
2000	8.0	46–90	100	100	4.0	8	70	90
2005	14.0	45–94	95	95	4.5	8	65	75

STB, stemphylium blight; BGM, botrytis gray mould

A study on the influence of climatic factors (temperature, relative humidity and rainfall) on the incidence of mungbean yellow mosaic virus was also studied (Anonymous 2002). The findings suggested that because of the higher atmospheric temperature and relative humidity during April to September, the disease incidence was higher in this period of the year. It was also observed that the white fly population was higher which is directly related to disease spread.

Due to changing climate and gradual warming there has been an increase in the average temperature and relative humidity. The amount and pattern of rainfall were also changed. In Bangladesh, unusual rainfall and warming towards the end of winter season is becoming a regular phenomenon during the recent years which created the apprehension that out break of many diseases causing severe crop damage might have influenced by the change in climatic factors.

23.5 Future Thinking to Combat Crop Loss Due to Diseases Mediated Through Changing Climate

Cultivation of chickpea and lentil in Bangladesh have been threatened by the terminal heat stress which has however been considered to changes of climatic factors. Gradual increase of temperature during grain filling stages of the crops has been major concern towards reducing crop yield and occurrence and increase of severity of diseases. To meet this challenge:

- Efforts should be made towards development of heat tolerant and disease resistant varieties of lentil and chickpea.
- Efforts should also be made to assess the existing pathotypes and pathogens of major chickpea and lentil diseases.
- A package of improved management practices for various pulses need to be developed for different agro-ecological zones.

References

Ahmed HU, Bakr MA, Alam KB (1981) Disease survey of major winter and summer pulses in Bangladesh. Proceedings of the National Workshop on Pulses held on August 18–19, 1981 at BARI, Joydebpur, pp 257–264

Anonymous (1977) Progress report on plant pathology section of co-coordinated pulses improvement project. Submitted BARC during 1977

Anonymous (2002) Annual Report. Plant Pathology Division, Bangladesh Agricultural Research Institute, Gazipur, India

Bakr MA, Ahmed F (1993) Integrated management of Stemphylium blight of lentil. Abst. No. 3.5.47. presented in the 6th Intl. Congress of Plant Pathology held in Montreal, Canada, 28 July–6 August 1993, p 361

Bakr MA, Rashid MH (2007) Strategic intervention of Pulse disease research at BARI. In: Bakr MA, Ahmed HU, Uadud Miah MA (eds) Proceedings of the National workshop on "Strategic

Intervention of Plant Pathological Research in Bangladesh. 11–12 February, 2007. BARI (Bangladesh Agricultural Research Institute), Joydebpur, Gazipur, p 344
Bakr MA, Zahid MI (1986) Stemphylium blight; a new foliar disease of lentil in Bangladesh. Bangladesh J Plant Pathol 2:69–70
Bakr MA, MacLeod WJ, Johansen C, Afzal MA, Knights EJ, Pande S, Bretag TW, Taylor PWJ, Mondal NA, Islam S, Siddiqu KHM (2004) Developing resistant varieties and integrated management package against botrytis grey mould of chickpea: an international effort. Bangladesh J Plant Path 20:77–80
deCandolle A (1883) Originedes plantes cultivees. Paris, pp 258–260
Rashid MH, Ali MO, Bakr MA, Sarker A (2008) Botrytis gray mold: a new foliar disease of lentil in Bangladesh. Bangladesh J Plant Pathol 23 (in press)
Talukder MJ (1974) Plant diseases in Bangladesh. Bangladesh J Agril Res 1:61–86
Vavilov NI (1926) Studies in the origin of cultivated plants. Leningrad, pp 129–238

Chapter 24
Climate Change and Sustainable Irrigation Management in Bangladesh

Md. Shirazul Islam and M. Harun-ur-Rashid

Contents

Paper for presentation in the international symposium on "Climate Change and Food Security in South Asia" held in Dhaka, Bangladesh from 25 to 30 August 2008.

M.S. Islam (✉)
Farm Machinery and Postharvest Process Engineering Division, Bangladesh Agricultural Research Institute, Gazipur 1701, Bangladesh
e-mail: sislambari@yahoo.com

M. Harun-ur-Rashid
Bangladesh Agricultural Research Council, Farmgate,
Dhaka 1201, Bangladesh
e-mail: harunbari52@yahoo.com

R. Lal et al. (eds.), *Climate Change and Food Security in South Asia*,
DOI 10.1007/978-90-481-9516-9_24, © Springer Science+Business Media B.V. 2011

Abstract Predictions from different studies reveal that the agricultural sector will face the acute challenges due to climate change. Models have shown that due to such changes, the rainy season will be wetter due to increased rainfall and dry season will be drier due to higher temperatures. This will cause more surface water bodies to dry out making less water available for irrigating crops in dry season. Presently, about 75% of total irrigated area is based on groundwater resources. Due to drying out of surface water bodies, additional pressure will be exerted on groundwater to meet crop water demand. High loss in irrigation systems will also invite more groundwater to pump. This extended withdrawal will threaten sustainable groundwater use. Reports are available on the declining trends of groundwater levels both in urban and some pocket rural areas in Dhaka, Joydebpur, Comilla, and Barind tracts of Bangladesh. Excessive withdrawal of groundwater is one of the principal reasons for water table declination. Though the water loss is much higher in paddy field, it is also quite remarkable in non-rice crop fields. Obviously, the system efficiencies are much lower for the crops. Therefore, it is important to devise ways and means to reduce water losses in conveyance, distribution and application methods. Innovative works are already in progress to address these issues. Alternate drying and wetting for rice irrigation, improved earthen channel, buried pipe water distribution system and hose pipe irrigation for both rice and non-rice crops and low cost simple drip irrigation system for high value horticultural crops (like tomato, watermelon, egg plants etc.) are becoming popular day by day. Drought is another regular phenomenon that has to be addressed properly because a severe drought reduces production of Aman rice grown in monsoon by about 90%. So, in the drought prone area, proper irrigation facility should be developed and at the same time efforts to be employed to develop drought tolerant crop varieties. In the coastal areas, sea level rise will increase salinity intrusion further inland to grasp non-saline agricultural lands. Therefore, suitable crop varieties to withstand soil salinity need to be developed. At the same time, improved method of cultivation to grow high value crops in the coastal saline areas of Bangladesh will also be developed.

Keywords Climate change • Disasters • Irrigation • Sustainability

Abbreviations

BMDA	Barind Multipurpose Development Authority
MPO	Master Plan Organization
SRDI	Soil Resources Development Institute
BPS	Buried Pipe Water Distribution System

24.1 Introduction

Climate change due to changing temperature from increased greenhouse gases in the atmosphere is a global concern. Increasing population, changing land use, advancing technology and growing industries are the major drivers of climate change.

The phenomenon involves changes in temperature, evaporation, air moisture, rainfall, snow melt, glacial melt, soil moisture, overland flow, river flow, sea level, etc. These changes, in turn, bring changes in dry season water availability, drought, fluvial flooding, river bank erosion and cyclonic storm surges.

Bangladesh will face greater challenges from the climate change. Among other sectors, its agriculture will face the deadliest experience from flood, drought, tornado, cyclone, tidal surge and soil salinity (Rashid and Islam 2007). The "World Watch Institute" published a report on 22 October 2007 realizing that 21 countries of the world including Bangladesh are under serious risk due to rise of sea level from climate change. Besides, climate change is likely to bring changes in the timing and strength of the monsoon wind which is the source of wet season rainfall in Bangladesh. Monsoon rainfall and river floodplain system play a vital role in agriculture, fisheries, forestry, public health, industry, waterway communication and sustenance of wetland ecosystems. The monsoon rainfall also causes damage to agricultural lands, crop productions, homesteads and infrastructures destabilizes the socio-economics of the country. These water related hazards are of great concern to the sustainable development of the country. These significantly disrupt the agro-environmental practices and socio-economic activities of the country (Ahmed and Roy 2007).

Agricultural activities and most of the environmental issues are related to water. So water availability and its proper utilization is a major concern for sustainable development of agriculture. Bangladesh is highly vulnerable to climate change as a vast area of the country has low altitude, located on the Bay of Bengal in the delta of the Ganges, Brahmaputra and Meghna rivers and densely populated. Its national economy depends on agriculture and natural resources that are sensitive to climate change and sea level rise (Koudstaal et al. 1999). Increased droughts and salinization in the dry season and prolonged inundation in the wet season will change the areas suitable for growing rice. Since agriculture is the mainstay for the economy, Bangladesh is very sensitive to growth of agricultural sector. Availability of irrigation water is the key issue in the production of important crops like wheat, maize, tomato, egg plants, potato, watermelon, cauliflower, cabbage etc. All are susceptible to water stress and produce fewer yields than the potentials under water shortage situation. During March through April water scarcity limits the cultivation of high yielding varieties of rice that accounts for approximately 36% of the total rice production in Bangladesh.

This paper, thus, includes discussions comprising water management scenario, effect of climate change on soil, water and crop production, salinity intrusion and use of improved techniques for soil and water management and possible effective measures for sustainable irrigation management in Bangladesh.

24.2 Water Management Scenario in Bangladesh

As reported by Selvaraju (2006), climate changes have impacts on Bangladesh agriculture, especially in the north western district. There are increased evapotranspiration resulting in more water demand and exploitation of groundwater, scarcity of surface and groundwater resources, increase in winter temperature affecting wheat and pulse cultivation, increased spikelet sterility and reduced crop yield, less animal comfort and reduction in milk yield and higher infestation of pest and diseases to crops. With the climate change, additional pressure will be exerted on the existing irrigation and water management systems.

Present food production of around 30 million tons per year in Bangladesh is mostly contributed by irrigated agriculture (56% on cropped area basis). For irrigation, deep tubewell, (DTW), shallow tubewell (STW), low lift pumps (LLP) and other indigenous equipment are used in addition to gravity flow methods. A status of equipment is given in Table 24.1 (MoA 2005).

Presently about 75% of total irrigated area is based on groundwater. The water distribution and application methods of these irrigation systems are mainly unlined channels. The water losses are very high in these channels. There are also some lined channels but these are not performing well due to improper managements. Only a few buried pipe water distribution stems are operating with higher potentials. The water loss scenarios of different systems are given in Table 24.2 (Hoque et al. 1986; Rashid et al. 1991).

Many farmers are using flexible pipes to convey water from pumps to crop fields. This method is suitable for those who have lands near the pump or tube well but not for those who have lands far from the source of water. This method can save 95% water loss (Islam 1991; Sarkar et al. 2006).

The overall system efficiency for rice and non-rice crops ranges from 30% to 35% (Sarkar et al. 2006) and 45–55%, respectively. Out of their ignorance about the ultimate effects of water loss on irrigation economy, the farmers are reluctant to be careful about maintaining well distribution systems.

Table 24.1 Summary of irrigation equipment for the year 2004–2005

Equipment	Number in operation	Area coverage (ha)	% of total irrigated area
DTW	27,117	6,54,189	13.67
STW	11,28,991	31,59,899	66.00
LLP	99,255	8,38,377	17.51
Gravity flow method	–	1,09,381	2.29
Traditional method	–	25,500	0.53

Source: Minor irrigation survey report 2004–2005, MoA, Bangladesh

Table 24.2 Water losses in different distribution systems

Distribution system	Water loss (l/s/100 m)
Buried pipe (BP)	0.35–0.53
Lined channel system	1.52–2.10
Unlined channel system	4.1–10.32
Flexible (hose) pipe	1.15–1.50

Withdrawal of regulation from irrigation appliances in 1992, a rapid growth in STWs has been observed. Having no restriction on tubewell installation, density of installation of shallow tubewells increased, but at the same time, command area or capacity utilization has decreased. This has made the tubewells underutilitzed. Inferences are there about only 0.5 ha irrigated area by a 14 l/s capacity STW.

However, better management is also found in some agency managed irrigated systems. Barind Multipurpose Development Authority (BMDA) project is functioning better in Rajshahi and Chapai Nawabgonj districts, the two less rainfed areas of the country.

The gravity flow irrigation projects have much less efficiency and cover only about 2.3% of irrigated area of the country (MoA 2005).

Apart from appliances, the management and crop selections considering soil properties and climatic factors are vital importance for sustainable irrigated agriculture. Farmers always have the tendency to cultivate rice on availability of water and apply much more water than required. Thus, the efficiency of rice irrigation is always less with low water productivity. Wheat and vegetables are mostly grown by irrigation with higher water productivity.

24.3 Effect of Climate Change on Soil, Water and Crop Production

24.3.1 Effect on Soil Characteristics

The main potential changes in soil forming factors directly resulting from global climate change would be in organic matter supply from biomass, soil temperature regime and soil hydrology, the latter due to shifts in rainfall zones as well as changes in potential evapotranspiration (Brinkman and Sombroek 2007). The probable effects of sea level rising due to climate change on soil characteristics of a gradual enstatic rise in sea level will vary from place to place depending on a number of local and external factors, and interactions between them (Brammer and Brinkman 1990). Other changes due to variations in temperature and precipitation are expected to be relatively well buffered by the mineral composition, the organic matter content or the structural stability of soil.

Besides, salinity intrusion towards inland, grasping more fresh agricultural lands into salinity, will stand as a large obstacle to food security. The natural phenomenon in the change of soil salinity implies that due to high rainfall in rainy season, the soil salinity goes down to below 2.0 dS/m in all the coastal belts of the country (SRDI 2000). The salinity starts increasing from November and gets maximum intensity in February or March (Mondal 1997; Islam et al. 2007). In many areas of Khulna, Satkhira and Noakhali, greater than 16 dS/m salinity restricts growing food crops leaving a vast fallow land in dry season.

24.3.2 Effect on Water Availability

Due to erratic climatic conditions, more area is expected to come under water shortages in the near future. Already some areas in Dhaka, Joydebpur and Comilla are showing rapid declines in groundwater levels. This leaves a huge number of suction mode tubewells inoperable (Bhuyan 1983; MPO 1987; Khan and Islam 1999). In most parts of Bangladesh water table shows natural annual fluctuation of about 3.6 m but failed to return to its previous year's level at the end of the rainy season (Hyde 1979; Haq and Sattar 1987).

Water use efficiency (WUE) from growers' perspective, is the crop yield output per unit of water applied. This is limited by the amount of water availability. Except in a few areas like High Barind tract, no problem has as far been found to get higher water use efficiency by applying the optimal water to crops. Water use efficiency of crops available in different reports of Irrigation and Water Management Division, BARI, and personal contact with the scientists of BRRI, are shown in Table 24.3.

With change of climate, the dry season water availability will further deteriorate. The result is the scarcity of both drinking and irrigation water creating acute health and food problems.

24.3.3 Effect on Crop Production

Shallow rooted plants will suffer most due to rise in temperature. Upper soil layer will lose moisture quickly leaving the plant root zone with inadequate moisture. With the rise of temperature, plants will require more water for evapotranspiration. Since, a major part of water used for irrigation comes from groundwater, additional pressure will be exerted on sustainable use of this precious resource with the climate change.

The impact of climate change on the existing crop cultivars and cropping patterns are expected more. Due to temperature and humidity changes, some crops will be

Table 24.3 Water use efficiency of crops based on crop water use and irrigation requirement

Crop	WUE (kg/ha-mm)
Rice	5–6
Wheat	12–15
Maize	15–18
Potato	35–40
Tomato	100–115
Brinjal	55–65
Onion	25–30
GN.	8–10

eliminated or produce less yields requiring that either new varieties of the existing cultivars are to be developed or the whole cropping patterns should be changed as per demand. Increased insect infestation and pest and disease infections of crops will appear as a great problem.

24.3.4 *Effect on Soil Salinity*

Bangladesh has, at present, about 1.0 mha of lands under various levels of salinities (Islam et al. 2007). In Bangladesh, soil salinity is only found in the coastal belts. Table 24.4 shows the areas under different mapping units and their salinity characteristics.

Coastal soils vary widely in nature of salinity, depth and fluctuation of groundwater and the seasonal variation in salinity of surface water (MPO 1986). The subsoil and sub-strata remain saline throughout the year and shallow groundwater also remains very harmful in these areas (SRDI 1991). Soil Resources Development Institute (SRDI) has developed five mapping units of the entire coastal zone (SRDI 2000). These are presented in Table 24.4.

Depending on the intensity of salinity, it has been categorized by S_1 (2.0–4.0 dS/m), S_2 (4.1–8.0 dS/m), S_3 (8.1–12.0 dS/m), S_4 (12.1–16.0 dS/m) and S_5 (>16.0 dS/m). The total land area under each category of 5 mapping units is given in Table 24.5. It is seen from the Table that about 232,240 ha of land are already

Table 24.4 Mapping units of the coastal regions of Bangladesh

Mapping units	Characteristics of mapping units	Locations
1	No saline to slightly saline	A large part of Satkhira, Khulna, Bagerhat, Narail, Jessore, Gopalgonj, Madaripur, Pirojpur, Jhalokathi, Barisal, Bhola, Patuakhali, Borguna, Laxmipur, Noakhali, Feni and Chittagong
2	Slightly saline to low saline	Many parts of Satkhira, Khulna, Bagerhat, Narail, Jessore, Madaripur, Pirojpur, Bhola, Patuakhali, Borguna, Laxmipur, Noakhali, Feni, Chittagong and Cox's Bazar
3	Low saline to medium saline	Many parts of Satkhira, Khulna, Bagerhat, Narail, Jessore, Madaripur, Pirojpur, Bhola, Patuakhali, Borguna, Laxmipur, Noakhali, Feni, Chittagong and Cox's Bazar
4	Medium to high saline	Many parts of Satkhira, Khulna, Bagerhat, Bhola, Patuakhali, Borguna, Noakhali, Chittagong and Cox's Bazar
5	High saline to very high saline	Many parts of Satkhira, Khulna, Bagerhat, Bhola, Patuakhali, Borguna, Noakhali, Feni and Cox's Bazar

Table 24.5 Category of soil salinity and saline areas of Bangladesh

Mapping unit	Saline area (ha)	Category of salinity (dS/m) S_1 2–4	S_2 4.1–8	S_3 8.1–12	S_4 12.1–16	S_5 >16.0
1	1,15,370	82,260	31,590	1,520	–	–
2	3,09,190	170,380	110,390	29,420	–	–
3	2,40,220	35,490	113,890	61,240	25,870	2,650
4	1,98,890	1,630	36,060	73,400	55,130	32,750
5	1,57,080	–	15,270	25,900	64,100	51,740
Total	10,20,750	289,760	307,200	191,550	145,100	87,140

Table 24.6 Intensity of drought, area covered and yield loss of transplanted Aman

Intensity of drought	Land area (ha)	Locations	Average yield (t/ha)	Yield loss due to drought (%)
Very severe	342,990	Rajshahi, Nawabgonj (Barind area)	1.7–2.5	70–90
Severe	737,028	Barind area, Gangetic alluvium	2–2.5	50–70
Moderate	3,154,950	Western, central and southern regions; Modhupur tracts, Kustia and Jessore	2.5–3.5	30–50
Slight	2,867,895	Teesta, Brahmaputra and Gangetic alluvium; alluvium soils of Meghna and Surma-Kushiara rivers	3–4	10–30

under high salinity (S_4 and S_5). Thus, the concern is that more agricultural lands will be lost to salinity unless appropriate technologies are available for cultivation of crops in saline soils.

24.3.5 *Effect on Drought*

Drought is a very well known natural disaster to the people of Bangladesh. Every year, 3–4 million ha of land are affected by droughts of different magnitudes. A survey by BARC shows 4.2 million ha of land are prone to droughts of different intensities. For drought, average rainfall becomes less during the critical growth stages of crops and they suffer from soil moisture deficits. Drought stress caused by higher evapotranspiration and reduced rainfall may override any growth benefits from the higher CO_2 levels, unless irrigation can be stepped up to compensate the deficit soil moisture (Warwick 2007). Out of 8.3 million hectares of cultivable land, 60% is used for Aman rice cultivation. But this rice is grown rain fed. During drought of different intensities, a heavy loss to Aman production affects the economy of the farmers. Table 24.6 gives the intensities of drought, area involved and the yield loss of transplanted Aman.

24.3.6 *Effect on Floods*

Approximately 25% of the country is flooded to varying degrees each year during May through September when over 60% of the cereals are produced. Recurrent flooding severely restricts the farmers' choices of cropping to traditional low yielding broadcast varieties of rice that can thrive in deep water.

Climate change will have a tremendous effect on flooding. The higher snowmelt will cause greater intensity floods to sweep all over Bangladesh damaging agricultural crops. Flash floods also will be frequent and dangerous for Sunamganj, Cox's Bazar and other areas of Bangladesh. In 2002, flash floods damaged shelters and crops of about 31,500 ha in 12 upazillas of Chittagong and Cox's Bazar districts. Floods from tide and ebbs are not so dangerous except in the rainy season when the strong southern wind causes swelling of tidal waves that retards the flooded river waters to flow to the sea at a normal rate. This results in the inundation of a vast coastal land by saline water making additional agricultural lands uncultivable.

Due to climate change, the distribution of rainfall and its intensity will be changed resulting in increased flooding. This will threaten economic development and food security of Bangladesh. As predicted, flood area will increase by about 6%. This increase in flooding might result in crop production loss of about 11% in the Ganges basin, 7% in Meghna basin and 2% in Brahmaputra basin. Also about 6,400 km embankments may face protection and safety problems (Hassan et al. 2008).

24.3.7 *Effect on Cyclones and Tornadoes*

Climate change will enhance frequent cyclones and tornadoes of high intensities. This will threaten Bangladesh agriculture most. From the experience of SIDR, the people of Bangladesh have learnt how cyclones can destroy human and cattle lives with huge standing crops.

24.4 Possible Measures for Sustainable Water Use in Crop Fields

24.4.1 *Optimization of Water Use in Crop Field*

Water is so precious and a scarce natural resource in many countries of the world that it has already become a burning issue worldwide to optimize the use of this natural resource. In Bangladesh the farmers are still in the dark about this information. They are using almost double the amount of water needed to grow crops especially rice. Already the drying and wetting method of water application has been proved to be very useful to save 50% water without any or a little sacrifice of yield (Sattar 2008). Researches have found that irrigation water application may be optimal to produce maximum benefit somewhere below the amount needed to produce maximum yield (Barret and Skogerboe 1980). On the other hand, farmers do not apply adequate water to many of the high value upland crops like, wheat, maize, tomato, potato etc. and get fewer yields. Obviously, the returns also become less than the optimal. Therefore, optimal or near optimal use of water to these crops should be ensured for both better yield and return.

24.4.2 Use of Drip Method for Irrigation

Apart from rice, crops like tomato, brinjal, potato, wheat, watermelon etc. can be irrigated under improved soil and water management techniques. For small land holdings, drip irrigation with raised bed and mulch can be more effective than any other traditional methods. For saline soils, this method can be a good choice for the coastal farmers (Islam et al. 2007). Table 24.7 shows a comparative crop production scenario in saline soils by drip irrigation in raised bed with mulch and traditional practice.

24.4.3 Channel Improvement

Channel improvement by adequate compaction has been shown to increase command area of about 10–30% over that of normally constructed channels. It is very costly to construct lined channels. These channels also can save water effectively when new but gradually suffer from cracking and other damaging processes. Eventually, these channels lose efficacy to transport water.

Table 24.7 Crop production in saline soils by improved and traditional methods

	Location: Noakhali			Location: Satkhira		
Sequence	Crop	Yield (t/ha)	Salinity (t/ha)	Crop	Yield (t/ha)	Salinity (dS/m)
Drip irrigation in raised bed with mulch	Tomato	50–70	10–12 (4.5–5.5)	Tomato	48–55	9–10 (4.75–6.15)
	Watermelon	40–50	10–12 (4.5–5.5)	Okra	10–11	9–10 (4.75–6.15)
Traditional method (pitcher irrigation in flat land)	Tomato	18–20	10–12	Tomato	16–19	9–10
	Watermelon	15–18	10–12	Okra	5–6	9–10

Figures in parenthesis indicate reduced soil salinity due to improved management practices

24.4.4 Use of Hose Pipes for Irrigation

Farmers using low capacity pumps or tubewells can use hose pipes for water conveyance. This has been proved to be an acceptable means of water delivery to crop fields even though farmers experience some difficulties in using these pipes. A good quality pipe can save 95% water loss. This practice to convey water should be advocated to the farmers extensively.

24.4.5 Use of Buried Pipe Water Distribution System

The most efficient water saving technique for irrigation is the buried pipe water distribution system (BPS). This is costly but an ensured water distribution method. It can handle the land undulation, land loss of about 1.8% due to construction of channels, irrigation cost and water distribution uniformity (Islam 1991). According to Michael (1987), the land loss by channels is much higher than this and it ranges from 2% to 4%. The land loss during channel construction can be avoided by using buried pipe water distribution systems (BPS). So, government should come forward to take necessary measures to install BPS by phases all over Bangladesh and get the money reimbursed from the irrigators.

24.4.6 Changing Cropping Pattern

In many locations of the country, tubewells are installed in light textured soils to irrigate rice crop. Almost every day, water is applied to crop fields and a huge amount of water is lost as seepage and percolation. This results in higher pumping duration and higher fuel cost thereby swelling up irrigation cost. But, a change in cropping pattern with upland non-rice crops can make irrigation much more profitable. Growing low water requiring crops like wheat, maize and vegetables, potato and horticultural crops in the cropping pattern more area can be brought under irrigated agriculture and farmers can be financially more benefited. However, based on soil, climate and socio-economics, the pattern wise irrigation practice should be introduced. A comparative water use by different crops based on their respective cultivated area during 2007–2008 is shown in Table 24.8.

Some other crops like pulses, oilseeds and spices crops are not usually irrigated in Bangladesh and are excluded from the table. In the calculation, the overall system efficiency for rice and other crops are taken as 40% and 55%, respectively. From Table 24.8, it appears that out of 99,065 Mm^3 water used in irrigating crops, the maximum amount (88,121 Mm^3) is required in Rabi season when the acute shortage of water is experienced. The groundwater contribution of the total requirement is about 74,299 Mm^3. Only 24,766 Mm^3 is contributed from surface water and rainfall. Rice (Boro, T. Aus and Aman) alone takes the major share, 86,354 Mm^3 (87%) of

Table 24.8 Annual water requirements of different crops

Name of crop	Area cultivated ('000' ha)	Seasonal crop water requirement (ha-mm)	Approx. Irrig. water requirement (Mm^3)
Rabi season (October–February)			
Boro rice	4,675	650	75,969
Wheat	370	300	2,018
Potato	400	280	2,036
Maize (w)[a]	346	450	2,831
Mustard	547	200	1,990
Groundnut	70	250	318
Vegetables (w)[a]	448	250	2,034
Onion	158	230	659
Garlic	64	230	266
		Sub-total	188,121
Kharif-I season (March–May)			
T. Aus[b]	1,154	240	6,924
Vegetables[b]	231	120	504
Maize[b]	30	100	55
		Sub-total	7,483
Kharif-II season (June–September)			
Aman rice[b]	1,154	120	3,461
		Sub-total	3,461
		Grand total	99,065

[a](w), winter
[b]Supplemental irrigation

the total irrigation requirement. The T. Aus, Aman, summer maize and summer vegetables need only supplemental irrigation. Thus, the rice-rice-fallow pattern can be replaced by introducing wheat, potato, horticultural crops, pulses or oilseeds in the pattern and the farmers can be more benefited from the new pattern.

24.4.7 Construction of Water Management Infrastructures

Since 1960, various types of hydraulic structures have been used. These structures provided flood protection to about 3.5 million ha of land out of about eight million ha of flood vulnerable areas in Bangladesh (Hossain and Siddiqui 2006). In the coastal areas over 12 million people are receiving the benefits of around 3,600 km embankment and rest 6,400 km embankment are in the non-tidal areas. It has been found that the intensity of salinity is much higher outside the embankment than that within the embankment in the coastal areas of Noakhali district. So, to retard saline water intrusion, necessary and well planned water management infrastructures are essential.

24.4.8 Maintaining Proper Spacing of Tubewells

A technical sub-committee formed by the Ministry of Agriculture has submitted a report on tubewell spacing for the country. This need to be taken into consideration and implement immediately to avoid underutilization of pumps/tubewells as well as excessive withdrawal of groundwater.

24.4.9 Conjunctive Use of Surface and Ground Waters

Since dry season availability of irrigation water is gradually becoming scarce, conjunctive use of both surface and ground waters may be the best option for sustainable irrigation management. Government has also emphasized on surface water utilization. This will reduce pressure on groundwater and help sustainable use.

24.4.10 Upgrading Farmers' Knowledge

Most of the farmers of the country are illiterate and do have a little or no idea about the consequence of mismanagement of irrigation water. So, they should be made aware of the water managements in irrigated agriculture through training, demonstration and group discussion. Unless they are aware of the fatal consequences of excessive water use to crops, they will be little concerned about the optimal water use and benefits.

24.5 Conclusions

From the above discussions, the following conclusions can be drawn in response to climate change effects on agriculture in Bangladesh:

- Water conveyance systems should be improved.
- Irrigation of diversified crops should be encouraged.
- Farmers should be made aware of the bad effects of water losses on their economy.
- Conjunctive use of water for irrigation should be encouraged.
- Drying and wetting method of irrigation for rice to be followed.
- Government should take necessary measures to install Buried Pipe Water Distribution Systems by phases all over Bangladesh.
- Existing rules of tubewell installation should be revived.
- Mass media should be used to propagate the above messages to general people.

References

Ahmed AMMM, Roy K (2007) Utilization and conservation of water resources in Bangladesh. J Dev Sustain Agric 2:35–44

Barret H, Skogerboe GV (1980) Crop production and the allocations and use of irrigation water. Agric Water Manage 3:53–64

Bhuyan SI (1983) Groundwater irrigation in Bangladesh: development at the crossroads. Paper presented at the Institution of Engineers, Bangladesh

Brammer H, Brinkman R (1990) Changes in soil resources in response to a gradually-rising sea level. In: Sharpenseel et al. (eds) Chapter 12, pp 1990

Brinkman R, Sombroek WG (2007) The effects of global change on soil conditions in relation to plant growth and food production. Land and Water Development Division, FAO, Rome, Italy

Haq KA, Sattar MA (1987) Groundwater table fluctuation and response to rainfall on a micro level basin. Bangladesh J Agric Eng 1:45–48

Hassan A, Chowdhury JU, Hoque A (2008) Impact of climate change on riverine flooding in Bangladesh. Paper Presented in the Final Workshop on Impact of Climate Change and Sea Level Change in Part of the Indian Sub-continent (CLAIC). Department of International Development (DFID), UK, BUET, Dhaka, Bangladesh

Hoque DMR, Idris DM, Samad MMA, et al (1986) Integrated irrigated command area development. Fifth Annual Report, Coordinated Irrigation and Water Management Studies, 1984–85. Agricultural Engineering Division. Bangladesh Agricultural Research Council, Farmgate, Dhaka, Bangladesh

Hossain ANHA, Siddiqui KU (eds) (2006) Options for flood risk and damage reduction in Bangladesh. University Press Ltd, Dhaka

Hyde LW (1979) Hydrology of Bangladesh. Consultancy Report of Groundwater Circle, BWDB, Bangladesh

Islam MS (1991) A study of deep tubewell irrigation with buried pipe distribution system. A Master's thesis. Department of Water Resources Engineering, Bangladesh University of Engineering and Technology, Dhaka, Bangladesh

Islam MS, Akanda MAR, Islam MN et al (2007) Development of integrated salinity management techniques for the coastal eco-system of Bangladesh. Irrigation and Water Management Division, Bangladesh Agricultural Research Institute, Joydebpur, Gazipur

Khan LR, Islam AKKMM (1999) Estimation of groundwater recharge for flood plain Aquifer in Dhunat Upazilla of Bogra District. J Inst Eng Bangladesh 26:19–27

Koudstaal R, Werners SE, Rahman A et al (1999) Considering adaption to climate change towards a sustainable development of Bangladesh. Prepared for. South Asian Region, The World Bank, Washington, DC

Michael AM (1987) Irrigation theory and practice. Indian Agricultural Research Institute, New Delhi, India

MoA (2005) Minor Irrigation Survey Report 2004–2005. Ministry of Agriculture. Government of the People's Republic of Bangladesh, Segunbagicha, Dhaka, Bangladesh

Mondal MK (1997) Management of soil and Water Resources for Higher productivity of the Coastal Saline rice land of Bangladesh. Ph.D. Thesis, UPLB, Philippines

MPO (1986) National Water Plan. Bangladesh Water Development Board, Dhaka, Bangladesh

MPO (1987) The groundwater resource and its availability for development. Technical Report 5. Ministry of Irrigation, Water Development and Flood Control, Government of Bangladesh, Dhaka, Bangladesh

Rashid MA, Islam MS (2007) Adaptation to climate change for sustainable development of Bangladesh Agriculture. Paper presented in the 3rd session of Technical Committee of Asian and Pacific Center for Agricultural Engineering and Machinery (APCAEM) on November 20–21, 2007. Beijing, China

Rashid MA, Das RK, Uddin MZ (1991) Deep Tubewell Project II. Water Management Study. 1990–91. Mott MacDonald International Limited in association with Hunting Technical Services Limited. Engineering and Planning Consultants, Dhaka

Sarkar KK, Islam MN, Sattar MA et al (2006) Water management for improvement of shallow Tubewell performance during boro rice cultivation. Bangladesh J Agric Eng 17:41–48

Sattar MA (2008) Irrigation and water management, agricultural mechanization and post harvest process engineering for sustainable agricultural development of Bangladesh. A key note paper presented in the Annual Paper Meet (Agricultural Engineering, 2008). Institution of Engineers, Bangladesh

Selvaraju D (2006) Livelihood adaption to climate change in drought prone areas. Climate Cell, Department of Environment, Component 4B. Comprehensive Disaster Management Programme (CDMP), Government of Bangladesh

SRDI (1991) Soil and water salinity monitoring report. Soil Resources Development Institute, Ministry of Agriculture, People's Republic of Bangladesh

SRDI (2000) Soil salinity in Bangladesh. Soil Resources Development Institute, Ministry of Agriculture, People's Republic of Bangladesh

Warwick (2007) Climate Change and its Impact on Water Use in Field Crops. Wave: (http://www2.warwick.ac.uk/fac/sci/whri/research/climatechange/cgwateruse)

Chapter 25
Cool Rice for a Warmer Environment: Concept, Progress and Prospect – Bangladesh Perspective

M. Sirajul Islam, Abdullah A. Mahbub, and M. Serajul Islam

Contents

Abstract Achieving higher yield of rice depends on increasing total biomass. Total crop biomass is determined mainly by crop photosynthesis and respiration losses, both of which are sensitive to temperature. Global mean surface air temperature is increased by about 0.6°C in the twentieth century. Over the twenty-first century a range of 1.1–2.9°C warming is predicted in a scenario with low emissions of greenhouse gases, and 2.4–6.4°C in a high-emissions scenario. Scientists have attempted to assess the effect of increasing temperature and high carbon dioxide in the atmosphere on the growth and yield of rice using simulation models. Direct studies on the effects of observed climate change on crop growth and yield could provide more accurate information for assessing the impact of climate change on crop production. When rice crop is exposed to temperatures higher than 35°C and

---

Paper presented at the International Conference on Global Climate Change and Its Effects... held on 25–30 August 2008 at Dhaka, Bangladesh.

M.S. Islam (✉)
Agronomy Division, Bangladesh Rice Research Institute, Gazipur 1701, Bangladesh
e-mail: salam_brri@yahoo.com

A.A. Mahbub and M.S. Islam
Plant Physiology Division, Bangladesh Rice Research Institute, Gazipur 1701, Bangladesh
e-mail: mrishaon@yahoo.com; rafiqshaon@ymail.com

R. Lal et al. (eds.), *Climate Change and Food Security in South Asia*,
DOI 10.1007/978-90-481-9516-9_25,

last for more than 1 h, injuries occur according to its growth stages. Especially high temperature during flowering can induce floret sterility and can limit grain yield. The direct cause of sterility is reduction in number of germinated pollen grains on the stigma. In many cases, poor pollination is the main cause of sterility induced by high temperature at flowering time. In Bangladesh, the highest monthly mean temperature is observed either in March or April or in May. In this time the maximum temperature is often more than 35°C in Rajshahi, Barisal and Khulna regions. The temperature trend analysis showed that the monthly mean of minimum temperature increased by at least 0.5°C during the period from 1986–1995 to 1996–2005 in these regions of Bangladesh. Spikelet panicle^{-1} was significantly lower when high temperature (38°C) was induced in spikelet differentiation stage and reduction division stage due to spikelet abortion. While the sterility percentage was significantly increased when high temperature was induced at heading stage, distinct varietal difference in rice was observed when high temperature (38°C) was induced at anthesis stage. Sterility percent ranged from 20–78% for susceptible variety like BR9, and ranged from 12–22% for moderately tolerant variety like BR1. There is genetic variation in both escape (time of flowering) and absolute tolerance (spikelet sterility). Besides breeding for high temperature-tolerant rice varieties (cool rice), basic understanding of the physiological processes of rice in warmer climate and suitable crop management strategies to cope with the high temperature problems due to global warming need to be developed immediately.

Keywords Rice production • Global warming • High temperature stress • Floral sterility • Temperature tolerant rice varieties

Abbreviation

BRRI Bangladesh Rice Research Institute

25.1 Introduction

Rice (*Oryza sativa* L.), the food that feeds almost half of the planet, has a special significance in Bangladesh, where about 140 million people consume it as a staple food. World rice production must increase by almost 1% annually to meet the growing demand for food that will result from population growth and economic development. Most of this increase must come from higher yield on existing cropland to avoid environmental degradation, destruction of natural ecosystems, and loss of biodiversity (Peng et al. 2004). Achieving greater yields depends on increasing total biomass. Total crop biomass is determined mainly by crop photosynthesis and respiration losses, both of which are sensitive to temperature (Peng et al. 2004). Extreme temperatures are destructive to plant growth and development. The critically low and high temperatures, normally below 20°C and above 30°C, vary from one growth stage to another. These critical temperatures differ according to variety, duration of critical temperature, diurnal changes, and physiological status of the plant (Yoshida 1981).

25.2 Materials and Methods

The present paper consists of the basic concept of global warming, its impact on rice, research progress of high temperature stress on rice in Bangladesh context and future prospect of research on high temperature stress for developing high temperature tolerant rice varieties (cool rice) to cope with the high temperature problems due to global warming. The recent available literature was reviewed and duly cited. The effect of high temperature on rice was studied under green house condition. Plants were grown in pot in normal environment and then transferred to a green house having a temperature of 38°C in different growth stages. Normal and high temperature treated plants were compared in terms of percent sterility.

25.3 Results and Discussion

25.3.1 Global Warming and Its Impact on Rice

The conclusions of the Fourth Assessment Report of the Intergovernmental Panel on Climate Change Intergovernmental Panel on Climate Change (IPCC 2007) leave no doubt that the global climate is changing in a manner unprecedented in the past 400,000 years. Global mean surface air temperature is increased by about 0.6°C in the twentieth century. Over the twenty-first century a range of 1.1–2.9°C warming is predicted in a scenario with low emissions of greenhouse gases, and 2.4–6.4°C in a high-emissions scenario. In the past century, daily minimum nighttime temperature increased at a faster rate than daily maximum temperature in association with a steady increase in atmospheric greenhouse gas concentration (Peng et al. 2004).

Scientists have attempted to assess the effect of increasing temperature and high carbon dioxide in the atmosphere on the growth and yield of rice using simulation models (Horie et al. 1996). Direct studies on the effects of observed climate change on crop growth and yield could provide more accurate information for assessing the impact of climate change on crop production. Peng et al. (2004) reported that the annual mean maximum and minimum temperature at IRRI, the Philippines have increased by 0.35°C and 1.13°C, respectively, during the period of 1979–2003 and a close linkage between rice grain yield and mean minimum temperature during the dry cropping season (January–April). Grain yield declined by 10% for each 1°C increase in growing-season minimum temperature in the dry season, whereas the effect of maximum temperature on crop yield was insignificant. It was suggested by the scientists that adoption of high temperature-tolerant

Table 25.1 Symptoms of high temperature stress at different growth stages of rice

Growth stage	Symptom
Vegetative	White leaf tip, chlorotic bands and blotches, reduced tillering, reduced height
Reproductive	White spikelets, white panicles, reduced spikelet number
Anthesis	Sterility
Ripening	Reduced grain filling

cultivars, popularly known as cool rice, is one of the most effective countermeasures to sustain high productivity and stability of rice under anticipated climate change (Horie et al. 1996).

25.3.2 High Temperature Stress

When rice crop is exposed to temperatures higher than 35°C and last for more than 1 h, injuries occur according to its growth stages (Table 25.1). Especially high temperature during flowering can induce floret sterility and can limit grain yield (Yoshida 1981; Matsushima et al. 1982). Clear differences exist in the tolerance of rice varieties for high temperature induced sterility. At 35°C, N22, an upland rice variety from India, has higher than 80% spikelet fertility, whereas BKN6624–46–2, a lowland selection from Thailand, has about 10% spikelet fertility (Yoshida 1981).

25.3.3 Mechanism of High Temperature Induced Floret Sterility

The mechanism responsible for high temperature induced floret sterility has been studied under controlled environment using phytotron and growth chambers. The direct cause of sterility was reduction in number of germinated pollen grains on the stigma (Matsui et al. 2001). In many cases, poor pollination is the main cause of sterility induced by high temperature at flowering time.

The driving force behind anther dehiscence is the rapid swelling of pollen grains in side the anther locules in response to the floret opening (Matsui et al. 1999). High temperature at flowering inhibits the swelling of pollen grains, and results in indehiscence of the anthers (Matsui et al. 2000) and poor release of pollen grains (Matsui et al. 2005). High temperature at flowering also makes the pollen grains sticky. The sticky pollen grains choke the basal dehiscence and remain in the anther.

25.3.4 High Temperature Stress – Bangladesh Scenario

In Bangladesh, a high percentage of spikelet sterility usually occurs in Transplanted Aus season (March–June) and sometimes in Transplanted Aman season (June–October) when the temperature exceeds 35°C at anthesis and lasts for more than 1 h. High temperature in Boro season (January–May) may also induce spikelet sterility and reduced grain yield. This detrimental effect of high temperature becomes more prominent when the relative humidity is low. In Bangladesh, the highest monthly mean temperature is observed either in March, April, or May. In this time the maximum temperature is often more than 35°C in Rajshahi, Barisal and Khulna regions (Tables 25.2 and 25.3). The temperature trend analysis showed that the monthly mean of minimum temperature increased by at least 0.5°C during the period from 1986–1995 to 1996–2005 in these regions of Bangladesh (Fig. 25.1). However, the monthly mean of maximum temperature did not show such

Table 25.2 Year wise (1986–2005) maximum temperature (°C) in April and May at Rajshahi, Bangladesh

Year	April	May	Year	April	May
1986	*35.1*	33.4	1996	*36.6*	*36.2*
1987	*35.3*	*35.8*	1997	31.6	*36.7*
1988	*37.1*	34.1	1998	33.3	34.5
1989	*37.8*	34.5	1999	34.1	34.3
1990	34.3	33.2	2000	33.9	33.8
1991	*36.2*	34.5	2001	*35.1*	33.4
1992	*38.1*	*35.0*	2002	32.8	32.6
1993	34.0	33.4	2003	*35.3*	*35.3*
1994	*35.5*	*35.9*	2004	*35.3*	*36.8*
1995	*37.7*	*37.5*	2005	34.9	34.9

Table 25.3 Year wise (1986–2005) maximum temperature (°C) in April and May at Barisal, Bangladesh

Year	April	May	Year	April	May
1986	33.3	34.3	1996	*36.1*	*35.7*
1987	34.0	34.9	1997	32.5	33.7
1988	*35.4*	33.9	1998	34.9	*35.3*
1989	34.6	34.4	1999	*36.1*	34.4
1990	33.6	*35.0*	2000	34.8	34.6
1991	*35.6*	33.9	2001	*36.7*	*35.0*
1992	*37.1*	*35.2*	2002	34.2	34.2
1993	*35.0*	34.0	2003	*35.5*	*35.8*
1994	*35.0*	*35.0*	2004	*35.0*	*36.9*
1995	*37.2*	*35.3*	2005	*37.0*	*36.6*

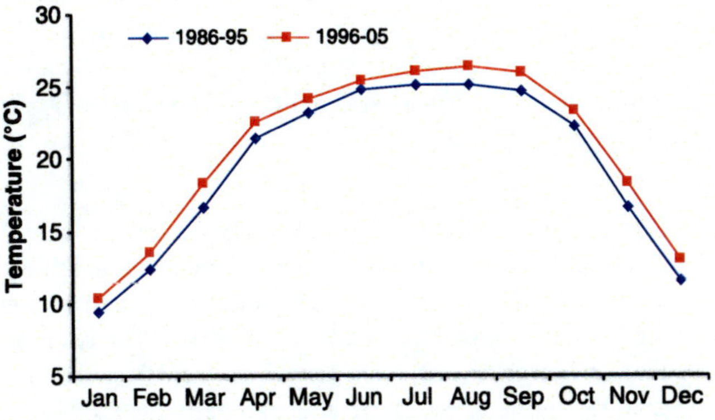

Fig. 25.1 Average minimum temperature in two decades at Rajshahi, Bangladesh

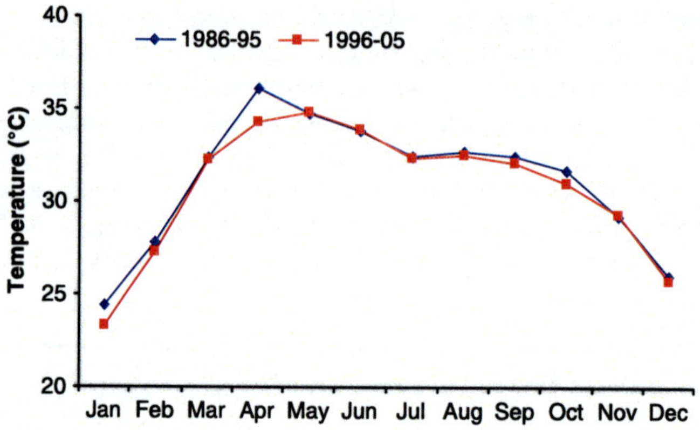

Fig. 25.2 Average maximum temperature in two decades at Rajshahi, Bangladesh

increase (Fig 25.2). During these two decades there were certain years when the mean maximum temperature of these months reached more than 35°C.

25.3.5 Research Progress on High Temperature Stress

It was observed that the spikelet panicle^{-1} was significantly lower when high temperature (38°C) was induced in spikelet differentiation stage (PI to 16 days before heading) and reduction division stage (7 days before heading to heading) due to spikelet abortion (BRRI (Bangladesh Rice Research Institute) 1980). While the sterility percentage was significantly increased when high temperature was induced at heading stage (Table 25.4). Distinct varietal difference in rice was observed when high temperature (38°C) was induced at anthesis stage. Sterility percent ranged from 20% to 78% for a susceptible variety like BR9, but it ranged from 12% to 22% for a moderately tolerant variety like BR1 (Table 25.5). Recent varieties and advanced line developed by BRRI also showed differences in tolerance with high temperature stress (Table 25.6).

25.3.6 Approaches and Prospects to Cope with the High Temperature Stress

There are two main approaches for breeding rice varieties adapted to warmer climates. The first is to breed for heat tolerance, i.e. rice with high fertility despite exposure to temperatures exceeding 35°C during the most sensitive growth stage of the rice plant, i.e., at anthesis or dehiscence of the anther. The second approach is to change the time of flowering, i.e., from usual between 10 am and noon (Yoshida 1981) to earlier in the morning when the temperature is lower. Both conventional

Table 25.4 Effect of high temperature on spikelet no. panicle^{-1}, grain size and percentage of sterility of BR1 at BRRI, Gazipur, Bangladesh during 1980

Treatment (stage)	Spikelet no. panicle^{-1}	100 grain weight (g)	Sterility (%)
Spikelet differentiation	149.7	2.02	18.02
Reduction division	151.4	2.16	23.03
Heading	172.6	2.07	46.54
control	174.6	2.13	19.31

Table 25.5 Effect of high temperature on BR1 and BR9 at BRRI, Gazipur, Bangladesh during 1980

	Spikelet no. panicle^{-1}		100 grain weight (g)		Sterility (%)	
Treatment (stage)	BR1	BR9	BR1	BR9	BR1	BR9
Spikelet differentiation	103.3	196.2	1.91	2.13	13.2	22.8
Reduction division	105.0	197.8	2.00	2.34	22.0	24.0
Heading	114.6	230.6	1.95	2.20	15.2	77.8
control	115.2	234.0	1.94	2.33	12.0	26.6

Table 25.6 Effect of high temperature on percent sterility of some advanced lines during 2004–2005 at BRRI, Gazipur

	Sterility (%)		
Variety/line	Control	Heat at RD stage	Heat at heading stage
BR6888-93-3-1	25.0	54.2	49.8
BR6895-22-4-7	41.9	60.0	30.9
BR6985-15-3-3	51.5	47.9	28.6
BR6985-50-1-2	*50.6*	*58.5*	*27.7*
BR6985-80-1-1	35.9	27.3	52.2
BR6988-20-2	51.8	47.1	52.5
BR7055-7-1	25.6	28.1	38.0
BR7056-6-1	21.6	31.4	52.3
BR7057-8-1	49.3	59.1	55.3
M2-13-26-UL	31.4	42.1	28.4
GUI-4XUAN	35.3	31.6	36.8
QI GUI ZHAN	*23.9*	*18.3*	*21.5*
BM9820	*20.9*	*15.8*	*27.9*
BR6028-34-3-2-1-HR1	29.6	32.5	43.9
BR1 (Check)	52.0	42.3	40.3
BR26 (Check)	41.1	46.5	12.7
BRRI Dhan28 (Check)	44.9	42.2	34.6

breeding and molecular breeding/genetic engineering methodologies may be employed for using these approaches (Redona et al. 2007).

Genetic variability for high temperature tolerance has been reported in rice as early as the late 1970s. Jennings et al. (1979) found varieties, such as Hoveyzeh from southern Iran, which remained fertile at temperatures higher than 45°C, when the other varieties were already completely sterile. Yoshida (1981) found IET4658, IR2006-P12-12-2-2, IET2684, Agbede, Carreon, Dular, N22, OS4, PI215936, IR1561 and Sintiane Diofor to be high temperature tolerant, with their respective fertilities ranging from 84–90% at 35°C. There are fewer reports on genetic variation for time of flowering. Yoshida (1981) noted that *Oryza glaberrima* flowers early in the morning and suggested that this trait can be incorporated into *O. sativa* through breeding.

25.4 Conclusion

Global mean surface air temperature has increased by about 0.6°C in the twentieth century. Over the twenty-first century a range of 1.1–2.9°C warming is predicted in a scenario with low emissions of greenhouse gases, and 2.4–6.4°C in a high-emissions scenario. The current research thrust of Bangladesh Rice Research Institute (BRRI) to cope with high temperature due to global warming is limited only with the screening of breeding materials. There is genetic variation in both escape (time of flowering) and absolute tolerance (spikelet sterility). These should be understood and used for breeding programs. More genotypes from home and abroad should bring in the screening program. Besides breeding for high temperature-tolerant rice varieties (cool rice), basic understanding of the physiological processes of rice in a warmer climate and suitable crop management strategies like high temperature and nutrients and water management to cope with the high temperature problems due to global warming need to be developed immediately.

References

BRRI (Bangladesh Rice Research Institute) (1980) Annual Report for 1979–1980. Bangladesh Rice Res. Inst. Joydebpur, Gazipur, p 187

Horie T, Matsui T, Nakagawa H, Omasa K (1996) Effect of elevated CO_2 and global climate change on rice yield in Japan. In: Omasa K, Kai K, Toda H, Uchijima Z, Yoshino M (eds) Climate change and plants in East Asia. Springer, Tokyo, pp 39–56

IPCC (Intergovernmental Panel on Climate Change) (2007) Climate Change 2007: the physical science basis (summary for policymakers) IPPC secretariat. WMO, Geneva, Switzerland

Jennings PR, Coffman WR, Kauffman HE (1979) Rice improvement. IRRI, Los Banos, Philippines

Matsui T, Omasa K, Horie T (1999) Mechanism of anther dehiscence in rice (*Oryza sativa* L.). Ann Bot 84:501–506

Matsui T, Omasa K, Horie T (2000) High temperature at flowering inhibits swelling of pollen grains, a driving force for the dehiscence in rice (*Oryza sativa* L.). Plant Prod Sci 3:430–434

Matsui T, Omasa K, Horie T (2001) The difference in sterility due to high temperature during the flowering period among japonica-rice varieties. Plant Prod Sci 4:90–93

Matsui T, Kobayashi K, Kagata H, Horie T (2005) Correlation between viability of pollination and length of basal dehiscence of the rice under a hot and humid condition. Plant Prod Sci 8:109–114

Matsushima S, Ikewada H, Maeda A, Honda S, Niki H (1982) Studies on rice cultivation in the tropics. 1. Yielding and ripening responses of the rice plant to extremely hot and dry climate in Sudan. Jpn J Tropical Agric 26:19–25

Peng S, Huang J, Sheehy LRC, Visperas RM, Zhong X, Centeno GS, Khush GS, Cassman KG (2004) Rice yield decline with higher night temperature from global warming. PNAS 101:9971–9975
Redona ED, Laza MA, Manigbas NI (2007) Breeding rice for adaptation and tolerance to high temperatures. In: Abstract. International workshop cool rice for a warmer world, 26–30 March, 2007. Huazhong Agricultural University Wuhan, P.R. China
Yoshida S (1981) Fundamentals of rice crop science. International Rice Research Institute, Los Banos, Philippines, p 269

Chapter 26
Soil Carbon Sequestration Under Rubber Plantations in North-East India

D. Mandal and K.R. Islam

Contents

Abstract Rubber plantations on degraded forest lands are expanding worldwide due to economic, social and environmental perspectives. The study was conducted to evaluate the temporal effects of rubber plantations on soil carbon (C) sequestration in the north-eastern Indian state of Tripura. Composite soil samples were randomly collected at 0–15 (surface) and 15–30 cm (sub-surface) depths from 0 (control), 5, 10, and 20 years old rubber plantations, 2-mm sieved, air-dried, and analyzed for oxidizable (CO_x), non-oxidizable (CNO_x), and total C (CT) concentration, bulk density (ρb), and selected properties. Results showed that soil C concentration, stock and sequestration varied significantly by rubber plantations' age. Averaged across soil depth, the CO_x concentration was increased by 32–105% with increasing rubber plantations' age. Likewise, the CT concentration increased by 14–62% over time. The increase in CO_x and CT concentration was more pronounced at surface soil. The proportion of CO_x in CT was increased (13%) with an associated decrease in CNO_x content. Averaged

D. Mandal
Rubber Research Institute of India, PO. Kunjaban, Agartala 79900, India

K.R. Islam (✉)
Soil and Water Resources, Ohio State University South Centers at Piketon,
1864 Shyville Road, Piketon, OH 45661
e-mail: islam.27@osu.edu

R. Lal et al. (eds.), *Climate Change and Food Security in South Asia*,
DOI 10.1007/978-90-481-9516-9_26,

across soil depth, the CO_x, CNO_x, and CT stocks increased quadratically over time. Soil profile-wise CT stock was increased by 14–57% which accounted for an accumulation of 34 Mg C ha^{-1} over a 20 year period. The CT sequestration increased quadratically, which accounted for 956, 919 and 852 kg C ha^{-1} $year^{-1}$ under 5, 10 and 20 years old rubber plantations, respectively. Results suggest that, if managed properly, rubber plantation on degraded tropical forest lands can be a C sink over time.

Keywords Bulk density • Degraded forests • Stock • Temporal • Tropical climate

Abbreviations

CT	Soil organic total carbon
CNO_x	Non-oxidizable carbon
CO_x	Oxidizable carbon

26.1 Introduction

Terrestrial ecosystem is an important component of the global C cycle. However, a substantial amount of C is emitted from the terrestrial reservoirs as CO_2 to the troposphere by human activities (Lal 2004; Richter et al. 2007; Bonan 2008). Currently, world-wide fossil fuel consumption and deforestation, biomass burning and soil cultivation are accounted for 8 and 1.6 billion Mg of annual C emissions, respectively (Koonin 2008). In India, widespread degradation and deforestation of forests are ongoing processes which accounted for an estimated annual C emission of 45 million Mg (Tg) (Ravindranath et al. 1997). In particular, the tropical forests in the north-eastern Indian state of Tripura are deforested annually at the rate of 1.12% (Ravindranath et al. 1997; Rao 1997). Deforestation not only transfers soil-plant ecosystems' C stock directly to the troposphere by combustion and decomposition, but it also destroys a valuable mechanism to control tropospheric CO_2 accumulation (Ravindranath et al. 1997; Rao 1997). The rate at which forests are being degraded is becoming an environmental concern in India.

Terrestrial vegetation and soil currently absorb about 40% of global CO_2 emissions from human activities (Adam 2001). Simulated models have predicted that the terrestrial biosphere acts as an overall C sink until 2050, but turns into a net source thereafter (Van Kessel et al. 2000). In response to accelerated deforestation, tropospheric CO_2 accumulation and the resulting adverse changes in the global climate, there is an urgent need to find ways for efficient management of biosphere as a net C sink in the future (Seneviratne 2002). Increasing flow of plant biomass through reforestation or plantation is considered one of the important mitigation options to enhance C sequestration and rejuvenate degraded forest lands.

Planting rubber is one of the management practices in north-east India to replace shifting cultivation and control land degradation, produce natural rubber and generate income, and improve site quality. The state of Tripura has been under rubber plantation on denuded forest lands since 1975. Consequently, a large portion of the degraded tropical rain forests is under managed rubber plantations, even though the climate and other conditions are not always favorable. Rubber plantation is basically a special type of secondary forest. Previously, research efforts were mainly focused on increasing rubber yield world-wide (Priyadarshan 2003). So far, few studies have been conducted to investigate the environmental and/or ecological aspects of managed rubber plantations on degraded forest soils under humid tropical climate (Osman et al. 1988; Dey et al. 1996; Sharma and Rai 2007). Specifically, the impact of more subtle land-use changes like rubber plantations on soil C sequestration is relatively unknown (Zhang et al. 2007). Since soils are not able to sequester C indefinitely, information on temporal C sequestration capacity of soil under rubber plantations may help us to develop suitable strategies for managing degraded tropical forests. The objectives of the study were to measure the concentration and stock of C pools and calculate the soil C sequestration capacity under different age groups of rubber plantations in north-eastern state of Tripura, India.

26.2 Materials and Methods

26.2.1 Description of the Site

The study was conducted on 0 (control), 5, 10 and 20 year old rubber plantations in Agartala, Tripura, India. The standard management practices including land clearing and plowing, fertilization, liming, and pest/weed control were performed before planting rubber seedlings. The climate of the region is subtropical with a hot and humid monsoon summer. The average maximum temperature is 33°C and the minimum temperature is 10°C. Relative humidity ranges between 67% and 85% with 160–200 cm annual rainfall. Maximum rainfall occurs during May to September. Average sunshine is slightly more than 8 h day^{-1}.

Soils at the study sites are Ultisols which developed on tertiary hill sediments of unconsolidated beds of sedimentary rocks. They are sandy clay loam/clay loam in texture, and moderately rich in Kaolinite and Illite with a high K fixation capacity. Accelerated erosion accounted for a substantial loss of silt, clay and organic matter from upland areas, which, in turn, were deposited in the valleys. Induced leaching and surface runoff of basic cation have caused the soils to become strongly acidic over time. Soil basic properties were pH 4.2–4.7, E_C 85–168 μS cm^{-1}, CEC 5.7–9.5 meq 100 g^{-1}, base saturation 47–57%, available N 26–29 mg kg^{-1}, available P 0.21–0.7 mg kg^{-1}, and extractable K 6.2–8.8 mg kg^{-1}, porosity 39–43%, sand 36–58%, silt 20–35, and clay 18–42%.

26.2.2 Soil Collection, Processing and Analysis

Soil core samples were randomly collected at 0–30 cm depth from replicated plots within each age group of the rubber plantation. Soil cores were segmented into 0–15 and 15–30 cm depth, 2-mm sieved, oven-dried at 105 ± 5°C for 24 h, and ground prior to analysis. Soil organic matter was determined by using the loss-on ignition method. Total organic C (CT) was calculated by dividing the soil organic matter content bye the factor of 1.724. Soil oxidizable carbon (CO_x) was determined by the modified Walkley–Black's wet oxidation method (Nelson and Sommers 1982) without multiplying by a factor of 1.9. Non-oxidizable C (CNO_x) was calculated by subtracting the CO_x from the CT. The stock of various C fractions was calculated by multiplying the C concentration with concurrently determined soil bulk density (ρb) as follows:

$$\text{Carbon stock}\,(\text{Mg ha}^{-1}) = \text{C concentration}\,(\text{g kg}^{-1}) * \rho b\,(\text{Mg m}^{-3}) * \text{soil depth}\,(\text{m}) * 10$$

The ρb values that were used to convert the C concentration into stock were determined from mass and volume relationship of the soil as follows:

$$\rho b\,(\text{g cm}^{-3}) = (\pi r^2 * L * n) w^{-1}$$

where r is the internal radius of the soil core sampler (0.9 cm), L is the length of soil cores (15 cm), n is the number of soil cores, and w is the total dry weight (g) of soil. Selected soil properties such as pH, cation exchange capacity, base saturation, available N, available P, extractable K, total porosity, sand, silt, and clay content were determined by following standard methods of analysis.

26.2.3 Statistical Analysis

Data were analyzed by using Analysis of Variance procedure of the SAS Institute (2008). The effects of rubber plantations' age on CO_x, CNO_x, and CT concentration, stock and sequestration were analyzed by following a completely randomized design in 4 (plantation age) × 2 (soil depth) factorial combination. Simple and interactive effects of rubber plantation and soil depth on dependent variables were separated and evaluated at $p \leq 0.05$ by the least significant different test, unless otherwise mentioned. Linear and non-linear regression models were used to calculate soil C sequestration rates.

26.3 Results and Discussion

26.3.1 Soil Organic Carbon and Bulk Density

Soil C concentration was significantly influenced by rubber plantations' age and soil depth without an interaction (Table 26.1). The concentration of soil oxidizable C (CO_x) has increased by 32–105% over time. A maximum increase in CO_x concentration was found at 5–10 years growth interval of rubber plantations. The increase in CO_x concentration was more pronounced at surface soil (0–15 cm depth). Likewise, the total C (CT) concentration increased (14–62%) over time. In contrast, the non-oxidizable C (CNO_x) concentration did not increase consistently (2–22%). The CO_x:CNO_x, CO_x:CT, and CNO_x:CT were significantly varied by rubber plantations age and soil depth (Table 26.1). The CO_x:CNO_x increased by 39–73% over time, and the highest increase (84%) was found at the soil surface. The CO_x:CT was increased by 13%, with an associated decrease in CNO_x concentration.

Inclusion of soil bulk density (ρb) to calculate stock has shown significant temporal effects of rubber plantations on C content (Table 26.2). The ρb was significantly reduced (0.0034 g cm^{-3} per year) with increasing soil C content, and the effect was more pronounced at surface soil (0.03 g cm^{-3} per year) than at subsurface soil (0.01 g cm^{-3} per year). Soil CO_x, CNO_x and CT stocks have increased asymptotically over time (Table 26.2 and Figs. 26.1–26.3). The CO_x stock has increased by 33–100%, CNO_x by 0–17%, and CT by 14–57% over a 20 year period. Summing

Table 26.1 Soil carbon concentration under different age groups of rubber plantations

Plantation age (Year)	Depth of soil (cm)	CO_x (g kg^{-1})	CNO_x	CT	Co_x:CNO_x^{-1} (g 100 g^{-1})	$CO_xC_T^{-1}$	$CNO_xC_T^{-1}$
0	0–15	7	5.9	12.9	0.63	54.4	45.6
	15–30	4.7	6.9	11.6	0.36	40.4	59.6
		5.9D	6.4B	12.3B	0.49C	47.4C	52.6A
5	0–15	9.7	5.7	15.4	0.9	63.2	36.8
	15–30	9.3	6.8	16.1	0.72	57.9	42.1
		7.8C	6.2B	14B	0.8B	54.8B	45.2B
10	0–15	11.2	6.2	17.4	0.95	64.3	35.7
	15–30	9.3	6.8	16.1	0.72	57.9	42.1
		10.3B	6.5B	16.8A	0.84A	61.1A	38.9C
20	0–15	14.7	6.9	21.7	1.12	68.1	31.9
	15–30	9.5	8.7	18.2	0.58	52.2	47.8
Soil depth effects		12.1A	7.8A	19.9A	0.85A	60.2A	39.8C
	0.15 cm	10.7a	6.2b	16.8a	0.9a	62.5a	37.5b
	15–30 cm	7.3b	7.3a	14.6b	0.53b	49.3b	50.7a

Co_x, soil oxidizable carbon; CNO_x, soil non-oxidizable carbon; CT, soil organic total carbon. Mean values separated by same upper or lower case letters within a column did not vary significantly at $p < 0.05$

Table 26.2 Soil carbon stock under different age groups of rubber plantations

Plantation age (year)	Depth of soil (cm)	ρb (Mg ha^{-1})	CO_x	CNO_x	C_T
0	0–15	1.61	16.9	14.1	31
	15–30	1.62	11.4	16.9	28.3
		1.62A	14.1D	15.5B	29.7D
5	0–15	1.60	23.4	13.7	37.1
	15–30	1.62	14.2	16.3	30.6
		1.62A	18.8C	15B	33.8C
10	0–15	1.55	26	14.4	40.4
	15–30	1.57	22.5	20.5	42.9
		1.56B	23.9B	15.2B	39.1B
20	0–15	1.54	34.1	16	50.1
	15–30	1.57	22.5	20.5	42.9
		1.55B	28.3A	18.2A	46.5A
Soil depth effects	0–15 cm	1.57a	25.1a	14.6b	39.7a
	15–30 cm	1.60a	17.5b	17.4a	34.9b

CO_x, soil oxidizable C; CNO_x, soil non-oxidizable C; CT, soil total C. Mean values separated by same upper or lower case letters within a column did not vary significantly at $p < 0.05$

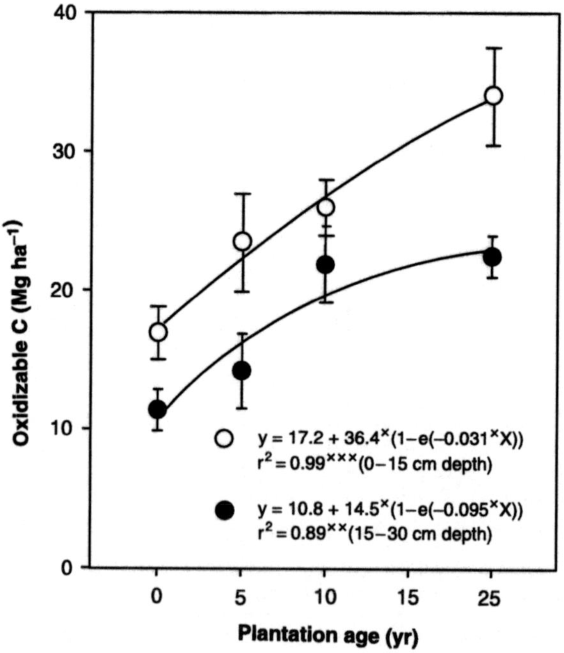

Fig. 26.1 Soil oxidizable carbon stock under different age groups of rubber plantations. *Bars* indicate standard error of mean (±)

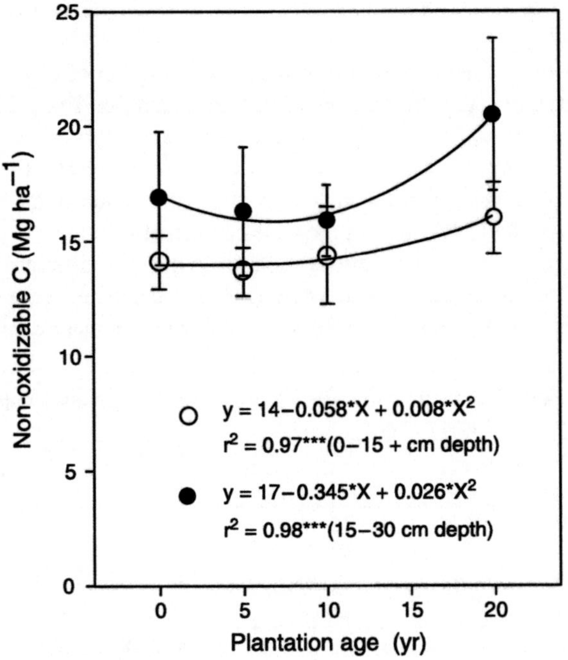

Fig. 26.2 Soil non-oxidizable carbon stock under different age groups of rubber plantations. *Bars* indicate standard error of mean (±)

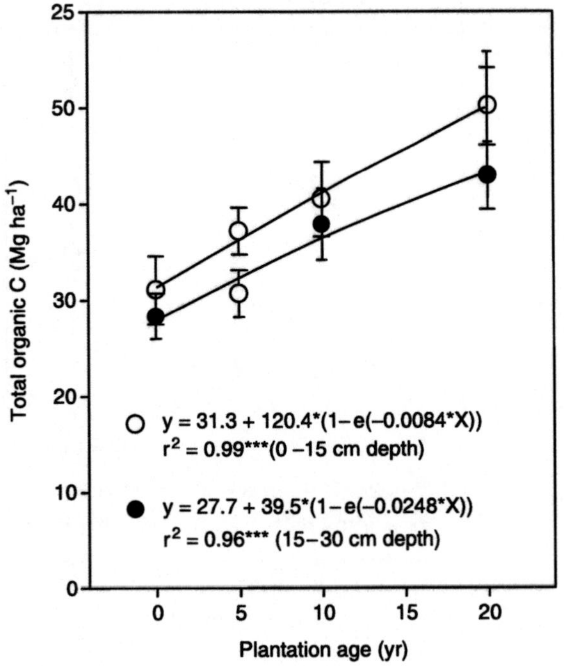

Fig. 26.3 Soil total organic carbon stock under different age groups of rubber plantations. *Bars* indicate standard error of mean (±)

the C stock over soil depth, the profile-wise accumulation of CO_x and CT (except CNO_x) increased in a quadratic manner (Table 26.3 and Fig. 26.4). The profile-wise CO_x stock increased by 33–100% which accounted for an accumulation of 28 Mg C ha^{-1} within 20 years of rubber plantation. Similarly, the CT stock has increased by 14–57%, which accounted for an accumulation of 34 Mg C ha^{-1} over time. In contrast, the CNO_x did not increase consistently.

Significantly higher concentrations of C in soil under different age groups of rubber plantations is most likely caused by the balance between organic matter input and decomposition over time. It is reported that during the mature phase, an amount

Table 26.3 Soil profile-wise C stock under different age groups of rubber plantations

Plantation age (year)	CO_x (Mg ha^{-1})	CNO_x	C_T
0	28.3D	31.1B	59.4D
5	37.7C	30.0B	67.7C
10	47.96B	30.3B	78.2B
20	65.6A	36.5A	93.1A

CO_x, soil oxidizable C; CNO_x, soil non-oxidizable C; CT, soil total organic C. Mean values separated by same upper or lower case letters within a column did not vary significantly at $p < 0.05$

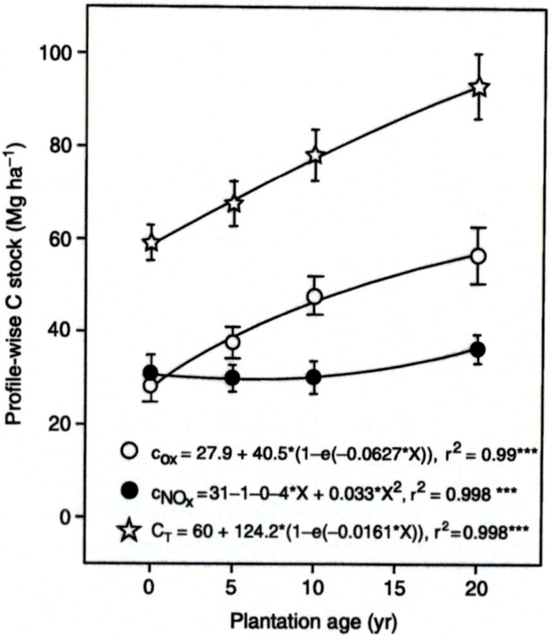

Fig. 26.4 Soil profile-wise oxidizable, non-oxidizable and total organic carbon stocks under different age groups of rubber plantations. *Bars* indicate standard error of mean (±)

of 5–7 Mg ha^{-1} of plant residues is added annually to the soil floor which on subsequent decomposition accumulated a considerable amount of C (Hutchinson et al. 2007). In general, the loss or accumulation of C depends on the standing C stocks in vegetation, litter-falls, and native soil organic matter (Sharma and Rai 2007). A greater soil C concentration with progressive increases in the age of rubber plantations is most probably related to a lack of disturbance and efficient recycling of plant residues (Osman et al. 1988, Sharma and Rai 2007). A reduced contact between plant litter-falls and partially anaerobic undisturbed soil surface under rubber plantations is most probably facilitated by a dominance of fungal food webs. Because of their filamentous growth, greater surface area, and inherent nature of slow decomposition of high C:N plant residues, fungi are the most predominant in the forest ecosystems for efficient metabolism of organic substrates (Fransson et al. 2004). Moreover, the fungal hyphae contribute to protect C in macroaggregation through physical enmeshing of microaggregates or soil primary particles, or releasing extracellular polysaccharides as binding agents (Chenu 1989). It is reported by Dey et al. (1996) that average soil C that could be sequestrated by rubber plantation in the northeastern region of India is about 99.7 Mg C ha^{-1}.

A higher content of organic matter (C) suggests a better soil environment developed under rubber plantations, which could be seen by a significant decrease in soil compaction (ρb). The low values of ρb suggest higher porosity under rubber plantations. It is expected that a continuous deposition of low density leafy biomass on soils under rubber plantations as sources of organic matter may have impacted to reduce ρb. Several studies have reported that due to its low bulk density and the ability to enhance soil aggregation, organic matter accounted for lower ρb, higher porosity, increasing water holding capacity, minimizing erosion, and subsequently improved soil physical quality (Ekwue 1992; Ohu et al. 1994). However, greater reduction in ρb at surface soil is most probably related to stratification of organic matter.

26.3.2 Soil Carbon Sequestration

Soil C sequestration under different age groups of rubber plantations varied significantly over time (Table 26.4 and Figs. 26.1–26.3). The rate of CO_x sequestration increased quadratically which accounted for 1,071 kg ha^{-1} $year^{-1}$ under the 5 year old plantation, 930 kg ha^{-1} $year^{-1}$ under the 10 year old plantation, and 729 kg ha^{-1} $year^{-1}$ under the 20 year old plantation (Fig. 26.1). Likewise, the CT sequestration was 956, 919 and 852 kg ha^{-1} $year^{-1}$ for 5, 10 and 20 years old plantations, respectively (Fig. 26.3). However, the rate of CNO_x sequestration increased exponentially which accounted for only 139 kg ha^{-1} $year^{-1}$ in the 20 year old plantation (Fig. 26.2). There was hardly any CNO_x sequestration within the first 10 years of the plantations. The CO_x (952 vs. 867 kg ha^{-1} $year^{-1}$) CNO_x (36 vs. –42 kg ha^{-1} $year^{-1}$), and CT (964 vs. 854 kg ha^{-1} $year^{-1}$) sequestration rates were higher at 0–15 cm soil depth than at 15–30 cm depth. Soil profile-wise, C sequestration was influenced significantly by rubber plantations age (Table 26.4 and Fig. 26.4). The CO_x sequestration increased

Table 26.4 Soil carbon sequestration under different age groups of rubber plantations

Plantation age (year)	Depth of soil (cm)	CO_x (kg ha^{-1} year^{-1})	CNO_x	C_T
5	15	1,045.3	−17.5	990.4
	30	1,096.5	−215.0	921.3
		1,070.8A	−116.3C	955.9A
10	15	970.3	22.5	970.0
	30	889.2	−85.0	867.6
		929.8B	−31.3B	918.8AB
20	15	840.9	102.5	931.0
	30	616.6	175.0	772.3
		728.8C	138.8A	851.7B
	0–15 cm	952.2a	35.8a	963.8a
Soil depth effects	15–30 cm	867.4b	−41.7b	853.7b

CO_x, soil oxidizable C; CNO_x, soil non-oxidizable C; CT, soil total organic C. Mean values separated by same upper or lower case letters within a column did not vary significantly at $p < 0.05$

quadratically, which accounted for 2,180, 1,886 and 1,447 kg ha^{-1} year^{-1} for 5, 10 and 20 year old rubber plantations, respectively. However, the CNO_x sequestration did not increase consistently.

Results suggested that degraded soils under managed rubber plantations have considerable potential for C sequestration. Lal et al. (1995) estimated that with resilient land management practices, degraded soils can sequester 100–1,000 kg C ha^{-1} year^{-1}. In a recent study, (Cheng et al. 2007) reported that degraded soils under rubber plantations have more C sequestration potential than by primary and secondary rain forests. However, soil C sequestration eventually achieves an equilibrium point and the ability of the soil to sequester C decreases over time as previously depleted stocks are replenished and the soil returns to an equilibrium condition. However, the time it takes for soils to return to equilibrium is highly variable. Similar results were found in the present study where soils under rubber plantations or protected forests had significantly higher C sequestration as compared to soils under fallow and degraded forests (Brown and Lugo 1982; Osman et al. 1988; Raghubanshi 1991; Homann and Grigal 1996; Islam et al. 2001).

26.4 Conclusions

After converting degraded tropical forest lands into well managed rubber plantations, there was a significant increase in soil C concentration and stock over time. The CT sequestration increased non-linearly which accounted for 956 kg ha^{-1} year^{-1} under the 5 year old plantation, 919 kg ha^{-1} year^{-1} under the 10 year old plantation, and 852 kg ha^{-1} year^{-1} under the 20 year old plantation, respectively. Results suggested that, if managed properly, rubber plantations can act as one of the important terrestrial C sinks to minimize the adverse effects of CO_2-driven global warming over time.

References

Adam D (2001) Royal society disputes value of carbon sinks. Nature 412:108–108

Bonan G (2008) Carbon cycle: fertilizing change. Nat Geosci 1:645, October

Brown S, Lugo AE (1982) The storage and production of organic-matter in tropical forests and their role in the global carbon-cycle. Biotropica 14:161–187

Cheng Chun-Man, Ru-Song W, Ju-Sheng J (2007) Variation of soil fertility and carbon sequestration by planting *Hevea brasiliensis* in Hainan Island, China. J Environ Sci 19:348–52

Chenu C (1989) Influence of a fungal polysaccharide, scleroglucan, on clay microstructures. Soil Biol Biochem 21:299–305

Dey SK, Chowdhury D, Vinod KK, Pothen J, Seturaj MR (1996) Estimation of biomass in Hevea clones by regression method. Ind J Nat Rubber Res 9:40–43

Ekwue EL (1992) Effect of organic and fertilizer treatments on soil physical properties and erodibility. Soil Till Res 22:199–209

Fransson AM, Valeur I, Wallander H (2004) The wood-decaying fungus *Hygrophoropsis aurantiaca* increases P availability in acid forest humus soil, while N addition hampers this effect. Soil Biol Biochem 36:1699–1705

Homann PS, Grigal DF (1996) Below-ground organic carbon and decomposition potential in a field-forest glacial-outwash landscape. Biol Fert Soils 23:207–214

Hutchinson JJ, Campbell CA, Desjardins RL (2007) Some perspectives on carbon sequestration in agriculture. Agric Forest Meteo 142:288–302

Islam KR, Ahmed MR, Bhuiyan MK, Badruddin A (2001) Deforestation effects on vegetative regeneration and soil quality in tropical semi-evergreen degraded and protected forests of Bangladesh. Land Degra Develop 12:45–56

Koonin SE (2008) The challenge of CO_2 stabilization. Elements 4:293–294

Lal R (2004) Soil carbon sequestration impacts on global climate change and food security. Science 304:1623–1627

Lal R, Kimble J, Levine E, Whitmore C et al (1995) Towards improving the global database on soil carbon. In: Lal (ed) Soils and global change. Lewis, Boca Raton, FL, pp 343–436

Nelson SD, Sommers LE (1982) Total carbon, organic carbon, and organic matter. In: Page AL et al (eds) Methods of soil analysis, vol 9, Part 2 2nd edn, Agron. Monogr. ASA and SSSA, Madison, WI, pp 539–579

Ohu JO, Ekwue EI, Folorunso OA (1994) The effect of addition of organic matter on the compaction of a vertisol from Northern Nigeria. Soil Tech 7:155–162

Osman KT, Islam KR, Hoque MN, Alim A (1988) Influence of topographic positions on the growth and yield of rubber planted in the hills of Ramu. Bang J Bot 13:97–102

Priyadarshan PM (2003) Breeding *Hevea brasiliensis* for environmental constraints. Adv Agron 79:351–400

Raghubanshi AS (1991) Dynamics of soil biomass-C, N, and P in a dry tropical forest in India. Biol Fert Soils 12:55–59

Rao KS (1997) Natural resource management and development in Himalaya – a recourse to issues and strategies. Environmental information system on Himalayan ecology, monograph 1. G.B. Pant Institute of Himalayan Environment and Development, Kosi-Katarmal, Almora, 263 643, India. http://gbpihed.gov.in/envis/HTML/monograph1/contents.html

Ravindranath NH, Somashekhar BS, Gadgil M (1997) Carbon flow in Indian forests. Climate Change 35:297–320

Richter D Jr, Callham MA Jr, Powlson DS, Smith P (2007) Long-term soil experiments: keys to managing earth's rapidly changing ecosystems. Soil Sci Soc Am J 71:266–279

SAS Institute (2008) SAS online doc 9.13. SAS Institute, Inc, Cary, NC

Seneviratne G (2002) Litter controls on carbon sequestration. Scientific correspondence. Curr Sci 82:130–131

Sharma P, Rai SC (2007) Carbon sequestration with land-use cover change in a Himalayan watershed. Geoderma 139:371–378

Van Kessel C, Horwath WR, Hartwig U, Harris D, Luscher A (2000) Net soil carbon input under ambient and elevated CO_2 concentrations: isotopic evidence after 4 years. Global Change Biol 6:435–444

Zhang H, Gan-Lin Z, Yu-Guo Z, Wen-Jun Z, Zhi-Ping Q (2007) Chemical degradation of a Ferralsol (Oxisol) under intensive rubber (*Hevea brasiliensis*) farming in tropical China. Soil Till Res 93:109–116

Chapter 27
Incidence and Severity of Rice Diseases and Insect Pests in Relation to Climate Change

Mainul Haq, M.A. Taher Mia, M.F. Rabbi, and M.A. Ali

Contents

Abstract Climatic factors especially temperature and relative humidity are the key factors influencing development of any insect pest and disease of rice. CO_2 is the key factor for global climate change, resulting increase in temperature. The intergovernmental panel on climate change predicted that with the current emission scenario, global mean temperature would rise between 0.9°C and 3.5°C by the year 2100. Under such condition frequency of precipitation, intensity of drought and UV-B radiation is predicted to increase, which might affect the structure of rice plant and intensity of insect pests and diseases. Climate change may change the pest–plant relationship resulting in positive or negative impact on incidence and severity of different diseases and insect pests. Sheath blight (*Rhizoctonia solani*), a minor disease in early 1970s, is now a most destructive disease of rice. Similar change has also been observed on the incidence and severity of some other diseases and insect pests. Ear-cutting caterpillar (*Mythimna separata*), a major pest of rice in Bangladesh in 1960s had only a few occurrences in the last few decades. Besides, leaf roller (*Cnaphalocrocis medinalis*, *Marasmia exigua*) that had lower ranks in the list of major pests has been coming at the top of the list since 1980s. In Bangladesh many interventions such as change in crop diversity, variety, cropping intensity, irrigation, fertilization, etc. along with climate change in the rice production system

M. Haq (✉) and M.F. Rabbi
Entomology Division, Bangladesh Rice Research Institute, Gazipur, Bangladesh
e-mail: brrihq@bdonline.com

M.A.T. Mia and M.A. Ali
Plant Pathology Division, Bangladesh Rice Research Institute, Gazipur, Bangladesh

R. Lal et al. (eds.), *Climate Change and Food Security in South Asia*,
DOI 10.1007/978-90-481-9516-9_27, © Springer Science+Business Media B.V. 2011

affected incidence and severity of insect pests and diseases. However, exact and individual contribution of such factors or interventions has not been worked out in Bangladesh or elsewhere. Probable change in the years to come on the status of insect pests, pathogens and natural enemies along with change in the host plants due to global climate change is discussed in this paper.

Keywords Insects and pests • Climate change • Disease and pest problems • Integrated nutrient management • Cropping systems

Abbreviations

BCA Biocontrol agents
RH Relative humidity
UV Ultraviolet

27.1 Introduction

Massive industrialization causes extensive emission of CO_2 is the key factor for global climate change. According to the intergovernmental panel on climate change (IPCC) the predicted changes include more frequent extreme high temperatures and less frequent extreme low temperatures. Similarly, intensity of precipitation events is increased. While water vapor, evaporation and precipitation are predicted to increase on average, predictions about increased or decreased precipitation are region specific. In southern and eastern Asia, precipitation is predicted to increase in summer. IPCC also predicted that with the current emission scenario, global mean temperature would rise between 0.9°C and 3.5°C by the year 2100 (IPCC 2001). Although the IPCC did not find any evidence of any change of tropical and extratropical storms, more recent analysis concluded that there have been changes in storm patterns in recent years (Emanuel 2005; Webster et al. 2005), which could influence the global movement of pathogens (Brown and Hovmoller 2002).

Incidence and severity of pests, especially insect pests and diseases are influenced by the environmental factors. Prevailing environmental conditions is a decisive factor especially for the occurrence and severity of diseases. In Bangladesh so far, 32 diseases and 175 insect pests have been reported to occur on rice. Among them, 10 diseases and 20 insect pests have the potentiality to cause economic damage to the crop and thus considered as major. However, status of pests is dynamic. Due to several factors, namely change in environmental conditions, change in varieties, change in cropping pattern and intensity, change in the pathogen and introduction of a new pathotype, etc. Climatic change would certainly bring change in cropping patterns and also cropping intensity. To cope up with changed climatic factors new varieties may also be released, which might have either positive or negative impact on severity of a particular disease or insect pest thus change its status. Climatic changes might also have a negative effect on delicate natural enemies such as hymenopteran parasitoides and small predators. This can also cause increase in population of certain pests. Evidence on the change already occurred on the incidence of some diseases and insect pests and predictions on the future change in insect and disease status under changing climatic conditions are discussed in this article.

27.2 Status of Diseases and Insect Pests Over the Year

Surveys on insect pest and disease incidence were made in different agro-ecological zones of Bangladesh since 1980. In addition, observations on the incidence were made from time to time field visits at different locations of the country. Changes observed on the pests status from such surveys and field visits are discussed below.

Talukder (1968) reported that sheath blight (*Rhizoctonia solani*) disease of rice has occurred in Bangladesh from a long time as a minor disease. However, over the course of time the status of this disease has been changed to a major one (Miah et al. 1985). High temperature and high relative humidity (RH) are very much conducive for rapid proliferation of this pathogen and spread of the disease. Replacement of indigenous rice varieties with modern varieties having thick canopy and responsive to higher dose of fertilizers especially nitrogenous fertilizers changed the microclimate of rice field in favor of sheath blight disease development. These factors along with climate change probably influenced the disease and changed its status from minor to a major disease. Presently, sheath blight disease is wide spread in the country especially in rainfed ecosystem and the severity has also increased further (Fig. 27.1). In some locations where Boro crop is sown late is vulnerable to this disease.

Blast disease was recorded for the first time in the Indian sub-continent in 1913 (Padmanabhan 1965). In late sixties and early seventies blast (*Pyricularia grisea*) was considered as a relatively unimportant disease with low incidence in limited areas (Anonymous 1977). But during 1980 and 1990 outbreak of blast on Boro rice was observed in different regions of Bangladesh. (Miah et al. 1980; Shahjahan et al. 1991). Cultivation of a susceptible rice variety Pajam in all the three consecutive seasons was

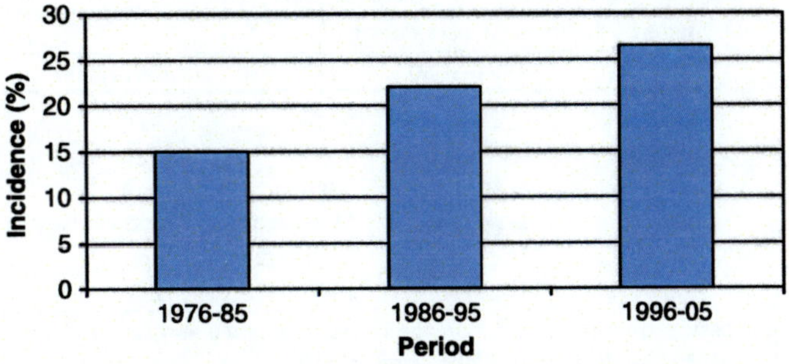

Fig. 27.1 Incidence pattern of sheath blight disease during 1976–2005

considered to be the main factor for blast outbreak in 1980. However, the probable reason of epidemic outbreak of this disease in 1990 was assumed to be due to higher level of both maximum and minimum RH, lower variation between maximum and minimum temperatures, higher number of rainy days and minimum temperature in the month of January to March. The northern parts of the country were free from this disease since the beginning but recently neck blast has been found to occur in T. Aman and Boro crops in Rajshahi, Joypurhat, Sirajgonj and Bogra districts. This disease is now widely spread in the country with medium to high incidences in Boro and T. Aman seasons.

Bakanae (*Fusarium moniliforme*) disease of rice used to occur mostly in Aus season and was of very little importance in Boro crop. However, since the year 2000 it has become a big concern for Boro rice especially in areas with single Boro rice. The reason of such shifting of the disease in a different season is not yet clear. The trend of daily minimum temperature from 1981 to 2005 showed a considerable variation. An increasing trend of decadal mean of daily minimum temperature was observed at Comilla from 1981 to 2005 (Fig. 27.2), which is a positive indication of global warming. Such rise in minimum temperature during winter might have increased the incidence of blast and increased appearance of bakanae in Boro season. Changes in status of some major diseases over time are shown in Table 27.1.

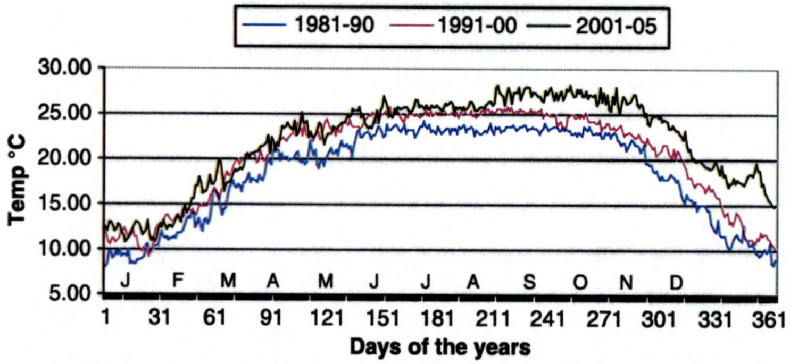

Fig. 27.2 Average of daily minimum temperature (°C) during three periods at Comilla

Table 27.1 Change in status of some major diseases of rice over time

Disease	Status of different diseases during				
	Before 1970	1971–1980	1981–1990	1991–2000	2001–2006
Tungro	High	High	High	High	High
Sheath blight	Low	Low	High	High	High
Blast	Low	Low	High	High	High
Bakanae	Low	Low	Moderate	Moderate	High
Brown spot	High	High	Moderate	Low	Low
BLB	High	High	High	High	High

The status was assigned based on the Damage Index (incidence and severity) of the disease as follows: Low = <30%; Moderate = 31–50%; High = >50%

Source: Survey reports of Plant Pathology Division, BRRI and Anonymous (1977)

In undivided Bengal, ufra disease (*Ditylenchus angustus*) was reported for the first time in deep-water ecosystem. Gradually this disease has spread in 18 districts of Bangladesh, both in rainfed and irrigated ecosystems. However, from the year 2000, the area under this disease has been shrinking remarkably.

Recently bacterial leaf streak (*X. campestris* pv *oryzicola*) emerged as an alarming proportion especially in south and southwest parts of the country. The incidence of the disease might be due to rice varieties and environmental factors especially rain and storm.

Moreover, incidence of five major insect pests of rice was monitored everyday at BRRI farm Gazipur. The pests were brown plant hopper, leaf roller, caseworm, green leaf hopper and stem borer. A Pennsylvanian light trap placed in a corner of the BRRI farm was operated dusk to dawn every day from January 1981 to December 2006. Total number of individual pest trapped in a year was considered as the severity of that pest in that particular year.

Data on the number of pests suggested that brown planthopper and green leafhopper increased whereas, leaf roller, caseworm and stem borer population decreased over the years at BRRI farm (Figs. 27.3–27.7).

Weather records collected from BRRI weather station at Gazipur indicated that maximum and minimum temperatures increased to a little extent during the observation period. Although it could not be related to the incidence trend with the temperature change, it seems that temperature might be one of the factors responsible for the change in pest incidence pattern.

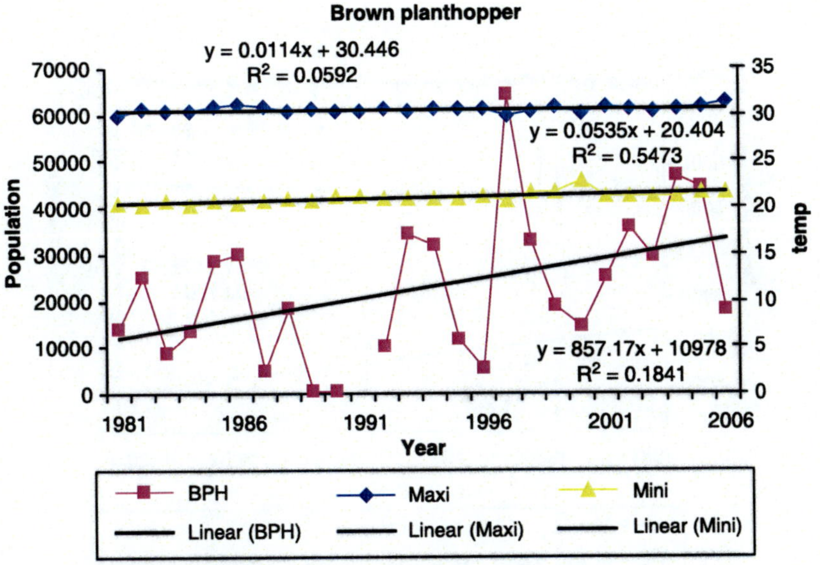

Fig. 27.3 Brown plant hopper population at Gazipur (1981–2006)

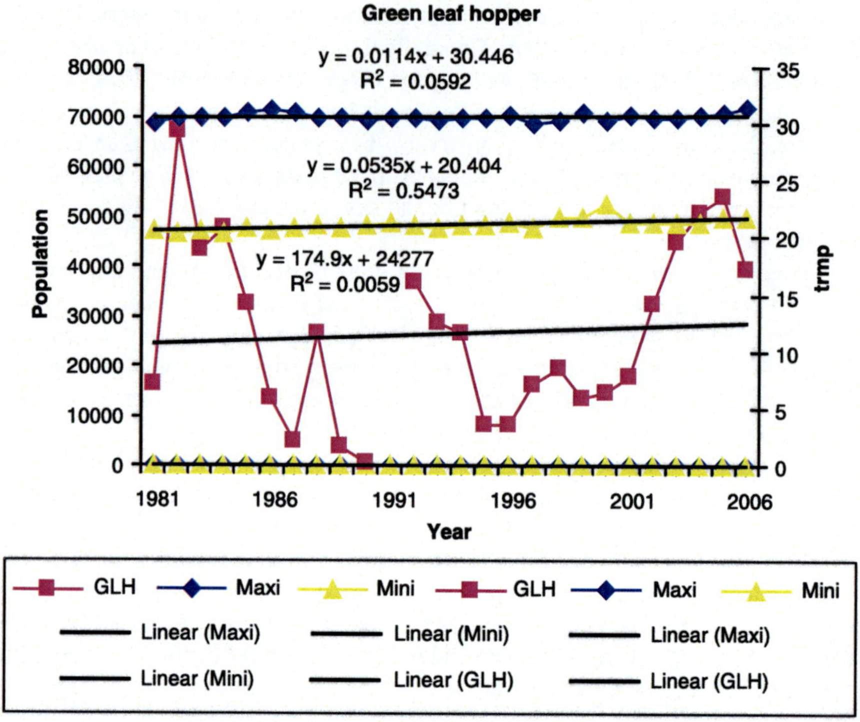

Fig. 27.4 Green leaf hopper population at Gazipur (1981–2006)

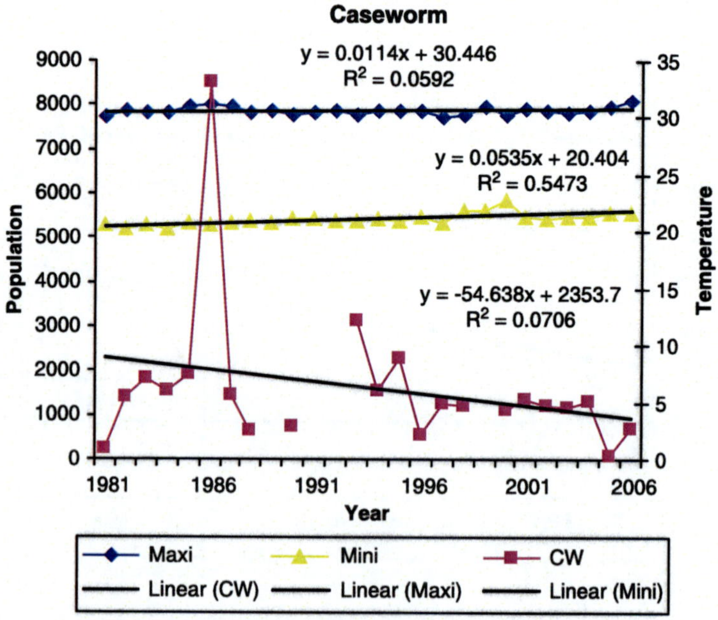

Fig. 27.5 Caseworm population at Gazipur (1981–2006)

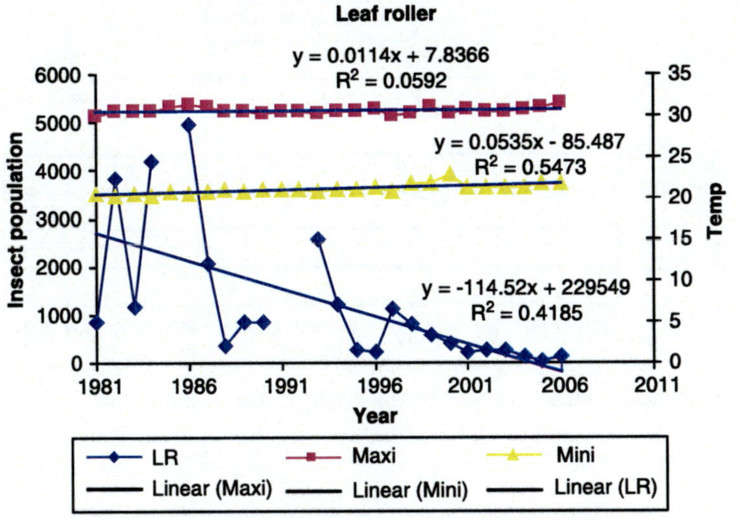

Fig. 27.6 Leaf roller population at Gazipur (1981–2006)

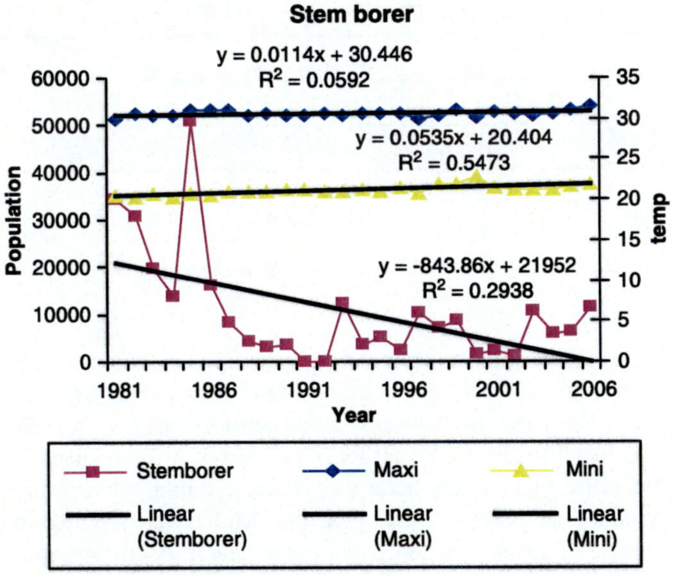

Fig. 27.7 Stem borer population at Gazipur (1981–2006)

Monthly incidence pattern of the above five insect pests for the last 10 years (1998–2007) were compared with those of the previous decade (Figs. 27.8 and 27.9). It was observed that the monthly incidence pattern of green leafhopper did not change in the last decade. Peak incidence was observed in April–May and October–November which was similar to the pattern reported earlier (Bangladesh Rice Research Institute 2008). For leaf roller, three peaks in March, May and September–October were reported

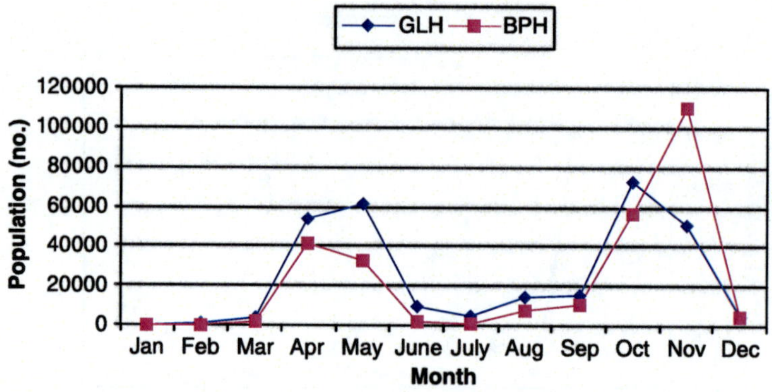

Fig. 27.8 Monthly incidence pattern of GLH and BPH at Gazipur (1998–2007)

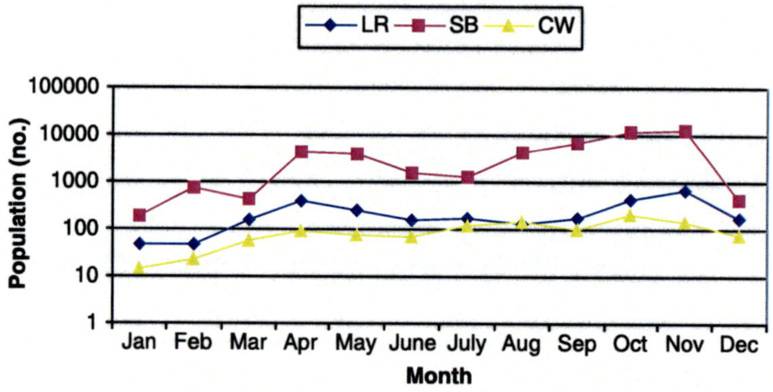

Fig. 27.9 Monthly incidence pattern of LR, SB and CW at Gazipur (1998–2007)

in the earlier reports. But in the present observation period, two peaks, one in April and another in October–November, were observed. Brown plant hopper was earlier reported to have two peaks in May–June and October–November. In 1998–2007, the first peak was found to occur a bit earlier (April–May). A sharp change in the incidence pattern of the yellow stem borer was observed during the reported period. In the earlier report, three peaks in April, June and October were reported for this pest. But in the period of 1998–2007, only two peaks, one in April-June and another in October–November, were found to occur. A major change also occurred in the incidence pattern of caseworm. Instead of one peak in August–September, three peaks (March–May, August and October) occurred during the observation period. The infestation in the field was observed from August to November (Haq et al. 2006). Fei et al. (1995) found that climatic factors like RH, total rainy days and temperature were responsible for greater caseworm incidence.

Besides light trap records, visual surveys from insect pest incidence were made in different agro-ecological zones of Bangladesh in 1980 and 1990. Survey records suggested changes in incidence trend of some insect pests of rice. Notable changes occurred in the case of ear-cutting caterpillar, rice leaf roller, gall midge and rice hispa. No severe incidence of ear-cutting caterpillar was found during this period although it had been considered as a notorious pest of rice in the 1960s. An increasing trend of rice leaf roller incidence was observed in the south-western districts of Bangladesh, although light trap records reported a decreasing trend at Gazipur. Upsurgence of gall midge has been observed in the North-Western Bangladesh, although earlier it had been considered as a pest of rain prone low land areas. Moreover, endemic range of rice hispa is increasing in Bangladesh. Although, the south-western districts of Bangladesh had been considered as the endemic area of rice hispa, new areas particularly in the north-western districts has become hispa-prone since 1990s (Bangladesh Rice Research 2005).

27.3 Consequences of Predicted Climate Change

Elevated CO_2 levels tend to result in changed plant structure viz., increased leaf area, increased leaf thickness, higher number of leaves, and higher total leaf area per plant with greater diameter (Pritchard et al. 1999). Changes in plant architecture may affect microclimate and thus risks of infection (Burdon 1987). Under such condition of plants, the microclimate of the rice field will be changed significantly especially RH and temperature would be increased, which predispose the plants to certain pathogens viz. *Rhizoctonia solani* and *Sclerotium oryzae.* Presently, sheath blight disease mostly occurs in rainfed upland and lowland rice ecosystems and rarely under irrigated ecosystem, except where the later is planted late due to cultivation of three rice crops or a third crop in between. Under expected changing climatic conditions, the winter temperature is likely to be increased that would be congenial for development of sheath blight in irrigated rice and may cause additional concern to the growers. Under elevated temperature there is a possibility of shorten the life span of rice varieties and thereby increased susceptibility to sheath blight as varieties with short duration is more susceptible to sheath blight disease (Sharma et al. 1995). Therefore, the risk of increased severity of sheath blight disease is most likely in the future.

In general, increased plant density will tend to increase leaf surface wetness and leaf surface wetness duration, and so encourage infections by foliar pathogens more likely (Harvell et al. 2002), especially by *Pyricularia grisea*, causing blast disease of rice. Climate change would increase the temperature in winter months and thus the chances of aggravating blast disease are more likely. However, further increase in temperature during T. Aman season may have a negative impact on blast disease development.

Abiotic stresses such as heat and drought may contribute to plant susceptibility to pathogens or it may increase resistance. Drought conditions often predispose plants to *Bipolaris oryzae*, causing brown spot disease of rice. Drought prevailed during T. Aman season in 1993 and 2004 at Rajshahi and Chuadanga, respectively was associated with very high incidence of brown spot disease (Personal observation). Therefore, incidence of these diseases likely to be increased under changed climatic conditions.

In the years to come, increases in UV-B radiation is likely due to depletion of ozone layer. Small decreases in the ozone layer may cause significant increases in UV-B can significantly increase the susceptibility of rice to blast disease (Olszyk and Ingram 1993). UV-B can alter plant morphology through reduction in plant height and leaf area and overall plant geometry may be changed (Barnes et al. 1988). Such conditions may also increased production of secondary plant metabolites that might change the pest–host relationship and also the status of individual pest (Renger et al. 1989; Cladwell et al. 1989). Elevated ozone concentrations can change the structure of leaf surface, altering the physical topography as well as the chemical composition of surfaces, including the structure of epicuticular wax. These changes in leaf structure may alter leaf surface properties such as leaf wettability and the ability of leaves to retain solutes, all influencing the ability of pathogens to attach to leaf surface and infect (Karnosky et al. 2002). Ozone exposure has been proposed to enhance attacks on plants by necrotrophic fungi (Sandermann 2000).

The changes in insect incidence pattern in rice might happen due to many factors. Change in temperature is one of the reasons to influence disease in rice plants. Temperature may affect the population trend and incidence pattern of insect pests in direct or indirect ways. Reproductive biology of an insect pest may be affected both positively and negatively which ultimately affect its population build-up (Chapman and Reiss 1995). Delicate natural enemies specially the small parasitoids possess very narrow temperature threshold range and an increase in temperature may be detrimental to such parasitoids and may affect natural enemy-pest relationship. Brown planthopper is at least 17 times more tolerant of 40°C than its predator *Cyrtorrhynus lividipennis*. In contrast, the wolf spider *pardosa pseudoannulata* was more tolerant to 40°C than brown plant hopper. This implies that global warming might result in a disruption of this predator prey relationship and may cause a change in the population build-up of the pest (Heong and Domingo 1992). A change in temperature may interact with the cropping pattern affecting arthropod diversity in a region. Moreover, rise in temperature in winter may help to continue the life cycle of some pests without disruption in Boro season.

27.4 Approaches to Mitigate the Predicted Problems

The predicted disease and pest problems under changed climate may be addressed through multi-dimensional approaches. The probable approaches are as follows:

27.4.1 Development of Resistant Varieties

As the change will not be sudden, therefore, the approaches should be continuous. The most economic and environment friendly approach is the development and deployment of crop varieties resistant to major diseases and insect pests. For the development of crop varieties resistant to major diseases, the pathogen population should be characterized to monitor if there are any changes in the pathogen population over time. However, varieties, breeding lines and germplasms needs to be screened against the mixed population to identify entry(ies) with partial resistance that should be durable. Searching for resistant source (s), development of partial resistant varieties and replacement of more susceptible varieties should be a continuous proccss. Similar activities should also be done for major insect pests.

27.4.2 Survey and Monitoring of Diseases and Insect Pests

Regular monitoring of diseases and insect pests should be done to create a database and monitor if there is any change on the occurrence of any pests and diseases.

27.4.3 Use of Bio-control Agents (BCA)

Although success of controlling rice insect pests and diseases using BCA is not much encouraging in Bangladesh or elsewhere due to shorter field duration of rice crop. Extensive search for thermophilic BCA should be taken up. Collecting antagonists/ parasitoids/ predator during hot summer from the areas where the prevailing temperature is high enough, preserved or rearing these and also efficacy study would be done at higher temperature. Use of already developed management practices until these remain effective.

27.5 Conclusions

Adaptation to climate change through choice of management practices is an important strategy to minimize the incidence of pests and pathogens. Important among management practices are complex cropping systems (e.g., sequential cropping of cereals with legumes, mixed cropping, relay cropping, agroforestry systems, integrating animals with crops and trees), new cultivars relatively tolerant to biotic and abiotic stresses, and innovative systems of soil and water management. Adopting strategies of integrated pest management (IPM) is a cost-effective and environment-friendly option. Growing crops with conservation tillage involving minimal soil disturbance and the liberal use of crop residue mulch, incorporation of leguminous cover crops within the rotation cycle, use of integrated nutrient management (INM) options based on biofertilizers including manures and compost integrated with mycorrhizae and rhizobium inoculation, and creating disease-suppressive soils through enhancement of soil biodiversity. The overall strategy is to enhance use efficiency of inputs by minimizing losses, recycling biosolids, and conserving natural resources.

References

Anonymous (1977) Literature review of Insect pests and diseases of rice in Bangladesh. Bangladesh Rice Research Institute, Joydebpur, Dacca, Bangladesh

Anonymous (1986) Adhunik Dhaner Chash. Bangladesh Rice Research Institute, Gazipur

Bangladesh Rice Research Institute (1976) BRRI Annual Report for 1976. Bangladesh Rice Research Institute, Gazipur

Bangladesh Rice Research Institute (2005) Proceedings of the workshop on Ecology and management of rice hispa in Bangladesh, 24–25 April, 2004, p 48

Bangladesh Rice Research Institute (2008) Adhunic Dhaner Chash. Publication No. 5, 2008. BRRI, Gazipur, p 63

Barnes PW, Jordan PW, Gold WG, Flint SD, Cladwell MM (1988) Competition, morphology, and canopy structure in wheat (*Triticum aestivum* L.) and wild oat (*Avena fatua* L.) exposed to ultraviolet-B radiation. Funct Ecol 2:319–330

Brown JKM, Hovmoller M (2002) Aerial dispersal of pathogens on the globe and continental scales and its impact on plant diseases. Science 297:534–541

Burdon JJ (1987) Diseases and Population Biology. Cambridge University Press, New York

Chapman JL, Reiss MJ (1992) Ecology principles and applications. Cambridge University Press, Cambridge, p 294

Cladwell MM, Teramura AH, Tevini M (1989) The changing solar ultraviolet climate and the ecological consequences for higher plants. Ecol Evol 4:363–366

Emanuel K (2005) Increasing destructiveness of tropical cyclones over the past 30 years. Nature 436:686–688

Fei H, Su Q, Zhang X, Fei H, Su Q, Zheng X (1995) Effects of immigrated population and climatic factor a field population dynamics of rice caseworm. Acta Phytophylacica Simica 22:193–197

Haq M, Haque NMM, Karim ANMR (2006) Incidence pattern of rice caseworm(*Nymphyla* sp.). J Agric Rural Dev 4:75–81

Harvell CD, Mitchell CF, Ward JR, Altizer S, Dobson AP, Ostfeld RS, Samuel MD (2002) Climate warming and disease risks for terrestrial and marine biota. Science 21:2158–2162

Heong KL, Domingo I (1992) Shifts in predator–prey ranges in response to global warming. IRRN 17:29–30

IPCC (2001) Climate Change 2001: the scientific basis. Contrin. Work. Grpup I Third Assess. Rep. Intergov. Panel Clim. Change. In: Houghton JT, Ding Y, Griggs DJ, Noguer M, van der Linden PJ (eds) Cambridge University Press, Cambridge

Karnosky DF, Percy KE, Xiang B, Callan B, Noormets A (2002) Interacting elevated CO_2 and tropospheric O_3 predisposes aspen (*Populus tremuloides* michx.) toinfection by rust (*Melampsora medusae* f. sp. *tremuloides*). Glob Change Biol 8:329–338

Miah SA, Shahjahan AKM, Haque MA (1980) Severe blast outbreak during 1979-80 Boro in Bangladesh. IRRN 5:10–11

Olszyk DM, Ingram KT (1993) Effects of UV-B and global climate change of rice production. The EPA/IRRI cooperative research plan. International Rice Research Institute, Philippines

Padmanabhan SY (1965) Breeding for blast resistance in India. In: The rice blast disease. Johns Hopkins Press, Baltimore, Maryland, pp 343–349

Pritchard SG, Rogers HH, Prior SA, Peterson CM (1999) Elevated CO_2 and plant structure: a review. Glob Change Biol 5:807–837

Renger G, Volker M, Eckert HJ, Fromme R, Hohm-viet S, Graber P (1989) On the metabolism of photosystem II deterioration by UV-B radiation. Phochem Photobiol 49:97–105

Sandermann H Jr (2000) Ozone/biotic disease interactions: molecular biomarkers as a new experimental tool. Environ Pollut 108:327–332

Shahjahan AKM, Ahmed HU, Mia MAT, Hossain MA, Sharma NR, Nahar NS (1991) Outbreak of leaf and neck blast in Boro rice crop in Bangladesh. IRRN 16:21

Sharma NR, Akanda SI, Shahjahan AKM (1995) Development of sheath blight in short, tall, early and late maturing rice cultivars. Bangladesh J Bot 24:143–146

Talukder MJ (1968) Diseases of rice in East Pakistan and their control. Mimeo. paper, Div. Mycol. Pl. Path. BARI, Dacca. In: Literature review of insect pests and diseases of rice in Bangladesh. Bangladesh Rice Research Institute. Joydebpur, Dacca, Bangladesh, 1977, p 36

Webster PJ, Holland GJ, Curry JA, Chang HR (2005) Changes in tropical cyclone number, duration and intensity in a warming environment. Science 309:1844–1846

Chapter 28
Assessing Impacts of Climate Change on Cereal Production and Food Security in Bangladesh

S. Ghulam Hussain

Contents

Abstract As a consequence of future climate change, agriculture is likely to be affected, which would also lead to risk of hunger and water resource scarcity with enhanced climate variability. Currently, the estimated population of Bangladesh stands at over 143.4 million and is likely to be 214.6 million in 2050. To keep pace with population growth and shrinking land resource base, the food production needs to be doubled by the year 2020. Cereal production more than doubled in the last 25 years and the production gains were achieved mainly due to yield increases.

A simulation study was conducted to assess the climate change related vulnerability of food grain production in Bangladesh. Four climate change scenarios (baseline, GFDL–TR = Geophysical Fluid Dynamics Laboratory Transient; HadCM2 = Hadley Centre Unified Model2 Transient ensemble and UKTR = UK Met. Office/Hadley

---

Presented at the International Conference on Global Climate Change and Its Effects, held during August 25–30, 2008 in Dhaka, Bangladesh.

S.G. Hussain (✉)
Planning & Evaluation Division, Bangladesh Agricultural Research Council,
Farm Gate, New Airport Road, Dhaka 1215, Bangladesh
e-mail: ghussain@agni.com

R. Lal et al. (eds.), *Climate Change and Food Security in South Asia*,
DOI 10.1007/978-90-481-9516-9_28, © Springer Science+Business Media B.V. 2011

Centre Transient) were used. Simulation runs were made for high yield varieties of rice for *Aus* (March–August), *Aman* (August–November), and *Boro* (February–July), the traditional growing seasons, using the CERES-Rice model. Simulation was carried out for wheat, which is grown from November through March, using the CERES-Wheat model. The detrimental effect of temperature rise was observed even with elevated CO_2 levels. Considerable spatial and temporal variations were also noted. Impact of these changes on food security was also assessed.

Keywords Agriculture • Bangladesh • Climate change • Food security • Rice • Wheat • CERES-Rice • CERES-Wheat • Simulation

Abbreviations

GDP	Gross Domestic Product
MoA	Ministry of Agriculture
GCM	Global Circulation Model
GFDL–TR	Geophysical Fluid Dynamics Laboratory Transient
HadCM2	Hadley Center Unified Model 2
UKTR	UK Met. Office/Hadley Center Transient
HYV	high yielding varieties
DSSAT	decision support system for agrotechnology transfer
CERES	Clouds and the earth's radiant energy system
BRRI	Bangladesh Rice Research Institute
BARI	Bangladesh Agricultural Research Institute
BARC	Bangladesh Agricultural Research Council
ICASA	International Consortium for Agricultural System Analysis
MAGICC	Model for the Assessment of the Greenhouse Gas Induced Climate Change
SCENGEN	SCENerio GENerator
HADC	Hadley Centre Unified Model 2
(HadCM2 HADC50, HADC70)	Transient Ensemble GCM Scenario

28.1 Introduction

As a consequence of future climate change agriculture is likely to be affected, which would also lead to risk of hunger and water resource scarcity with enhanced climate variability and more rapid melting of glaciers (Cruz et al. 2007). The Fourth Assessment Report of the IPCC considers agriculture as one of the most vulnerable systems to be affected by climate change in the south Asian region. Bangladesh being one of the countries of the region, it is very likely that the climate change vulnerability would increase. Currently the share of agriculture to the total GDP is about 21.77% and the overall GDP growth rate is around 4.49%. The share of this sector is 4.9% to the total export earnings. While the share of crop sector in GDP is 12.19% (2005–2006 Fiscal Year), the contribution of fisheries has increased from 4.5% to 4.86% during the same period. Although the contribution of agriculture is not very lucrative, it employs about 63% of the labor force. Farming in the country is mainly of subsistence in nature. Therefore, livelihood of the majority of the population is directly or indirectly dependent on agriculture.

Rice is the staple of Bangladeshi people. The country produces about 6.89% of the world's paddy from 11.20 M ha of land. In the year 2006, Bangladesh produced 43.73 M t of rice paddy equivalent to 30.61 M t of clean rice. About 0.13% of the global total of wheat is also produced in the country. During 2006 the production of wheat was 0.77 M t from 0.48 M ha (FAO 2008).

The climate of Bangladesh is generally sub-tropical in the north to hot humid in the south. Southwest monsoon influences the climate during June to October, and during the winter the climate is controlled by the northeast monsoon from November to March. The summer is hot and humid and the winter is mild. The country is vulnerable to many environmental hazards, including frequent floods, droughts, cyclones, and storm surges that damage life, property, and agricultural production.

Bangladesh, a developing country in South Asia, is primarily a deltaic floodplain, and elevations in most of the country do not exceed 10 m. The country has a humid tropical climate. Rains start in March–April and end in October, and average 1,500 mm year^{-1} in drier regions to 5,000 mm year^{-1} in wetter regions. Winter in Bangladesh is from November to February, when the average minimum and maximum daily temperatures are about 9.7°C and 26.6°C, respectively. Summer is the hottest period of the year, with temperatures generally above 35.4°C but rarely exceeding 40.0°C. Maximum solar radiation occurs during March, April, and May and ranges from 16 to 22 MJ m^{-2} day^{-1}; minimum solar radiation occurs during December and ranges from 12 to 14 MJ m^{-2} day^{-1}. Bangladesh is highly vulnerable to severe weather events – floods and droughts are frequent and damaging. The crop most affected by floods and droughts is rice, which contributes 95.48% of the total cereal production in the country, followed by wheat (2.65%) and other cereals (1.88%) (computed from MoA 2007).

Mirza and Dixit (1997), Khan et al. (2000), and Mirza (2002) reported that during the period from 1985 to 1998 an increasing trend of about 1°C in May and 0.5°C in November has already occurred. Decadal rain anomalies above long term averages since the 1960s were also reported. In many countries of Asia, production of major cereals like rice, maize and wheat during the past few decades has declined due to increasing water stress arising partly from increasing temperature, increasing frequency of El Niño and reduction in the number of rainy days (Wijeratne 1996; Aggarwal et al. 2000; Fischer et al. 2002; Tao et al. 2003; Tao et al. 2004). Peng et al. (2004) reported in a study at the International Rice Research Institute that rice yield decreased by 10% for every 1°C increase in growing-season minimum temperature. Climate change could make it more difficult than it is already to accelerate the agricultural production to meet the growing demands in developing countries in Asia.

This study considers the possible impacts of climate change on the production of rice and wheat. It looks at the effects of three GCMs scenarios (GFDL–TR = Geophysical Fluid Dynamics Laboratory Transient; HadCM2 = Hadley Centre Unified Model2 Transient ensemble and UKTR = UK Met. Office/Hadley Centre Transient) and baseline, in six locations for rice, and in three locations for wheat. These locations represent the major rice and wheat growing regions of the country. Effort was also made to assess the impact of climate change on food security in Bangladesh.

28.2 Materials and Methods

28.2.1 Study Locations and Crops

Six sites of the country that represent the drier (Dinajpur and Jessore), wetter (Sylhet and Mymensingh), and coastal regions (Chittagong and Barisal) were chosen for the rice study, and three sites (Jessore, Mymenshing, and Dinajpur) were selected for the wheat study (Fig. 28.1). A simulation study was conducted for three high yield variety (HYV) rice crops, grown during three seasons: *Aus* (pre-monsoon-season paddy, March–August,), *Aman* (monsoon-season paddy, July–November), and *Boro* (dry-season paddy, December–May). Three widely grown varieties *BR3, BR11,* and *BRRI Dhan29* were selected for Aus, Aman, and Boro, respectively. In the simulation experiments, Aus rice was planted on 01 April, Arnan rice on 15 July, and Boro rice on 01 February. *Shourav* was the selected wheat variety which was planted on 10 November in Dinajpur, Jessore, and Mymensingh.

28.2.2 Crop Models

The simulation runs for rice and wheat were made by using CERES-Rice and CERES-Wheat crop models of Decision Support System for Agrotechnology Transfer (DSSAT) version 4.0 (Hoogenboom et al. 2003). Eighteen experiments (three crops at six locations) for rice and three experiments for wheat (one crop at three locations) were conducted. Each experiment was composed of seven treatments. Several inputs are required by these crop models. In this section, how these inputs were generated have been described.

Genetic Coefficients Both CERES models use as inputs coefficients that account for differences in growth and development among cultivars. These coefficients are referred to as genetic coefficients. Genetic coefficients allow the model to simulate performance of different varieties under diverse weather and management conditions. Therefore, to obtain reasonable outputs from the crop simulation models, it is necessary to have the appropriate genetic coefficients for the selected varieties. Since the genetic coefficients for the varieties considered in this study were not available, those were estimated following the method as described by Ritchie et al. (1998).

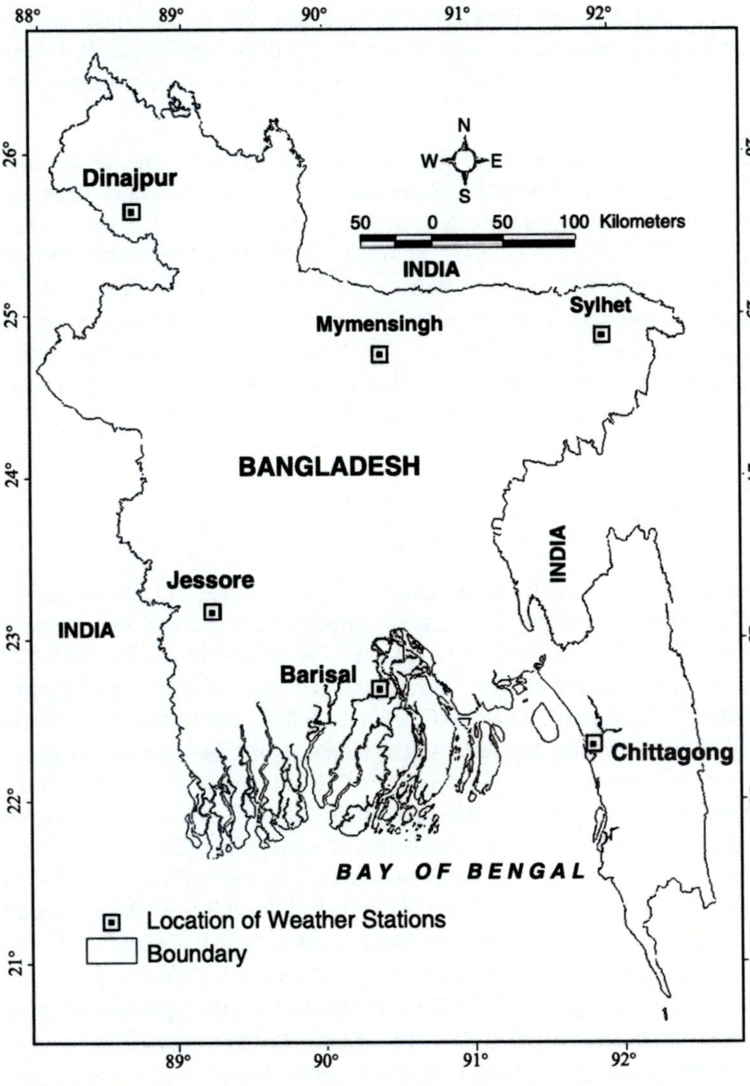

Fig. 28.1 Selected weather stations for this study

Soils Soil information is contained in a soil data file. The file contains information collected for the soil profile at a specified site along with supplemental information extracted from soil survey reports for a soil with similar taxonomic classification. Soil profiles in DSSAT v4.0 input file format were created from information compiled from various sources (Department of Soil Survey 1975) and were added to the SOILSOL file of DSSAT v4.0.

Crop Data The required crop data on development and growth characteristics and cultural practices for the selected varieties were collected from publications of

BRRI (2007) for rice, and BARI (1999) for wheat. The development and growth characteristics information was used to estimate genetic coefficients of the selected varieties. The cultural practices information was used for creating different simulation experiments.

Weather Data Twenty years of (1986–2005) daily weather data for the six sites (Fig. 28.1) was obtained from the Bangladesh Meteorological Department, which is stored at the BARC (Bangladesh Agricultural Research Council) Computer Center as ASCII files. This data set included maximum and minimum temperature (°C), rainfall (mm), and bright sunshine hour (hour) or cloud cover (Octas). Since the model requires solar radiation data (MJ m^{-2} day^{-1}), these were calculated based on the method described by Allen et al. (1998). All the weather data were then formatted as per model requirement into ICASA (International Consortium for Agricultural System Analysis) format.

28.2.3 Climate Change Scenarios

Seven scenarios were used: one baseline, and three climate change scenarios each for two intervals. The extent of change in climatic parameters or the 'delta values' of mean temperature and precipitation were generated by using MAGICC 2.4 (Model for the Assessment of Greenhouse-gas Induced Climate Change) and SCENGEN 2.4 meaning SCENario GENerator (CRU, 2000). The following model parameters for Emission Scenario IS92a (IPCC Third Assessment Report 2007) were used: According to the best guess of IPCC emission scenario IS92a, the concentrations of CO_2 for 2050 and 2070 are 515 ppmv and 575 ppmv respectively. These values were used for the simulation of crop yields. Delta values for mean temperature and precipitation were generated for the GCMs – GFDL–TR = Geophysical Fluid Dynamics Laboratory Transient, HadCM2 = Hadley Centre during 1995 and 1996 using the Second Version of the United Kingdom Meteorology Office's Unified Model UKTR = UK Met. Office/Hadley Centre Transient with respect to the base year of 1990. Default MAGGIC model parameters and high climate sensitivity were used for two time intervals, 2050 and 2070.

The range of change in mean temperature with respect to the base year (1990) was different for the three GCMs. GFDL–TR predicted delta values ranging from 1.1°C to 1.7°C for 2050, while it varied from 1.5°C to 2.3°C for 2070. Whereas, UKTR generated mean temperature increase was 1.5–2.1°C in 2050 and 1.2–2.7°C in 2070. According to HadCM2 generated delta value s, temperature would increase 1.3–2.9°C in 2050 and 1.7–4.0°C in 2070.

From Table 28.1 it can be seen that different GCMs predict different sets of values for rainfall increase (or decrease). Among the three GCMs, GFDL-TR predicted milder changes while HadCM2 suggested severe changes and the values increased with time. In case of rainfall, GFDL-TR and UKTR both predicted a decreasing tendency in the future during winter season (DJF). But, HadCM2 suggested that there would be higher precipitation (35.6% increase in 2050 and 48.9% increase in 2070) during the winter season (DJF).

Table 28.1 Delta values for mean temperature and rainfall generated for various Global Circulation Models for 2050 and 2070 using MAGICC 4.2/SCENGEN 4.2

	Global circulation models											
	GFDL–TR				HadCM2				UKTR			
	Mean temperature (°C)		Rainfall (%)		Mean temperature (°C)		Rainfall (%)		Mean temperature (°C)		Rainfall (%)	
Seasons	2050	2070	2050	2070	2050	2070	2050	2070	2050	2070	2050	2070
DJF	1.5	2.1	−1.4	−1.9	1.9	2.6	35.6	48.9	2.0	2.2	−5.2	−7.2
MAM	1.4	1.9	7.6	10.4	1.8	2.5	−1.8	−2.4	1.7	2.1	6.8	9.3
JJA	1.3	1.8	18.5	25.5	1.9	2.7	−0.2	−2.2	1.7	1.7	24.0	33.0
SON	1.3	1.8	7.4	10.3	2.1	2.9	−8.1	−5.7	1.7	1.9	−0.3	−0.4

GFDL–TR, Geophysical Fluid Dynamics Laboratory Transient; HadCM2, Hadley Centre during 1995 and 1996 using the Second Version of the United Kingdom Meteorology Office's Unified Model; UKTR, UK Met. Office/Hadley Centre Transient; DJF, December–January–February; MAM, March–April–May; JJA, June–July–August; SON, September–October–November

Table 28.2 Projected population from 1995 to 2050

Year	Population[a] (million)
1995	120.611
2000	131.155
2005	153.281
2010	166.574
2015	179.995
2020	193.129
2025	205.689
2030	217.429
2035	228.098
2040	237.575
2045	245.697
2050	252.306

[a]The World Bank HNPStats – Population Projection Tables by Country and Group.htm, URL: http://go.worldbank.org/072F5QBOC0)

28.2.4 Demography and Land Use

28.2.4.1 Population

During the period from 1961 (50.84 million) to 2001 (124.36 million) five population censuses were conducted in Bangladesh (BBS 2007)According to these census results, the population has almost doubled in less than 30 years since 1961 and is likely to be 214.6 million in 2050. Currently, the estimated population stands at over 153 million. The projections made by the World Bank (2008) show that the population would be about 205.9 million and 252.3 million under Medium Fertility Scenario and Standard Mortality Scenario in 2025 and 2050 (Table 28.2). The input data used for the projections include a base year (mid-2005) population estimate by age and sex, and base period (2005–2010) estimates of mortality, fertility, and migration (The World Bank HNPStats-Population Projection Tables by Country and Group.htm). To keep pace with the population growth and shrinking land resource base, the food production needs to be doubled by the year 2020. Sustainable growth and development of the agricultural sector is the important issue to the government for meeting the future challenge of increased production.

28.2.5 Land Use Under Major Cereals

28.2.5.1 Food Grain Production

Cereal production almost doubled in the last 25 years and the production gains were achieved mainly due to yield increases. Total production of food grains increased from 15.025 M t (million metric tons) in 1980–1981 to 27.805 M t in 2005–2006. Rice production increased by 91.10% in 2005–2006 from that of 1980–1981 (13.883 M t). On the other hand, wheat production decreased from 1.092 M t in 1980–1981 to 0.735

Table 28.3 Area and production of major cereals in Bangladesh

	2004–2005		2005–2006	
Crops	Area (M ha)	Production (M t)	Area (M ha)	Production (M t)
Local	0.574	0.638	0.517	0.664
HYV	0.451	0.862	0.517	1.081
Total Aus	1.025	1.5	1.034	1.745
Local	1.88	2.668	1.73	2.714
HYV	2.906	6.693	3.193	7.505
Total Aman	4.786	9.361	4.923	10.219
B. Aman	0.494	0.458	0.505	0.591
Total Aman	5.28	9.819	5.428	10.81
Local	0.188	0.391	0.174	0.347
HYV	3.876	13.446	3.892	13.628
Total Boro	4.064	13.837	4.066	13.975
Total rice	10.369	25.156	10.528	26.53
Wheat	0.558	0.976	0.479	0.735
Maize	0.067	0.356	0.098	0.523
Other cereals	0.038	0.018	0.027	0.017
Total cereals	11.032	26.506	11.132	27.805

Source: Bangladesh Bureau of Statistics (BBS) quoted from the Handbook of Agricultural Statistics, Ministry of Agriculture (MoA 2007) Government of Bangladesh

M t in 2005–2006 (Table 28.3). However, there was no significant change in the cultivated area. Area increased by only 2.15% for rice, while the area under wheat was reduced by about 19%. Notable increase in the area as well as the production of maize was made during the period from 1980–1981 to 2005–2006.

28.2.5.2 Food Grain Requirement

The availability of food grain in Bangladesh showed substantial improvement during the past three to 4 years. The consumption of rice, wheat, spices, edible oil, fruits, potato and fish has increased considerably, but other food items like vegetables, meat and milk consumption is still very low. The requirement of most of the energy comes from 454 g day^{-1} head^{-1} of rice and wheat. The population was 120.61 million in 1995 and 153.28 million in 2005 then the food grain requirements were 19.99 M t and 28.342 M t, respectively (Table 28.4).

28.3 Results and Discussion

28.3.1 Impact of Climate Change on Crop Yields

28.3.1.1 Rice

Under the Baseline scenario, Aus rice yields varied from a minimum of 3.08 t ha^{-1} in Jessore to a maximum of 5.08 t ha^{-1} in Dinajpur. Similarly, minimum (4.82 t ha^{-1}) and

Table 28.4 Food grain and production requirements from 1995 to 2050

Year	Food grain requirement[a] (M t)	Production requirement[b] (M t)
1995	19.986	22.301
2000	21.734	24.250
2005	25.400	28.342
2010	27.603	30.799
2015	29.827	33.281
2020	32.003	35.709
2025	34.085	38.032
2030	36.030	40.202
2035	37.798	42.175
2040	39.369	43.927
2045	40.714	45.429
2050	41.810	46.651

[a]Food grain requirement is calculated @ 454 g day^{-1} $head^{-1}$
[b]Production requirement is calculated by adding 11.58% for seed, feed and wastage with food grain requirement as per the study on 'Seed, Feed and Post Harvest losses' Ministry of Food (MOF), Government of Bangladesh

maximum (6.01 t ha^{-1}) of Aman rice yields were obtained for Barisal and Dinajpur. Maximum (6.58 t ha^{-1}) and minimum (4.66 t ha^{-1}) Boro rice yields were recorded in Sylhet and Jessore, respectively. The percentages of yield difference as compared to baseline are presented for Aus, Aman, and Boro rice at the six study sites in Table 28.5.

Aus Rice With different climate change scenarios it was evident that in the case of Aus rice under GFDL50 scenario compared to Baseline 3.8% and 15.7% yield increases were noted in Jessore and Barisal, respectively. Under GFDL70 scenario yield increases were recorded for all six locations, which varied from 3.2% in Dinajpur to 14.8% in Jessore. It is evident from Table 28.5 that Aus rice under HADC50 and HADC70 scenarios yield reduction occurred in all locations and the percentages varied from 13.5 in Jessore to 43.6 in Dinajpur. In most cases the yield reductions were around 25%. For Aus grown under UKTR50 scenario, the yield differences compared to baseline varied between +3.4% in Jessore and –15.4% in Mymensingh and Chittagong.

Aman Rice For Aman rice under GFDL50 scenario, the percent change in yields over Baseline varied between –4.9 and 1.4 in Dinajpur and Barisal, respectively. Similarly, under GFDL70 scenario the percent change in yields over Baseline varied between –5.2 and 0.1 in Dinajpur and Barisal, respectively. In the case of HADC50 and HADC70 scenarios, yield reductions were more for the latter scenario than the former. For HADC50 scenario the yield differences were between –13.5% and +1.9% in Jessore and Barisal, respectively. Yield reductions were observed in all locations under HADC50 scenario, maximum for Dinajpur (–7.7%) and minimum (–1.4%) in Jessore. The effect of UKTR50 and UKTR70 on Aman rice yields was negative in all six locations. The effect of the latter scenario was harsher than the former. Yield reductions ranged from 3.2% in Jessore to 14.2% in Dinajpur.

Table 28.5 Rice and wheat yields under different climate change scenarios

	Sylhet			Mymensingh			Dinajpur		
Simulation	Aus	Aman	Boro	Aus	Aman	Boro	Aus	Aman	Boro
Baseline (t ha^{-1})	498.9	5,439.5	6,581.5	4,169.0	5,425.9	5,664.8	5,080.3	6,006.8	5,771.8
Percent change over baseline									
GFDL50	−6.1	−1.7	6.1	−7.8	0.0	7.3	−7.0	−4.9	1.1
GFDL70	6.7	−1.7	10.7	6.1	−2.3	15.5	3.2	−5.2	9.7
HADC50	−25.7	−1.0	3.6	−26.5	−0.9	7.5	−43.6	−13.5	−5.7
HADC70	−24.0	−3.8	5.8	−25.8	−4.6	11.0	−25.8	−7.7	10.0
UKTR50	−13.4	−5.2	2.2	−15.4	−6.4	5.7	−14.1	−9.2	1.6
UKTR70	−3.9	−9.8	4.4	−6.7	−11.7	11.2	−6.3	−14.2	9.3
	Jessore			Chittagong			Barisal		
Simulation	Aus	Aman	Boro	Aus	Aman	Boro	Aus	Aman	Boro
Baseline (t ha^{-1})	3,077.3	5,115.2	4,658.7	4,169.0	5,968.9	6,059.0	3,450.6	4,819.2	6,019.3
Percent change over baseline									
GFDL50	15.7	3.3	2.5	−7.8	−0.2	0.2	3.8	1.4	3.0
GFDL70	14.8	0.6	9.3	6.1	−0.4	4.7	9.9	0.1	7.3
HADC50	−13.5	1.9	1.3	−26.5	−1.6	−0.9	−19.0	0.1	1.3
HADC70	−24.9	−1.4	6.6	−25.8	−2.6	1.0	−24.1	−1.7	3.3
UKTR50	3.4	−3.2	−0.6	−15.4	−4.6	−3.6	−5.7	−3.4	−0.6
UKTR70	−2.4	−9.0	4.7	−6.7	−7.9	−1.7	−4.3	−7.3	1.3

Table 28.6 Wheat yields under different climate change scenarios

Simulation	Dinajpur	Mymensingh	Jessore
Baseline (t ha^{-1})	3.01	2.67	2.72
Percent change over baseline			
GFDL50	20.4	28.6	29.0
GFDL70	−18.4	−15.8	−17.3
HADC50	11.2	23.8	23.0
HADC70	−23.1	−23.7	−25.6
UKTR50	17.0	24.8	24.0
UKTR70	−22.8	−21.5	−24.0

Boro Rice In general, it was noted from Table 28.5 that there is a beneficial effect of climate change in the case of Boro rice. Significant yield increase of Boro were noted for GFDL70 scenario, irrespective of locations. Under this scenario, highest yield increase was recorded in Mymensingh (15.5%), followed by Sylhet (10.7%). However, only in a few cases was a negative effect seen; –5.7% in Dinajpur for HADC50 scenario, in Chittagong –3.6 % for UKTR50, –1.7% for UKTR70 and –0.9% HADC50; and –0.6% for UKTR50 in Jessore and Barisal.

Wheat Under the Baseline scenario wheat yields varied from a minimum of 2.67 t ha^{-1} in Mymensingh to a maximum of 3.01 t ha^{-1} in Dinajpur. Simulation results in Table 28.6 show that all the scenarios for 2050 had a positive effect on wheat yield. Highest relative yield increases were around 29.0% in Jessore and Mymensingh and 20.4% in Dinajpur. Conversely, all the scenarios for 2070 had a negative effect on wheat yield. Highest relative yield decreases were 25.6% in Jessore and 23.0% in Dinajpur and Mymensingh. The reductions ranged from 15.8% to 25.6%.

28.3.2 Effect of Climate Change on Aggregated Rice and Wheat Production

Production of rice and wheat for the entire country was aggregated by crop and by season (Table 28.7). The observations are discussed in the following sections.

Aus Rice The average aggregated production of Aus rice in Bangladesh was 1.72 M t during 2005–2006. Productions were reduced by 1.5–25.8% under different climatic scenarios. Only with GFDL70 scenario was a 7.8% increase in production was recorded.

Aman Rice The aggregated production of Aman rice for 2005–2006 was 10.81 M t. Simulation results show that under all scenarios, production was lowered from 0.4% for GFDL50 scenario to 10.0% for UKTR70.

Boro Rice The baseline production of Boro rice for 2005–2006 was 13.98 M t. Under all climate change scenarios, production increased in the range of 1.2% for HADC50 and 9.5% for GFDL70 scenarios.

Table 28.7 Aggregated rice and wheat production of Bangladesh under different climate change scenarios

Simulation	Aus (M t)	Change (%)	Aman (M t)	Change (%)	Boro (M t)	Change (%)	Wheat (M t)	Change (%)	Total (M t)
Baseline	1.75	–	10.81	–	13.98	–	0.74	–	27.27
GFDL50	1.72	–1.5	10.77	–0.4	14.45	3.4	0.93	25.98	27.86
GFDL70	1.88	7.8	10.65	–1.5	15.31	9.5	0.61	–17.19	28.44
HADC50	1.30	–25.8	10.54	–2.5	14.14	1.2	0.88	19.35	26.86
HADC70	1.31	–25.1	10.42	–3.7	14.85	6.3	0.56	–24.14	27.13
UKTR50	1.57	–10.1	10.24	–5.3	14.08	0.8	0.90	21.94	26.78
UKTR70	1.66	–5.1	9.73	–10.0	14.65	4.9	0.57	–22.79	26.61

Wheat The aggregated production of wheat for 2005–2006 was 0.735 M t. It is interesting to note that with all the scenarios for 2050, the wheat production increased in the tune of 19.35–25.98%. While, under scenarios for 2070 the production decrease ranged from 17.19% for GFDL70 to 24.14% for HADC70 scenarios.

28.3.3 Effect of Climate Change on Crop Growing Season

The growing period for Aus, Aman, and Boro rice and wheat under Baseline conditions were 130, 145, and 165 and 110 days, respectively. In general, all the climate change scenarios enhanced the crop growing season by 2–10 days compared to baseline. Under all scenarios for 2070, the growing season was reduced by 8–10 days and for 2050 the reduction was 5–7 days.

Yield reductions of rice and wheat are attributed to the effect of high temperature during Aus and Aman seasons and dry spells. During Boro season the temperature increase under all scenarios varied between 1.5°C for GFDL50 scenario and 2.6°C for HADC70. These temperature increases did not exceed the threshold value of 35.0°C for rice. Wheat is sensitive to high temperature, but increase of wheat yields was observed. This may be attributed to the fact that when the temperature regime is not detrimental and wheat being a C3 crop, it is likely that at atmospheric CO_2 concentrations of 515 and 575 ppmv CO_2 fertilization might have occurred.

The existing literature suggests similar findings: Reilly et al. (1996) reported that in Asia, including Bangladesh, under different GCM scenarios and sites, the yields of cereals varied with CO_2 effect. The range of variations for rice, maize, and wheat were −22 to +28, −65 to −10, and −61 to +67, respectively. Squire and Unsworth (1988) indicated that determinate crops with discrete elements to their life cycle develop faster in higher temperatures, and so the stage of seed filling is shortened, limiting the benefits of elevated CO_2. When CO_2 concentration doubled, but daily weather data from a typical year were used, the potential wheat yield was 27% larger than in a control run at 340 m mol/mol CO_2 (11.5 and 9.0 t ha^{-1}, respectively). The date of maturity was unchanged. When daily temperatures were increased by 3°C and CO_2 was doubled, more rapid development of the crop shortened the growing season, and the potential grain yield (10.4 t ha^{-1}) was only 15% larger than in the control, but maturity occurred 30 days earlier.

Simulation studies conducted by Karim et al. (1996, 1999) showed that rice and wheat production in Bangladesh decreased due to rise in temperature. Wheat was more susceptible than rice. It was found that a 4°C increase in temperature had severe impact on food grain production, especially on wheat production. On the other hand, CO_2 fertilization would facilitate the food grain production. A rise in temperature will cause significant decrease in production, some 28% and 68% for rice and wheat, respectively. The apparent increase in yield of Boro and other crops might be constrained by moisture stress. A 60% moisture stress on top of other effects might cause as high as 32% decline in Boro yield instead of having an overall 20% net increase.

28.3.4 Food Security

Bangladesh is highly vulnerable to floods and droughts. Droughts of different intensities occur, especially in the dried regions of the country, due to uneven distribution of rainfall and a higher evapotranspiration rate accelerated by increased temperature, low humidity, and high wind speed. Wet season drought severely affects transplanted Aman rice, reducing the annual production by about 1.5 M t of rice. On the other hand, dry season drought affects the production of wheat, potato and Aus paddy (Karim et al. 1990). Drought related vulnerability can be minimized by supplemental irrigation. Attempts should be made to develop drought resistant varieties of crops for drought prone areas of the country.

One of the obvious effects of climate change may be changes in pests and weeds in agriculture. This is because this change will likely affect the pest–weed–host relationship in one (or more) of three ways: by affecting the pest–weed population; by affecting the host population; and by affecting the pest–weed–host interaction. These would make food security more vulnerable.

If the production of major cereals is hampered due to climate change then food security would be in jeopardy, as out of a recommended 2122 calories day^{-1} person^{-1}, the most part is met from intake of 454 g of food grains (rice and wheat). Most of the estimates show that a huge amount of additional food grain will be required to feed the growing population of the country. Hossain and Shahabuddin, (1999) estimated that at a 4.5% national income growth per annum, cereal requirement will be 30.57 and 35.39 M t in 2010 and 2020, respectively. At a 6.5% national income growth per annum cereal requirement in 2010 and 2020 would be 32.82 and 35.14 M t, respectively.

World Bank's estimate based on the assumption that NRR (net reproductive rate) would be lower from present level of about 1.2 to unity by 2025–2030, indicates a population of 131 million, 167 million and 217 million during 2000, 2010 and 2030 respectively.

Currently, the estimated population stands at over 153 million and by the year 2030 and 2050 it is projected to be 217 and 252 million, respectively (World Bank 2008 To feed this huge population, the food grain requirement of the country would be 36.03 M t in the year 2030 and 41.81 M t in the year 2050.

Therefore, to become self-sufficient in food grain production in the years 2030 and 2050, compared to the production of 2005–2006 (27.805 M t), an additional 8.22 and 14.00 M t respectively would be required. Besides, as per estimation by the Ministry of Food of Bangladesh about 11.58% additional food grain would be required for seed, feed and wastage. Therefore, the production requirement would be 40.2 M t in 2030 and 46.65 M t in 2050. Although food grain production is an important part of food security, it is not the only strategy. With the income growth the preference for food is shifted from carbohydates to more protein, vegetables and fruits. This will moreover reduce the pressure on cereals in the future.

28.4 Conclusion

Under the business as usual scenario, with the available technologies Bangladesh has the potential to meet demand for cereals up to 2025 and even beyond. However, it is difficult to predict the consequences of climate change on agriculture as a whole. Cereal demand will be difficult to meet as the population is very likely to be more than 252 million by the year 2050. However, under changed scenarios, new technologies need to be developed to combat climate change and sea level rise, coupled with the great pressure of population increase.

To predict the consequences of climate change on agriculture, limited efforts have been made to generate climate change scenarios using various Global Climate Models (GCMs) and DSSAT (Decision Support System for Agrotechnology Transfer) based Crop Simulation models of International Consortium for Agricultural Systems Applications (ICASA). Due to unavailability of high resolution GCMs for Bangladesh as well as for the South Asia region, reliable scenarios cannot be generated. On the other hand, for running different crop models, various required minimum datasets for crops, management, and climatic parameters are not collected adequately or are incomplete. Another aspect of these efforts is validation of the output from these models, which is very important. To make reliable predictions the model outputs need to be thoroughly substantiated with evidence. National facility should be developed to conduct phytotron (plant growth chamber) experiments under simulated climate change scenarios.

Since climate change and sea level rise will have adverse effect on the ecosystems, many important germplasms are likely to be lost. Research should be initiated to preserve the vulnerable species of crop, livestock, etc. Genetic tolerance to adverse environment/improvement must be achieved in order to have sustainable food security and development. Research should be undertaken for screening of genotypes for withstanding adverse edaphic, environmental, and climatic conditions. As a consequence of climate change, drier regions would become more arid in the winter and the possibility of growing rainfed crops would be diminished. Therefore, low moisture consuming crops/cropping patterns need to be developed and disseminated for wider adoption by the farmers.

References

Aggarwal PK, Bandyopadhyay SK, Pathak H, Kalra N, Chander S, Kumar S (2000) Analysis of yield trends of the rice-wheat system in north-western India. Outlook Agr 29:259–268

Allen RG, Pereira LS, Reas D, Smith M (1998) Crop evapotranspiration: guidelines for computing crop water requirements. FAO Irrigation and Drainage Paper 56. Food and Agriculture Organization of the United Nations, Viale delle Terme di Caracalla, 00100 Rome, Italy

BARI (Bangladesh Agricultural Research Institute) (1999) "Krishi Projukti Hatboi" (Handbook on agro-technology) in Bangla. In: Rashid MM, Ali SME, Wahab MA, Amin MS, Quyuum MA, Alam MS (eds) Bangladesh Agricultural Research Institute, Joydebpur, Gazipur-1701

BBS (Bangladesh Bureau of Statistics) (2007) Statistical pocket book, Bangladesh 2006. Statistics Division, Ministry of Planning. Government of the People's Republic of Bangladesh

BRRI (Bangladesh Rice Research Institute) (2007) "Adhunik Dhaner Chash" (Modern rice cultivation) In: Mustafi BAA, Salam MA, Mondal M, Quashem MA (eds) Bangladesh Rice Research Institute, Joydebpur, Gazipur-1701

CRU (Climatic Research Unit) (2000). MAGICC (Model for the Assessment of Greenhouse-gas Induced Climate Change)/SCENGEN (SCENario GENerator), Version 2.4. University of East Anglia, Norwich, UK

Cruz RV, Harasawa H, Lal M, Wu S, Anokhin Y, Punsalmaa B, Honda Y, Jafari M, Li C, Huu Ninh N (2007) Asia Climate Change 2007: impacts, adaptation and vulnerability. In: Parry ML, Canziani OF, Palutikof JP, van der Linden PJ, Hanson CE (eds) Contribution of Working Group II to the Fourth Assessment Report of the Intergovernmental Panel on Climate Change. Cambridge University Press, Cambridge, UK, pp 469–506

Department of Soil Survey (1975) Reconnaissance soil survey report. Government of the People's Republic of Bangladesh

FAO (Food and Agriculture Organization of the United Nations) (2008) FAOSTAT. FAO statistics division. (http://faostat.fao.org/site/567/DesktopDefault.aspx?PageID=567), accessed 01 July 2008

Fischer GM, Shah M, vanVelthuizen H (2002) Climate change and agricultural vulnerability. Preprints. World Summit on Sustainable Development, Johannesburg, p 160

Hoogenboom G, Jones JW, Porter CH, Wilkens PW, Boote KJ, Batchelor WD, Hunt LA, Tsuji G (eds) (2003) Decision support system for agrotechnology transfer version 4.0. vol 1: Overview. University of Hawaii, Honolulu, HI

Hossain M, Shahabuddin Q (1999) Sustainable agricultural development in Bangladesh: challenges and issues. In: Bhuya SI, Karim ANMR (eds) Increasing rice production in Bangladesh: challenges and strategies. International Rice Research Institute, Los Baños, Philippines

IPCC (2007) Summary for policymakers. In: Solomon S, Qin D, Manning M, Chen Z, Marquis M, Averyt KB, Tignor M, Miller HL (eds) Climate Change (2007) The physical science basis. Contribution of Working Group I to the Fourth Assessment Report of the Intergovernmental Panel on Climate Change. Cambridge University Press, Cambridge

Karim Z, Ibrahim AM, Iqbal A, Ahmed M (1990) Drought in Bangladesh agriculture and irrigation schedules for major crops. Bangladesh Agricultural Research Council, Soils Publication No: 34, p 11

Karim Z, Hussain SG, Ahmed M (1996) Assessing impacts of climate variations on foodgrains production in Bangladesh. J Water Air Soil Pollut 92:53–62

Karim Z, Hussain SG, Ahmed AU (1999) Climate change vulnerability of crop agriculture. In: Huq S, Karim Z, Asaduzzaman M, Mahtab F (eds) Vulnerability and adaptation to climate change for Bangladesh. Kluwer, Dordrecht, The Netherlands, pp 29–53

Khan TMA, Singh OP, Rahman MDS (2000) Recent sea level and sea surface temperature trends along the Bangladesh coast in relation to the frequency of intense cyclones. Marine Geodesy 23:103–116

Mirza MQ (2002) Global warming and changes in the probability of occurrence of floods in Bangladesh and implications. Global Environ Change 12:127–138

Mirza MQ, Dixit A (1997) Climate change and water management in the GBM Basins. Water Nepal 5:71–100

MoA (Ministry of Agriculture) (2007) Handbook of agricultural statistics. Agro-Economic Research Branch, Ministry of Agriculture, Government of the People's Republic of Bangladesh

Peng S, Huang J, Sheehy JE, Laza RE, Visperas RM, Zhong X, Centeno GS, Khush GS, Cassman KG (2004) Rice yields decline with higher night temperature from global warming. P Natl Acad Sci USA 101:9971–9975

Reilly J, Baethgen W, Chege RE, van de Geijn SC, Erda L, Iglesias A, Kenny G, Patterson D, Rogasik J, Rötter R, Rosenzweig C, Sombroek W, Westbrook J (1996) Agriculture in a changing climate: impacts and adaptation. In: Changing climate: Impacts and response strategies, Report of Working Group II of the Intergovernmental Panel on Climate Change. Chapter 13. Cambridge University Press, Cambridge, UK

Ritchie JT, Singh U, Godwin DC, Bowen WT (1998) Cereal growth, development and yield. In: Tsuji GY, Hoogenboom G, Thornton PK (eds) Understanding options for agricultural production. Kluwer, Dordrecht, pp 79–98

Squire GJ, Unsworth MH (1988) Effects of CO_2 and climate change on agriculture. University of Nottingham, England, Report to U.K. Department of the Environment

Tao F, Yokozawa M, Hayashi Y, Lin E (2003) Changes in agricultural water demands and soil moisture in China over the last half-century and their effects on agricultural production. Agr Forest Meteorol 118:251–261

Tao F, Yokozawa M, Zhang Z, Hayashi Y, Grassl H, Fu C (2004) Variability in climatology and agricultural production in China in association with the East Asia summer monsoon and El Niño South Oscillation. Climate Res 28:23–30

Wijeratne MA (1996) Vulnerability of Sri Lanka tea production to global climate change. Water Air Soil Poll 92:87–94

World Bank (2008) The World Bank HNPStats – population projection tables by country and group.htm. (http://go.worldbank.org/072F5QBOC0), accessed 01 July 2008

Chapter 29
Climatic Variations: Farming Systems and Livelihoods in the High Barind Tract and Coastal Areas of Bangladesh

Md. Badirul Islam, Md. Yusuf Ali, Mohammad Amin, and Sheikh Mostafa Zaman

Contents

Abstract Geographically, Bangladesh is highly vulnerable to climate change. In particular, impacts of climate variability on agriculture and consequences on other sectors are already evident in the drought prone High Barind Tract and coastal regions. The agriculture and fisheries sectors in the High Barind Tract (HBT) and southern coastal region are very likely to face significant yield reduction in the future due to climate change. Global circulation model results revealed that higher

M.B. Islam (✉) and M.Y. Ali
On-Farm Research Division, Bangladesh Agricultural Research Institute, Gazipur 1701, Bangladesh
e-mail: ofrdjoy@yahoo.com

M. Amin
On-Farm Research Division, Bangladesh Agricultural Research Institute, Noakhali 3800, Bangladesh
e-mail: ofrdnoa@gmail.com

S.M. Zaman
On-Farm Research Division, Bangladesh Agricultural Research Institute, Pabna 6600, Bangladesh
e-mail: sm_zaman@yahoo.com

R. Lal et al. (eds.), *Climate Change and Food Security in South Asia*,
DOI 10.1007/978-90-481-9516-9_29, © Springer Science+Business Media B.V. 2011

temperature and water stress due to heat results an in decline in vegetation and agricultural production, especially in the drought affected HBT. While the coastal region would suffer from increased degradation of land, salinity intrusion, river bank erosion, siltation, water logging, tidal surge and floods. Drought delays the timely planting of T.Aman rice, the main crop of HBT, while drought in September and October drastically reduces the yield of said crop, and the chance of sowing/planting of different rainfed rabi crops markedly decreases. Nationwide rice production losses due to drought in 1982 were about 50% more than losses due to flood in the same year, particularly in the HBT, >80% T.Aman rice production was lost. Moreover, the ground water table of HBT is continuously going down in the dry season due to over exploitation by deep-tube well. The 1997 drought caused a reduction of around1 million tons of food grain, of which about 0.6 million tons was T.Aman rice, entailing a loss of around $ 500 million. A cyclone in 1970 resulted in 300,000 deaths, and another in 1991 led to the loss of 138,000 lives. These effects are likely to be exacerbated by climate change as peak intensity of cyclones is projected to increase by 5–10%, and precipitation rates may increase by 20–30%. The strength of SIDR and economic losses was caused by the major hurricane in 2007 fit into this trend. Even before the new impacts of SIDR, about 1.2 million hectares of arable land were already affected by varying degrees of soil salinity, tidal flooding during wet season, direct inundation by saline water and upward and lateral movement of saline ground water during the dry season. Inundation of brackish water for shrimp farming is a key cause for secondary salinization of coastal lands. The severity of salinity problem has increased over the years and is expected to increase in the future due to rise of sea level. Even in non-cyclonic situations, higher mean sea-levels are going to increase the problem of coastal flooding and salinization, causing significant pressure on livelihood activities. Thus, climate change effect has a large negative impact on the farming systems and livelihoods of rural people of HBT and coastal area and on the overall economy of Bangladesh.

Keywords Salinity • Coastal belt • Cropping patterns • Drought mitigation • Adaptation measures • Livelihood

Abbreviations

B.Aus	Broadcasted Aus
BARI	Bangladesh Agricultural Research Institute
BMDA	Barind Multipurpose Development Authority
D.Aus	Dibbling Aus
DAE	Department of Agricultural Extension
GO	Government organization
HBT	High barind tract
HL	Highland

HYV	High yielding variety
ICM	Integrated Crop Management
IPM	Integrated Pest Management
IPNS	Integrated Plant Nutrient System
MHL	Medium highland
NGO	Non-government organization
OFRD	On-Farm Research Division
SRDI	Soil Resources Development Institute
T.Aman Rice	Transplanted Aman Rice
T.Aus rice	Transplanted Aus rice

29.1 Introduction

Climatic variations have a profound effect on agriculture and livelihoods of rural communities in Bangladesh. The effect of climatic variation is more pronounced in the drought prone High Barind Tract (HBT) and coastal areas of Bangladesh. Among the drought affected areas, HBT is one of the largest areas (Brammer 1999). In Bangladesh, the HBT of north-west Rajshahi division is different from other parts of the country due to its undulating topography having compact and low fertile soils. The region experiences high temperatures with limited soil moisture storage along with low and erratic rainfall (1,075 ± 325 mm) (BMDA 2006). Moreover, no river/large water bodies are present within the HBT. Also, vegetation is scanty compared to other parts of the country. Moisture holding capacity of HBT soil is poor due to low organic matter content and low infiltration of water (Ali 2000; Ali et al. 2007a). These situations make the area drought-prone, along with poor crop productivity. Moreover, recent reports suggest that ground water level of Chapai Nawabganj district (within HBT) is rapidly falling due to over exploitation through deep-tube well (Selvaraju et al. 2006). About 55% area of HBT is rainfed (Ali 1998), and the deep-tube wells have far less command area than expected. Also, irrigation cost is higher than that of other areas. T.Aman rice is the major crop which suffered regularly due to early or late drought and planting of post-rainy crop. Thus, the areas, livelihood is often vulnerable to climate change, particularly to drought (Ali et al. 2007b). The 1997 drought caused a reduction of around 1 million tons of food grain, of which about 0.6 million tons was T.Aman rice, entailing a loss of around $ 500 million. A cyclone in 1970 resulted in 300,000 deaths, and another in 1991 led to loss of 138,000 lives. These effects are likely to be exacerbated by climate change as the peak intensity of cyclones is projected to increase by 5–10%, and precipitation rates may increase by 20–30%. The strength of SIDR and economic losses caused by the major hurricane in 2007 fits into this trend. Even before the new impacts of SIDR, about 1.2 million ha of arable land were already affected by varying degrees of soil salinity, tidal flooding during wet season, direct inundation by saline water and upward, and lateral movement of saline ground water during dry season.

Salinity causes unfavorable environment and hydrological situation, restricting the normal crop production throughout the year (Amin et al. 2008), and few crops/cultivars can survive in severe saline soils. In Bangladesh, more than 30% of net cultivable area is in the coastal belt, of which about 833,000 ha consist of saline soils (Karim and Iqbal 2001), and have already increased up to 1.2 million ha. Salinity gradually increased from south-eastern coastal belt to south-western coastal belt, mainly due to very low flow of upstream water during November to May because of commissioning of the disastrous Farraka Barrage on the Ganges river in the West Bengal of India since 1974. The coastal area of Bangladesh has less cropping intensity due to rise of salinity during dry season and other constraints such as river erosion, flood, poor drainage condition, heavy clay basin, scarcity of fresh water, low soil fertility, exposure to cyclones, and risk of early and late rainfall. In Khulna region (secondary salinity), crop lands are converted to shrimp production (gher), thus rendering the non-saline crop lands to saline soils. As a result, lands are becoming unsuitable for rice or other crop production. In general, non-saline lands are becoming saline due to flooding by saline water during storm-surge or by breaking of a coastal embankment, drought in the pre-monsoon and post-monsoon season, and saline water used for irrigating crops (Brammer 1999). With the intrusion of highly saline water, the production of crops and fruit trees are seriously affected, and as a result the livelihood and economy of the farmers of those areas are seriously hampered. Moreover, Global Circulation Model results predicted that due to rise of temperature, drought, salinity, cyclones and storm surge in Bangladesh could increase at an alarming rate in the near future. In this context, trials were conducted by OFRD, BARI at coastal and drought areas to overcome the adverse effects of climate change. Some findings of three vulnerable areas in Bangladesh are furnished in the report, along with some primary data and information on climatic effect on agriculture and livelihoods.

29.2 Methodology

Crops and cropping pattern with in drought prone (HBT) and coastal areas (Noakhali, Patuakhali and Khulna) were selected for data collection. A survey was conducted at HBT, Godagari, Rajshahi and for coastal belt of six locations viz. Bagerhat, Dacope of Khulna, Noakhali sadar, Subarnochar, Noakhali, Dumki and Kalapara of Patuakhali to evaluate the effect of climatic variations on the farming systems and livelihoods of the people during May–July 2008. Primary information was collected from elderly farmers of above 50 years of age who could compare between the present and past agricultural systems, livelihoods and climatic change through a pre-designed and pre-tested questionnaire. Collected data were verified though Focal Group Discussion and discussion with key informants. A total of 200 samples were randomly collected from seven Upazilas. Among them for the drought prone HBT, 50 samples were collected from eight villages of Godagari Upazila of Rajshahi district. For coastal and saline areas, 150 samples were collected from six Upazilas viz. Bagerhat, Dacope of Khulna, Noakhali sadar, Subarnochar, Noakhali, Dumki and Kalapara of Patuakhali. For each above mentioned Upazilas, 25 samples were collected at random. Meteorological information from Bangladesh Meteorological Department, Dhaka, data also are presented in graphical forms. Necessary secondary information was also collected from DAE, BARI, SRDI & BMDA reports and through literature review.

29.3 Results and Discussion

29.3.1 High Barind Tract (HBT), Rajshahi

In High Barind Tract, the predominant cropping patterns were Fallow-Fallow-T.Aman rice, B.Aus rice-T.Aman rice & Wheat-T.Aman rice for highland and Fallow-Fallow-T.Aman rice, Chickpea-T.Aman rice, Chickpea + Linseed-T.Aman rice, Mustard/Linseed-T.Aman rice for medium highland under rainfed condition in the 1980s. During that time, the local T.Aman rice varieties, namely Raghusail, Dadkhani, Magurshail, Batraj, Chengul, Chita, etc, and high yielding varieties of T.Aman rice, namely BR10, BR11 and BR4, were cultivated under rainfed condition. In 1985, most of the crops produced the lower yield with local varieties. Due to erratic rainfall, poor management practices, lack of irrigation facilities and use of low yield crop varieties. But in 2007, most of the crops produced higher yield due to use of high yielding varieties, development of irrigation system and use of improved management practices. In 1985, cropping intensity of HBT was 117% and 166% in 1992–1993, whereas present cropping intensity of this area is 216% (BMDA 2006).

Experimental results showed that persistent drought prevailed during the flowering stage of T.Aman rice. As a result, an 11–34% yield loss occurred for local T.Aman rice and a 43–50% loss for recent varieties. Supplemental irrigation by collection run off in a mini pond having an area of 300 m^3 (10 × 10 × 3 m) can save the yield loss for an area of one hectare of land (Annonymous 1991). In recent years, the irrigation facilities have further expanded, which creates the opportunity to cultivate high yielding varieties of crops with modern methods of cultivation, which resulted in an increase in yields. Under the changing circumstances, different *rabi* crops like wheat, boro rice, tomato, potato, mustard, hybrid maize, etc, have been also cultivated under irrigated condition.

Drought is the major climatic hazard in the HBT. About 82% farmers responded crop failure (particularly T.Aman rice) due to low rainfall and acute scarcity of water. However, they have not identified that intensity of drought has increased or decreased. Eighty-two percent of the farmers reported that water was held in perennial ponds year round in the 1980s. However, 66% responded that it has reduced due to drought, low/inadequate rainfall, siltation and increased use of pond water for crop production (Appendixes 29.1–29.3). Seventy-two percent of the farmers had the opinion that more ponds should be excavated, and 16% of them felt the need of pond re-excavation (Appendix 29.4). Moreover, cropping pattern has changed mainly due to development of irrigation facility along with availability of high yielding varieties and modern technologies (Appendixes 29.5 and 29.6). Livelihood of the farmers has improved through the intervention of modern

Appendix 29.1 Present and past water holding duration of pond in the High Barind Tract, Godagari, Rajshahi

Present duration of water holding	% farmer responded	Past (10 years before) duration of water holding	% farmer responded
July–November	8	July–November	14
July–January	10	July–January	20
Round the year	66	Round the year	82

Appendix 29.2 Causes of decreasing pond water in the HBT, Godagari, Rajshahi

Causes	No. of respondent	%
i. Drought	41	82
ii. Irrigation	29	58
iii. Decreasing pond depth due to siltation	27	54
iv. Inadequate/low rainfall	9	18
iv. Leaching loss	3	6

Appendix 29.3 Utilization of pond water in the HBT, Godagari, Rajshahi

Type of utilization	No. of respondent	%
i. Domestic purpose	41	82
ii. Irrigation in crop field and home garden	39	78
iii. Fish culture	28	56
iv. Use for livestock and poultry rearing	15	30

Appendix 29.4 Farmers' perception on necessity of more ponds in the HBT, Godagari, Rajshahi

Farmers perception	No. of respondent	%
More needed	30	72
No need	11	28
Management/ re-excavation of existing pond	9	16

Appendix 29.5 Cropping patterns of present day (2007) and mid eighties in the HBT, Godagari, Rajshahi

Land type	Present cropping pattern	% farmer practiced	Land type	Past cropping pattern	% farmers practiced
HL	Fallow-Fallow-T.Aman rice	42	HL	Fallow-Fallow-T.Aman rice	62
	Boro-T.Aman rice	31			
	Mustard-T.Aman rice	18			
	Wheat-T.Aman rice	16			
	Tomato-T.Aman rice	24			
MHL	Chickpea-T.Aman rice	26	MHL	Fallow-Fallow-T.Aman rice	80
	Boro-T.Aman rice	40		Chickpea/linseed-T.Aman rice	15
	Wheat-T.Aman rice	16			
	Potato-T.Aman rice	10			
	Mustard-T.Aman rice	14			
	Mustard-Boro-T.Aman rice	6			
	Linseed-T.Aman rice	6			

Appendix 29.6 Causes of changing cropping pattern/fish culture/livestock's rearing/plantation, HBT, Godagari, Rajshahi

Causes	No. of respondent	%
i. Development of modern technologies in agriculture like new crop varieties	46	82
ii. Increase irrigation facilities	23	46
iii. Increase their demands	14	28
iv. Availability of treatment facilities for livestock's	13	26
v. Increase awareness on crops and cropping and tree plantation	9	18

Appendix 29.7 Livelihood change of changing of cropping pattern/Fish culture/livestock's rearing/plantation in HBT

Changing indicators	No. of respondent	%
i. Introduce modern crop varieties, i.e. increased yield	41	82
ii. Increase irrigation facilities	23	46
iii. Increase fish culture	14	28
iv. Increase livestock's rearing	13	26
v. Increase food security	10	20
vi. Create employment opportunity	9	18
vii. Increase percent literacy	9	18

agricultural technologies, along with the development of better communication (Appendix 29.7). However, long term effect of increased ground water lifting should be monitored carefully. Maximum and minimum mean temperature have increased slightly (Fig. 29.1). Rainfall was erratic around 800–2,200 mm, but at the latter it was exceptionally high for HBT (Fig. 29.2).

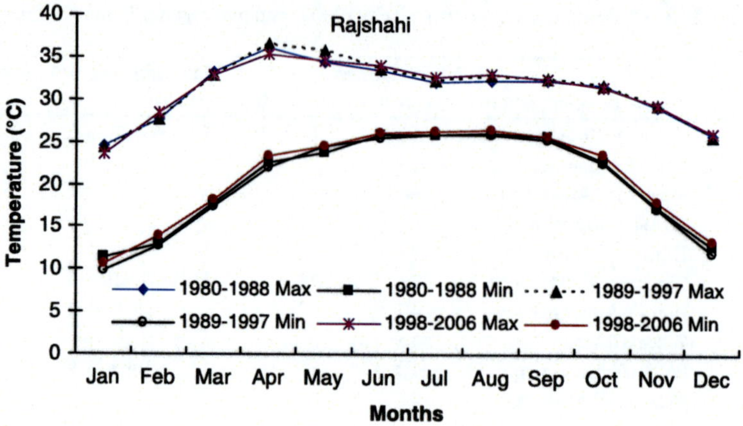

Fig 29.1 Mean monthly maximum and minimum temperature in Rajshahi during 1980–2006

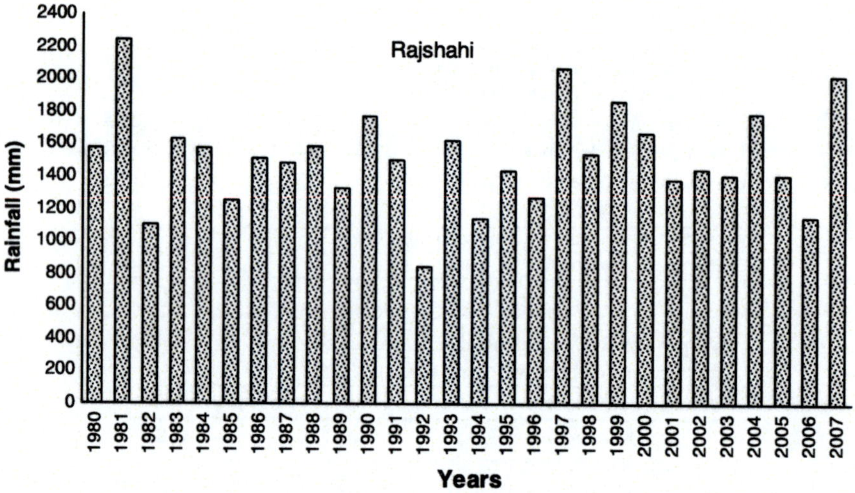

Fig 29.2 Yearly total rainfalls in Rajshahi during 1980–2007

29.3.2 Noakhali

In Noakhali, the predominant cropping patterns in the 1980s were Fallow-Fallow-T.Aman rice & relay Khesari – T.Aman rice for medium lowland and Groundnut-T.Aman rice, Sweet potato-T.Aman rice, Cowpea-T.Aman rice & Chilli – B.Aus/D.Aus rice – T.Aman rice for medium highland. Cropping intensity, as well as yields of different crops, increased due to use of improved varieties, quality of seeds and use of recommended production practices. In the 1980s, the major crops were T.Aman rice, khesari, groundnut and cowpea. These crops produced lower yields due to higher salinity and lack of irrigation. But in 2007, major crops had higher yields due to introduction of modern varieties and use of improved management practices.

Salinity of Noakhali decreased over time by different natural ways such as leaching by rain water and further control of intrusion of saline water (Appendix 29.8). Crop diversification markedly increased along with adoption of high yielding varieties and technologies for ensuring food security to growing population which resulted in increased food production (Appendix 29.10). However, fodder production was reduced (Appendix 29.11) due to bringing of fallow land under cultivation and adoption of modern rice cultivars, which gave a reduced amount of straw. Also, production of vegetables, fruits, timber tree and poultry birds increased for meeting the demand and higher earning. However, cattle/buffalo/goat population was reduced due to fodder/feed crisis and for lack of grazing land. Due to climatic effect and intensive cropping, cost of cultivation was increased, higher incidence of insect, pest and diseases, and decrease of soil fertility of the area was observed (Appendix 29.9). Open water fish availability was also reduced. Overall livelihoods of the farmers improved, ensuring enhanced food security and improvement of other social indicators. Maximum and minimum

Appendix 29.8 Reasons for decrease of salinity in Noakhali and Patuakhali

	Farmers' responded (%)	
Reasons	Noakhali	Patuakhali
1. Wash out by rain water	92	88
2. Infiltration/run off etc	54	–
3. Crop removal	36	–
4. Further no entrance of saline water	78	76
5. Establish of embankment	–	100

Appendix 29.9 Effect of climatic change on the human livelihoods in Bagerhat, Dacope (Khulna), Noakhali and Patuakhali

	% Farmers responded			
Reasons	Bagerhat	Dacope	Noakhali	Patuakhali
1. Timely sowing of rabi crops hampered due to cyclone, tornado, SIDR etc.	40	72	56	80
2. Increase the disease/insect incidence	100	36	72	74
3. Cost of cultivation increased	72	28	100	100
4. Soil fertility decreased	20	–	100	100
5. Number of trees decreased	12	–	–	–
6. Uncertain human death	–	32	–	–
7. Fish availability decreased	–	–	100	100

Appendix 29.10 Reasons for livelihood change in Bagerhat and Dacope (Khulna), Noakhali and Patuakhali

Reasons	% Farmers responded			
	Bagerhat	Dacope	Noakhali	Patuakhali
1. Farmers income increased due to shrimp cum rice culture	100	–	–	–
2. Farmers income increased due to adoption of modern technologies	–	–	100	100
3. Increase of cropping intensity and also gradual reduction of salinity	–	–	66	72
4. Food security increased	–	–	70	70
5. Knowledge and education increased	–	–	78	82
6. Off-Farm income increased	–	–	54	44
7. Semi pacca housing constructed to avoid water logged	72	–	–	–
8. Fuel shortage increased due to decreased the number of trees	72	72	–	–
9. The cost of cultivation increased due to crop, fisheries, livestock system change	–	56	–	–
10. Semi pacca housing constructed due to crop, fisheries, livestock system change	–	40	–	–
11. Improved sanitation system due to crop, fisheries, livestock system change	–	16	–	–

Appendix 29.11 Reasons for lower fodder production in Bagerhat and Dacope (Khulna), Noakhali and Patuakhali

Reasons	% Farmers responded			
	Bagerhat	Dacope	Noakhali	Patuakhali
1. Due to increased of salinity	100	64	–	–
2. Lack of fallowland for fodder production	88	56	100	100
3. Increase of cropping intensity caused no seasonal fallowland	–	–	72	88
4. Less straw production due to cultivation of HYV T.Aman rice rice	88	68	66	80
5. Due to unavailability of pure drinking water	12	–	–	–
6. Lack of highland	–	100	–	–
7. Due to change of cropping pattern	–	100	–	–
8. Incidence of cyclone/SIDR/sea water damage the rice straw	–	–	48	54

temperature (Fig. 29.3) increased slightly and rainfall variation was 2,000–4,800 mm (Fig. 29.4).

29.3.3 Bagerhat and Khulna

The cropping patterns which were practiced by the farmers of Dacope, Khulna in the 1980s were Fallow-Fallow-T.Aman rice, Lentil-T.Aman rice & Sesame-T.Aman rice in medium highland and Fallow-Fallow-T.Aman rice in medium lowland.

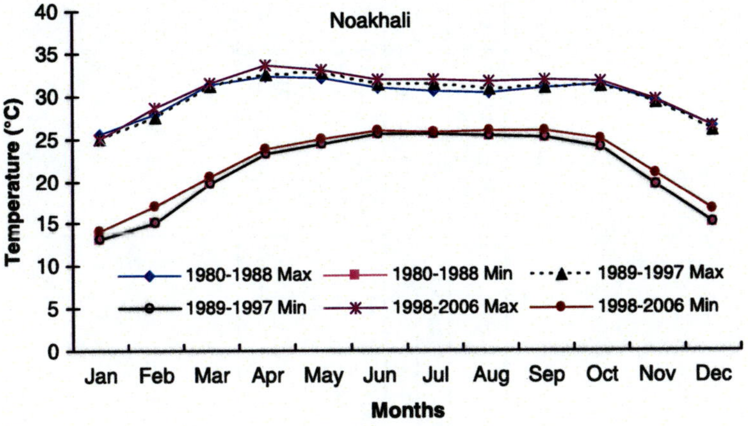

Fig 29.3 Mean monthly maximum and minimum temperature in Noakhali during 1980–2006

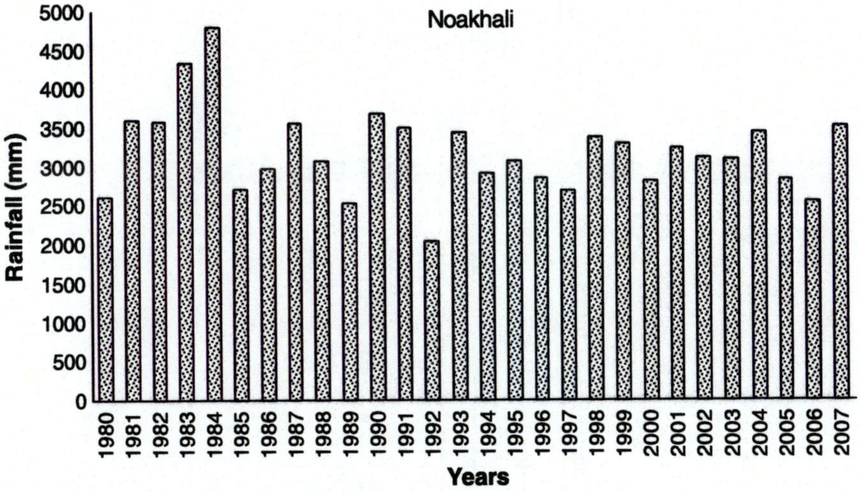

Fig 29.4 Yearly total rainfalls in Noakhali during 1980–2007

The cropping patterns have been changed due to use of salt tolerant modern rice varieties and to making Gher for shrimp culture. In the 1980s, major crops produced lower yields due to high salinity and lack of irrigation. But in 2007, major crops yielded higher due to introduction of modern varieties and use of improved management practices. The yields of local T.Aman rice and sesame have been decreased due to an increase in the salinity level in recent years (Table 29.1).

Salinity of Bagerhat and Khulna (Ganges Tidal floodplain) is mainly related to decreased water flow from the upward of the Ganges and its tributaries due to Farraka barrage, resulting in increased intrusion of saline water from the Bay.

Table 29.1 Comparison of soil salinity status between 1973 and 2000 in coastal areas of Bangladesh

District	Salt affected area (000'ha)		Salinity class								Salinity increase over three decades	
			S_1 2.0–40 dS/m		S_2 4.1–8.0 dS/m		S_3 8.1–16.0 dS/m		S_4 > 16.0 dS/ma		Area	
	1973	2000	1973	2000	1973	2000	1973	2000	1973	2000	(000' ha)	%
Khulna	120.04	145.25	3.90	28.83	92.54	37.32	13.00	59.49	10.60	19.61	25.21	21.00
Bagerhat	107.98	125.13	8.30	35.66	77.08	41.50	2.60	41.23	20.00	6.74	17.15	15.88
Satkhira	146.35	147.08	16.50	27.03	85.60	38.01	33.50	60.03	10.75	22.01	0.73	0.50
Patuakhali	115.10	139.35	68.50	40.11	46.60	43.62	0	46.10	0	9.52	24.25	21.07
Noakhali	49.6	53.55	6.3	13.04	39.90	16.93	3.40	15.83	0	7.75	3.95	7.96

Source: SRDI (2003)

However, secondary salinity (for making shrimp Gher) is also another cause of increasing soil and water salinity. Shrimp culture has become a major cause of increased salinity in Bagerhat. Eighty-four percent of the farmers at Dacope reported that intrusion of saline water within the embankment (by breaking) is the major reason of increasing salinity along with shrimp culture (20% of farmers) (Appendix 29.12). In Bagerhat, crop diversification has reduced due to culture of shrimp. Whereas at Dacope, crop diversification has markedly increased through the intervention of modern technology, as shrimp cultivation is still in a small area. In Bagerhat, 12% of farmers reported that tree population has decreased due to increase of salinity and year round inundation of land for making shrimp gher (Appendix 29.13). While at Dacope, a change of the agricultural systems is mainly due to intrusion of saline water through embankment breakage (like through SIDR), and lack of fresh water for irrigation in dry season (Appendix 29.13). In both locations, the yield of T.Aman rice has been reduced due to increased salinity and late planting, though yield of modern variety was more than local variety (Appendix 29.14). Increased salinity has a profound negative effect on the livestock health due to decrease of fodder production because of salinity, from shrimp Gher and long inundation, modern rice cultivation and polluted water. Increased salinity and climatic hazards like SIDR have ill effect on the health of human, along with increased disease incidence and sudden death (Appendix 29.9). Through shrimp Ghers, farmers of Bagerhat increased their income and improved their livelihood. However, people of Dacope failed to improve their livelihood through increased crop diversification

Appendix 29.12 Reasons for increasing salinity in Bagerhat and Dacope, Khulna

	% Farmers responded	
Reasons	Bagerhat	Dacope, Khulna
1. Entrance of saline water within the gher area	100	84
2. Rice cum fish culture	40	–
3. Due to shrimp culture	–	20

Appendix 29.13 Reasons for changing crop, fisheries, trees and livestock system in Bagerhat and Dacope, Khulna

	% Farmers responded	
Reasons	Bagerhat	Dacope
1. Salinity level increased over time	100	100
2. Due to shrimp culture	100	–
3. Stagnation of water	96	–
4. Stagnation of water due to constriction of embankment	76	64
5. Lack of irrigation	20	80
6. Number of trees decreased due to increased of salinity	12	–
7. Lack of repair and maintenance of sluice gate	–	24
8. Weather and climatic change	–	24

Appendix 29.14 Reasons for changing productivity in Bagerhat and Dacope (Khulna), Noakhali and Patuakhali

	% Farmers responded			
Reasons	Bagerhat	Dacope	Noakhali	Patuakhali
1. Decreased the yield of T.Aman rice in gher area	72	–	–	–
2. Rice cum shrimp culture increased the income	72	–	–	–
3. Increased the yield of HYV Boro	60	–	–	–
4. Yield of local T.Aman rice decreased due to late planting	–	84	–	–
5. Increased the yield of HYV T.Aman rice modern variety	–	64	100	100
6. Yield of local T.Aman rice increased due to improve management	–	–	72	68
7. Increased the yield of vegetable	–	4	78	82
8. Decreased the number of trees	–	88	86	–
9. Increased the number of trees/fruit trees	–	–	–	88
10. Decreased the number of poultry	–	12	–	–
11. Increased the number of poultry	–	–	60	78
12. Cattle/buffalos/goat decreased	–	–	100	100

because of escalation of production cost, and frequent natural hazards, as it is closer to the Bay. Mean monthly maximum and minimum temperature increased slightly (Fig. 29.5), while rainfall ranges from 1,100–2,600 mm (Fig. 29.6).

29.3.4 Patuakhali

In the 1980s, major cropping patterns in Patuakhali were Fallow-Fallow-T.Aman rice, Chilli-T.Aman rice under medium highland situation and Sweet potato-T.Aman rice, Groundnut-T.Aman rice, Sesame-T.Aman rice, T.Aus rice-T.Aman rice, Relay Khesari-T.Aman rice & Mungbean/Sesame-T.Aman rice under medium lowland condition. Utilization of fallow land, use of modern crop varieties, and appropriate production technologies have helped in changing the cropping systems and simultaneously increased the crop yields. Traditionally, Patuakhali was less saline affected. Further salinity was reduced through construction of embankments, wash out by rain water and no further intrusion of saline water (Appendix 29.8). Cropping intensity was increased through introduction of new crops along with modern varieties and technologies, which resulted in increased crop production and food security (Appendix 29.10). However, fodder production was reduced due to absence of fallow land and cultivation of modern rice varieties (Appendix 29.11).

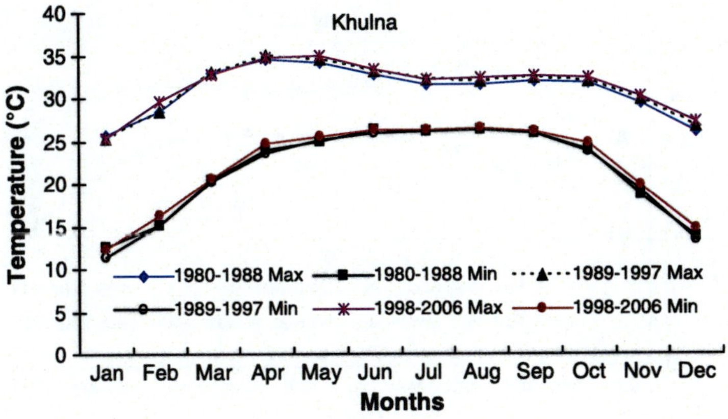

Fig 29.5 Mean monthly maximum and minimum temperature in Khulna during 1980–2006

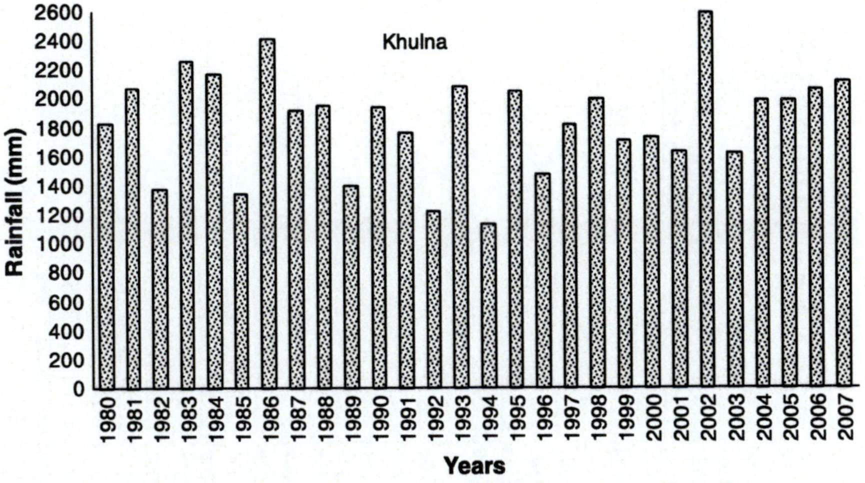

Fig 29.6 Yearly total rainfalls in Khulna during 1980–2007

Due to climatic effect and intense cropping, cost of cultivation was increased, along with a higher incidence of insect, pest and diseases, and decrease in soil fertility (Appendix 29.9). Open water fish availability was also reduced. Overall livelihood of the farmers improved through adoption of modern agricultural technologies along with improvement of other social indicators (Appendix 29.10). Maximum temperature increased slightly while minimum temperature showed a bit of a reduction (Fig. 29.7). Year to year rainfall variation (1,800–4,300 mm) was large (Fig. 29.8).

Soil salinity status between 1973 and 2000 in Noakhali, Satkhira, Bagerhat, Khulna and Patuakhali are presented in Table 29.1. It was observed that the salinity level of different locations increased in the year 2000, except in Patuakhali.

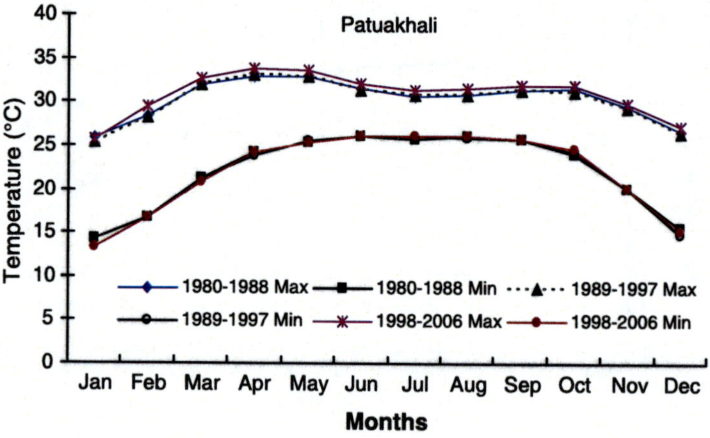

Fig 29.7 Mean monthly maximum and minimum temperature in Patuakhali during 1980–2006

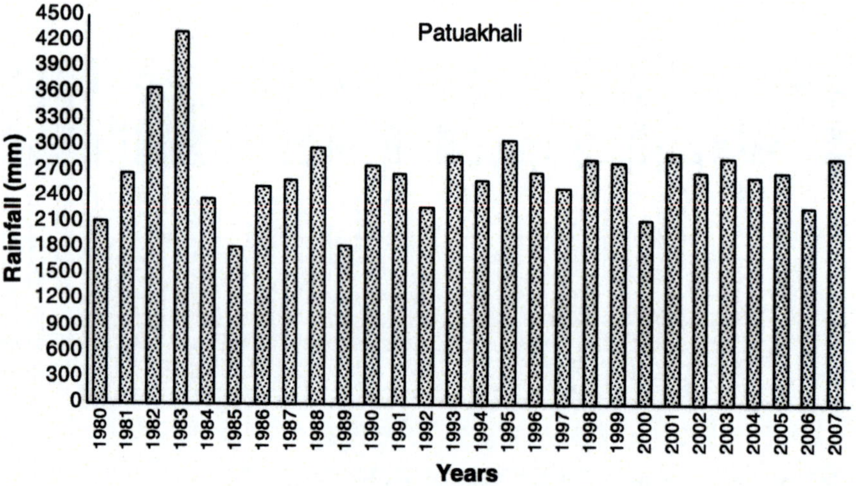

Fig 29.8 Yearly total rainfalls in Patuakhali during 1980–2007

In the Khulna area, 82% farmers responded that they had sown their rabi crops timely, but in 2007, only 18% responded (Table 29.2). The incidence of disease and pest was decreased in the 1980s by 92%, where in 2007, only 8% responded. The crop production affected only 2% in 1980, but it affects 98% in 2007. In the 1980s, 70% farmers opined that they had sown their rabi crops timely but in 2007, only 30% responded in Noakhali areas (Table 29.2). About 98% of farmers responded that incidence of disease and pests decreased in the 1980s, where in 2007, only 2% responded. Only 70% of farmers responded that salinity affected crop production in the 1980s, but in 2007, only 30% responded. In the 1980s, the

Table 29.2 Effect of climate change on the livelihoods in Dacope (Khulna), Noakhali and Patuakhali

	% Farmers responded					
	Dacope (Khulna)		Noakhali		Patuakhali	
Factors/reasons	1980	2007	1980	2007	1980	2007
Sowing rabi crops timely	82	18	70	30	75	25
Incidence of disease and pest increase	17	83	14	86	10	90
Incidence of disease and pest decrease	92	8	98	2	87	13
Cost of crop production increase	0	100	0	100	0	100
Cost of crop production decrease	100	0	100	0	100	0
Soil fertility increase	98	2	100	0	95	5
Soil fertility decrease	3	97	0	100	3	97
No. of trees increase	73	27	8	92	16	84
No. of trees decrease	22	78	90	10	87	13
Water availability increase	86	14	28	72	85	15
Water availability decrease	12	88	70	30	14	86
Farmers' livelihood increase	20	80	15	85	13	87
Farmers' livelihood decrease	85	15	80	20	90	10
Cropping pattern increase	30	70	6	94	8	92
Cropping pattern decrease	68	32	95	5	90	10
Salinity increase	0	100	100	0	74	26
Salinity decrease	98	2	0	100	30	70
Salinity affects crop production	2	98	70	30	55	45

livelihood of the local farmers was increased 15%, but 85% of farmers reported in 2007. In Patuakhali, rabi crops were sown timely by 75% in the 1980s, whereas only 25% responded in 2007 (Table 29.2). The incidence of disease and pest decreased in the 1980s, whereas in 2007 only 13% responded. The salinity affected crop production by 55%, but in 2007, 45% responded. The livelihood pattern of the farmers was increased by 13%, whereas in 2007 it increased up to 87%. Bangladesh Agricultural Research Institute has already carried out some study to adapt the drought situation of High Barind Tract and screening of crops/variety for saline areas.

29.3.5 *Technologies to Adapt Drought Situation of HBT*

- On-farm experimental results revealed that in many years drought prevailed during flowering and grain filling stage of T.Aman rice. As a result, an 11–34% yield loss occurred for local and 43–50% for modern T.man rice varieties. To adapt the situation, supplemental irrigation by harvesting run off water in a mini pond (10 × 10 × 3 m) can save the yield loss of one hectare of land (Annual Progress Report, OFRD, BARI 1991) (Photo 29.1).
- Through on-farm trials, seed priming of chickpea increased the seed yield of 20–25% over non-priming under moisture stress situation of HBT.

Photo 29.1 On-farm experimental results revealed that in many years drought prevailed during flowering and grain filling stage of T.Aman rice. As a result, an 11–34% yield loss occurred for local and 43–50% for modern T.Aman rice varieties. To adapt the situation, supplemental irrigation by harvesting run off water in a mini pond (10 × 10 × 3 m) can save the yield loss of one hectare of land (Annual Progress Report, OFRD, BARI 1991)

- Chickpea, barley, sesame, linseed, kangkong, stem amaranth, coriander and mungbean were found promising for moisture stress/drought situation.
- Besides these, improved crop management practices viz., mulching and minimum tillage were also found to mitigate the challenges of moisture stress/drought of HBT.

29.3.6 Screening of Crops/Variety for Saline Areas

- Tolerance of rabi field crops to salinity is approximately as follows:
  - Sweet potato > cowpea > groundnut > millets > soybean > triticale > mungbean > sesame > barley > wheat > mustard.
- For vegetables, spices and fruits the ranking would look like this:
  - Batishak > chilli > kangkong > garlic > Indian spinach > okra > water melon > red amaranth.

- Besides these, improved crop management practices viz., raised bed with mulch, mulching, minimum tillage, sorjan method, etc were also found to mitigate the challenges of salinity and rise of temperature.

In summary the following points may be considered for impact assessment on the basis of climate change scenarios.

29.3.7 Climate Influence

- Crop production
- Population distribution
- Vegetation
- Soil, water and animal resources
- Temperature, rainfall and natural hazards like cyclone, flooding, drought, salinity changes and erosion
- Human health
- Industry, energy, infrastructure and communication
- Urbanization
- Food demand and supply

29.3.8 Impact on Agriculture

- Agricultural crops are highly vulnerable to climate events.
- Occurrence of unprecedented floods.
- Occurrence of flash floods.
- Occurrence of droughts (kharif I and later part of kharif II and rabi).
- Emissions of nitrous oxide, carbon monoxide and nitrogen oxides from agril. residues.
- Excessive emission of methane gas has been contributing to the global warming.
- The residual effects of climate changes have been destroying the content of organic matter in the soil and also the protein sources.

On the above fact, the following adaptation measures may be taken:

29.3.9 Adaptation Measures

- Improved irrigation efficiency
- Crop diversification (introduction of salt and drought tolerant variety for coastal and Barind area)

- Conjunctive use of surface and ground water irrigation
- Change in fertilization techniques (deep placement of fertilizer, hole system etc.)
- Coastal green belt forestry through GO–NGOs collaboration
- Agro-forestry development
- Homestead vegetable and agroforestry development
- Community forest development through GO–NGOs partnership

29.3.10 Institutional Adaptation

- Improved of agricultural extension services and proper linkage with farmers for adoption of new technology for those area
- Enhance training program and dissemination activities
- Research and development of salinity and drought tolerant crops and high yielding varieties
- Re-excavation of ponds/cannel in Barind area
- Change practices (cultural and other management practices such as tillage, sorjan, relay, ICM/IPM/IPNS concept)
- Expanded access to incentive for conservation agriculture
- Guidelines to incorporate climate change in future planning
- Development of reserved/protected areas in different agro-ecological zones (coastal areas)
- Co-operative social forestry support services with the involvement of vulnerable women and men

References

Ali MY (1998) Optimum management of water and nutrients in High Barind Tract, Bangladesh. A discussion paper presented in the Mini-Symposium on production system oriented research and development in the High Barind Tract of Bangladesh, 3 December, 1998, ICRISAT, Patancheru, India

Ali MY (2000) Influence of phosphorus fertilizer and soil moisture regimes on root system development, growth dynamics and yield of chickpea. PhD thesis. Bangabandhu Sheikh Mujibur Rahman Agricultural University, Salna, Gazipur. p 56

Ali MY, Rahman MA, Quayyum MA, Islam MB (2007a) Improved adaptive capacity to climate change for sustainable livelihoods in the agricultural sectors – the case of High Barind Tract: BARI-FAO collaborative program. On-Farm Research Division, Bangladesh Agricultural Research Institute, Gazipur-1701, Bangladesh

Ali MY, Ahmed S, Johansen C, Harris D, Kumar Rao JVDK (2007b) Root traits of different crops under rainfed conditions in the High Barind Tract of Bangladesh. J Plant Nutr Soil Sci 170:296–302

Amin M, Rahman MA, Zaman SM, Howlader MIA, Hossain MZ, Islam MI (2008) Cropping system in coastal and Barind area. Paper presented in National workshop on Multiple Cropping, held at BARC, Farmgate, Dhaka on 23–24 April, 2008

Annonymous (1991) Annual Progress Report for 1990–91. Farming System Research Programme, On-Farm Research Division, Bangladesh Agricultural Research Institute, Barind Station, Rajshahi

BMDA (2006) Barind authority-past present. Barind Multipurpose Development Authirity, Rajshahi, pp 11, 103, 262

Brammer H (1999) Agricultural disaster management in Bangladesh. The University Press, Dhaka, pp 273–74, 331

Karim Z, Iqbal A (2001) Impact of land degradation in Bangladesh: changing scenario in agricultural land use. Bangladesh Agricultural Research Council, Farmgate, Dhaka, p 106

Selvaraju R, Subbiah AR, Baas S, Juergens I (2006) Livelihood adaptation to climate variability and change in drought-prone areas in Bangladesh. Asian Disaster Preparedness Centre and Food and Agriculture of the United Nations, Rome, pp 2–21

Chapter 30
Influence of Climatic Changes on the Abundance of Major Insect Pests of Sugarcane

Md. Abdullah and Md. Abdul Mannan

Contents

Abstract The abundance of major sugarcane insect pests like top shoot borer, stem borer and rootstock borer were investigated for the last 28 years from 1980 to 2007 at Ishurdi, Pabna. The incidence of these major insect pests occurs in the months of March to October. The abundance of top shoot borer, *Scirpophaga excerptalis* Walker, was positively correlated with rainfall (r = 0.448*) in the month of June, where temperature and humidity did not affect significantly on their incidence. Stem borer, *Chilo tumidicostalis* Hampson, was found positively correlated with average maximum temperature in July (r = 0.399*), and negatively correlated with average minimum temperature in October (r = –0.379*) where humidity and rainfall did not affect significantly on their incidence. The abundance of rootstock borer, *Emmalocera depressella* Swinhoe, was positively correlated with average maximum temperature (r = 0.525**), and negatively correlated with rainfall in the month of May (r = –0.563**) and September (r = –0.380*). Top shoot borer and rootstock borer infestation were increased by 2.09 and 1.43%, respectively, during the last 28 years. Maximum and minimum temperature was increased by 0.18% and 0.22%, respectively, in the month of May during the last 28 years.

Keywords Climatic changes • Abundance • Insect pests • Sugarcane

Md. Abdullah and Md.A. Mannan (✉)
Bangladesh Sugarcane Research Institute, Ishurdi 6620, Pabna, Bangladesh
e-mail: bsri@bsri.gov.bd

R. Lal et al. (eds.), *Climate Change and Food Security in South Asia*,
DOI 10.1007/978-90-481-9516-9_30, © Springer Science+Business Media B.V. 2011

Abbreviation

BSRI Bangladesh Sugarcane Research Institute

30.1 Introduction

Agriculture is strongly interrelated with climatic factors. Temperature, which is one of the main factors of climate, is closely associated with agricultural production. Global warming is the increase in the average temperature of the Earth's atmosphere and oceans that has been observed in recent decades. The scientific opinion on climate change is that much of the recent change may be attributed to human activities. Carbon dioxide and other greenhouse gases released by the burning of fossil fuels, land clearing, agriculture, among other human activities, are the primary sources of the human-induced component of warming. Observational sensitivity studies and climate models referenced by the IPCC (Intergovernmental Panel on Climate Change) predict that global temperatures may increase by 1.4°C to 5.8°C between 1990 and 2100. An increase in global temperatures can in turn cause other changes, including rises in sea level and changes in the amount and pattern of precipitation. These changes may increase the frequency and intensity of extreme weather events, such as tropical cyclones or floods.

The tropical and subtropical countries are more vulnerable to the potential impact of global warming through the effects on crops, soils, insects, weeds, and diseases. Bangladesh is in the subtropical region. Therefore, the agriculture of this country may be affected. The effects of climate change are already evident in the agro-ecosystem of the country. Climate change may act as a factor for sea level rise in the coastal regions of Bangladesh. This may cause an increase in salinity in water and soil of the coastal regions. Growth of standing crops (like rice, jute, sugarcane, etc.) may be affected due to soil salinity, and this may limit overall crop production in the coastal regions as well as make the soil unsuitable for many potential crops. Climate change may lead to an increase in world hunger unless population growth rates in developing nations are much smaller than currently projected, and farmers obtain adequate assistance.

It is expected that due to climate change, humidity, wind flow, and temperature in Bangladesh may be changed. These three climatic mechanisms, in changing conditions, cause an increase in insects, pests, diseases and microorganisms in agriculture, and accordingly, crop production may be decreased.

Increase in range of a species is considered to be limited by two broad categories of physical environment, namely geographic barriers and climatic restrictions. In agriculture, geography is rapidly losing importance because of the activities of man. Because of this, the bioclimatic aspect of insect distribution and abundance has drawn considerable attention in recent years (Messenger 1959). Temperature has been considered as the most dominant factor of environment, influencing the development, survival, feeding, fecundity, dispersal, distribution and abundance of insects. Insects generally grow rapidly in warmer conditions. Within the zone of effective temperature, the rate of development bears a linear relationship with temperature. Optimum humidity varies not only for each of the insect species but also for every stage in its life cycle. These two physical factors show high level of interaction when affecting insect activity, development and survival simultaneously, as it happens in nature. O'Malley (2005) reported that global warming of only 1°C is likely to slash wheat yields by 10% and to encourage wheat, cotton, and sugarcane pests (Collected from internet: articles from News Limited's Australian papers). It seems obvious that any significant change in climate on a global scale should impact local agriculture, and therefore, affect the world's food supply.

So far about 70 species of insect pests have been identified and reported to feed on sugarcane in Bangladesh, (Anonymous 1973–1978, 1992) (Table 30. 1). Among various factors, insect pests inflict considerable losses, which are estimated to be around 20% in cane yield and 15% in sugar recovery (Avasthy 1983). It is reported that the most damaging pests are top shoot borer, stem borer and root stock borer. Losses in yield incurred due to borer attack were estimated to be 21–48% by top shoot borer (Karim and Islam 1977), 8.2–12.6% by stem borer (Khanna et al. 1957), 8.55% (Miah et al. 1986) and up to 10% (Gupta and Avasthy 1952) by rootstock borer. Alam et al. (2006) reported that 10% infestation by top shoot borer caused 12.29% yield loss and 80% infestation caused 35.01% yield loss. He reported that 10% infestation by stem borer caused 12.01% yield loss and 80% infestation caused 27.81% yield loss. He also reported that 10% infestation by rootstock borer caused 10.44% yield loss and 60% infestation caused 27.76% yield loss. Abdullah et al. (2006) reported that stem borer caused 28.73% yield loss and 15.93% recovery loss.

Much work has been done on the influence of various climatic factors on top borer activity and incidence in India (Gupta 1956). Low maximum temperature is generally considered favourable for its build-up. Gupta (1959) has deduced it to be below 37.8°C. The optimum relative humidity for borer activity varies 60–80% (Singh et al. 1957; Gupta 1959). High rainfall is also regarded as a contributory factor for profuse multiplication of the borer (Gupta, 1959). Rainfall appears to favour stem borer multiplication. However, under drought conditions, profuse egg laying take place in July. High atmospheric humidity during August also favours rapid build-up of the borer (Gupta and Avasthy 1952). The incidence of the pest is significantly high in heavy soils and under water logged and flooded conditions (Khanna et al. 1957). The root borer has been observed to be active at high temperatures and moderate humidity levels, and appears to be tolerant to rain to an extent of 45 cm, after which its population declines (Gupta 1953). Root borer incidence and population are generally high in unirrigated fields and in sand or sandy loam soils (Gupta and Avasthy 1952). Therefore, investigation was made to observe the effect of climatic factors on the abundance of major insect pests in sugarcane.

Table 30.1 List of sugarcane insects, mites and nematode pests (Anonymous 1973–1978, 1991)

Si. no.	Common and scientific name		Sugar mills areas where recorded	Status
1.	Top shoot borer	*Scirpophaga excerptalis* (Lep., Pyralidae)	All sugar mills	Major
2.	Internode borer	*Chilo tumidicostalis* Hampson (Lep., Pyralidae)	All sugar mills	Major
		Proceras indicus Kapur (Lep., Pyralidae)	MKSM & SRI	Minor
		Chilo auricilius Dudgeon (Lep., Pyralidae)	JSM, PSM & KCSM	Major
3.	Rootstock borer	*Emmalocera depressella* Swinhoe (Lep., Pyralidae)	All sugar mills	Major
4.	Shoot borer	*Chilo infuscatellus* Snellen (Lep., Pyralidae)	All sugar mills	Major
		Sesamia inferens Walk. (Lep., Noctuidae)	All sugar mills	Major
		Baris sp. (Coleop., Curculionidae)	TSM, SRI	Minor
5.	Leaf eating caterpillar	*Psalis pennatula* F. (Lep., Lymantridae)	All sugar mills	Minor
		Mythimna separata Walk (Lep., Noctuidae)	TSM, PSM, JSM, SRI	Minor
		Spodoptera pecten Gn. (Lep., Noctuidae)	SRI	Minor
		Ramesa testa Walk (Lep., Noctuidae)	SRI	Minor
		Phalera combusta Walk (Lep., Noctuidae)	SRI	Minor
6.	Leaf cutting weevil	*Tanymecus hispidus* Mshl. (Celeop., Curculionidae)	All sugar mills	Minor
		Myllocerus discolour var. *Variegatus* Bohem (Coleop., Curculionidae)	SRI, KSM, RSM	Minor
		Tanymecus sciurus (Oliv.) (Coleop., Curculionidae)	All sugar mills	Minor
		Lepropus chrysochlorus (Wied.) (Coleop., Curculionidae)	STSM	Minor
		Xanthochelus faunus (Oliv.) (Coleop., Curculionidae)	NBSM	Minor
7.	Pyrilla hopper	*Pyrilla perpusilla pusana* Dist. (Hemip., Lophopidae)	All sugar mills	Major
8.	Spittle bug	*Clovia* sp. (Hemip., Aphrophoridae)	All sugar mills	Major
		Aprophora sp. (Cercopidae, Homoptera)		

(continued)

Table 30.1 (continued)

Si. no.	Common and scientific name		Sugar mills areas where recorded	Status
9.	Hispa	*Dorcathispa* cusp/data Maulik (Coleop., Chrysomelidae)	SRI, TSM, KCSM, PSM, JSM	Major
		Dactylispa atkinsoni Gst. (Coleop., Chrysomelidae)	PSM	Minor
10.	Click bettle	*Heteroderes lenis* Candeze (Coleop., Elateridae)	All sugar mills	Major
11.	White fly	*Neomaskellia* sp. (Homop., Aleurodidae)	SRI, TSM	Minor
		Aleurolobus barodensis (Mask.) (Homop., Aleurodidae)	SRI	Minor
12.	Termites	*Odontotermes* probably *lokanardi* (Isop., Termitidae)	Carew & Co.	Major
		Microtermes sp. (Isop., Termitidae)	SRI	Major
		Odontotermes sp. (Isop., Termitidae)	KCSM	Major
		Microtermes obesi Holm (Isop., Termitidae)	SHSM	Major
		Odontetermes parvidens Holm & Holm		
13.	Leaf minor	*Downesia tarsata* Baly (Colep., Chrysomelidae)	SRI, NBSM	Minor
14.	Proutista leaf hopper	*Proutista moesta* Westw. (Hemip., Derbidae)	SRI, most of the sugar mills	Minor
15.	Leaf roller	*Marasmia suspicalis* Walk (Lepidop., Pyralidae)	Carew & Co. & most of the sugar mills.	Minor
16.	Dynastid beetle	*Pentodon bengalens* Arrow (Coleop., Scarabaeidae)	SRI, all sugar mills	Minor
17.	Whitegrubs	*Holotrichia* sp. (Coleop., Scarabaeidae, Melolonthinae)	STSM	Locally major
		Anomala siliguha Arrow (Coleop., Scarabaeidae, Rutelinae)	STSM	Minor
		Holotrichia seticollis Moser (Coleop., Scarabaeidae, Melolonthinae)	STSM	Minor
		(Coleop., Scarabaeidae, Rutelinae)	STSM	Minor
		Holotrichia serrate (F) (Coleop., Scarabaeidae)	Pabna	Minor
		Brahmina sp. (Coleop., Scarabaeidae, Melolontfinae)	STSM	Locally major

(continued)

Table 30.1 (continued)

Si. no.	Common and scientific name		Sugar mills areas where recorded	Status
		Maladera sp. (Coleoptera; Scarabaeidae, Melolonthinae)	STSM	Minor
		Lepidiota sp. *sticticotera* Blanchard (Coleoptera; Scarabaeidae, Melolonthinae)	STSM	Minor
		Anomala polita Branch. (Coleoptera; Scarabaeidae, Rutelinae)	STSM	Minor
		Adoretus versutus Harold (Coleoptera; Scarabaeidae, Rutelinae)	STSM	Minor
		Anomala bengalensis Blanch. (Coleoptera; Scarabaeidae, Rutelinae)	STSM	Minor
		A. biharenis Arrow (Colepptera; Scarabaeidae, Rutelinae)	STSM	Minor
		A. varicolor (Gyllenhal) (Colepptera; Scarabaeidae, Rutelinae)	STSM	Locally major
		A. sp. nr. *varicolor* (Gyllenhal) (Colepptera; Scarabaeidae, Rutelinae)	STSM	Locally major
		Mimela sp. nr. *anopunctata* (Burmeister) (Colepptera; Scarabaeidae, Rutelinae)	STSM	Minor/ major
		Adoretus lasiopygus Burmeister (Colepptera; Scarabaeidae, Rutelinae)	STSM	Minor
		Alissonotum impressicolle Arrow (Coleoptera; Scarabaeidae, Dynastinae)	KCSM	Minor
		Genocephalum sp. (Coleoptera; Tenebrionidae)	STSM	Minor
18.	Grass hopper	*Autarches Miliaris* L. (Orthop., Acrididae)	STSM	Locally major
19.	Shield bug	*Agonoscelis nubila* (Fabr.) (Hemip; Pentatomidae)	SRI	Minor
20.	Thrips	*Baliothrips* serrate (Kabus) (Thysanop, Thripidae)	All sugar mills	Major

(continued)

Table 30.1 (continued)

Si. no.	Common and scientific name		Sugar mills areas where recorded	Status
21.	Mealybug	*Dysmicoccus* sp. (Homop., Pseudococcidae)	SRI	Minor
		Kiritshenkella sacchari Gr Minor (Homop., Pseudococcidae)	All sugar mills	Minor
22.	Scale insect	Ac/orda sp. *takahashi* Kuw. (Homop., Aclerdidae)	All sugar mills	Minor
		Me/anasp/s *glomerata* Green (Homop., Diaspididae)	All sugar mills	Major
23.	Black leaf hopper	*Eoeurysa* flavocap/tata Muir. (Homop., Delphacidae)	All sugar mills	Major
24.	Woolly aphis	*Ceratovacuna lanigera* (Zehnt.) (Homop., Aphididae)	All sugar mills	Major
25.	Aphid	*Longiunguis sacchari* (Zehnt.) (Homop., Aphididae)	RSM	Minor
		Rhopalosiphum maidis (Fitch) (Homop., Aphididae)	SRI	Minor
26.	Zebra hopper	Ricania *Zebra* (Dist.) (Homop., Ricanidae)	SRI	Minor
27.	Mite (Red)	*Oligonychus indicus* (Hirst) (Acarina, Tetranychidae)	SRI, Careew & Co.	Minor
28.	Icerya	*Icerya* sp. (Margorodidae, Homoptera)	SRTI	Minor
29.	Root aphid (ground pearl)	*Geoica lucifuga* (Zehntner) (Aphididae; Homoptera.)	Careew & Co.	Minor
30.	Mite (white)	*Schizotetranychus andropogoni* (Hirst) (Tetranychidae; Acarina.)	Dhaka/SRTI	Minor
31.	Nematodes	*Helicotylenchus* sp. (Tylenchida; Tylenchidae)	KSM, TSM, STSM	Major
		Hop/o/a/mus sp. (Tylenchida; Tylenchidae)	KSM, TSM, STSM	Major
		Tylenchorhynchus sp. (Tylenchida; Tylenchidae)	KSM, TSM, STSM, SRTI	Major
		Pratylenchus sp. (Tylenchida; Tylenchidae)	NRS, TSM, KSM	
		Aphelenchoides sp. (Tylenchida; Aphelenchoididae)	KSM	Minor
		Meloidogynae sp. (Tylenchida; Heteroderidae)	NRS, SRTI, STSM	Major
		Xiphinema sp. (Dorylaimida, Longidoridae)	NRS, STSM, TSM	Minor
		Hemicriconemoides sp. (Tylenchida; Criconematidae)	STSM	Minor

30.2 Materials and Methods

The investigation was made at the experimental farm of Bangladesh Sugarcane Research Institute (BSRI), Ishurdi, Pabna for the last 28 years from 1980 to 2007. An area of 0.5 ha was planted with sugarcane varieties/clones through conventional sett placement in the trenches. Fertilizer application, irrigation, weeding, mulching and earthing-up were done as per normal cultural practices. Sugarcane varieties/ clones were subjected to natural infestation. No pest control measure was applied.

Data on the incidence of top shoot borer (*Scirpophaga excerptalis* Walker), stem borer (*Chilo tumidicostalis* Hampson) and rootstock borer (*Emmalocera depressella* Swinhoe) were taken from March to October. The percentage of infestation was calculated by counting the total and infested canes. In case of rootstock borer, up-rooted stocks were dissected to observe their infestation. Data on climatic factors viz., maximum temperature, minimum temperature, relative humidity (%) and rainfall were recorded by the biometry section of the institute. Since these pests are more active in the months of March to October, average maximum and minimum temperature, average relative humidity and total rainfall were calculated month-wise from the 28 years. Correlations of climatic factors with borer infestation were calculated. Growth rate of temperature and pest infestation is calculated by using Discrete formula:

Growth Rate r: ((A/P)1/t − 1), where P = value of the variable at beginning of period, A = value of the variable at end of period, t = number of periods including first and last.

Growth rate is also calculated using Microsoft Excel: ((A/P)^(1/t))*100.

30.3 Results and Discussion

Climatic factors like temperature, humidity and rainfall play an important role in the abundance of pest population. Investigation conducted at BSRI showed that top shoot borer infestation varied from 0.3% to 88.66% (Fig. 30.1), stem borer varied from 5.36% to 84.40% (Fig. 30.2) and rootstock borer infestation varied from 4.12% to 44.57% (Fig. 30.3) during the last 28 years from 1980 to 2007. During the study period, year wise (January–December) average maximum temperature and average minimum temperature varied from 30.03°C to 32.29°C, average minimum temperature varied from 20.21°C to 21.47°C, average relative humidity ranged from 78.81% to 91.25% and total rainfall ranged from 863.48 to 2145.91 mm. Since these pests are more active in the months of March to October, therefore, year-wise (March–October) average maximum and minimum temperature from 1980 to 2007 are shown in Figs. 30.4–30.11.

Twenty-eight years of data of individual months showed that average maximum temperature varied from 31.84°C to 35.59°C and average minimum temperature varied from 18.56°C to 26.41°C, where the month of April had the highest average

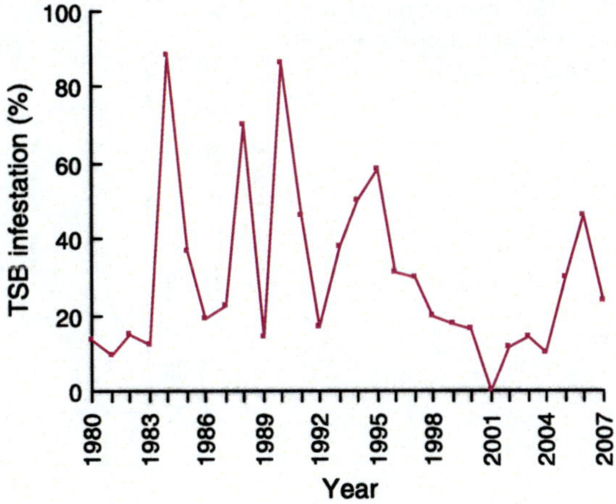

Fig. 30.1 Top shoot borer infestation in different years

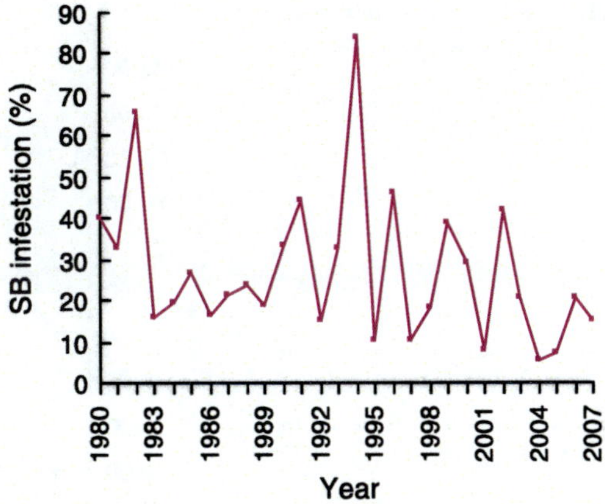

Fig. 30.2 Stem borer infestation in different years

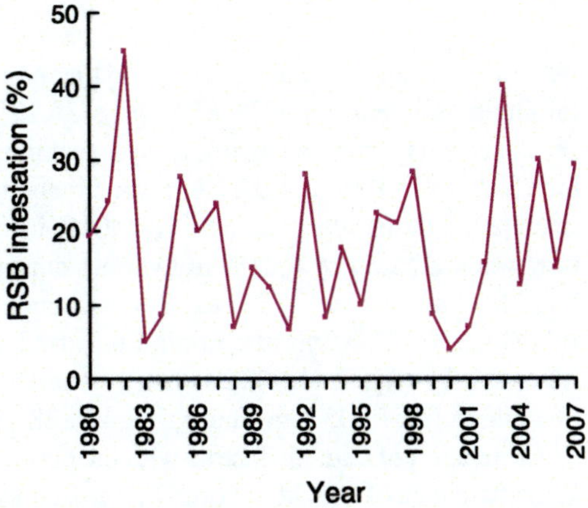

Fig. 30.3 Rootstock borer infestation in different years

Fig. 30.4 Average maximum and minimum temperature in March (28 years)

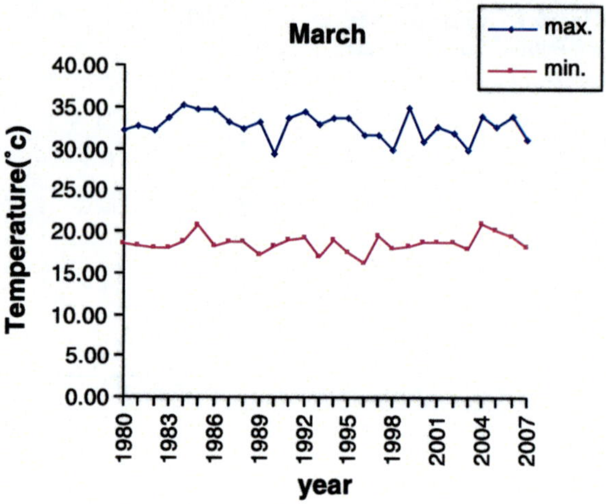

Fig. 30.5 Average maximum and minimum temperature in April (28 years)

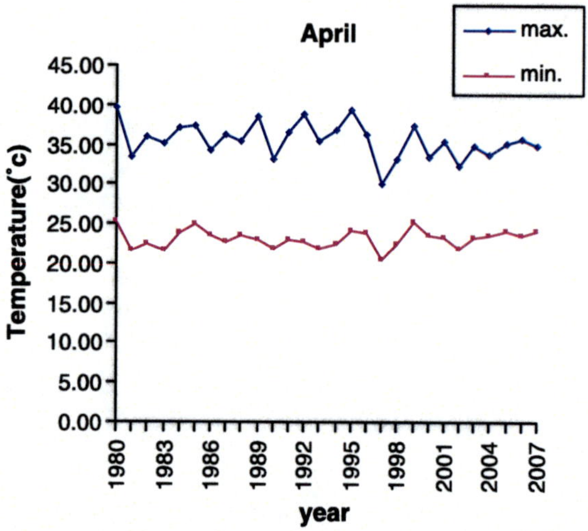

maximum temperature of 35.59°C and the month of August had the highest average minimum temperature of 26.41°C (Fig. 30.12). From growth rate analysis of temperature, it is observed that average maximum and minimum temperature was increased by 0.18% and 0.22%, respectively, in the month of May, average maximum temperature was increased by 0.04% in June and average minimum temperature was increased by 0.01% in October during the last 28 years. Average relative humidity varied from 74.3% to 89.81% from March to October with the highest in July (Fig. 30.13). Minimum rainfall occurred in the month of March (928.75 mm) followed by April (1,902.25 mm), whereas the highest rainfall occurred in September (7,713.05 mm) during the last 28 years (Fig. 30.14).

Correlation of climatic factors with the pest infestation was evaluated (Table 30. 2). The abundance of top shoot borer was positively correlated with rainfall (r = 0.448*)

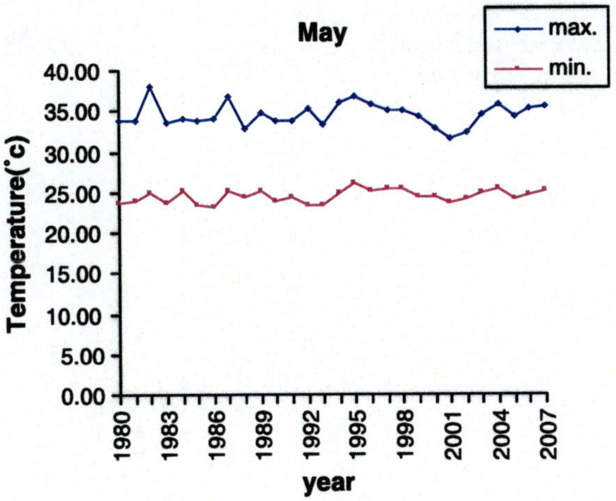

Fig. 30.6 Average maximum and minimum temperature in May (28 years)

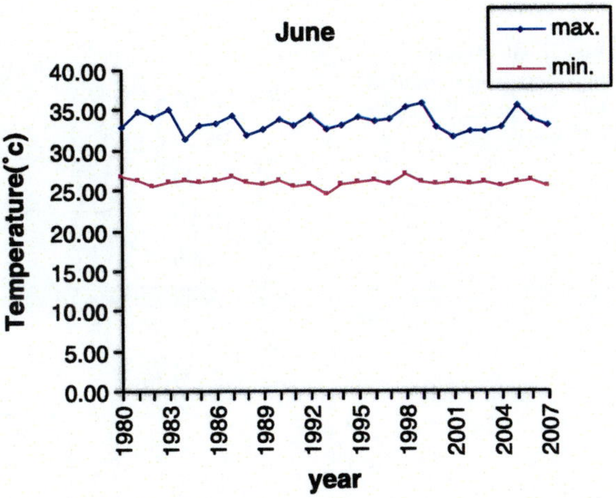

Fig. 30.7 Average maximum and minimum temperature in June (28 years)

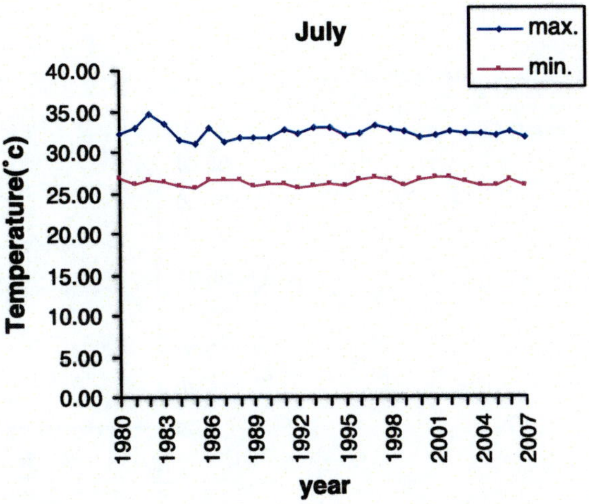

Fig. 30.8 Average maximum and minimum temperature in July (28 years)

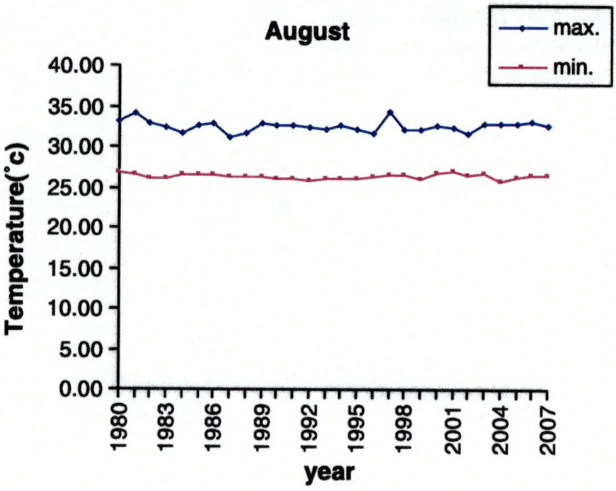

Fig. 30.9 Average maximum and minimum temperature in August (28 years)

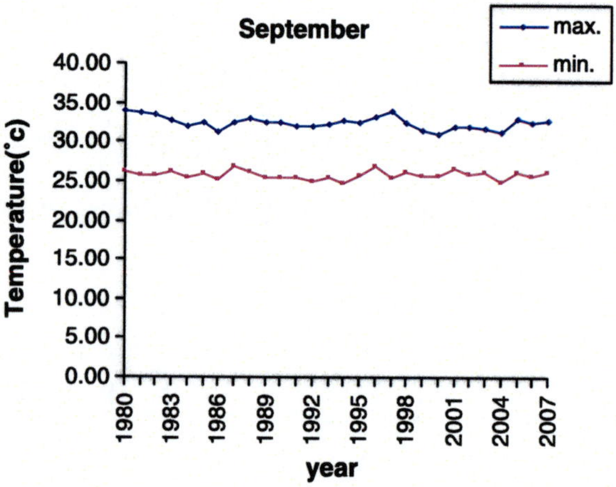

Fig. 30.10 Average maximum and minimum temperature in September (28 years)

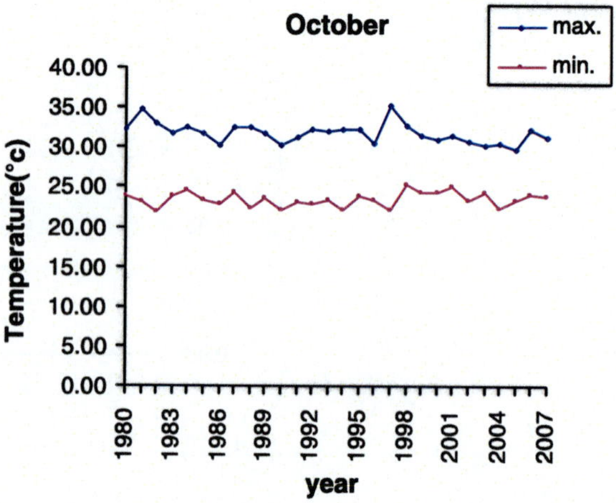

Fig. 30.11 Average maximum and minimum temperature in October (28 years)

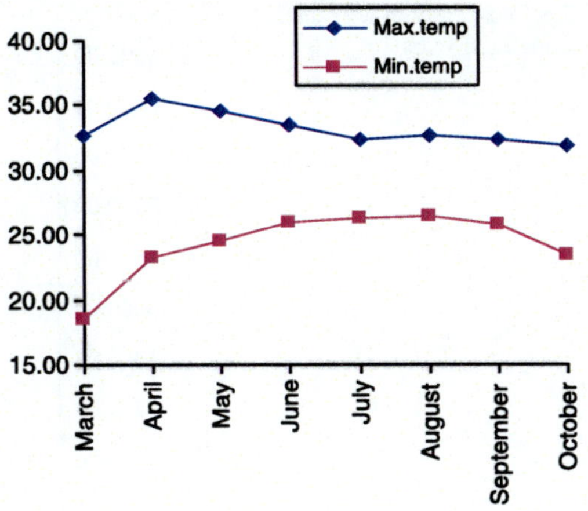

Fig. 30.12 Month-wise average maximum and minimum temperature (28 years)

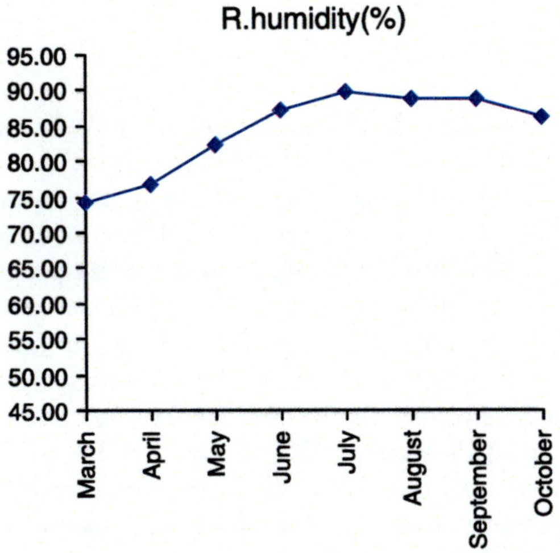

Fig. 30.13 Month-wise average relative humidity (28 years)

in the month of June where temperature and humidity did not affect significantly on their incidence. Rainfall has a positive influence on the population of *S. excerptalis*. It multiplies profusely during the rainy season from July to September, though its activity starts from April and continues upto October (Srivastava and Singh 1957). High rainfall is also regarded as a contributory factor for profuse multiplication of the borer (Gupta 1959). The role of rainfall during July in influencing the population build-up has also been established by Bhardwaj et al. (1981). This result has similarity with the rainfall effect on top shoot borer infestation in the present study. In March 1984, the average maximum temperature was 35.19°C (Fig. 30.4) where the top shoot borer infestation was 88.66%. In Uttar Pradesh, the climatic conditions during the monsoon months, with average maximum temperature ranging from 32.2°C to 35°C and mean relative humidity from 75% to 85% are conducive for its multiplication (Gupta 1954). These results support the effects of temperature on top

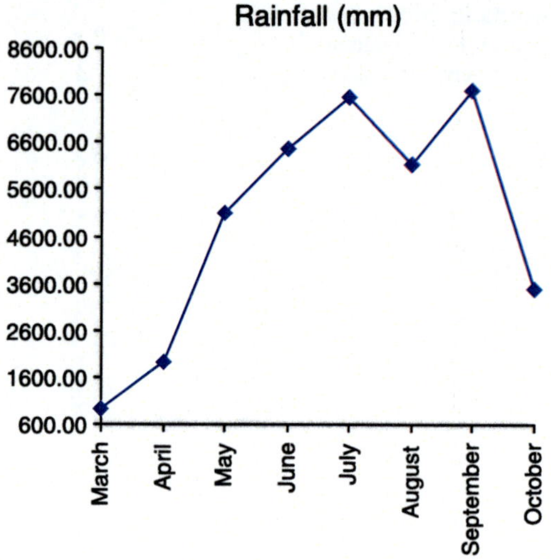

Fig. 30.14 Month-wise total rainfall (28 years)

shoot borer infestation. So, these factors alone cannot have a role in influencing top borer activity (Avasthy 1969). In view of this, the role of biotic factors like natural enemies, varieties under cultivation and others on top borer activity under various agro-climatic conditions have differential incidence of the pest.

Stem borer infestation was observed positively correlated with average maximum temperature in July (r = 0.399*) and negatively correlated with average minimum temperature in October (r = –0.379*), whereas humidity and rainfall did not affect significantly on their occurrence. Increased soil moisture level, waterlogged conditions and heavy soils are found to favour the multiplication of *C. tumidicostalis* (Khanna et al. 1957; Singh et al. 1981). The abundance of rootstock borer was positively correlated with average maximum temperature (r = 0.525**) and negatively correlated with rainfall in the months of May (r = –0.563**) and September (r = –0.380). In 1982, the average maximum temperature was 37.97°C in the month of May, for which the infestation of rootstock borer was high (44.57%). At the same time, the rainfall was too low (81.28 mm) and the average relative humidity was 77.57%, favoring the rootstock borer infestation. Gupta (1954, 1959) reported that rootstock borer is adversely affected by rains. The root borer has been observed to be active at high temperatures (Gupta 1953). The borer flourishes well at high temperatures 31.7–42.1°C and low humidity conditions in north India. Generally, borer incidence and population are generally high in unirrigated fields and in sandy or sandy loam soils (Gupta and Avasthy 1952). All these reports have similarities with the effects of climatic factors on rootstock borer in the present study. Data computed for growth rate analysis showed that Top shoot borer and rootstock borer infestation were increased by 2.09% and 1.43%, respectively, during the last 28 years. At the same time, maximum and minimum temperature was increased by 0.18% and 0.22%, respectively, in the month of May during the last 28 years.

Table 30.2 Correlation (r) of climatic factors with borer incidence

Month	Climatic factors	Top shoot borer	Stem borer	Rootstock borer
March	Av. max. temp. (°C)	0.060	–0.041	–0.293
	Av. min. temp. (°C)	0.008	–0.233	0.016
	Av. r. humidity (%)	–0.080	0.043	0.187
	Total rainfall (mm)	0.048	0.127	0.351
April	Av. max. temp. (°C)	0.125	0.160	–0.041
	Av. min. temp. (°C)	0.027	–0.059	–0.017
	Av. r. humidity (%)	–0.165	0.048	0.063
	Total rainfall (mm)	–0.047	0.195	0.076
May	Av. max. temp. (°C)	0.034	0.179	0.525**
	Av. min. temp. (°C)	0.154	–0.091	0.114
	Av. r. humidity (%)	–0.022	0.081	–0.263
	Total rainfall (mm)	0.149	0.038	–0.563**
June	Av. max. temp. (°C)	–0.225	–0.009	0.300
	Av. min. temp. (°C)	0.011	–0.105	0.138
	Av. r. humidity (%)	–0.059	0.014	–0.108
	Total rainfall (mm)	0.448*	0.139	–0.254
July	Av. max. temp. (°C)	–0.309	0.399*	0.223
	Av. min. temp. (°C)	–0.264	0.160	–0.006
	Av. r. humidity (%)	–0.256	–0.299	–0.054
	Total rainfall (mm.)	0.052	–0.369	0.002
August	Av. max. temp. (°C)	–0.241	–0.053	0.276
	Av. min. temp. (°C)	–0.209	–0.156	0.084
	Av. r. humidity (%)	–0.177	0.134	–0.238
	Total rainfall (mm)	0.180	–0.080	–0.240
September	Av. max. temp. (°C)	0.038	0.221	0.354
	Av. min. temp. (°C)	–0.225	–0.204	0.192
	Av. r. humidity (%)	–0.264	0.008	–0.349
	Total rainfall (mm)	–0.006	–0.344	–0.380*
October	Av. max. temp. (°C)	0.009	0.089	0.075
	Av. min. temp. (°C)	–0.228	–0.379*	–0.154
	Av. r. humidity (%)	–0.259	–0.099	–0.201
	Total rainfall (mm)	–0.117	–0.200	0.001
	5%	1%		
Tabulated r′	0.374	0.479	df = 26	

*significant; ** highly significant

References

Abdullah M, Alam MA, Ahmed T, Siddiquee MNA (2006) Estimation of losses caused by stem borer, *Chilo tumidicostalis* Hampson in sugarcane. Planter Kuala Lumpur 82:33–38

Alam MA, Biswas MM, Abdullah M, Jabber MA (2006) Loss assessment due to attack of borers in sugarcane. Bangladesh J Sugarcane 28:19–23

Anonymous (1973a) Bangladesh sugarcane research institute annual report, 1973–78. Sugarcane Research and Training Institute, Ishurdi, Pabna, pp 123–142

Anonymous (1973b) Bangladesh sugarcane research institute annual report, 1973–78. Sugarcane Research and Training Institute, Ishurdi, Pabna, pp 138–140

Anonymous (1991) Bangladesh sugarcane research institute annual report, 1991. Sugarcane Research and Training Institute, Ishurdi, Pabna, pp 88–91

Anonymous (1992) Bangladesh sugarcane research institute annual report, 1992, Ishurdi, Pabna, pp 87–96

Avasthy PN (1969) The top borer of sugarcne, *Scirpophaga nevella* (F.). In: Williams JR, Metcalfe RW, Mungomery, Mathes R (eds) Pests of sugarcane. Elsevier, New York, pp 189–205

Avasthy PN (1983) Insect pest management for sugarcane in India. In: Balasubramanian M, Solayappan AR (eds) Sugarcane pest management in India. Tamil Nadu Co-operative Sugar Federation, Madras, pp 71–77

Bhardwaj SC, Gupta JN, Jain BK, Yadav SR (1981) Comparative incidence of stalk borer, *Chilo auricilius* Ddgn in autumn and spring planted and ratoon crops of sugarcane. Indian J Agric Res 15:135–140

Gupta BD (1953) Resume of work done under the insect pests scheme during 1946–47 to 1950–51. Indian Cent Sug Cane Comm, New Delhi, p 111

Gupta BD (1954) Consolidated report of the work done under the insect pests scheme during 1952–53. Indian Cent Sug Cane Comm, New Delhi, p 32

Gupta BD (1956) Consolidated report of the work done in India under the insect pests scheme. Indian Inst Sug Cane Res, Lucknow, p 43

Gupta BD (1959) Insect pests of sugarcane in India-III. The top borer, *Scirpophaga nivella* Fabr. Indian Sug 9:127–149

Gupta BD, Avasthy PN (1952) Biology of sugarcane borer, *Emmalocera depressella* Swinh., in the Uttarpradesh. J Agric Anim Husb U P 2:19–25

Karim MA, Islam MN (1977) Sugarcane insect pests and their natural enemies recorded in Bangladesh. Proceeding annual convention Bangladesh society sugarcane technologist, 1974–76, pp 32–68

Khanna KL, Nigram LN, Puri VD (1957) *Chilo tumidicostalis* Hampson, a serious stem borer pest of sugarcane in Bihar. Proc Indian Acad Sci B 46:75–95

Messenger PS (1959) Bioclimatic studies with insect. A Rev Ent 4:183–206

Miah MAH, Kundu R, Mannan A, Abdullah M (1986) Assessment of cane damage caused by root stock borer. Emmalocera depressella Swinhoe and effects of pesticides on the pest. Bangladesh J Sugarcane 8:1–4

O'Malley B (2005) Collected from internet: articles from News Limited's Australian papers

Singh H, Kalra AN, Sidhu AS (1957) Incidence of sugarcane top borer in relation to relative humidity and temperature. Indian J Sug Cane Res Dev 1:81–84

Singh, AK, Prasad AR, Gupta SC (1981) Brief resume of work done on stem borer pests of sugarcane in Bihar (1935–80). Proc natn Symp Stalk borer, Karnal, p 161–164

Srivastava AS, Singh RP (1957) Bionomics and control of *Scirpophaga nivella* Fabr, top shoot borer of sugarcane. Proc A Conv Sug Technol Assoc India 24:161

Chapter 31
Soil, Water and Climate Related Constraints for Crop Production in Bangladesh

M. Shahabuddin Khan, Ranjit Sen, Shamsun Noor, Habib Mohammad Naser, and Md. Khairul Alam

Contents

Abstract The crop production situation in Bangladesh is becoming worse due to a range of soil, water and climate related constraints. Global warming and climate change phenomena exacerbate this situation. Declining soil fertility, accelerated soil erosion and soil salinity are the major soil related constraints to crop production. About 0.87 million hectares (Mha) of land is affected by different degrees of salinity and about 1.74 Mha is prone to soil erosion. The rise of 3°C atmospheric temperature would cause an 11% decrease in soil organic matter (SOM) content. Judicious use of chemical fertilizer and increasing use of organic manure with legume based cropping pattern can restore soil fertility. Shortage of water in the

M.S. Khan (✉), R. Sen, S. Noor, H.M. Naser, and M.K. Alam
Soil Science Division, BARI, Gazipur 1701, Bangladesh
e-mail: dmrahman@agni.com

R. Lal et al. (eds.), *Climate Change and Food Security in South Asia*,
DOI 10.1007/978-90-481-9516-9_31, © Springer Science+Business Media B.V. 2011

dry season and abundance of water in the rainy season are the main water-related constraints. Winter (boro) season rice cultivation using a huge amount of ground water creates arsenic contamination in soil and water. Less water consumptive crops should be selected in the rabi season. Climatic changes results in natural hazards like floods, drought and cyclones. Short duration and increasing temperature are the recent trends of the winter season in Bangladesh. Therefore, appropriate crop variety (heat and salt tolerant), improved crop management, fertilizer use, water management and agronomic practices are needed to alleviate these constraints.

Keywords Climate change • Soil fertility decline • Soil erosion • Soil salinity • Flood and drought

Abbreviations

As	Arsenic
B	Boron
BARI	Bangladesh Agricultural Research Institute
BR 22	Bangladesh Rice 22
BR 23	Bangladesh Rice 23
CO_2	Carbon di-oxide
Fe	Iron
GHGs	Green House Gases
HYVs	High Yielding Varieties
IPNS	Integrated Plant Nutrient System
Mha	Million hectares
Mt	Metric Ton
Na	Sodium (Natrium)
NCA	Net Cultivated Area
OM	Organic Matter
P	Phosphorus
S	Sulphur
SOM	Soil Organic Matter
T.aman	Transplanted Aman
T.wet	Transplanted wet
WUE	Water Use Efficiency
Zn	Zinc

31.1 Introduction

Global climate, an extremely fragile phenomenon, has been kept it in a perfect balance for thousands of years, and has enabled this planet to sustain diverse life forms. But recent emissions of green house gases (GHGs) and other environmental degradation done by anthropogenic activities are affecting this balance with drastic consequences. The concentrations of GHGs in the atmosphere is equivalent to around 388 ppm CO_2 compared with 280 ppm before the industrial revolution. Concentrations have increased mainly because of rapid population increase and subsequent pace of high mass consumptive lifestyles of human population. This concentration has already caused the world to warm by >0.5°C, and will also increase by another 0.5°C over the next few decades because of inertia in the climate systems (Fairbanks 1989). Thus, crop productivity is projected to increase slightly at mid to high latitudes for local mean temperature increases of up to 1–3°C, depending on the crop cultivars, and then decrease beyond that in some regions. Adaptations such as altered cultivars and planting times allow low and mid to high latitude cereal yields to be maintained at or above baseline yields for modest warming. Increases in the frequency of droughts and floods are projected to affect local production negatively, especially in subsistence sectors at low latitudes.

However, crop production is a complex function of many factors like crop, soil, environment and production management practices. Bangladesh is situated in a climatically vulnerable position, and is impacted frequently by natural calamities. Due to recent global atmospheric changes, abrupt extreme climatic phenomena are appearing repeatedly over the last few years. Adverse climatic conditions severely affect Bangladesh's agriculture. Different catastrophic environmental hazards like uneven rainfall, extreme temperatures, flood, drought, hailstorm, storm, tidal-surges, saline water-intrusion, etc., impact this country frequently. This review synthesizes the available literature on risks of soil degradation by climate change and proposes adaptation strategies.

31.2 Principle Constraints to Crop Production

Review of various reports and publications show the following major constraints to crop production: soil fertility decline, soil erosion and land loss, soil salinity, loss of water/inefficient use of water, and climatic stresses.

31.2.1 Soil Fertility Decline

In Bangladesh, major food crops remove about 2.98 m tons (Mt) of nutrients annually against a total addition of 0.72 Mt. According to an appraisal report on Bangladesh's soil resources, soils on about 6.10 Mha contain low (less than 1%) organic matter (OM), 2.15 Mha contain low (1–2%) OM and the remaining 0.90 Mha contain high (more than 2%) OM. The state of decline is also reflected in deficiency of N in almost all soil types, and widespread deficiency of P, Zn, S, B, etc. As much as 5 Mha of land suffer from S and 2 Mha from Zn deficiency. Boron deficiency has lately been reported to cause sterility in mustard, wheat and chickpea. Boul et al. (1990) made a quantitative evaluation of the effect of a temperature increase on the soil OM. They predicted that 3°C rise in atmospheric temperature would cause an 11% decrease in soil OM to 30-cm depth, which will adversely affect plant growth (Tables 31.1 and 31.2).

Table 31.1 Changes in soil organic matter content with time in Major AEZs (Karim et al. 2004)

Name of AEZ	Land type	Organic matter (%) 1969–1970	1989–1990	1999–2000	Depletion of OM (%) (1970–2000)
Madhupur Tract	High Land	1.78 (1.32–2.4)	1.20 (0.06–1.7)	1.02 (0.05–1.63)	42.7
Barind Tract	High Land and Med. High Land	1.45 (1.06–2.0)	1.15 (0.9–1.4)	0.76 (0.40–1.3)	47.6
Old Himalayan Piedmont Plain	High Land	1.32 (1.0–1.65)	1.2 (0.8–1.5)	1.04 (0.70–1.45)	21.2
Tista Meander Floodplain	High Land and Med. High Land	1.55 (1.46–1.6)	1.23 (0.8–1.5)	0.97 (0.67–1.45)	37.4
Northern and Eastern Hills	High Land	2.04 (1.49–2.46)	1.32 (1.0–1.5)	1.05 (0.68–1.28)	48.5
Old Meghna Estuarine Floodplain	High Land	2.16 (1.92–2.61)	1.17 (1.0–1.5)	0.91 (0.58–1.40)	57.9
High Ganges River Floodplain	High Land	1.21 (0.64–1.61)	0.98 (0.31–1.4)	0.83 (0.55–1.29)	62.4
Old Brahmaputra Floodplain	Medium High Land	1.56 (1.09–2.16)	1.23 (0.9–1.5)	1.15 (0.85–1.39)	26.3

Table 31.2 Different types/areas of land degradation and their extent in Bangladesh (BARC 1999)

Types of land degradation	Areas (in Mha) affected by different degrees of degradation: Light	Moderate	Strong	Extreme	Total area (Mha)
1. Water Erosion	0.1	0.3	1.3	–	1.7
– Bank erosion	–	1.7	–	–	1.7
2. Wind Erosion	–	–	–	–	–
3. Soil Fertility Decline	3.8	4.2	–	–	8.0
– P deficient (for HYV rice)	5.3	3.2	–	–	8.5
– P deficient (for Upland crops)	3.1	2.5	–	–	5.6
– K deficient (for HYV rice)	4.0	3.4	–	–	7.4
– K deficient (for Upland crops)	2.1	5.4	–	–	7.5
– S deficient (for HYV rice)	4.4	3.3	–	–	7.7
– S deficient (for Upland crops)	4.1	4.6	–	–	8.7
Soil Organic Matter depletion	1.94	1.56	4.05	–	7.55
4. Water logging	0.69	0.008	–	–	0.7
5. Salinization	0.29	0.43	0.12	–	0.84
6. Pan formation	–	2.82	–	–	2.82
7. Acidification	–	0.06	–	–	0.06
9. Active floodplain	–	–	–	–	1.53
10. Deforestation	–	0.3	–	–	0.3
11. Barind	–	–	–	–	0.773

Table 31.3 Month wise soil loss at different hill slope (average of 1998–2000) (Khan 2001)

Months	Soil loss ($t\ ha^{-1}$) at different hill slopes: 10%	20%	30%
April	0.30	0.46	0.60
May	1.64	2.10	2.56
June	6.02	8.18	10.0
July	4.18	7.06	9.11
August	3.06	5.12	6.33
September	1.28	2.05	2.82
October	0.92	1.36	1.68

31.2.2 Soil Erosion and Land Loss

High seasonal rainfall, low OM content in soils, poor soil management and steep slopes particularly in hilly areas contribute to accelerated soil erosion. It is reported that about 1.74 Mha of land is prone to soil erosion. Land degradation results in part from deforestation, both in hills and plains, occurring now at 5–6% annually (Table 31.3).

Table 31.4 Effect of saline water on decrease in grain yields of rice and wheat (FAP 1993)

Yield decrease (%)	EC on rice ($\mu S\ cm^{-1}$)	EC on wheat ($\mu S\ cm^{-1}$)
0	2,000	4,000
10	2,600	4,900
25	3,400	6,400
50	4,800	8,700

31.2.3 Soil Salinity

Of the 2.85 Mha of coastal areas (some 30% of the total arable land of the country), about 0.87 Mha are affected by different degrees of soil salinity. The problem has been exacerbated in recent years, especially in Khulna-Jessore region as a result of reduced dry season water flow below the Farakka Barrage in India. The cropping intensity in the coastal saline areas range from only 62% in Chittagang region to 114% in Potuakhali region. The low intensity is mainly due to soil salinity and unavailability of a high quality water for irrigation in the dry period. Besides, shrimp farming is causing damage to the environment, leading to deforestation and salinity. Local transplant of wet season (aman) rice is the principal crop cultivated here with an average yield of 1.27 t ha^{-1}. Potential cropping patterns for different degrees of salinity against land types are available, which need to be adopted at the field level for optimizing agricultural production. Water conductivity of 750 m mhos is damaging to human health, while a rating of 2,000 m mhos affects yields of irrigated rice and many other crops. Yield decrease of rice and wheat in saline irrigation water are illustrated in Table 31.4.

31.2.4 Inefficient Use of Water

Water will become the most limiting factor of all natural resources with crop intensification. Despite this, a considerable amount of water is wasted every year that can be attributed mainly to a faulty flood system. The water use efficiency in Bangladesh will hardly exceed 30%. The quality of ground water is deteriorating due to the accumulation of different types of salts, chloride, Fe, Na and arsenic that are affecting not only the operation of irrigation equipment but also the public health. The salinity level in both surface and ground water is also rising, especially in the dry season.

31.3 Climate Stresses

Climate remains the most important factor in crop productivity, although the impact of climate on productivity is often not appreciated. Changes in climate have a profound effect on gram crops. A rise of 1–2°C in combination with lower radiation causes sterility in rice spikelets. High temperature reduces yield of HYVs of summer, wet and winter season rice in all study locations and in all especially if the ambient temperature exceeds 40°C.

31.3.1 Rainfall

Crop production in Bangladesh is predominantly rainfed. Hence the onset, duration, amount and the periodic aberrations of rainfall dictate the nature and type of crops to be grown and the sequence of farming activities. The annual average rainfall of the country ranges from 1,194 to 3,454 mm. The occurrence of rainfall before and during summer season helps to cultivate crops normally escaping any drought. On the other hand, the occurrence of excessive monsoon rainfall (80% of the total rainfall concentrates during July to September) causes unusually early, high or late floods, which damage crops. The uneven distribution of monsoon rains in space and time over different parts of the country may lead to periodic drought and flood situations. Therefore, high variability of rainfall is the single environmental factor which influences the fluctuations of crop yields in different parts of the country.

31.3.2 Flood

Flood of different types, namely flash flood, river water flood and rainwater flood or combinations of them are occurring in Bangladesh mainly due to the occurrence of excessive rainfall. Nearly 1.23 and 5.05 Mha of crop lands are severely and moderately flood prone, respectively. The record floods of 1988 inundated about 2.6 Mha. Recently, it has been estimated that about 7.2 Mha, about half of the total land area, are flood prone. Floods cause colossal losses of crops, poultry, and animals. In some places, reduction of soil fertility is caused mainly by sand deposit.

Late T. wet season varieties like BR 22 and BR 23 may be cultivated with aged seedlings of increased number $hill^{-1}$. Planting split tillers of rice after flood damage could be a good alternative for getting higher field coverage of T. wet season rice at the time of acute seedling shortage and thus higher crop production may be ensured. After receding of flood water, pulses, potato and mustard may be cultivated without tillage and vegetables can be produced to recover the severe crop loss and have alternative food supply within a shorter period. Dapog or floating seedbed for raising seedlings of rice may be practiced in areas where most of the land goes under water during rainy season, so that winter (boro) rice can be planted in time just after receding of floodwater. Short duration modern winter (boro) rice varieties may be developed and cultivated instead of long duration varieties in the low lying areas to avoid yield loss due to the early flash flood (Table 31.5).

Table 31.5 Natural hazards and affected area (Mahatab and Karim 2002)

Hazards	Area affected (Mha)
Floods (flash, rain water, river water and tidal floods)	1.32 of net cultivated area (NCA) severely affected
	5.05 of NCA moderately affected
Droughts	2.32 of crop affected in summer season
	1.20 affected in winter season
Salinity and coastal surges	0.87 is affected
Cyclones and wind	2.80 of coastal area are subjected to damaging cyclones

31.3.3 Drought

Droughts of different intensities occur throughout the country, which severely affect annually about 2.3 Mha in the summer season and 1.2 Mha in the dry (winter and pre-summer) seasons. During summer season, transplant wet season (aman) rice, the principal crop of the country, is mainly affected resulting in the reduction of about 1.5 Mt of rice which is about 8% of the total annual production. During the dry (winter and pre-summer) season mustard, potato, wheat, pulses, winter and broadcast summer rice crops are affected.

Drought resistant modern crop varieties need to be developed and included in the existing cropping patterns for efficient drought management. Adjusting planting time (early or late) of the crops may be a better way for avoiding drought effects, supplemental irrigation during the early stage of summer (Aus) rice or at the later stage of wet season (Aman) rice may increase the yield to a significant level. Winter (Boro) rice and other (Rabi) crops are to be irrigated for their higher production. Much emphasis needs to be given for cultivating Rabi crops, including winter vegetables, pulses and oilseed crops instead of winter (Boro) rice for economic water use and higher profit in areas where irrigation facilities are scarce. Moreover, addition of OM, use of mulches, practicing cover cropping, judicious intercropping etc. help to minimize the adverse effect of drought (Table 31.6).

31.3.4 Temperature

Temperature in the country ranges from 36.7°C to 40.6°C and from 10°C to 27.5°C during summer and winter, respectively. The winter temperature sometimes can be below 7.2°C. High temperature accompanied by scanty or no rainfall accelerates the severity of drought, especially in the later part of the summer (kharif) season, which results in crop loss. On the other hand, sterility or failure in grain setting is a common feature for wheat and rice (early boro and late aman) due to prevalence of high and low temperatures, respectively, at the period of grain setting. Cold tolerant modern T.aman and Boro rice varieties need to be developed to mitigate yield loss.

Late planting of short duration Boro rice and early planting of long duration T.aman rice varieties may be grown for avoiding cold injury. Shorter winter period might be the effect of global warming, which adversely reduces the wheat yield and this necessitated developing and cultivating wheat varieties with the potential to give stable yields at a relatively higher temperature.

Global warming will increase the yield potential of rice in colder regions and decrease yield in warmer regions where temperature is already above optimum or marginal. Most of the present varieties are sensitive to high temperature, with yields decreasing linearly with increase in day time temperature above 33°C (Satake and Yoshida 1978). Various studies indicate that high temperature reduces the yields of HYVs of Aus, Aman and Boro rice during all the seasons throughout Bangladesh (Table 31.7).

31.3.5 *Cyclone and Tidal Surges*

Cyclone and tidal surges are also common in the coastal area and they cause severe loss to the crops and lives. Salinity and tidal submergence tolerant rice varieties and appropriate cropping pattern with proven soil reclamation methods need to be developed to increase the cropping intensity of the coastal saline area.

Table 31.6 Intensity of winter (Rabi) and Pre-summer (Kharif) droughts and yield reduction of crops (Mahatab and Karim 2002)

		Percent of yield reduction of crops			
Drought classes	Area (Mha)	Wheat	Potato	Mustard	B. Aus
Very severe	0.37	60–70	>70	>50	>40
Severe	0.86	50–60	60–70	40–50	30–40
Moderate	3.28	40–50	50–60	30–40	20–30
Less moderate	1.46	30–40	40–50	20–30	10–20
Slight	4.35	<40	30–40	<20	<10

Table 31.7 High temperature effect on key development stage of major agricultural crops (Acock and Acock 1993)

Crops	Effect	References
Wheat	Temperature > 30°C for more than 8 h can reverse vernalization	Evans et al. (1975)
Rice	Temperature > 35°C for more than 1 h at anthesis causes high percentage spikelets sterility	Yoshida (1981)
Maize	Pollen begins to lose viability at temperatures > 36°C	Decker et al. (1986)
Potato	Temperature > 20°C depress tuber initiation and bulking	Prange et al. (1990)
Soybean	It has great ability to recover from temperature stress	Shibles et al. (1975)

31.3.6 Hailstorms

Hailstorms are usually associated with violent thunderstorms. In Bangladesh, hailstorms mainly occurs during March to May and cause severe damage to standing crops like boro rice, wheat, jute, mungbean, summer vegetables, fruits etc. Crops damaged due to hailstorm in April, 1997 were 12667 tons of boro rice, 185 tons of wheat, 861 tons of jute and 46 tons of mung. In May 1997, 63.31 tons of boro rice on 18473 ha of land area were damaged. Resistant varieties with desired plant type and capability of quick recovery need to be developed and cultivated to reduce the yield loss due to the occurrence of hailstorms, cyclone, storm, tidal surges, etc. Research in these themes must be strengthened.

31.4 Opportunities of Adaptation to the Constraints

It is important to face the challenges of the coming years by developing and applying technologies that can help increase yield to feed the growing population. This is to be achieved in ways that are economically viable and environmentally sustainable. Some of the yield increases must come from the available technologies and the remaining from the new and frontier research, mainly in the area of plant breeding (genetic improvement), and soil and crop management.

31.5 Crop Management

Crop improvement depends not only on genetic improvement, but also on crop management. In crop management, increases in productivity can be achieved through the efficient use of fertilizers and water resources and by the management of pests, diseases and climate.

31.6 Nutrient/Fertilizer Management

Improving N efficiency, which is particularly low in irrigated crops and tropical countries, must be a major future intervention to increase productivity and reduce production cost. Various methods to improve N efficiency such as split application, deep placement of urea, urea super granules and time of application have been developed specially for rice and wheat. These kinds of nutrient management techniques should also be initiated or promoted in other crops like oilseeds, pulses, vegetables and tubers in the future decades. Research on green manuring must be given emphasis on a regular basis to improve OM content of soils.

It is recognized that neither mineral fertilizers nor organic matter alone can increase yields on a sustainable basis. Thus, balanced use of fertilizers with OM, that is, integrated plant nutrient system (IPNS), is essential to achieving the twin objective of increasing and sustaining crop productivity. Another approach towards N efficiency would be utilizing N from the air through non-leguminous crops. Methods have to be developed to identify non-leguminous species capable of utilizing atmospheric N. Recently, a positive response of B fertilization (better grain formation) has been reported in mustard, chickpea and groundnut.

31.6.1 Water Management

Water use efficiency (WUE) with the present flood irrigation system is around 30%. Thus, efficient use of water at farm level must be a major goal and studies to this end need to be undertaken or strengthened. The minimum/zero tillage technique that is now being practiced in wheat, potato, mustard, etc., must be refined and used commercially to increase WUE and reduce production cost. Research on the scheduling of irrigation based on water requirement and appropriate growth stages of crops is also imperative in the next decades. It may be worthwhile to mention that one irrigation at the crown root initiation stage in wheat is more efficient than applying several irrigations at other growth stages. Harvesting of water and its conservation (that is at present practically neglected) must receive priority attention in future years for supplementary irrigation to T.aman rice and irrigation to dry season crops. Photosynthesis and other physiological processes are affected by different degrees of water stress. But research in this area is scanty. Therefore, initiation of studies to understand how water stress affects these physiological processes and thereby causes reduction in yields are suggested.

31.6.2 Climate

Another issue of research that must receive adequate attention to meet the challenges of future years is the changes in climate (i.e. rise in temperature and CO_2 and reduction of radiation). Detrimental effects of high temperature on grain crops have been documented by several studies. What is necessary in the future, however, is to study the interaction of temperature, CO_2 and radiation on individual crops or production systems as a whole. Simultaneously, the effects of temperature, humidity and radiation on the incidence of pests and diseases and micro-organisms must be investigated.

31.7 Conclusion

Global warming and climate change have a detrimental impact on soil fertility and crop productivity. Soil OM is decreasing due to rise of soil temperature. Extent and severity of natural disasters like flood, drought, cyclones and tidal surges will be greater in coming years. Increased drought and salinity, prolonged inundation and excessive soil erosion will reduce the crop covering area and field. Appropriate crop management practices must be followed in the climate change affected areas. Selection of appropriate crop species/variety must be chosen for specific area.

Acknowledgement Authors acknowledge the support from the library of Bangladesh Agricultural Research Institute, Bangladesh Rice Research Institute, Bangladesh Agricultural Research Council, Department of Environment and Meteorology. Field research was done on soil erosion at Hill Agriculture Research Station, BARI, Ramgargh. Data were collected from secondary sources and in consultation with the respective resource personnel.

References

Acock B, Acock MC (1993) Modeling approaches for predicting crop ecosystem response to climate change. In: International crop science, vol 1. Crop Science Society of America, Madison, Wisconsin, USA, pp 299–306

BARC (1999) Land degradation situation in Bangladesh. Soils Division, Bangladesh Agricultural Research Council

Boul SW, Sanchez PA, Weed SB, Kimble JM (1990) Predicted impact of climate warming on soil properties and use. In: Impact of carbon dioxide, trace gases and climate change on global agriculture, special publication no. 53, pp 71–82

Decker WL, Jones VK, Achutuni R (1986) The impact of climate change from increased atmospheric carbon dioxide on American agriculture. Washingtion DC, U.S. Department of Energy, Carbon Dioxide Research Division. DOE/NBB-0077, p 44

Evans LT, Wardlaw IF, Discher RA (1975) Wheat. In: Evans LT (ed) Crop physiology. Chambridge University Press, London, pp 101–149

Fairbanks RG (1989) A 17,000-year glacio-eustatic sea level record; influence on glacial melting rates on the Younger Dryas event and deep-ocean circulation. Nature 342:637–642

FAP (1993) Flood Action Plan Report, p 17

Karim Z, Miah MMU, Razia S (2004) Fertilizer in national economy and sustainable environment development. Asia Pac J Environ Dev 1:48–67

Khan MS (2001) Conservation of soil and water for crop production in the hills. Contract Research Project (BARC). BARI. Joydebpur, Gazipur, Bangladesh

Mahatab U, Karim Z (2002) Natural hazards in Bangladesh, First edition, A book published by Bangladesh Agricultural Research Council (BARC), p 11

Prange RK, McRae KB, Midmore DJ, Deng R (1990) Reduction in potato growth at high temperature. Role of photosynthesis and dark respiration. Am Potato J 67:357–369

Satake T, Yoshida S (1978) High temperature induced sterility in indica rice at flowering. Jpn J Crop Sci 47:6–17

Shibles RM, Anderson IC, Gibson AH (1975) Soybean. In: Evans LT (ed) Crop physiology. Cambridge University Press, Cambridge, pp 151–189

Yoshida S (1981) Fundamentals of rice crop science. The International Rice Research Institute (IRRI), Los Banos, Philippines, p 269

Chapter 32
Phenomena Leading to Food Chain Contamination to Intensify Climate Change Effects and Adaptation in Bangladesh

A.M.M. Maruf Hossain and S.F. Elahi

Contents

A.M.M. Maruf Hossain* (✉) and S.F. Elahi
Department of Soil, Water and Environment, University of Dhaka, Bangladesh

*Current address: Department of Environmental Science & Engineering,
Room # 208, Gwangju Institute of Science & Technology,
261 Cheomdan-gwagiro (Oryong-dong), Buk-gu, Gwangju 500-712,
Republic of Korea
e-mail: mueed_abd@yahoo.com

R. Lal et al. (eds.), *Climate Change and Food Security in South Asia*,
DOI 10.1007/978-90-481-9516-9_32,

Abstract Effects of climate change regarding food chain contamination are well-pronounced. Prevalence of environmental contaminants and chemical residues in food chain is one of them. To address such climate change effects sustainable public and environmental health are the prime concerns. Due to be of flood-plain physiography, Bangladesh is more vulnerable to the effects which can be intensified by regional eco-toxicity phenomena that are getting newer dimensions. Two of such dimensions are phenomena of chromium and lead contamination of food chain. In Bangladesh, among the routes of chromium eco-toxicity, feeds and fertilizer production from tanned skin-cut wastes (SCW) has been found to be the most direct one leading to food chain contamination. To study *heavy metal* status of production of protein-concentrates from SCW as fertilizer and feed input, extensive chromium pollution has been detected. Protein-concentrate production process was comprehensively studied and chromium concentration was found to be as high as 2.49%. In this way a huge migration of chromium can happen into poultry products, fish and vegetables, and further bio-magnify into food chain. The target population of this phenomenon is also huge. Another investigation was to study the environmental fate of intense auto-exhaust lead pollution in Bangladesh which is only recently been controlled. Regional bioconcentration has been investigated to assess the nature and extent of ecotoxicity of these two heavy metals. Single poultry egg from the sampling of 96 eggs covering eight districts of central Bangladesh was found to contain a mean chromium content of 23.3809 μg, which exceeds adequate daily dietary intake of children up to 8 years of age as well as corresponds to major part for other age groups. The mean regional bioconcentration of lead was found to be 8.1611 ppm which is about 80 times higher than the maximum permissible limit. As these are most likely to have become ecotoxicity phenomena, these will also involve other living species that constitute the ecosystem and can also exert effects on climate change through shaping altered ecological niche. Moreover, these phenomena will intensify the climate change effects on food chain contamination in this region and hence complicate the adaptation.

Keywords Ecotoxicology • Food chain contamination • Climate change • Adaptation

Abbreviations

SCW	Skin-cut wastes
AAS	Atomic Absorption Spectrophotometry
DRIs	Dietary reference intakes
UL	Upper intake level
RDA	Recommended dietary allowance
AI	Adequate intakes
APM	Airborne Particulate Matter
WHO	World Health Organization

32.1 Introduction

Climate-related extreme events, ocean warming and changes in surface temperature and humidity can result in food safety chemical hazards. There are many pathways through which global climate change and variability may impact environmental contamination and chemical hazards in foods. According to the Third Assessment Report of IPCC, South Asia is the most vulnerable region of the world to climate change impacts (McCarthy et al. 2001). The international community also recognizes that Bangladesh ranks high in the list of most vulnerable countries on Earth. With special regard to climate change effects on food chain contamination, Bangladesh's high vulnerability is due to its geographical location in South Asia, the flat deltaic topography with very low elevation, and extreme climate variability which is governed by monsoon. In this perspective, any phenomena leading to food chain contamination can strongly affect in intensifying climate change effects and adaptation in the region. In recent times regional eco-toxicity phenomena are getting newer dimensions which are addressed to explore in this work. The chromium and lead ecotoxicity are becoming such phenomena. The chromium problem originates from recycling of tanned skin-cut wastes (SCW) into protein-concentrates to be used in feeds and fertilizer production for more than a decade. These activities are mostly concentrated in Hazaribagh, situated in south-western part of Dhaka city. And the lead problem may have its possible origin in intense atmospheric lead pollution from auto-exhaust emission and successive deposition on agricultural soils.

32.2 Materials and Methods

32.2.1 Samples

- The tanned SCW are sliced cut pieces of hides produced at the end of tanning processes to maintain a definite thickness. The SCW are protein-rich and are unscientifically used for the production of poultry and fish feeds, and organic fertilizer. A total of 18 samples were collected to study the *heavy metal* status of protein-concentrate production in the Hazaribagh tanning area at various stages from the source to the final product, and to compare this with the imported ones. These were collected from three spots at Sonatengar of Hazaribagh, feed mill producing protein-concentrate in Hazaribagh, and Nimtoli poultry market, Kotwali Thana, Dhaka. Samples collected for studying protein-concentrate production process can be identified as:
  - **Class I** (sample numbers 1–4) samples were sampled from Sonatengar-I
    - Sample 1: SCW type I (with wet-blue treatment)
    - Sample 2: SCW type II (without wet-blue treatment)

- Sample 3: SCW type I + boiled + dried
- Sample 4: SCW type II + boiled + dried

- **Class II** (sample numbers 5–10) samples were sampled from Sonatengar-II
  - Sample 5: SCW type I (greenish-blue colored)
  - Sample 6: SCW type I + boiled + dried
  - Sample 7: Boiled animal guts
  - Sample 8: SCW type I + boiled + mixed with boiled animal guts + dried
  - Sample 9: Boiled skin-cuts, ear, and tail-cuts
  - Sample 10: Faded silver colored SCW (probable waste from alum tanning)
- **Class III** (sample number 11–14) samples were sampled from Sonatengar-III
  - Sample 11: SCW type I (greenish-blue colored)
  - Sample 12: SCW type II (faded greenish-blue colored)
  - Sample 13: Mixture of SCW type I and type II + boiled
  - Sample 14: Mixture of SCW type I and type II + boiled + dried

  The protein-concentrates were collected from

1. Local protein-concentrate from feed mill at Hazaribagh, Dhaka
2. Imported protein-concentrate from Nimtoli – 1, Kotwali, Dhaka
3. Local protein-concentrate from Nimtoli (supplied from Hazaribagh)
4. Imported protein-concentrate from Nimtoli – 2, Kotwali, Dhaka

The study area is shown in Fig. 32.1, where protein-concentrate production is being done from SCW.

- To study regional bioconcentration in eggs, the central Bangladesh was selected covering eight districts including capital Dhaka. The other seven districts are Narsingdi, Kishoreganj, Mymensingh, Tangail, Gazipur, Narayanganj, and Munshigonj. Twelve commercially produced chicken eggs were randomly sampled from each district's egg stock market. From Dhaka, Gazipur, and Tangail both brown and white colored eggs were found and six eggs per type were sampled. The rest five districts included only brown colored eggs. The study area is shown in Fig. 32.2 (gray color).

32.2.2 Sample Pretreatment

For SCW: The wet and moist samples were sun-dried to remove any sensible moisture. Then these were oven-dried at 105°C until difference in weights was found negligible.

For egg samples: The egg samples were boiled in deionized water and after complete boiling albumen and yolk were separately oven-dried at 80°C to remove all moisture. Each of oven-died albumen and yolk were treated and analyzed separately.

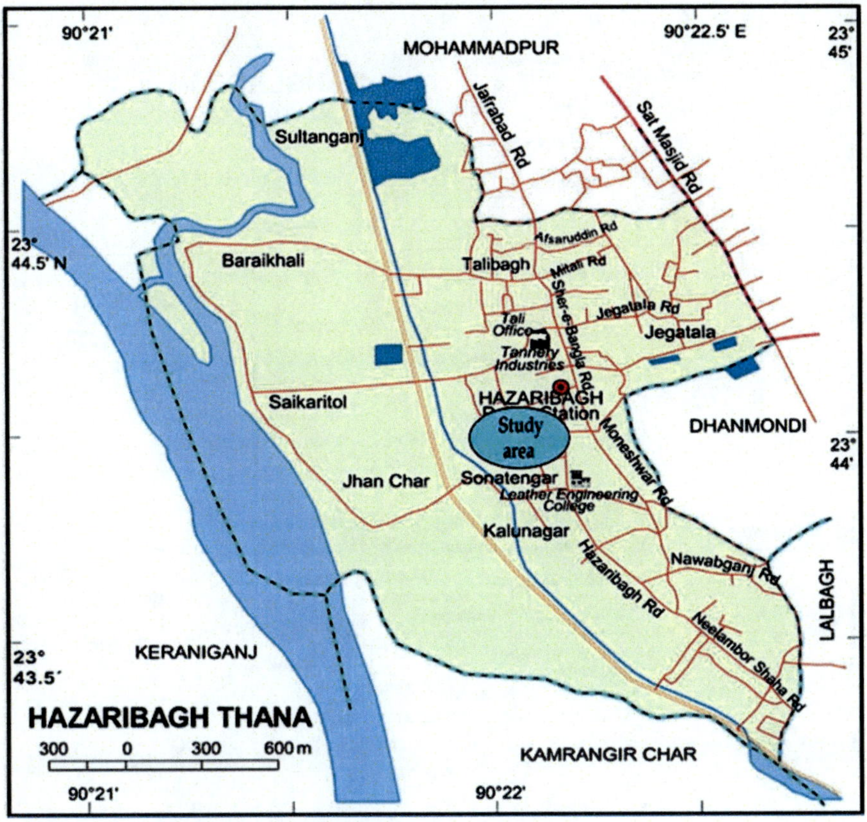

Fig. 32.1 Map of Hazaribagh Thana comprising the study area

32.2.3 Sample Preparation Method

The samples were prepared by using HNO_3–$HClO_4$ digestion (Kebbekus and Mitra 1998). Since the samples were of organic origin with a very high organic content, HNO_3–$HClO_4$ digestion was preferred over the more common HNO_3 extraction for the determination of heavy metals. This strongly oxidizing digestion decomposes organics quickly and efficiently.

32.2.4 Sample Analysis

Analysis of all prepared samples was performed through atomic absorption spectrophotometry (AAS). During the sample preparation for arsenic determination, hydrochloric acid (at least 11.6 ml concentrated HCl for 100 ml total sample) and

Fig. 32.2 Area selected to study regional bioconcentration of chromium and lead (*gray color*)

KI (at least 1%) were added. For the other heavy metals no special treatments were done. Hydride vapor generation technique was used in the determination of arsenic and mercury. Chromium, cadmium, lead, and arsenic determination were done in air-acetylene flame whereas mercury determination was done in cold vapor. Determination was carried out in *SpectrAA 220* Atomic Absorption Spectrometer, Varian, Canada, with using BDH standard solutions.

32.3 Results and Discussion

32.3.1 *Heavy Metal Concentrations in Protein-Concentrates*

The Hazaribagh tannery area, situated in south-western part of Dhaka city, comprise some 90% of tannery industries of Bangladesh on a 25 ha of land. The tanning industries of Hazaribagh process some 220 t of hide per day with an associated release of 600–1,000 Kg of tanned SCW per ton processed raw hide (Zahid et al. 2004). Being of protein origin, these wastes are converted to protein-concentrate to be used as poultry feed, fish feed, and in production of organic fertilizers with some treatment. Large amounts of chrome powder are used during tanning process. After absorption into skins under treatment some 47% of the collagen used as well as 85% of the chemicals enters the waste streams as effluent (UNIDO 2000).

There are many environmental hazards associated with the chemicals used in the tanning processes. But the hazards can come out in two ways. One is through the waters of canals and rivers after mixing with effluents. Substantial work has been done to study on this. But the other way, which is only recently been studied by Hossain et al. (2007), is the entrance of harmful chemicals into food chain through the use of SCW as feed stuff. This is a recent phenomenon happening at large extent for the last several years.

Results indicating chromium content in the three sample classes as well as in the protein-concentrate samples in dry weight basis are presented in Fig. 32.3.

Maximum chromium content of SCW is 3.2037% (sample 6). Data analysis shows that boiling and drying treatments brought no significant change in chromium levels in samples collected from the three different spots of sampling. In one case the concentration becomes higher while in other it is lowered. So it can be concluded that the source of SCW for any specific spot was not fixed rather the wastes were collected from various tanning units of Hazaribagh. That is why, for the chromium level what is at the dumping of solid waste is not the same which is being boiled and also which is sun-dried after boiling. As the sampling was done at a time from the dumping of solid waste and the waste after boiling and drying treatments, no regular change in chromium level among the treatments could be found. Since no extract is eliminated from the boiling of SCW, it can be said that no change in chromium level other than change in oxidation state is expected during the process of boiling with water.

The local protein concentrates (protein-concentrates 1 and 3) contained as high as 2.49% and 1.94% chromium, while the imported protein-concentrates (protein-concentrates 2 and 4) contained only 0.35–0.48%. Cadmium, lead, arsenic, and mercury were detected within the maximum and minimum concentration ranges of 3.888 and 0.991 ppm for cadmium, 30.114 and 7.577 ppm for lead, 2.212 and 0.099 ppm for arsenic, and 13.916 and 0.166 ppm for mercury. The comparative abundance of chromium on the other four heavy metals is shown in Fig. 32.4.

The results show that protein-concentrates produced from SCW are found to contain chromium at levels as high as 2.49%. This suggests that a huge migration

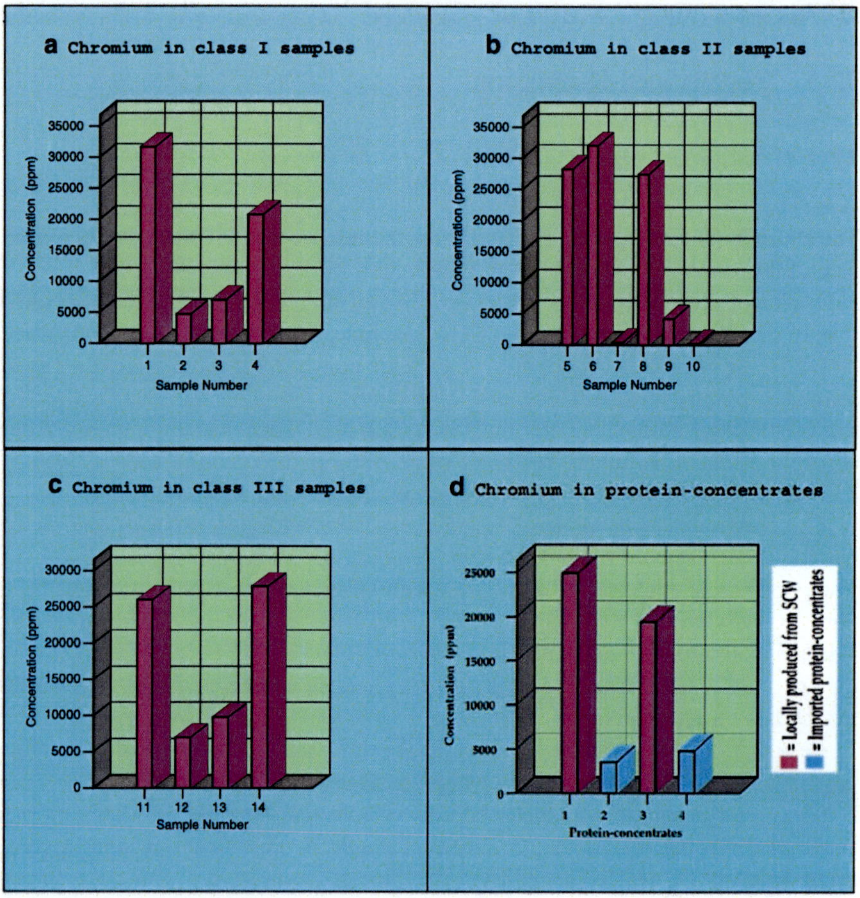

Fig. 32.3 Chromium concentration in the samples. (**a**) Chromium in class I samples; (**b**) chromium in class II samples; (**c**) chromium in class III samples; (**d**) chromium contents of protein-concentrates

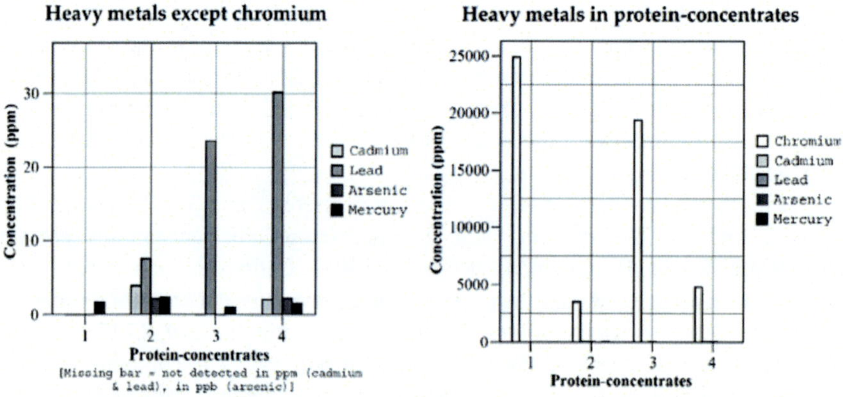

Fig. 32.4 *Heavy metal* concentration in the protein-concentrates. *Left*: all heavy metals analyzed; *Right*: heavy metals except chromium

of chromium can take place into poultry products, fish and vegetables through use of the protein-concentrates, and further bio-magnify into food chain. Zahid et al. (2004) have reported that chromium concentrations in effluents and canal water was 4.06 and 0.443 ppm, respectively; whereas suspended materials in effluents and canal water contained as high as 2.88% and 2.02%, respectively. These waters can also have multiplier indirect effects for chromium toxicity.

32.3.2 *Regional Bio-concentration of Chromium in Poultry Eggs*

Some average physical parameters were estimated from all samples so that the regional bio-concentration can be determined in a comparative way (Table 32.1).

Calculation of regional chromium bio-concentration is shown in Table 32.2 (average dry weight basis).

The average statistics for all egg samples, irrespective of albumen and yolk separately counted are presented in Table 32.3.

The comparison between albumen and yolk for their likelihood to contain chromium is shown in Fig. 32.5.

In all sample classes except Gazipur white eggs, the chromium content is higher in albumen than in yolk. From these results, it can be suggested that albumen is more likely to contain chromium than do yolk for chicken eggs as far as bio-concentration is concerned.

32.3.3 *Dietary Reference Intakes for Chromium*

In 1989, the National Academy of Sciences established an "estimated safe and adequate daily dietary intake" range for chromium. For adults and adolescents the range was 50–200 mcg (NRC 1989). In 2001, DRIs (Dietary Reference Intakes) for

Table 32.1 Average physical parameters for comparative analysis of regional bio-concentration

Physical parameter	Average values	Standard error
Whole weight of eggs, g	55.5098	0.7590
Shell: albumen: yolk ratio	1:4.8672:1.9775	NA
Albumen raw weight, g	34.4401	–
Albumen moisture content,%	84.0603	0.1512
Albumen dry weight, g	5.4900	–
Yolk raw weight, g	13.9930	–
Yolk moisture content,%	51.3656	0.1616
Yolk dry weight, g	6.8054	–

Table 32.2 Calculation of regional chromium bio-concentration in eggs

Sample class	Sample size and standard error	Cr in egg albumen (μg/g) (dry weight)	Cr in total egg albumen (μg)	Cr in egg yolk (μg/g) (dry weight)	Cr in total egg yolk (μg)	Total Cr in each egg (albumen + yolk) (μg)	Concentration in total edible egg (albumen + yolk) (μg/g)
Brown egg	6	1.5621	8.5759	1.1354	7.7269	16.3028	1.3259
(Dhaka)	Std. Error	0.1950	–	0.2149	–	–	–
White egg	6	2.0858	11.4510	1.3410	9.1260	20.5771	1.6736
(Dhaka)	Std. Error	0.6073	–	0.4133	–	–	–
Brown egg	12	1.0671	5.8584	1.0383	7.0660	12.9244	1.0512
(Narsingdi)	Std. Error	0.2070	–	0.2597	–	–	–
Brown egg	12	2.1827	11.9830	1.2334	8.3938	20.3768	1.6573
(Kishoreganj)	Std. Error	0.3379	–	0.1995	–	–	–
Brown egg	12	2.9545	16.2202	1.9318	13.1467	29.3669	2.3884
(Mymensingh)	Std. Error	0.3636	–	0.3741	–	–	–
Brown egg	6	1.9104	10.4881	1.7997	12.2477	22.7358	1.8491
(Tangail)	Std. Error	0.3401	–	0.3797	–	–	–
White egg	6	1.7281	9.4873	0.6670	4.5392	14.0265	1.1408
(Tangail)	Std. Error	0.2592	–	0.2476	–	–	–
Brown egg	6	2.3930	13.1376	1.5584	10.6055	23.7431	1.9311
(Gazipur)	Std. Error	0.4060	–	0.4111	–	–	–
White egg	6	1.4699	8.0698	1.9289	13.1269	21.1967	1.7240
(Gazipur)	Std. Error	0.1971	–	0.4274	–	–	–
Brown egg	12	3.3685	18.4931	2.2028	14.9909	33.4840	2.7233
(Narayanganj)	Std. Error	1.4602	–	0.2418	–	–	–
Brown egg	12	3.3878	18.5990	1.2682	8.6306	27.2296	2.2146
(Munshigonj)	Std. Error	1.5026	–	0.2536	–	–	–

Table 32.3 Statistics of all egg samples

Total sample number	192
Mean concentration (μg/g)	1.9016
Std. Error of Mean	0.1502
Minimum (μg/g)	ND[a]
Maximum (μg/g)	19.8051
Average total content in single egg (μg)	23.3809

[a]Not detected

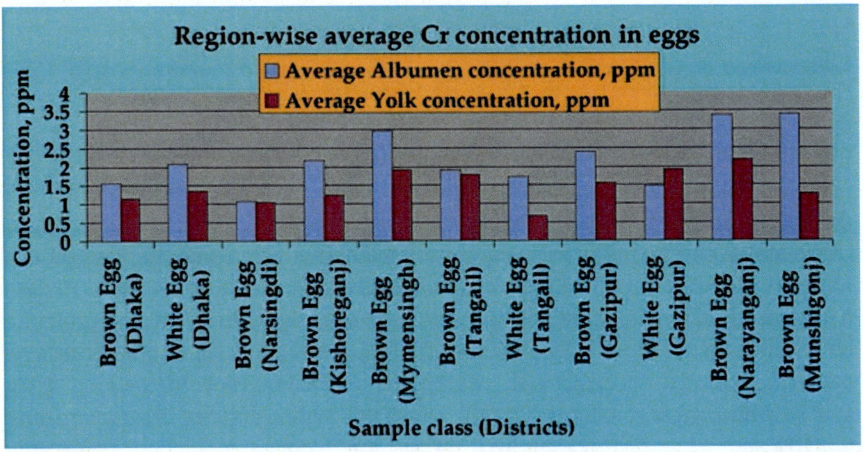

Fig. 32.5 Comparative chromium concentration in egg albumen and yolk

Table 32.4 Adequate Intakes (AIs) for chromium (Institute of Medicine 2001)

Age	Infants and children (μg/day)	Males (μg/day)	Females (μg/day)	Pregnancy (μg/day)	Lactation (μg/day)
0–6 months	0.2	–	–	–	–
7–12 months	5.5	–	–	–	–
1–3 years	11	–	–	–	–
4–8 years	15	–	–	–	–
9–13 years	–	25	21	–	–
14–18 years	–	35	24	29	44
19–50 years	–	35	25	30	45
>50 years	–	30	20	–	–

chromium were established. The research base was insufficient to establish RDAs (Recommended Dietary Allowances), so AIs (Adequate Intakes) were developed based on average intakes of chromium from food as found in other available studies (Institute of Medicine 2001). Chromium AIs are provided in Table 32.4.

32.3.4 Tolerable Upper Intake Level (UL) for Chromium

Due to few serious adverse effects being associated with high intakes of chromium, the Institute of Medicine did not established a Tolerable Upper Intake Level (UL) for this mineral (Stoecker 2001; Institute of Medicine 2001). A UL is the *maximum* daily intake of a nutrient that is unlikely to cause adverse health effects. It is one of the values [together with the RDA (Recommended Dietary Allowance) and AI (Adequate Intakes)] that comprise the DRIs for each nutrient.

32.3.5 Essentiality and Toxicity of Chromium

Chromium is a mineral that humans require in trace amounts, although its mechanisms of action in the body and the amounts needed for optimal health are not well defined. Chromium as heavy metal has no adverse effect. The trivalent form of chromium is considered as essential for normal carbohydrate and lipid metabolism (NRC 1980). Chromium is known to enhance the action of insulin (Mertz 1998), a hormone critical to the metabolism and storage of carbohydrate, fat, and protein in body. Little toxic effect is attributed to trivalent chromium when present in large concentrations. However, Cr (III) is ubiquitous in nature, occurring in air, water, soil, and biological materials. Cr (VI) has much higher toxicity in comparison to Cr (III), and the most important toxic effects, after contact, inhalation, or ingestion of hexavalent chromium compounds include dermatitis, allergic and eczematous skin reactions, skin and mucous ulcerations, perforation of the nasal septum, allergic asthmatic reactions, bronchial carcinomas, gastro-enteritis, hepatocellular deficiency, and renal oligo-anuric deficiency (Baruthio 1992).

32.3.6 In vivo Carcinogenicity of Chromium

Chromium occurs most commonly in valance states of +3 (III) and +6 (VI). Chromium III is the most stable oxidation state (Greenwood and Earnshaw 1997) and presumably is the form in the food supply due to the presence of reducing substances in foods. A dose of 5 mg chromium VI can be reduced to chromium III in 0.5 L of orange juice (Kuykendall et al. 1996), and endogenous reducing agents within the upper gastrointestinal tract and the blood also serve to prevent systemic uptake of chromium VI (Kerger et al. 1997). But in their in vivo toxicity the intermediates of Cr (V) and Cr (IV), or indirectly the reduced Cr (III) can be ultimately responsible.

In the carcinogenic behavior of chromium, chromate (CrO_4^{2-}) (which is a strong oxidizing agent) is reduced intracellularly to Cr^{5+} inside biological system and reacts with nucleic acids and other cell components to produce mutagenic and carcinogenic effects on biological systems (Clark 1994; McLean and Beveridge 2001).

But Cohen and Costa (2000) mentioned about derived trivalent form of chromium to be ultimately responsible for the carcinogenic effect. As the authors describe, the Cr (VI) ion is readily taken up into eukaryotic cells by anion-carrying proteins, after which it is reduced to Cr (III) by a number of cytoplasmic reducing agents. The final cellular form of chromium, Cr (III), becomes trapped intracellularly because it has low cell membrane permeability. This shift from Cr (VI) to Cr (III) allows a concentration gradient to be established such that a continual influx of Cr (VI) ions raises intracellular chromium levels until lethal burdens are achieved. While both valence states of chromium are able to interact with DNA, Cr (III) ions are responsible for decreasing the fidelity of DNA replication. In addition, both Cr (III) and Cr (VI) exhibit clastogenic potency; however, Cr (VI) possesses the greater activity and is also a powerful mutagen in many prokaryotic and eukaryotic cell systems. These properties of Cr (VI) support the claim that hexavalent compounds are likely to be active carcinogens, although it is more likely that the ultimate species responsible for the carcinogenic/mutagenic effects observed in vivo is the intracellularly derived trivalent form.

32.3.7 Assessment of Chromium Ecotoxicity

From the data, it is clearly seen that the mean chromium content of a single egg is 23.3809 μg, which exceeds adequate daily dietary intake of children up to 8 years of age as well as corresponds to major part for other age groups. As the *Institute of Medicine, USA* did not established a UL for chromium due to its adverse health effects observed in cases of high intakes, any intake exceeding the adequate daily dietary intake level is being considered undesirable. If a single poultry egg can contain more than the amount which should be taken from all dietary intakes in one day, then along with the chromium present in other food menus consumed in a day must turn the total amount exceeding the safety limit, especially for the children. The scenario is presented in Fig. 32.6.

32.3.8 The Target Population

The target population of tannery based chromium eco-toxicity is huge. Only in case of poultry, some 48% of our farmers use hand-mixed feeds which in most cases incorporate the various protein concentrates produced from tannery solid waste as protein supplement. In 2005, about 52% of poultry feed had been produced by feed mills and the rests by the farmers themselves. In the same year about 90% broiler

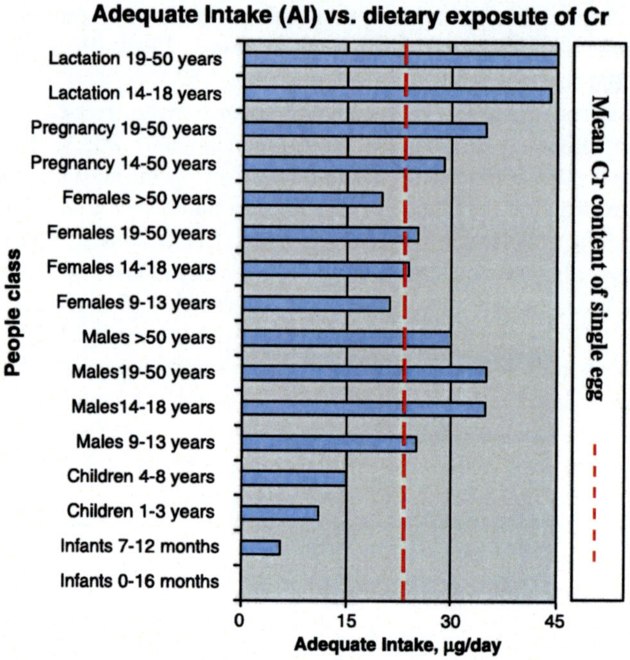

Fig. 32.6 Dietary exposure scenario of chromium from poultry egg

feeds had been produced by the feed mills, whereas only 18% layer feeds had been produced by feed mills (Poultry Business Directory 2007). This means, most of layer farmers used hand-mixed feeds which represents more severe health risk from eggs. In the same way, the products are supplied and added as primary protein supplements in fish culture and as organic fertilizer in vegetable farming.

32.3.9 Release of Lead (Pb) into Bangladesh Environment

The problem of Pb pollution in the capital city, Dhaka, Bangladesh, was identified in as early as 1980 (Khan et al. 1980) at the Atomic Energy Centre, Dhaka. In a survey of trace elements in whole blood of the adult population of Dhaka, the average Pb level was found to be 550 ± 180 µg/L in a selected group of 100 adults. The high level of Pb could not be understood as the automobile exhaust was not considered seriously at that time. At the same time, in food and drinks no abnormally high levels of lead were found (Ali et al. 1985; Khan et al. 1989). The problem remained an enigma until 1991, when high levels of Pb were found in suspended particles in air.

In 1999, unleaded gasoline was introduced to control auto-exhaust Pb pollution in Bangladesh. To assess the impact of the use of unleaded gasoline in Bangladesh, Airborne particulate matter (APM) was studied in Dhaka during the periods of

1994 and 1997–2000. The yearly average Pb concentration reached a maximum value of 370 ng/m^3 in the particulate matter with an aerodynamic diameter < 2.5 μm fraction in 1998. In 2000, the concentration decreased to approximately one-third (106 ng/m^3) of the high earlier values after the introduction of unleaded gasoline in the year 1999 (Biswas et al. 2003). However, there is a substantial amount of accumulated Pb in dust from earlier depositions. In 1999, some case studies around Dhaka-Aricha highway convincingly proved the migration of Pb to plants through deposition. The lead content in soil decreased with distance from highway as well as soil depth. Fruits and tree-leaves grown within 100 m of highway contained elevated concentration of lead (19.90–35.77 ppm). The average Pb concentrations in adjacent top-soil and sub-soil were 112.26 ppm and 74.65 ppm which was very high in comparison with the WHO recommended value (20 ppm). The lead accumulation by fruits in dry season was about two to six times higher than that of wet season (Ahmad et al. 2003).

Pb is still being used in an uncontrolled manner in paints and industry. It is, therefore, essential to continue the fight for lowering Pb level through removal of Pb in paints and control of emissions from suspected industries, especially Pb-based battery industries.

32.3.10 Regional Bio-concentration of Lead in Poultry Eggs

Using average physical parameters from Table 32.1, calculation of regional lead bioconcentration in eggs is shown in Table 32.5 (average dry weight basis).

The average statistics for all egg samples, irrespective of albumen and yolk separately counted are presented in Table 32.6.

The maximum permissible Pb concentration in chicken meat and fat is 0.1 μg/g (FAO/WHO 2008). So the mean regional bioconcentration of Pb is about 80 times higher (Table 32.6) than the maximum permissible limit. The possible reason may be the massive use of locally produced plant products by the layer farmers, where Pb might have been bioaccumulated through deposition on soil as well as through cycling into food chain.

32.3.11 Toxicity of Lead

Lead compounds may have a variety of targets within the nervous system causing neurotoxicity (Bondy 1988). Lead has multiple serious health effects. Acute lead poisoning in humans causes severe dysfunction in the kidneys, reproductive system, liver, and the brain and central nervous system; causing sickness or death. Lead poisoning from environmental exposure is thought to have caused mental retardation in many children. Mild lead poisoning causes anemia. The victims may have headaches and sore muscles and may feel generally fatigued and irritable (Manahan 1997).

Table 32.5 Calculation of regional lead bio-concentration in eggs

Sample class	Sample size and standard error	Pb in egg albumen (μg/g) (dry weight)	Pb in total egg albumen (μg)	Pb in egg yolk (μg/g) (dry weight)	Pb in total egg yolk (μg)	Total Pb in each egg (albumen + yolk) (μg)	Concentration in total edible egg (albumen + yolk) (μg/g)
Brown egg	6	5.5905	30.6918	18.3190	124.6680	155.3599	12.6356
(Dhaka)	Std. Error	1.1043	–	4.2439	–	–	–
White egg	6	11.1847	61.4041	6.4769	44.0777	105.4818	8.5790
(Dhaka)	Std. Error	1.7396	–	2.1627	–	–	–
Brown egg	12	10.4818	57.5453	15.0153	102.1849	159.7302	12.9911
(Narsingdi)	Std. Error	2.3770	–	3.1458	–	–	–
Brown egg	12	9.4945	52.1248	10.6030	72.1574	124.2822	10.1080
(Kishoreganj)	Std. Error	0.5917	–	0.7616	–	–	–
Brown egg	12	9.4769	52.0279	5.7009	38.7969	90.8248	7.3869
(Mymensingh)	Std. Error	1.1593	–	0.6024	–	–	–
Brown egg	6	11.0565	60.7000	5.4060	36.7898	97.4898	7.9290
(Tangail)	Std. Error	4.4361	–	1.8938	–	–	–
White egg	6	11.3587	62.3594	0.7730	5.2602	67.6197	5.4996
(Tangail)	Std. Error	2.4524	–	0.4713	–	–	–
Brown egg	6	6.8734	37.7349	1.6755	11.4023	49.1372	3.9964
(Gazipur)	Std. Error	2.0868	–	0.6178	–	–	–
White egg	6	5.9167	32.4825	2.2874	15.5666	48.0491	3.9079
(Gazipur)	Std. Error	2.1738	–	1.2855	–	–	–
Brown egg	12	0.7702	4.2285	5.4222	36.9002	41.1288	3.3451
(Narayanganj)	Std. Error	0.4071	–	1.9852	–	–	–
Brown egg (Munshigonj)	12	11.3360	62.2346	8.8177	60.0079	122.2426	9.9421
	Std. Error	2.3649	–	2.7654	–	–	–

Table 32.6 Statistics of all egg samples

Total sample number	192
Mean concentration (μg/g)	8.1611
Std. Error of Mean	0.5253
Minimum (μg/g)	ND[a]
Maximum (μg/g)	34.5637

[a]Not detected

High levels of Pb exposure are known to be fatal. Many people had been reported to die having their blood Pb levels elevated to 130 μg/dL (CDC 1991).

32.3.12 Food Safety Implications of Climate Change

There are many pathways through which global climate change and variability may impact environmental contamination and chemical hazards in foods. Climate-related extreme events, ocean warming and changes in surface temperature and humidity can result in food safety chemical hazards.

32.3.12.1 Through Contaminating Soils

Soils can be contaminated through remobilization of contaminated river sediments subsequently deposited on the flooded areas. Upstream contaminated terrestrial areas such as industrial sites, landfills, sewage treatment plants, etc. can contaminate soils through contaminating river waters and subsequent floods on the soils. Investigations after the Elbe and Mulde flooding in Central Europe in 2002 revealed presence of very high levels of polychlorinated dibenzo-p-dioxins and dibenzofurans (PCDD/Fs) in periodically flooded riverside pasturelands, and a significant transfer of these contaminants into livestock milk grazed on the floodplains (Umlauf et al. 2005). Sources of chemical contamination of floodwater following Hurricane Katrina included oil spills from refineries and storage tanks, pesticides, metals and hazardous waste (Manuel 2006). Several chemicals, such as hexavalent chromium, lead, manganese, p-cresol, toluene, phenol, 2,4-D, nickel, aluminum, copper, vanadium, zinc, and benzidine were detected in flood water while concentrations of most contaminants were within acceptable short-term levels, except for lead and volatile organic compounds in some areas (Pardue et al. 2005).

Arsenic is a well-recognized environmental problem in Bangladesh. Arsenic-contaminated groundwater is widely used for irrigation in Bangladesh during the dry season leading to increased levels of arsenic (As) in soils. On the contrary, monsoon flooding may lead to a reduction of topsoil-As contents almost to levels existing before irrigation. However, there are indications that As concentration in soils are increasing over-time because of irrigation with As-contaminated water. Though it is still unclear under what specific conditions and over what period of time As is accumulating in the soil, concentration of this contaminant in rice has been reported to increase over time because of prolonged input of contaminated irrigation water (Islam et al. 2005; Heikens 2006; Dittmar et al. 2007).

32.3.12.2 Through Contaminating Waters and Ocean Warming

Pollutants (pesticides, fertilizers, organic matter, heavy metals, etc.) are predicted to be increasingly washed from soils to water bodies in regions where intense rainfall is expected to increase (Environment Canada 2004). Higher runoff is expected to mobilize fertilizers and pesticides to water bodies in regions where their application time and low vegetation growth coincide with an increase in runoff (Soil and Water Conservation Society 2003). Because of compaction, heavy rainfall after drought can result in more severe runoff and an increased risk of certain types of contamination, thus alternating periods of floods and drought due to climate variability and change may aggravate the problem.

The Aral Sea area, which was once the fourth-largest lake in the world has become one of the world's largest environmental disasters. Agricultural mis-management (e.g. cotton monoculture, over irrigation, pesticide abuse) along with accelerated desertification due to both environmental degradation and climate change, have resulted serious contamination of soil, water and local foods with high levels of POPs and dioxins, leading to critical health and socio-economic impacts to local populations (Muntean et al. 2003).

Ocean warming facilitates methylation of mercury, and subsequent uptake of methyl mercury had been found to increase by 3–5% in fish and mammals for each 1°C rise in water temperature, thus increasing human dietary exposure (Booth and Zeller 2005).

32.3.13 Recent Outbreak of Lead Intoxication in Senegal

In explaining a death of 18 children living in the NGagne Diaw neighborhood of Thiaroye sur Mer, Dakar, Senegal, investigation by health and environmental authorities revealed that the area was contaminated with lead from an informal recycling of lead batteries (WHO 2008). In addition, siblings and mothers of the deceased children were reported to have extremely high blood lead concentrations, in many cases above 1,000 μg/L (in children, a concentration above 100 μg/L may impair neurological development, and 700 μg/L is considered to require immediate action). A number of children also showed evidence of neurological damage and environmental investigations resulted in very high concentrations of lead in soil both outside and inside of people's homes.

32.3.14 *Role of Metal Pollution in Affecting Climate Change Through Altering Ecological Niche*

The abundance of heavy metals in environment can have significant effects on community composition, and growth and metabolism of organisms in ecosystems. Some experimental studies report that, Pb treatment has a stimulating effect on soil enzymatic activities and microbial biomass carbon (C_{mic}) at low concentration and an inhibitory influence at higher concentration. When Pb treatment was raised to the level of 500 mg/kg, ecological risk appeared both to soil microorganisms and plants (Zeng et al. 2007). The results also revealed a consistent trend of increased chlorophyll contents and rice biomass initially, maximum at a certain Pb treatment, and then decreased gradually with the increase in Pb concentration. In a word, soil microbial biomass and community structure, therefore, may be sensitive indicators reflecting environmental stress of Pb (Zeng et al. 2007).

In experimental studies on response of *Saccharomyces cerevisiae* to elevated lead concentration it was reported that growth of *S. cerevisiae* shows a lag phase much longer than that in the absence of Pb (Chen and Wang 2007). Pb^{2+} at a concentration of 5 μM inhibited the microbial growth by approximately 30% with regard to control. The microbial assimilation of ammonium ion was also inhibited by the presence of Pb^{2+} in the medium with some 50% inhibition for Pb^{2+} concentration of 10 μM (Chen and Wang 2007). In the same way, higher levels of chromium can also have some definite role in altering community composition and metabolism of organisms. In addition, evolution can also take place by the introduction of chemical elements in new forms and present day change in chemical environment can also lead to such evolution (Williams 1997). As the chromium and lead problems are most likely to have become ecotoxicity phenomena in Bangladesh, these will also involve other living species that constitute the ecosystem and can also exert effects on climate change through shaping altered ecological niche. These effects can include emergence of new health problems or with factors driving climate change.

32.3.15 *Adaptation of Intensified Climate Change Effects in Bangladesh Perspective*

In Bangladesh, water-related impacts of climate change warrant the highest attention for adaptation due to the severity of such impacts and their socio-economic implications. High intensity flood scenario is of special importance for transport of environmental contaminants. If contaminants from high concentration source become more and more widespread, the ecotoxicity will tremendously increase because of the multiplier effect of cycling of chemicals through food chain. Moreover, climate change also has some direct food safety implications with various types of harmful microorganisms and their toxic metabolites. The widespread use of pesticides, especially organophosphate and carbamate compounds with impurities can also affect food safety in a changed climatic scenario in Bangladesh. With the other food safety problems such as arsenic, the newer dimensions of chromium and lead will certainly make the scenario worse. The various types of organic compounds released to environment in Bangladesh lack in their adequate assessment for ecotoxicity. The congregational effects of all of these food safety problems can be proven to be severe public health disaster in future, which will require special efforts to adapt with.

32.4 Conclusion

The current trajectory for warming and more extreme and unpredictable weather events could have catastrophic effects on South Asia region, especially Bangladesh. Other than the direct effects that chromium and lead can impose on public health through food chain, these can also bring about changes in species composition and interactions which can augment the emergence of unexpected changes including new health problems. Climate change can turn our living world subject to newly emerging diseases, while most of the developing countries are already subject to enormous disease burden (FAO 2008). Comprehensive risk assessment programs will have to be conducted regarding these new challenges this region is currently facing. The remediation options must be considered along with the indirect factors such as flood control. All possible entry of these metals beyond safety levels should be controlled by formulating comprehensive standards for all pathways towards food chain. Therefore any phenomenon allowing heavy metals to enter into food chain beyond those limits can readily be regarded as unsafe and will be directed for mitigation strategies. Adaptation with the impacts of new climatic conditions should also be emphasized.

References

Ahmad JU, Sayedur Rahman M, Islam R, Rahman MS (2003) Status of lead pollution in Bangladesh: case studies around Dhaka-Aricha highway. Bangladesh J Environ Sci 1:5–13

Ali M, Biswas SK, Akhter S, Khan AH (1985) Multielement analysis of water residue: a PIXE measurement. Fresenius Z Anal Chem 322:755–760

Baruthio F (1992) Toxic effects of chromium and its compounds. Biol Trace Elem Res 32:145–153

Biswas SK, Tarafdar SA, Islam A, Khaliquzzaman M, Tervahattu H, Kupiainen K (2003) Impact of unleaded gasoline introduction on the concentration of lead in the air of Dhaka Bangladesh. J Air Waste Manag Assoc 53:1355–1362

Bondy SC (1988) The neurotoxicity of organic and inorganic lead. In: Bondy SC, Prasad KR (eds) Metal neurotoxicity. CRC, Boca Raton FL

Booth S, Zeller D (2005) Mercury, food webs, and marine mammals: implications of diet and climate change for human health. Environ Health Persp 113:521–526

CDC (Centers for Disease Control and Prevention) (1991) Preventing lead poisoning in young children: a statement by the Center for Disease Control. CDC, Atlanta, USA

Chen C, Wang J (2007) Response of *Saccharomyces cerevisiae* to lead ion stress. Appl Microbiol Biotechnol 74(3):683–687

Clark DP (1994) Chromate reductase activity of *Enterobacter aerogenus* is induced by nitrite. FEMS Microbiol Lett 122:233–238

Cohen MD, Costa M (2000) Chromium. In: Lippmann M (ed) Environmental toxicants: human exposures and their health effects, 2/eth edn. Wiley, New York, pp 173–191

Dittmar JA, Voegelin LC, Roberts SJ, Hug G, Saha MA C, Ali A, Badruzzaman KR (2007) Spatial distribution and temporal variability of arsenic in irrigated rice fields in Bangladesh. Env Sci Tech Am Chem S 41(17):5967–5972

Environment Canada (2004) Threats to water availability in Canada. NWRI Scientific Assessment Report No 3, Prowse and ASCD Science Assessments series No. 1. National Water Research Institute, Burlington, Ontario

FAO (2008) Expert meeting on bioenergy policy, markets and trade and food security and global perspectives on fuel and food security, options for decision makers. FAO Headquarters, Rome, Italy. http://www.fao.org/fileadmin/user_upload/foodclimate/presentations/EM56/OptionsEM56.pdf (accessed on August 24, 2008)

FAO/WHO (2008) Codex general standard for contaminants and toxins in foods. CODEX STAN 193–1995, Rev.2–2006. http://www.codexalimentarius.net/web/ index_en.jsp (accessed on August 24, 2008)

Greenwood NN, Earnshaw A (1997) Chemistry of the elements, 2nd edn. Butterworth-Heinemann, Oxford

Heikens A (2006) Arsenic contamination of irrigation water, soil and crops in Bangladesh: risk implications for sustainable agriculture and food safety in Asia. RAP publication 2006/20. FAO, Regional Office for Asia and the Pacific

Hossain AMMM, Monir T, Rezwan-ul-Haque AM, Kazi MAI, Islam MS, Elahi SF (2007) Heavy metal concentration in tannery solid wastes used as poultry feed and the ecotoxicological consequences. Bangladesh J Sci Ind Res 42(4):397–416

Institute of Medicine Food and Nutrition Board (2001) Dietary reference intakes for vitamin A, vitamin K, arsenic, boron, chromium, copper, iodine, iron, manganese, molybdenum, nickel, silicon, vanadium, and zinc. National, Washington, DC

Islam MR, Jahiruddin M, Rahman GKMM, Miah MAM, Farid ATM, Panaullah GM, Loeppert RH Duxbury JM, Meisner CA (2005) Arsenic in paddy soils of Bangladesh: levels, distribution and contribution of irrigation and sediments. In: Behavior of arsenic in aquifers, soils and plants (Conference Proceedings) Dhaka

Kebbekus BB, Mitra S (1998) Environmental chemical analysis. Blackie Academic and Professional, London

Kerger BD, Finley BL, Corbett GE, Dodge DG, Paustenbach JP (1997) Ingestion of chromium (VI) in drinking water by human volunteers: Absorption, distribution, and excretion of single and repeated doses. J Toxicol Environ Health 50:67–95

Khan AH, Khaliquzzaman M, Zaman MB, Hussain M, Abdullah M, Akhter S (1980) Trace Element Composition of Blood in Adult Population in Bangladesh. J Radioanal Chem 56:157–167

Khan AH, Tarafdar SA, Ali M, Biswas SK, Akhter S, Saha DK, Islam A, Billah M, Hadi DA, Maroof FBA (1989) The status of trace and minor elements in Bangladeshi foodstuff. J Radioanal Nucl Chem 134:367–381

Kuykendall JR, Kerger BD, Jarvi JE, Corbett GE, Paustenbach DJ (1996) Measurement of DNA-protein cross-links in human leukocytes following acute ingestion of chromium in drinking water. Carcinogenesis 17:1971–1977

Manahan SE (1997) Environmental science and technology. CRC/Lewis, Boca Raton, FL

Manuel J (2006) In Katrina's wake. Environ Health Persp 114(1):A32–A39

McCarthy JJ, Canziani OF, Leary NA, Dokken DJ, White KS (eds) (2001) Climate change 2001: impacts, adaptation, and vulnerability, inter-governmental panel on climate change (IPCC), Work Group II Input to the Third Assessment Report. Cambridge University Press, Cambridge

McLean J, Beveridge JT (2001) Chromate reduction by a pseudomonad isolated from a site contaminated with chromated copper arsenate. Appl Environ Microbiol 67:1076–1084

Mertz W (1998) Interaction of chromium with insulin: a progress report. Nutr Rev 56:174–177
Muntean N, Jermini M, Small I, Falzon D, Fürst P, Migliorati G, Scortichini G, Forti AF, Anklam E, von Holst C, Niyazmatov B, Bahkridinov S, Aertgeerts R, Bertollini R, Tirado C, Kolb A (2003) Dietary exposure assessment to some persistent organic pollutants in the republic of Karakalpakstan of Uzbekistan. Environ Health Perspect 111:1306–1311
NRC (National Research Council) (1980) Mineral tolerance of domestic animals. National Academy Press, Washington, DC
NRC (National Research Council) FDB (Food and Nutrition Board) (1989) Recommended dietary allowances, 10th edn. National Academy Press, Washington, DC
Pardue JH, Moe WM, McInnis D, Thibodeaux LJ, Valsaraj KT, Maciasz E, van Heerden I, Korevec N, Yuan ZQ (2005) Chemical and microbiological parameters in New Orleans floodwater following Hurricane Katrina. Environ Sci Technol 39:8591–8599
Poultry Business Directory (2007) Poultry Business Directory 2007. Poultry Khamar Bichitra, Dhaka, Bangladesh
Stoecker BJ (2001) Chromium. In: Bowman B, Russell R (eds) Present knowledge in nutrition, 8/e. ILSI Press, Washington, DC, pp 366–372
Umlauf G, Bidoglio G, Christoph E, Kampheus J, KrügerF LD, Schulz AJ, Schwartz R, Severin K, Stachel B, Dorit S (2005) The situation of PCDD/Fs and Dioxin-like PCBs after the flooding of river Elbe and Mulde in 2002. Acta Hydroch Hydrob 33(5):543–554
UNIDO (2000) Regional programme for pollution control in the tanning industry in south-east Asia: chrome balance in leather processing. UN Industrial Development Organization, Vienna, Austria
Soil and Water Conservation Society (2003) Soil erosion and runoff from cropland. Report from the Soil and Water Conservation Society, USA
WHO (2008) Senegal: outbreak of lead intoxication in Thiaroye sur Mer, update 1–06 August 2008. World Health Organization, Geneva, Switzerland. http://www.who.int/environmental_health_emergencies/events/Senegal2008_update/en/index.html (accessed on August 24, 2008)
Williams RJ (1997) The natural selection of the chemical elements. Cell Mol Life Sci 53(10): 816–829
Zahid A, Balke KD, Hassan MQ, Flegr M (2004) Distribution of heavy metals in tannery effluent and their influence on sediments of Hazaribagh leather processing zone, Dhaka. In: Hassan MQ (ed) Water resources management and development in Dhaka City. Goethe-Institut Dhaka
Zeng LS, Liao M, Chen CL, Huang CY (2007) Effects of lead contamination on soil enzymatic activities, microbial biomass, and rice physiological indices in soil-lead-rice (Oryza sativa L.) system. Ecotoxicol Environ Saf 67(1):67–74

Part VIII
Policy Imperatives

Chapter 33
Metrics for Mainstreaming Adaptation in Agriculture Sector

Sivapuram Venkata Rama Krishna Prabhakar and Ancha Srinivasan

Contents

Abstract Agriculture is one of the most vulnerable sectors to climate change and mainstreaming climate change adaptation considerations at the sectoral level is paramount. The current available climate projections are not sufficient to take decisive actions on the ground. Hence, a framework is required for mainstreaming adaptation concerns at sectoral level. Stakes in adaptation are high since several countries are vulnerable and huge amounts of funds would have to be spent for adaptation. Identifying and institutionalizing a set of adaptation metrics helps in tracking the progress in adaptation. In this paper, a framework was presented for mainstreaming adaptation concerns in agriculture sector, some adaptation metrics and challenges and opportunities were identified to operationalize adaptation metrics. It has been concluded that operationalization of adaptation metrics require a context such as

S.V.R.K. Prabhakar (✉)
Institute for Global Environmental Strategies, 2108–11, Kamiyamaguchi, Hayama, Kanagawa 240–0115, Japan
e-mail: sivapuram.prabhakar@gmail.com

A. Srinivasan
Institute for Global Environmental Strategies, Hayama, Japan
and
Southeast Asia Department, Asian Development Bank, Manila, Philippines

R. Lal et al. (eds.), *Climate Change and Food Security in South Asia*,
DOI 10.1007/978-90-481-9516-9_33,

adaptation benchmarks and targets and main hurdle in identification of adaptation metrics is deciding between a set of qualitative and quantitative metrics using multi-criteria analytical approaches.

Keywords Adaptation • Mainstreaming • Effective adaptation options • Bali Action Plan • Risk management

Abbreviations

BAP	Bali Action Plant
GEF	Global Environmental Facility
GHG	Greenhouse gas
HDI	Human Development Index

Adaptation to climate change is important since the impacts of climate change will continue to be felt even if we bring down the greenhouse gas (GHG) emissions to the levels suggested by the Intergovernmental Panel on Climate Change (IPCC) in the next couple of years. However, to date, the global actions on climate change has largely been confined to mitigating the GHG emissions and the global efforts on adaptation have largely been slow and confined to few financing options available under Global Environmental Facility (GEF) of World Bank or financing by some multilateral and bilateral donor agencies.

Adaptation costs are going to be huge in the range of tens and hundreds of billion dollars (UNFCCC 2007a, b). Since stakes involved are high, there is a need to make sure that the adaptation actions identified and implemented help in effectively reducing the climate vulnerability of nations, sectors and communities at risk. This is where measuring adaptation comes into picture as an important component of adaptation framework in the climate regime that succeeds the Kyoto Protocol Regime after 2012.

33.1 Climate Change Adaptation in Agriculture Sector

Adaptation is important in all the sectors of society. However, it is important to recognize that some sectors are more vulnerable to climate change than others and hence there is a need for relatively higher emphasis on these vulnerable sectors, sectors those are directly influenced by changes in weather and climate. Agriculture is one of the climate vulnerable sectors (Fischer et al. 2002; Salinger et al. 2005; Butt et al. 2006). The inherent vulnerability of agriculture sector to climate owes to its direct linkage with different climatic and atmospheric variables such as temperature (of soil, water, atmosphere etc.), rainfall, humidity (in the micro-climate and macro-climate), and atmospheric carbon dioxide concentration which are directly impacted by the climatic change (IPCC 2007; Adams et al. 1998). Climate change could impact agriculture through droughts, floods, changes in soil temperature, change in seasonal patterns, early cessation or onset of cooling and warming periods which are associated with various stages of crop growth such as germination, vegetative growth, flowering, grain filling and maturity (Kumar 2006; Morison 1996).

Adaptation of agriculture sector to climate change is important for three reasons. Firstly, large proportion of population is dependent on agriculture in Asian region (about 47%) with relatively high percentage in Southeast (55.9%) and South Asia (55.7%) (FAO 2005). Secondly, large percent of this population live in rural areas which have poor access to various resources that would make them resilient to climatic vagaries (e.g. credit facilities, infrastructure such as roads, education and medical facilities etc.). Thirdly, the impacts of climate change would not stop with the agriculture sector itself but would also have rippling effects on the larger economy through multiple pathways and magnification effects (Fig. 33.1). For example, the loss of income from agriculture would affect the farmers' purchasing power and his dependence on natural resources such as forests (including natural and community forests) leading to destruction of natural environment. Repeated droughts have known to degrade the environment in terms of air quality as well as loss of vegetation further enhancing the vulnerability of rural population which is dependent on their physical as well as ecological services (Laurance and Williamson 2001; Wilhite 1993). Hence, the factors that make agriculture to fail would also be responsible behind the failure of interlinked sectors. One important aspect to be considered while identifying the agriculture vulnerability to climate change is complex linkages agriculture has with the rest of the society and the possible difficulties in isolating agricultural impacts from rest of the world. This also calls for clear definition of what 'agriculture' constitutes for the purpose of clear identification of impacts in the larger world.

33.2 Mainstreaming Adaptation in Agriculture Sector: A Framework

Mainstreaming adaptation refers to integrating the adaptation concerns in a sectoral or national level policy making. It typically involves broad range of activities and decisions ranging from consciously identifying the significance of adapting to climate change to taking up various actions such that the adaptation is realized. Figure 33.2 presents the sequence of steps to be taken for mainstreaming climate change adaptation in agriculture sector planning (Modified from Prabhakar et al. 2009). The scheme involves the following steps:

1. Identification of climate change impacts from regional and national studies
2. Ranking the impacts in the of their significance
3. Cross-checking local history with the broad regional trends to see if the location under question has tendency to follow the same trend as that of the regional trends
4. Identification of local strengths and weaknesses
5. Overlaying the identified possible local impacts over the local strengths and weaknesses

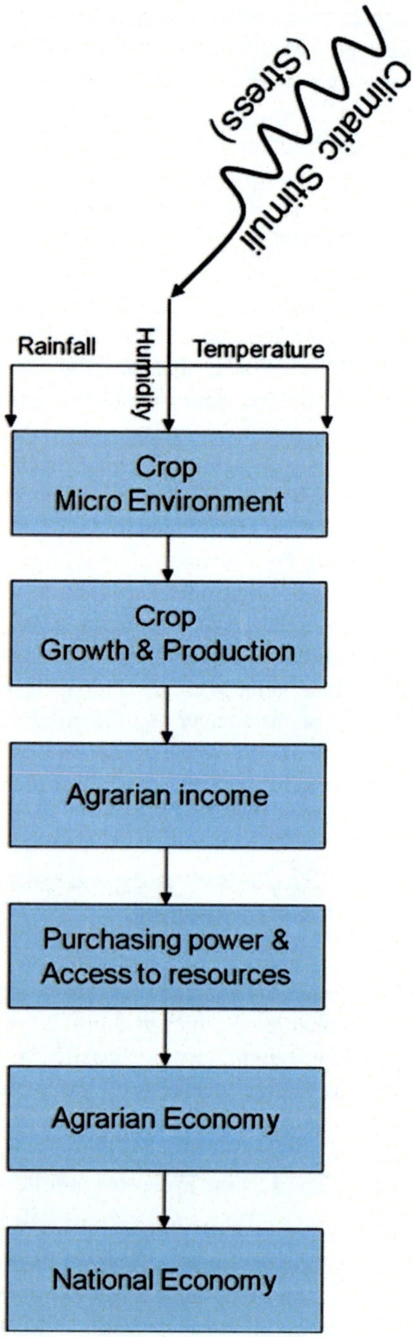

Fig. 33.1 A simplified scheme of climate change impact pathway on agriculture and its macro-economic linkages

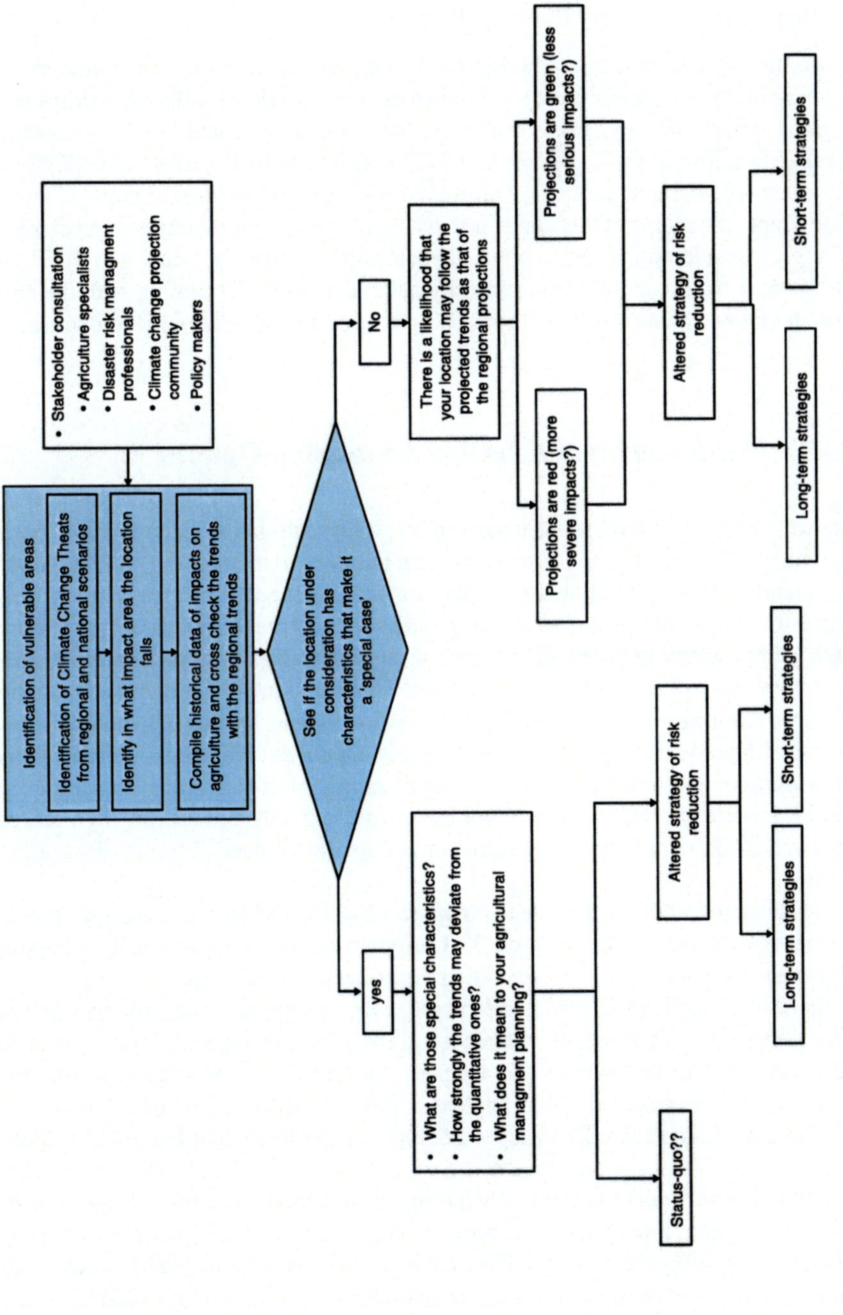

Fig. 33.2 Schematic diagram showing steps involved in mainstreaming climate change considerations in agriculture sector

6. See where the region under question falls in a range of categories such as still risky, out of risk, uncertain etc.
7. Design strategies to address the risks based on the identified broad categories supported by necessary resource allocation

For agriculture sector, this scheme also means identification of vulnerable crops (in addition to vulnerable areas), management practices those make these crops and regions vulnerable (e.g. excessive dependency on rainfall and lack of irrigation facilities), maladaptation practices, and existing strengths in terms of biophysical and socio-economic conditions. This model is scalable from local to national level. One simple advantage of this model is that it facilitates in easy mainstreaming climate change considerations even while dependable climate projects are lacking. However, what it cannot do is to identify suitable options for dealing with the perceived threats and how to monitor the progress in climate change risk reduction.

33.3 Identification of Effective Adaptation Options

So far under international climate regime, only mitigation actions have been measured. The objective of the measurement has been to monitor the progress made in climate change mitigation and to plan for future actions. The measurement was made through identifying and defining atmospheric greenhouse gasses, assessing their global warming potential and measuring their levels through establishing an inventory system among various countries. This inventory was established among Annex I countries which agreed for mitigation targets in the first commitment period of Kyoto Protocol (The Kyoto Protocol stipulates GHG emission reductions for a group of developed and industrialized countries referred as Annex I countries in the Protocol text). Measurement was necessary not only because of high stakes involved in GHG mitigation but also due to huge investments made for mitigation actions.

At this point, it is worthwhile to compare what enabled the mitigation actions to be measured at the first place. Table 33.1 provides a quick comparison of mitigation and adaptation and identifies areas where measurement had significant use.

In addition to the differences listed in the table, adaptation deals with the systems that are at different levels of vulnerability and adaptive capacity making it even more complex to understand and manage. Similarly, several adaptation options differ in their effectiveness when applied in different socio-economic contexts.

The Bali Action Plan (BAP) agreed during the Conference of Parties 13 in Bali, Indonesia in 2007, categorically stated the need for enhanced actions on adaptation through various means and especially through prioritization of actions, integration of adaptation actions into sectoral and national planning, specific programs and projects. The BAP also talks about incentivizing the implementation of adaptation actions and other means to foster climate-resilient development and reduce vulnerabilities.

How to prioritize and incentivize adaptation actions is the biggest question that would draw attention in discussions and negotiations towards a post 2012 regime.

Table 33.1 Adaptation metrics: mitigation vs. adaptation

Mitigation	Adaptation
Has a protocol (KP) that governs	No 'protocol' to govern adaptation
There are GHG reduction targets to meet with coordinated efforts	There are no 'adaptation targets' to meet
There are ways and means to *measure* the impact of collective actions	*No streamlined measurement system* for adaptation
Global actions and global benefits (more organized at global level)	Mostly local actions and local benefits (with some undeniable global spillover benefits)
Physical principles that govern mitigation	At nascent stages: complex interaction of biophysical and socioeconomic elements

Prioritization is possible by knowing where we want to go (adaptation targets), by setting a time frame for actions, and by knowing how much 'adaptation' we want to achieve at each stage in a series of stages drawn to meet the set target. Such a scheme is possible only when adaptation framework at global level is designed such a way that it includes the essential elements of adaptation targets and a measurement system that complements the adaptation targets. Measuring adaptation is an important means of achieving these multiple objectives of adaptation set by the Bali Action Plan. Measuring adaptation is also important for

1. The reason that adaptation has higher stakes now than ever since rapid climate change impacts are expected
2. The reason that huge amounts of funds would have to be invested in adaptation requiring accountability in how they are spent and how risks are mitigated
3. Prioritizing the adaptation actions according to their potential to reduce climate risks
4. Measuring the progress against an agreed benchmark (e.g. adaptation benchmark or baseline)

In addition, measurement is also necessary to check if adaptation concerns are appropriately and sufficiently 'mainstreamed' into the sectoral and national planning processes.

33.4 Measuring Adaptation

33.4.1 Underlying Principles/Pre-requisites for Adaptation Metrics

From the previous section, we determined that the climate change has multiple impacts on agriculture and that these impacts are not in isolation from the other changes happening at the global and local scales. From this perspective, it is important that any adaptation to happen should essentially be broad based. Several strategies have been proposed to design a system that is resilient to different changes and external stresses. Some of these strategies include facilitating self learning and evolving organizations (TERI and IISD 2006; Pelling and High 2004), and inculcating the attitude of looking at the local level players as innovators rather than as implementers (Prabhakar et al. 2009), which would collectively and eventually enable systems and societies to overcome the larger uncertainty involved in local level impacts of future climate change. Hence, measuring social learning and institutional evolution forms the basis for identifying effective adaptation options.

Adaptive management and transition management are found to be essential for dealing with the high levels of risk and uncertainty related to future social, technical, environmental and economic possibilities and outcomes (Foxon et al. 2008). While these two frameworks were developed independently, cross-framework learning was advocated to be essential for an effective adaptation process (van der Brugge and van Raak 2007). Both approaches emphasize the system learning and path dependent nature of change.

In order to measure adaptation, it is essential that the determinants of adaptation are identified. The ability of a system to adapt to external stimuli or forces causing it to change depends on a set of factors termed as determinants of adaptation. These factors could include economic resources, technology, information and skills, infrastructure, institutions, and equity. Figure 33.3 shows the relation between these determinants and climatic vulnerability.

Figure 33.3 shows that impacts depend on whether or not a system satisfies these determinants to the level of criticality. The criticality levels vary depending on the severity of climatic impacts and the system in question. This is the reason why

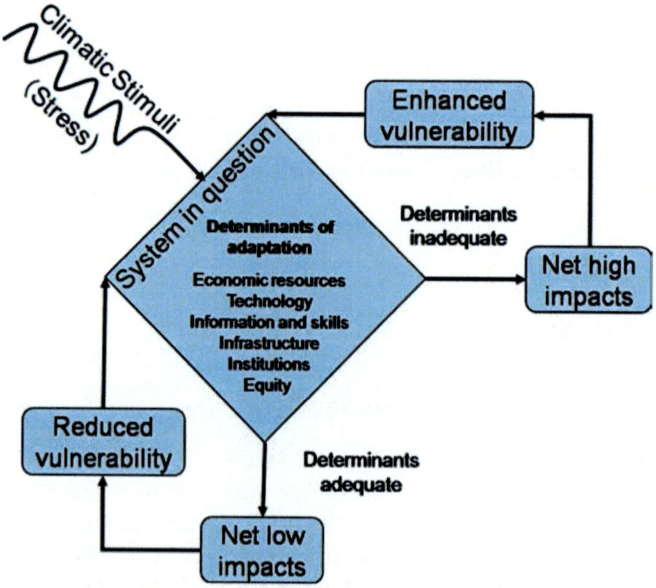

Fig. 33.3 Schematic diagram showing the relative relationship between determinants of adaptation with system vulnerabilities

the developed countries are known to have higher capacity to withstand climate change impacts than their developing country counterparts.

33.4.2 Methodologies for Identification of Adaptation Metrics

Different methodologies were used to assess the adaptation options and adaptive capacity of the system under question. Table 33.2 provides an overview of methodologies applied in the available literature. While literature on adaptation in agriculture was reviewed, literature from other subject fields such as natural disasters, human health and natural resources was also provided.

It appears that the methods could be as sophisticated as using simulation models and GIS based integration methods (Swanson et al. 2007) to simple methods such as cost-benefit analysis or expert elicitation using empirical (e.g. probit models) or non-empirical approaches (e.g. workshops). Multi-criteria evaluation methodologies also found to be useful as against single criteria approaches such as benefit-cost analysis or cost-effectiveness analysis (Dolan et al. 2001). In some of these studies, the availability and quality of data has been found as one of the limitations in approaching the problem at the local level (Swanson et al. 2007; Lemos 2007; Adger and Vincent 2005; Australian Government 2005; Yohe and Tol 2002) which is more pronounced in the developing countries where certain data (e.g. economic loss) is collected by agencies that are not trained in those subjects (e.g. health and humanitarian departments) (Yohe and Tol 2002).

33.4.3 Suggested Adaptation Metrics

Identification of specific impacts of climate change on agriculture would be necessary to identify suitable metrics to assess the adaptation actions since adaptation actions are designed to counteract the specific effects. The IPCC Fourth Assessment Report has identified some of the possible impacts of climate change on agriculture (Easterling et al. 2007). Some of the impacts reported include negative yield impacts on major cereals leading to food insecurity (Easterling 2007); complex impacts on smallholder and subsistence farmers, pastoralists and artisanal fisher folks (Easterling et al. 2007); reduced water flows from droughts and monsoon failures (Fischer et al. 2007), and most often these impacts may result in some of the countries importing more food than they used to (Easterling et al. 2007; Alcamo et al. 2007). Climate change is reported to impact all the four elements of food security i.e. availability, stability, utilization and access with a potential to threaten the food security of an additional 5–170 million people on the planet by 2080 (Schmidhuber and Tubiello 2007). Some others have also reported impacts such as enhanced soil degradation (ENFO 2004), CO_2 fertilization effects, incidence of pests and diseases, and higher irrigation water requirements (Tubiello et al. 2007).

Table 33.2 Different methodologies employed in assessing the adaptive capacity through metrics approach

Subject field	Methodology	Scope	Source
Agriculture	Benefit-cost analysis	Local, national and regional scales	Rosenzweig and Tubiello (2006), Rosenzweig and Tubiello (2007), Tubiello and Rosenzweig (2008)
	Cost-effectiveness analysis	Local, national and regional scales	Rosenzweig and Tubiello (2006), Rosenzweig and Tubiello (2007), Tubiello and Rosenzweig (2008)
	Multi-criteria analysis	Local, national and regional scales	Dolan et al. (2001), Rosenzweig and Tubiello (2006), Rosenzweig and Tubiello (2007), Tubiello and Rosenzweig (2008)
	Expert consultation (workshops)	Local, national and regional scales	Rosenzweig and Tubiello (2006), Rosenzweig and Tubiello (2007), Tubiello and Rosenzweig (2008)
	Dynamic crop models	Local, national and regional scales	Rosenzweig and Tubiello (2006), Rosenzweig and Tubiello (2007), Tubiello and Rosenzweig (2008)
	Modelling (simple regression to multi-variable predictive models) relationship between stressor and outcome variables	Local	Luers et al. (2003)
	GIS based index based on normalization and aggregation of determinants	Sub-national	Swanson et al. (2007)
	Historical trend analysis and constructing conceptual models	Sub-national	Allison and Hobbs (2004)
Natural resources	Case study approach	Community	Thomas et al. (2005)
Agriculture and water resources	Case study approach	Sub-national and national	TERI and IISD (2006)
Natural disasters	Delphi surveys and iterative Delphi surveys with expert judgment	National	Brooks et al. (2005)
	Building an indicator for coping capacity from the determinants of adaptive capacity	National	Yohe and Tol (2002)
Human health	Conjoint choice questions in expert judging engaging probit models	Society, individual, national	Alberini et al. (2005)

The literatures suggests that vulnerability of smallholder and subsistence agriculture farmers in developing Asia is mostly due to their location in tropical areas, and due to various socioeconomic, demographic, and policy trends limiting their capacity to adapt to change (Morton 2007). These farming systems area also difficult to model, which is essential in drawing useful adaptation programs, due to the difficulty in defining these farming systems due to wide variation among them, intrinsic characteristics of these systems in aspects such as complexity, location specificity, and their integration of livelihood and non-livelihood strategies, and their vulnerability to a wide range of climate related stressors. Similarly, assessing the impacts of climate change on agriculture in the developing Asia and suggesting possible adaptation measures become complicated due to the wide uncertainty in the projected socio-economic conditions in the region (Schmidhuber and Tubiello 2007). Similarly, there are suggestions that the impact of climate change could be much smaller than the impacts of socio-economic development in the region (Tubiello and Fischer 2007). This draws us to the fact that any adaptation metrics identified should essentially consider the climate change impacts as well as the larger global changes including socio-economic changes. Different metrics for measuring adaptation in agriculture sector is presented in Table 33.3.

Table 33.3 Adaptation metrics proposed in the available literature

Metric/s	Reference	Description on availability and limitations (includes authors judgment)
Mean and variability of yield and production, income, aggregate of value added	Tubiello and Rosenzweig (2008)	Measured and computed metrics. Available at local, national, regional and international levels in many countries. The aggregate of value added may need to be computed at the local level as such statistics will not be readily available.
Nutrition index	Tubiello and Rosenzweig (2008)	Computed metric (sum of local production and net imports divided by total food demand). Can be computed at national and regional level.
Yield estimates (remotely sensed), yield variability, highest relative yield/yield percentile	Luers et al. (2003)	Estimates could help in filling the gaps in the existing yield data, validating the measured yield data etc. Accuracy could be an issue when resolution of remote sensing is low.
Agricultural export, farm income, out-migration from farming, emergency payments	Venema (2006)	Agricultural exports and out-migration of farming are mostly applicable at the macro-economic level, while data on rest of the metrics (emergency payments) could be sparingly available.
Sources of income, livestock number, source of fertilizer	Brooks and Adger (2004)	It was not clear on how many sources of income is considered as optimal, and also the number of cattle. However, it is suggested that the higher the sources of income, with more diversification into non-farm sources, the higher the adaptive capacity.

33.4.4 Link Between Adaptation Metrics and Other Indicators

The metrics we mentioned in the previous section are not independent in themselves and they often influence or have correlation with other national and global indicators which may not be directly based on the metrics themselves. We checked the correlation between yileds of rice crop with the Human Developmnet Index (HDI) of United Nations (it should be noted that crop yileds are not considered in calculatoin of HDI) (refer to Fig. 33.4). The scatter plot of HDI values against rice yileds (t/ha) in some of the countries in Asia Pacific (Australia, Bangladesh, Bhutan, Cambodia, Chiina, India, Indonesia, Japan, Republic of Korea, Lao PDR, Myanmar, Nepal, Pakistan, Philippines, Sri Lanka, Thailand, and Vietnam) indicated a reasonably strong correlation between both the parameters and high positive correlation coefficient ($r=0.83$) indicating a direct dependence of one on another.

One important precaution to be taken is to check if all the metrics are applicable to all the countries and contexts. For example, in the case of Bangladesh, there is a close correlation between weather (as represented by rainfall) and GDP (Fig. 33.5) which is quite different from other countries such as India (Fig. 33.6) in the same region. Hence, choosing metrics should also be context specific.

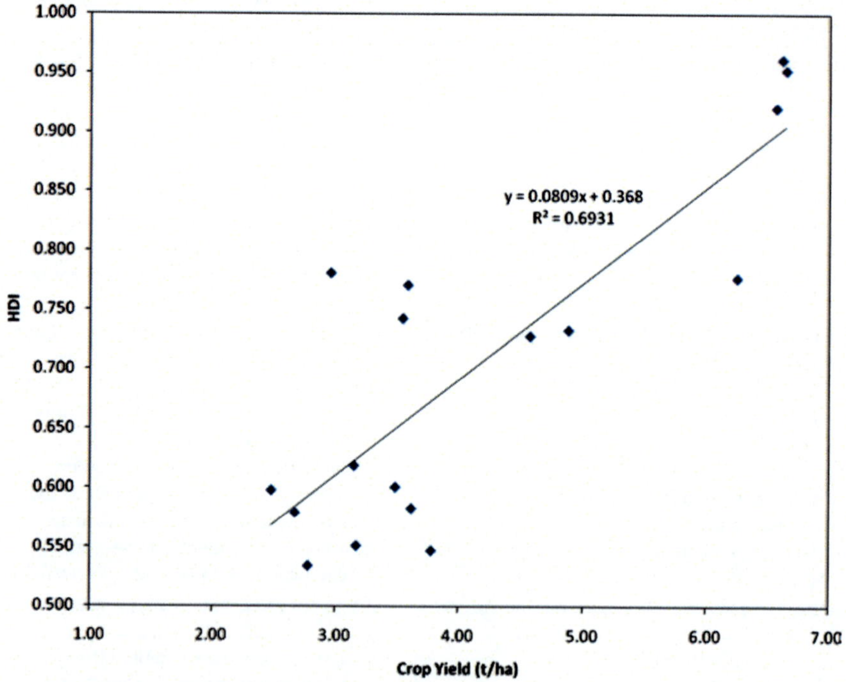

Fig. 33.4 Correlation between HDI and national level yields (t/ha) of rice crop across different countries (HDI figures from UNDP Human Development Report 2008 and yield figures from FAO Food and Agriculture Statistics)

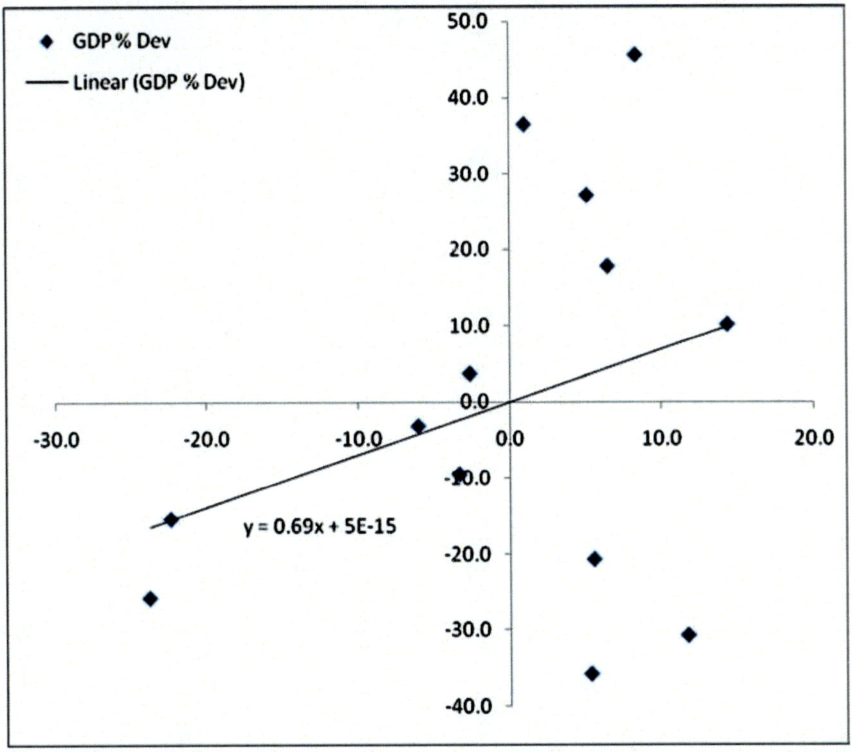

Fig. 33.5 Link between development and weather patterns (Bangladesh) (Source of data: GDP data from Bangladesh Bureau of Statistics, Rainfall data from Bangladesh Meteorological Department)

33.4.5 Measuring Adaptive Capacity of People and Policies

Assessing adaptive capacity is necessary for the purpose of checking the progress in adaptation as well as for implementing the adaptation strategies (Brooks and Adger 2004). Adaptive capacity would have to be assessed both for the people as well as for the policies as policies could undermine the progress made or the ability of the people to adapt to long-term stresses. Measuring adaptive capacity cannot be done using a blanket approach as the nature of adaptation, adaptation process and capacity to be adaptive differs at various scales. Similarly, adaptive capacity is context based. For example, the underlying factors determining adaptive capacity to floods differ from the adaptive capacity factors for economic changes or environmental changes such as deforestation and loss of related livelihoods.

Adaptive capacity at national level was found to be dependent on health, governance, political rights, and literacy, and economic well-being (Adger and Vincent 2005). However, there exists lot of uncertainty in understanding these factors as well as in their quantification which could lead the adaptation in different

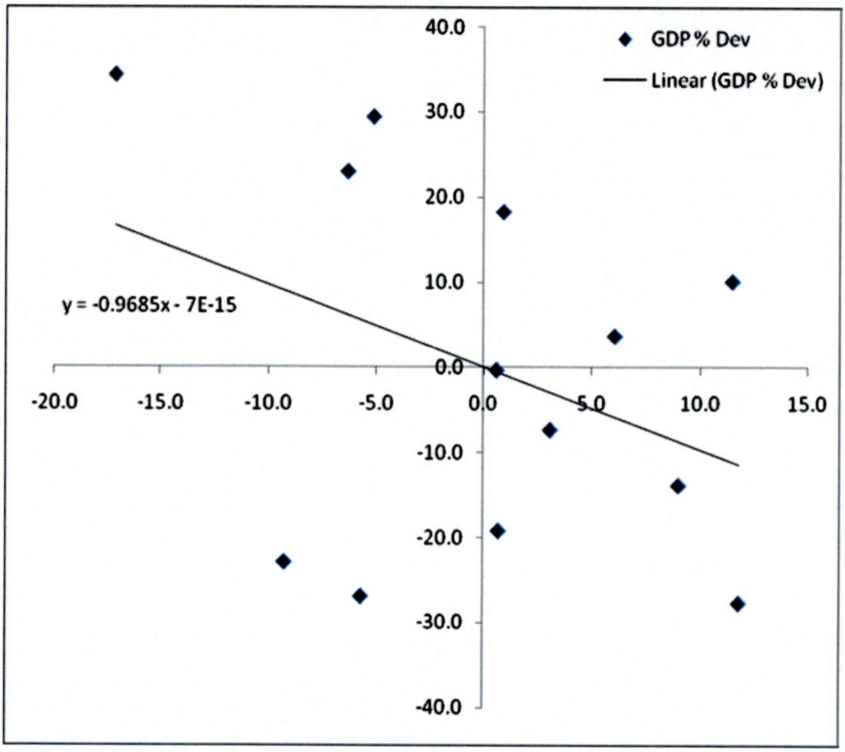

Fig. 33.6 Link between development and weather patterns (India) (Source of data: GDP data from Ministry of Statistics and Program Implementation, government of India, Rainfall data from Indian Meteorological Department)

direction than desired. Hence, it was suggested that the nature of uncertainty be understood through a traceable theoretical account since mere presence of adaptive capacity doesn't necessarily mean that the system would take a desirable decision which is intern dependent on the process of decision making and that there are no Newtonian laws guiding understanding of adaptation processes or elements (Adger and Vincent 2005).

Typical measures of adaptive capacity at national level could include nations' wealth, technology, education, information, skills, infrastructure, access to resources, and management capabilities (Haddad 2005). However, these are broad indicators and don't necessarily reflect the adaptive capacity of the agriculture sector in specific or of the people associated with it. Hence, sector specific adaptation metrics are necessary. Some work has already been taken up elsewhere on measuring adaptation or adaptive capacity in the agriculture sector (Dolan et al. 2001; Rosenzweig and Tubiello 2007; TERI and IISD 2006; Swanson et al. 2007; Allison and Hobbs 2004; Tubiello and Rosenzweig 2008; Rosenzweig and Tubiello 2006; Fischer et al. 2007; and Tubiello and Fischer 2007).

33.5 Conclusions and Way Forward

From this paper we discern that there is a possibility to identify a set of adaptation metrics those could be applicable at reasonably broader scales enabling operational use. However, there is much more that deserves attention in terms of methodological and capacity related issues which needs a concerted action from all the stakeholders.

There is an immediate need for adaptation metrics to be operationalized. These metrics should be policy-relevant, scalable, transferable, context specific and comparable. Methodological challenges needs to be overcome through concerted actions at all levels. A mix of qualitative and quantitative metrics is important rather than a single type. This is because the reason that all regions in the world, especially developing countries, may not have data available for reliable estimates on adaptation. It is also necessary to look into the cost implications of such data intensive methodologies. Hence, a mix of qualitative and quantitative methodologies would be most appropriate. Metrics in agriculture and water sectors could be complex due to complex interaction of social, economic and environmental factors; metrics should be able to not only help measure peoples adaptive capacity but also the adaptive capacity of policies and institutions; operationalization of metrics has a capacity angle which needs to be considered with priority; and any emphasis on metrics should not lead to overemphasis and lead to administrative and bureaucratic hurdles.

Operationalizing adaptation metrics is an important next step after identification of metrics. Operationalizing means that the data collection, analysis, and decision making are integrated with the existing decision making systems at all levels, from ground level to the state and national level policy decisions. This requires seamless integration for an effective and efficient outcome. Since adaptation funds most come from international sources, efforts should also be made to harmonize these procedures across different countries. A certain kind of convergence is highly appropriate which can be made possible by establishing proper framework that brings various governments and stakeholders together.

References

Adams RM, Hurd BH, Lenhart S, Leary N (1998) Effects of global climate change on agriculture: an interpretative review. Climate Res 11:19–30

Adger WN, Vincent K (2005) Uncertainty in adaptive capacity. Comptes Rendes Geoscience 337:399–410

Alberini A, Chiabai A, Muehlenbachs L (2005) Using expert judgment to assess adaptive capacity to climate change: evidence from a conjoint choice survey. Sustainability indicators and environmental valuation. September 2005. Fondazione Eni Enrico Mattei, Italy

Alcamo J, Dronin N, Endejan M, Golubev G, Kirilenko A (2007) A new assessment of climate change impacts on food production shortfalls and water availability in Russia. Global Environ Chang 17:429–444

Allison HE, Hobbs RJ (2004) Resilience, adaptive capacity and the lock-in trap of the Western Australian agricultural region. Ecol Soc 9:3

Australian Government (2005) Climate change risk and vulnerability: promoting an effective adaptation response in Australia. Report to the Australian Greenhouse Office, Department of Environment and Heritage

Brooks N, Adger N (2004) Assessing and enhancing adaptive capacity. In: Lim B, Spanger-Siegfried E (eds) Adaptive policy frameworks for climate change: developing strategies, policies and measures. United Nations Development Program and Cambridge University Press, pp 165–181

Brooks N, Adger WN, Kelly PM (2005) The determinants of vulnerability and adaptive capacity at the national level and the implications for adaptation. Global Environ Chang 15:151–163

Butt TA, McCarl BA, Kergna AO (2006) Policies for reducing agricultural sector vulnerability to climate change in Mali. Climate Policy 5:583–598

Dolan AH, Smit B, Skinner WM, Bradshaw B, Bryant CR (2001) Adaptation to climate change in agriculture: Evaluation of options. Occasional Paper No. 26. Department of Geography, University of Guelph, Candida

Easterling WE (2007) Climate change and the adequacy of food and timber in the 21st century. Proc Natl Acad Sci USA (PNAS) 104:19679

Easterling WE, Aggarwal PK, Batima P, Brander Km, Erda L, Howden SM, Kirilenko A, Morton J, Soussana JF, Schmidhuber J, Tubiello FN (2007) Food, fibre and forest products. In: Change ML, Parry OF, Canziani JP, Palutikof PJ, van der Linden, Hanson CE (eds) Climate change 2007: impacts, adaptation and vulnerability. Contribution of Working Group II to the Fourth Assessment Report of the Intergovernmental Panel on Climate. Cambridge University Press, Cambridge, UK, pp 273–313

ENFO (2004) Climate change: impacts. The Environmental Information Service, Dublin, Ireland

FAO (2005) Selected indicators of food and agriculture development in Asia-Pacific region. Food and Agriculture Organization, Rome, p 21

Fischer G, Shah M, Velthuizen HV (2002) Climate change and agricultural vulnerability. Report prepared for the World Summit on Sustainable Development, Johannesburg. International Institute for Applied Systems Analysis, Vienna, Austria

Fischer G, Tubiello FN, Velthuizen HV, Wiberg DA (2007) Climate change impacts on irrigation water requirements: effects of mitigation, 1990–2080. Technol Forecast Soc Chang 74: 1083–1107

Foxon TJ, Reed MS, Stringer LC (2008) Comparing adaptive management and transition management approaches to governing long-term social-ecological change. Ökologisches Wirtschaften 2:20–23

Haddad BM (2005) Ranking the adaptive capacity of nations to climate change when sociopolitical goals are explicit. Global Environ Chang 15:165–176

IPCC (2007) Climate change 2007: synthesis report. In: Pachauri RK, Reisinger A (eds) Contribution of Working Groups I, II and III to the Fourth Assessment Report of the Intergovernmental Panel on Climate Change. IPCC, Geneva, Switzerland, p 104

Kumar S (2006) Climate change and crop breeding objectives in twenty first century. Curr Sci 90:1053–1054

Laurance WF, Williamson GB (2001) Positive feedbacks among forest fragmentation, drought, and climate change in Amazon. Conserv Biol 15:1529–1535

Lemos MC (2007) Drought, governance, and adaptive capacity in North East Brazil: a case study of Ceara. Human Development Report 2008/2008, United Nations Development Program

Luers AL, Lobell DB, Sklar LS, Addams CL, Matson PA (2003) A method for quantifying vulnerability, applied to the agricultural system of the Yaqui Valley, Mexico. Global Environ Chang 13:255–267

Morison JIL (1996) Climate change and crop growth. Environ Manage Health 7:24–27

Morton JF (2007) The impact of climate change on smallholder and subsistence agriculture. PNAS 104:19680–19685

Pelling M, High C (2004) Social learning and adaptation to climate change. Paper read at International Work-Conference on Challenges of Complexity in Coping with Climate-Related Disasters. Wageningen Disaster Studies, 14–15 June 2004, Dreibergen

Prabhakar SVRK, Srinivasan A, Shaw R (2009) Climate change and local level disaster risk reduction planning: need, opportunities and challenges. Mitig Adapt Strateg Glob Change 14:7–33

Rosenzweig C, Tubiello FN (2006) Developing climate change impacts and adaptation metrics for agriculture. Presented at Global Forum on Sustainable Development on the Economic Benefits of Climate Change Policies, 6–7 July 2006, Paris, France

Rosenzweig C, Tubiello F (2007) Metrics for assessing the economic benefits of climate change policies in agriculture. 20 December 2007. Working party on global and structural policies, Organization for Economic Cooperation and Development, Paris, France

Salinger MJ, Sivakumar MVK, Motha R (2005) Reducing vulnerability of agriculture and forestry to climate variability and change: Workshop summary and recommendations. Clim Change 70:341–362

Schmidhuber J, Tubiello FN (2007) Global food security under climate change. PNAS 104:19703–19708

Swanson D, Hiley J, Venema HD (2007) Indicators of adaptive capacity to climate change for agriculture in the prairie region of agriculture: an analysis based on Statistics Canada's Census of Agriculture. IISD Draft Working Paper for Adaptation as Resilience Building. April 2007. International Institute for Sustainable Development, Canada

TERI and IISD (2006) Designing policies in a world of uncertainty, change, and surprise: adaptive policy-making for agriculture and water resources in the face of climate change. Phase I Research Report. The Energy and Resources Institute, New Delhi and International Institute for Sustainable Development, Canada

Thomas D, Osbahr H, Twyman C, Adger N, Hewitson B (2005) Adaptive: adaptations to climate change amongst natural resource-dependant societies in the developing world: across the southern African climate gradient. Technical Report 35, Tyndall Centre for Climate Change Research

Tubiello FN, Fischer G (2007) Reducing climate change impacts on agriculture: Global and regional effects of mitigation, 2000–2080. Technol Forecast Soc Chang 74:1030–1056

Tubiello FN, Rosenzweig C (2008) Developing climate change impact metrics for agriculture. Integr Assess J 8:165–184

Tubiello FN, Soussana JF, Howden SM (2007) Crop and pasture response to climate change. PNAS 104:19686–19690

UNFCCC (2007a) An assessment of the funding necessary to assist developing countries in meeting their commitments relating to the Global Environment Facility replenishment cycle. Subsidiary Body on Implementation, UNFCCC, November 2007, Bonn, Germany

UNFCCC (2007b) Investment and financial flows to address climate change. Background paper on analysis of existing and planned investment and financial flows relevant to the development of effective and appropriate international response to climate change, UNFCCC, 2007, Bonn, Germany p 273

van der Brugge R, van Raak R (2007) Facing the adaptive management challenge: insights from transition management. Ecol Soc 12:33

Venema HD (2006) The Canadian prairies: Biophysical and socio-economic context. In: TERI and IISD (eds) Designing policies in a world of uncertainty, change, and surprise: adaptive policy making for agriculture and water resources in the face of climate change. TERI, New Delhi and IISD, Canada

Wilhite DA (1993) Drought assessment, management, and planning: theory and case studies. Kluwer, USA

Yohe G, Tol SJ (2002) Indicators for social and economic coping capacity – moving toward a working definition of adaptive capacity. Global Environ Chang 12:25–40

Chapter 34
A Framework for Regional Cooperation on Integration of Disaster Risk Reduction and Climate Change Adaptation in South Asia

Sanjay K. Srivastava

Contents

Abstract Climate change has created newer grounds for disaster vulnerability in South Asia. Two third of the natural disasters the region experiences are having climate origins. In the recent times, their intensity, frequency and unpredictability are on the rise. The political leadership of the region has been sensitive to this emerging challenge. The South Asian Association for Regional Cooperation (SAARC) as a regional platform has made efforts to address the disaster vulnerability emanating from climate change collectively. The 29th Session of SAARC Council of Ministers, the SAARC Ministerial Meeting on Climate Change on July 3, 2008 at Dhaka signals a new beginning by adopting the SAARC Action Plan on Climate Change.

SAARC Disaster Management Centre (SDMC), the product of a collective vision to harness regional cooperation and enhance the resilience of vulnerable

S.K. Srivastava (✉)
SAARC Disaster Management Centre (SDMC), NIDM Building, IIPA Campus, 5-B, I.P. Estate, MG Road, New Delhi 110002, India
e-mail: sanjay.saarc@gmail.com

R. Lal et al. (eds.), *Climate Change and Food Security in South Asia*,
DOI 10.1007/978-90-481-9516-9_34, © Springer Science+Business Media B.V. 2011

South Asia to the natural disasters, has placed priorities on building regional cooperative mechanism addressing Climate Change Risks, Impacts and Adaptation and also developing a conceptual framework integrating Disaster Risk Reduction (DRR) strategies into Climate Change Adaptation (CCA).

There are practices in the region which bring in convergence between DRR and CCA strategies, while there are also areas identified, especially on institutional and programmatic fronts, responsible for the existing divergence. Harnessing the advances in Science & Technology and the recent progress made in Early Warning Systems as well as networking of the knowledge institutions enable integrating the methods, tools & techniques of DRR with the multi-sectoral dimensions of CCA in the operational arena. Proof-of-the-concept could be seen if climate related disaster risk 'hotspots' are targeted for designing risk reduction strategies and integrating climate, weather & EWS information in decision makings for the causes and purposes common to both DRR & CCA. Following the participatory approach wherein the concerns and voices of all SAARC member countries find expressions, SDMC is evolving a road map addressing the growing vulnerability of region due to climate change to support the SAARC Member States.

Keywords Climate change adaptation • Disaster risk reduction • SAARC disaster management centre • Hyogo framework of action • Bali action plan

Abbreviations

AWS	Automatic Weather Stations
CCA	Climate Change Adaptation
DRR	Disaster Risk Reduction
DWR	Doppler Weather Radars
GLOF	Glacial Lake Outburst Floods
SAARC	South Asian Association for Regional Cooperation
SCZMC	SAARC Coastal Zone Management Centre
SDMC	SAARC Disaster Management Centre
SMRC	SAARC Meteorology Research Centre
UNFCC	United Nations Framework Convention on Climate Change

34.1 Introduction

South Asia, the home for one fifth of the humanity, has perennially been a disaster-prone region. Two thirds of the disasters the region experiences are climate related and there have been phenomenal increase in their frequency, severity and unpredictability in the recent times. The severest impacts have been visualized in terms of sea level rise leading to submergence of low-lying coastal areas and depletion of Himalayan glaciers threatening the perennial rivers that sustained food, water, energy and environment security of the region. The climate change is surely creating grounds for newer and more severe risks of disasters in the region in the coming years (Allen and Ingram 2002; Lal et al. 2001a, b; McCarthy et al. 2001; O'Brien et al. 2004; Gosain et al. 2006).

Further, layers of vulnerabilities in the region – poverty, illiteracy, mal-nutrition and social inequities – are aggravating the risks from stresses on water, agriculture and environment and creating recipes of more disasters. With climate sensitive agrarian economies, all the countries of the region would be facing serious crisis unless the rising temperature of the globe and the region are checked and new technologies, practices and life styles are developed and adapted according to the changing climate scenarios (Richards 2003; Christoplos et al. 2001; Bruce et al. 1996). Therefore, climate change mitigation and adaptation have emerged as important tools for disaster risk reduction for all the countries in the region.

So far climate change and disaster management communities of the region have been working in relative isolation, with the former focusing more on long term modeling and projections of climate scenarios and their possible impacts and the later concentrating on short term preparedness and response to disaster events. The time has come when the implications of future climate projections for the current risks and vulnerabilities are understood and accordingly these are factored into the policies and programmes developed for reducing the risks of disasters. Surely there should be greater dialogue and interaction between the two communities so that the limited efforts for climate change analysis and adaptation and disaster risk reduction in the region can be integrated to the extent this possible and new innovative tools and methodologies developed for such integration in development projects and practices. Although, efforts to bring together stakeholders in climate change and disaster management have begun to create an opportunity for integration, challenges would lie not only in harmonizing diverse institutional structures and distinct sectoral planning and policies etc. but also in translating the common grounds into projects on the grounds (Smit and Pilifosova 2001; Smit et al. 2000).

Increasing trends of natural disasters and their threatening impacts on lives and livelihoods have resulted in a paradigm shift in disaster management in all the countries of South Asia – from one post disaster relief and rehabilitation to holistic management of management of disasters covering all phases of disasters (National Disaster Management Act 2005). The focus is clearly on Disaster Risk Reduction (DRR) – preparedness, mitigation and prevention. Many of the risk reduction measures particularly those related to hydro-meteorological disasters, such as drought proofing, flood protection, saline embankment and bio-shields, alternative livelihood development etc. have similarities with Climate Change Adaptation (CCA) programmes (Lim and Spanger-Siegfried 2005). Therefore synergies between DRR and CCA would be necessary not only to avoid duplicities and derive optimal benefits from scarce resources but also to add value to the projects through lessons learnt from the respective perspectives (Thomalla et al. 2006).

Factoring climate change issues in disaster risk mitigation projects would enrich the projects and make them more relevant to the emerging concerns just as risk management tools would assess climate change from the perspectives of risks and vulnerabilities over time and the cost-benefit of alternative strategies of adaptation.

34.2 Global Efforts

At global level, the *Hyogo Framework for Action 2005–2015: Building the Resilience of Nations and Communities to Disasters* adopted at the World Conference on Disaster Reduction in Kobe Japan lays emphasis on scaling up the use of disaster risk reduction tools viz., vulnerability and risk assessments, early warning systems, land-use planning, techno-legal regime for development practices, besides enhancing the institutional and legal capacities. The focus has also been placed on the integration of knowledge on disaster risk reduction and application of such knowledge to bridge the gaps in management of disaster risks. Each of these action areas can have significant bearings in climate sensitive sectors (UNISDR Report 2007 and UNISDR Report 2008a).

The United Nations Framework Convention on Climate Change (UNFCC) recognized that the climate system is a shared resource whose stability can be affected by industrial and other emissions of carbon dioxide and other greenhouse gases. The Convention adopted the Kyoto Protocol in the third conference of parties in Kyoto Japan in 1997 which commits the developed countries to stabilize greenhouse gases and provide a tool for such stabilization through the promotion of Clean Development Mechanisms in the developing countries. In the second most important initiative after Kyoto Protocol, the Convention adopted the Bali Action Plan in December 2007 which calls for enhanced action on adaptation, considering in particular risk management and risk reduction strategies, including risk sharing and transfer mechanisms such as insurance and other disaster reduction strategies. For the first time ever disaster risk reduction was included as a tool for climate change adaptation, which will guide the negotiations for a post Kyoto climate change agreement from 2012, This has opened up a range of possibilities for integration of climate change adaptation in disaster risk reduction strategies (UNISDR Report 2008b).

34.3 Regional Efforts

The need for regional cooperation addressing the concerns for environmental degradation in South Asia was voiced way back in 1987 during the Third SAARC Summit. The trans-boundary linkages of natural disasters with environment in the region were recognized for regional cooperation. The SAARC initiated a 'Regional Study on the Causes and Consequences of Natural Disasters and the Protection and Preservation of Environment' in 1991 and another study on 'Greenhouse Effect and its Impact on the Region' in 1992, which recommended regional measures in sharing experiences, scientific capabilities and information on climate change, sea level rise, technology transfer etc. (Regional Studies on Causes and Consequences of Natural Disasters, Reproduced by SDMC 2008). As a follow-up to these studies, SAARC Plan of Action on Environment was adopted in 1997. The Action Plan provided for the establishment of Regional Centers of Excellence. The SAARC

Meteorology Research Centre (SMRC) was established in Dhaka in 1995; the SAARC Coastal Zone Management Centre (SCZMC) was set up in Male in 2004; SAARC Disaster Management Centre (SDMC) came up in New Delhi in 2007 and the SAARC Forestry Center has come into the existence in Bhutan recently. All these SAARC Regional Centers provide credible institutional support for taking up climate change and disaster risk reduction issues in the region.

The Fourteenth SAARC Summit held in New Delhi in 2007 expressed 'deep concern' over the global climate change and called for pursuing a climate resilient development in South Asia. The member countries pledged for immediate collective action and stronger regional co-operation for the conservation and utilization of SAARC shared resources towards addressing the negatives of climate change. Further, the SAARC Council of Ministers, at their 29th Session held in New Delhi in December 2007, adopted the SAARC Declaration on Climate Change which reflects the collective vision of South Asia. On behalf of the SAARC H.E. the President of Maldives presented the Declaration in the UNFCC meeting at Bali in December 2007.

The SAARC Ministerial Meeting on Climate Change held on July 3, 2008 in Dhaka adopted the SAARC Action Plan on Climate Change. The emphasis has been laid on intensifying the regional cooperation on climate change adaptation. The effort is also placed on moving the SAARC from a declaratory to implementation phase with the SAARC Regional Centres to play the important role therein. The SAARC Meteorological Research Centre, the SAARC Coastal Zone Management Centre, SAARC Disaster Management Centre and SAARC Forestry Centre are tasked to contribute synergistically with their respective mandates in enhancing the SAARC climate change resilience by pursuing SAARC Action Plan on Climate Change.

The 15th Summit Meeting of Heads of States or Governments of SAARC countries held in Colombo on 2–3 August 2008 has endorsed the SAARC Action Plan and Declaration on Climate Change adopted by the Environment Ministers at Dhaka on 3 July 2008.

The SAARC Action Plan on Climate Climate Change and July 2008 stresses that the primary responsibility of implementing the Action Plan, proposed for an initial period of 3 years, rests with the National Governments. With regard to the regional cooperation, the Action Plan envisages that a mechanism should be agreed upon to effectively use the existing institutional arrangements of SAARC by giving clear directions and guidance.

SAARC Disaster Management Centre (SDMC) attaches a very high priority on implementing the SAARC Action Plan on Climate Change. In fact, SDMC, in its strategy to evolve the road maps on various themes, has taken into account the integration of disaster risk reduction into climate change adaptation as one of its priority areas of action. The SAARC Workshops on Science & Technology Applications in Disaster Risk Reduction in January 2008 in New Delhi and Coastal and Marine Risks in May 2008 in Goa emphasized on exchange of information and research on the linkages between climate change adaptation and disaster risk reduction in the region (Climate Change and Disasters, SDMC Publication 2008).

34.4 National Efforts – Emerging Trends

While there are efforts in South Asian countries to directly address climate change adaptation issues, through the development of National Adaptation Plans of Action (NAPAs), National Action Plan for Climate Change, their integration to disaster risk reduction need specific priority. In order to address adaptation concerns as part of their national development plans, the explicit focus on disaster risk is seen only in few cases (Prabhakar and Shaw 2008). For example, the Safe Island programme of Maldives is an integrated effort on addressing vulnerability through strategic planning for climate change adaptation (Box 34.1). Similarly, coastal zone management efforts in India, Pakistan and Sri Lanka are yet another example in this direction.

Except few cases in the arena of coastal zone management (Box 34.2) and also in case of integrated watershed development programs (Box 34.3), there is a clear disconnect between the institutional and legislative systems developed to address disaster risk and those developed to address climate change. The emphasis is to be laid on climate-related development outcomes – in areas such as agriculture, water resources, food security, health, the environment and livelihoods – that are sensitive to both climate variability and change (Smithers and Smit 1997).

Box 34.1 Safe Islands Programme of Maldives

The December 2004 tsunami exposed the risk Maldives is facing by virtue of having its unique geography and topography. Low elevation above sea level, perennial beach erosion, and dispersal of population across very small islands, remoteness and inaccessibility of islands, concentration of economic activities on tourism, high dependence on imports and high diseconomies of scale have added layers of coastal vulnerability in the country. Climate change and associated risks add to the growing vulnerability further. Maldives has developed the Safe Islands Programme, focusing on the development of the larger islands with better economic opportunities, high environmental resilience, and incentives for voluntary migration to these islands. To mitigate future risk from disasters, land use plans of the safer islands have been developed incorporating features of high resilience: with a wider environmental protection zone, elevated areas for vertical evacuation in case of floods, establishment of alternative modes of communication and energy and detailed disaster management plans. Currently five islands have been identified for the programme and development plans prepared in consultation with people. Challenges for the programme include geographical population dispersion, difficulties of access to islands, logistical difficulties, a high unit cost of delivery of construction material, inadequate human resource to manage projects and above all unpredictable weather and rough seas. But the Maldives are working to reduce the underlying risk and vulnerability factors that at the moment make them among the most "at risk" countries in the world

[Source: Report on Implementation of The Hyogo Framework for Action (HFA): Asia, ISDR/Gp/2007/Inf.5]

In South Asia where both climate-related hazard and vulnerability levels are likely to be drastically affected by climate change, it is necessary, based on the regional cooperation among South Asian countries, to establish systematic integration between the institutional frameworks, policies and strategies to address disaster risk with those related to adaptation to climate change.

Box 34.2 Coastal Zone Management for risk reduction in India: Highlights of Prof MS Swaminathan Report.

A National Coastal Zone Management Action Plan with an objective "to protect the coastal zone with people's participation, the livelihood security of the coastal fisher and other communities, and the ecosystem, which sustains the productivity of the coastal areas, while promoting sustainable development that contributes to nation's economy and prosperity." The salient features of the Committee's recommendations are as follows:

The Committee has recommended the reclassification of the coastal zone into four Zones i. e.:

1. Coastal Management Zone-I - consists of areas designated as Ecologically Sensitive Areas such as Mangroves, Coral reefs, Sand Dunes, Inland tide/water bodies such as estuaries, lakes, lagoons, creeks & straits, Mudflats, Marine parks and sanctuaries, Coastal forests & wildlife, Coastal fresh water lakes, Salt Marshes, Turtle nesting grounds, Horse shoe crabs habitats, Seagrass beds, Sea weed beds, Nesting grounds of migratory birds.
2. Coastal Management Zone-II - consists of areas identified as Areas of Particular Concern such as economically important areas, high population areas and culturally/strategically important areas. The administrative boundaries of these areas would be boundaries of CMZ-II.
3. Coastal Management Zone-III - consists of all other open areas including the coastal seas but excluding those areas classified as CMZ-I, CMZ-II and CMZ -IV.
4. Coastal Management Zone-IV - consists of Islands of The Andaman and Nicobar and Lakshadweep.

For the purpose of managing the above areas, the Committee has suggested Integrated Coastal Zone Management Approach. After taking into account the recent Tsunami, the Committee has laid emphasis on demarcation of vulnerability line all along the coastal areas and has suggested developmental activities to be regulated on the seaward side of the vulnerability line. Since the coastal management is a multi-disciplinary subject, the Committee has suggested a National Sustainable Coastal Zone Management Institute along with organisational structure to address issues relating to policy, law, conflict resolution and to creating public awareness. The Committee has laid special emphasis on developing bio shields all along the coastal areas by intensive plantation of mangroves, casuarina, etc.

[Source: Report of Ministry of Environment and Forest, Govt of India, 2005]

Box 34.3 Watershed development for CCA and DRR

India lives in villages, particularly where there are large tracts of arid and semi-arid areas with poor farmers battling with low productivity and sub-standard living conditions. Most of these farmers depend heavily on rainfall for agricultural production and sustenance. An innovative program of participatory watershed development project (Sujala in Karanataka State in Southern part of India) is implemented in five drought prone districts covering an area of around 0.5 Mha, and benefiting more than 400,000 households. Remote sensing & GIS products have been operationally used in Sujala project from the early stages of watershed prioritization, database and query system development to project action plan generation. The unique feature of the project is the way remote sensing, GIS and the Management Information System (MIS) are dynamically linked with the impact assessment both in terms of development of natural resources as well as socio-economic indicators. The approach of integrating these tools and techniques has been participatory through community themselves. Climate 'proofing' to watershed development is factorized in the action plan.

The mid-term assessment on the impact of the Sujala Watershed Development Project carried out has indicated very encouraging trends. The average crop yields have increased by 24% over the baseline. The average ground water level has increased by 3–5 ft. Shift to agro-forestry and horticulture, and reduction in non-arable lands has also been observed. Annual household income from employment, income generating activities and improvements in agricultural productivity has increased by 30% from a baseline. The 'extra mile' was prototyping a system ensuring greater transparency, social mobilization, inclusive growth and capacity building at the grassroots.

[Source: Ranganath et al. 2006]

A key challenge, in this context, is to strengthen regional capacities to manage and reduce risks associated with existing climate variability. To achieve this, closer linkages need to be forged between the policy arenas of climate change and disaster risk reduction, at national, regional and international levels.

Further, at global level, the implementation of the Hyogo Framework needs to be more clearly recognized as a primary tool to achieve the adaptation goals of the UN Framework Convention on Climate Change (UNFCCC). The reflection of such integration assumes greater importance and urgency in the climate risk hotspot of South Asia through regional cooperation under the SAARC Framework of Disaster Management.

34.5 Integration of Disaster Risk Reduction in Climate Change Adaptation

Climate change adaptation envisages climate proofing to the development programmes. At community level, there is autonomous adaptation which depends on the inherent adaptive capacity. Capacity could be enhanced further by more investment and absorption of technologies. It's a basically a planned adaptation and most of the interventions at policy and investment levels are made to realize this. Further, there are externalities, like Bali Action Plan, which bring priorities on strengthening both planned as well as autonomous adaptation (Burton 2004; Ivey et al. 2004; Moench and Dixit 2004) (Fig. 34.1).

Disaster risk reduction is similar to climate change adaptation in several ways but not the exact. There is autonomous resilience which brings in coping at the community level. Preparedness for risk reduction calls for applications of technologies and also investments. There are externalities like Hyogo Framework of Action which place focus on disaster risk reduction by strengthening autonomous resilience as well as preparedness. What is therefore required is the integration of disaster risk reduction to climate change adaptation and vice versa (Pearce 2003; Pittock and Jones 2000).

Integration of Disaster Risk Reduction (DRR) into Climate Change Adaptation (CCA) would be one of the challenges of risk management in South Asia. The task can be addressed by identifying those areas which create divergence between DRR and CCA processes, as also those which create convergence between the two. The forces that create divergence are the following:

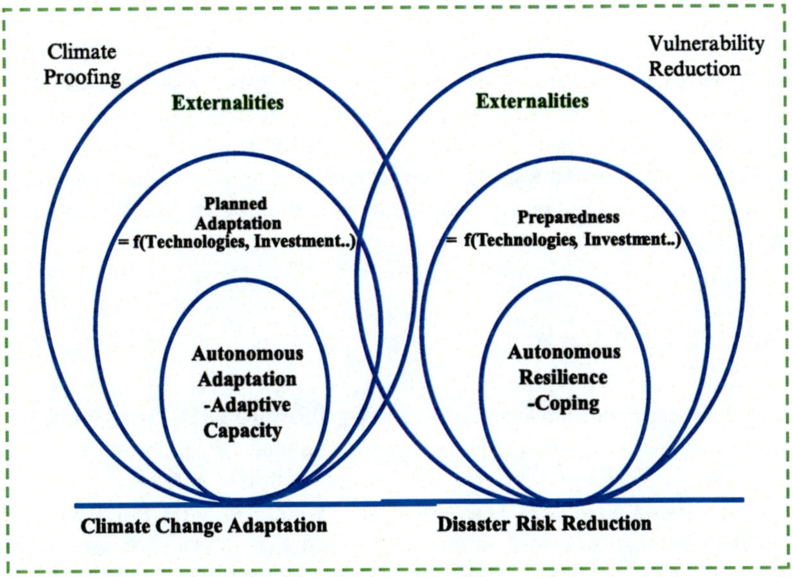

Fig. 34.1 Twin processes of Climate Change Adaptation and Disaster Risk Reduction

(a) **Diverse Institutional Structure** The institutional arrangements that exist in South Asian countries are such that DRR and CCA experts and functionaries are usually different, respond to different needs and to the different constituencies and do not have authority to implement policy decisions in the areas other than their specific responsibilities. In fact, such structural barriers also exist at international and regional levels.
(b) **Disconnected Policies, Planning and Programs** DRR and CAA policies, planning and programs often take place in isolation without sharing the respective goals, methodologies and objectives.
(c) **Lack of Relevant Information** Information concerned with DRR and CCA are inherently complex which cannot be packaged easily for integration into respective concerns. DRR related info, for example, often does not describe environmental and socio-economic information of underlying risk factors which are required in support of pursuing CCA.
(d) **Ad-hoc Short-term Approaches** For most of DRR projects, risks to investments are not considered for the full life-time of the project and thus ignores climate change risks, impact and adaptation factors.

The convergence between DRR and CCA processes has been observed in certain types of projects which need to be recognized for scaling-up and replications in the region, especially through regional cooperation. These are

(a) Integrated Coastal Zone Management
(b) Participatory Watershed Development Programme
(c) Land Use Planning in areas sensitive to climate and disaster risks
(d) River-basin Floodplain Management
(e) Integrated Drought Mitigation (Fig. 34.2)

The tools and techniques used for DRR such as early warning systems, hazard, risk and vulnerability analysis, risk assessment and monitoring, risk mitigation as well as response strategies need to be integrated with CCA strategies in the critical sectors like human health, food, water and environmental security, agriculture, forestry, tourism, etc. There are success stories and good practices demonstrating such integration, which should be replicated and further scaled up (Govt of India 2006; Kerr et al. 2002).

34.5.1 Policy Advisory

There are enabling mechanisms for integrating DRR and CCA through integration of appropriate technologies like ICTs, Space, Automatic Weather Stations (AWS), Doppler Weather Radars (DWR) etc. Similarly, networking of DRR and CCA institutions at national, regional and global levels coupled with multi-stakeholder communication and dialogues as well as exchange of information and expertise may catalyze such integration. From the 'conceptual framework' as outlines above to 'actionable strategies', the following steps are suggested:

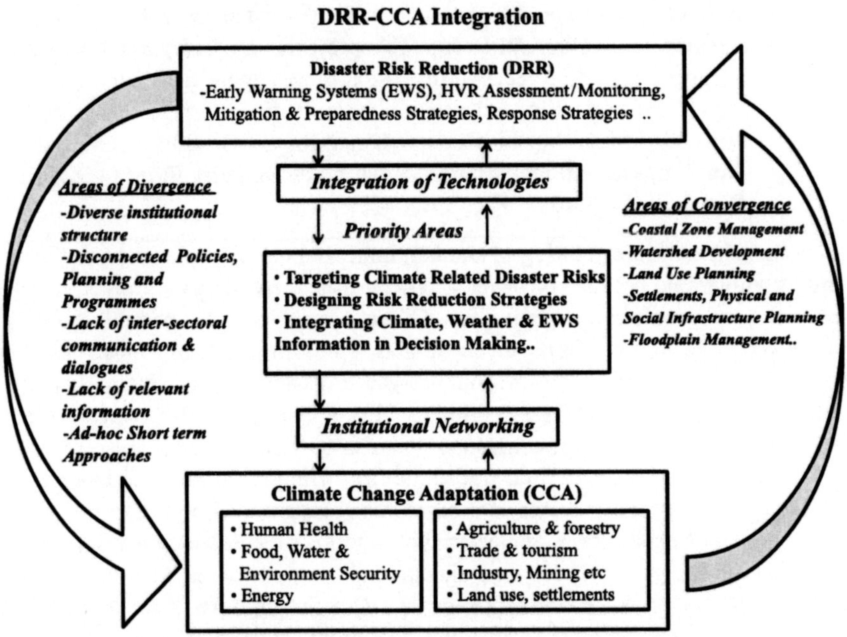

Fig. 34.2 Converging and diverging factors of climate change adaptation and disaster risk reduction

Step I: *Targeting Climate Related Disaster Risks* Most of the Hazard, Vulnerability and Risk (HVR) Assessment efforts are based on the frequency of occurrence of disasters in spatial and temporal domains. Climate risks are not captured well and also the simulated climate change scenarios are not factorized to target the climate related disaster risks especially in the 'hotspots' of South Asian region. While the strategy calls for recasting HVR mapping efforts, such efforts enable closer integration of DRR and CCA in the operational domain of end-to-end project implementation.

Step II: *Designing Risk Reduction Strategies* Designing Risk Reduction Strategies for hydro-meteorological risks must essentially be based on using the knowledge of climate risks. For instance, if it is to develop an effective and people's centred EWS to provide 'actionable' information about a climate hazard to a vulnerable population, the assessment of climate risk should form the key inputs. Further, the strategies must be dynamic and in tune with the changing practices and conditions such as depletion of the ecological foundation of the natural resources such as coral reefs and mangrove forests may aggravate risks; further effective insurance and micro-finance initiatives to transfer risks and provide additional resources may reduce risks.

Step III: Integrating Climate, Weather & EWS Information in Decision Making Besides implanting DRR in CCA projects, it is important to utilize advanced climate forecast information in managing risks from the existing climate variability and also utilize results from climate change models especially where known climate change impacts lead to a certain direction viz., glaciers retreat and Glacial Lake Outburst Floods (GLOF) in Himalayan region.

As climate hazards are growing in number, more and more people in the region are turning vulnerable because of poverty, powerlessness, population growth, and the movement of people to marginal areas. Climate change has the potential to derail the poverty alleviation efforts in the region, punishing first and most, the very people least responsible for greenhouse-gas emissions – and increasing their vulnerability to the natural disasters further. Concerted national efforts are necessary in support of climate change adaptation and disaster risk reduction.

Uniquely, with the inherited traditional knowledge, South Asia has got the civilizational heritage in terms of indigenous coping and community resilience. These heritages need further empowerment in terms of technology and knowledge to withstand the potential climatic shocks and their extremes. Further, with the growing climate risk, the adaptive capacity in South Asia is to be enhanced by providing the necessary financial resources, access to technology and knowledge, and by enhancing the institutional capacity. For example, the capital-intensive agricultural systems are less sensitive to climate, perhaps because they can control so many more inputs. Agriculture, water management, land use practices etc. in South Asia are therefore to harmonize with changing climate regimes.

34.5.2 Thematic Areas for Road Map

In order to evolve the roadmap, the workshop discussed the following thematic areas of the Regional Action Plan on Climate Change (Climate Change and Disasters , SDMC Publication 2008) on which SDMC was expected to play its role in promoting action for national action and regional cooperation.
Thematic Area One: Adaptation to Climate Change

- Adaptation to climate change impacts and risks in vulnerable communities, locations and ecosystems
- Adaptation in sectors (e.g. water, agriculture, fisheries, health and biodiversity)
- Adaptation to extreme climate events (e.g. flood, cyclone, glacial lake outburst, droughts and heat and cold waves)
- Adaptation to climate change impact (e.g. sea level rise, salinity intrusion, glacial melt and coastal and soil erosion)
- Adaptation suited to urban settlements, coastal structures and mountain terrain

Thematic Area Six: Management of Impacts and Risks Due To Climate Change

- Climate risk modeling and capacity building in the region on impact assessment of climate change.
- Sharing of information and capacity building in the management of climate change impacts and risks through cooperation among SAARC member states in early forecasting, warning and adaptation measures
- Cooperation amongst the SAARC member states in exchange of information on climate and climate change impacts (e.g. sea level rise, glacial melts, droughts, floods, etc.)
- Cooperation and sharing of good practices in disaster management

Priority Action Plan

- Exchange of information on disaster preparedness and extreme events
- Capacity building and exchange of information on climate change impacts (e.g. sea level rise, glacial melting, biodiversity and forestry)

34.6 Road Map for Implementation of SAARC Action Plan on Climate Change by SAARC Disaster Management Centre

Towards integrating DRR in CCA, the SDMC has adopted the following strategies and road map for the implementation of the Action Plan.

(i) *Adaptation to Climate Change* The experiences gained and lessons learnt from the existing and past initiatives on Disaster Risk Reduction (DRR) in the different countries of the region should be systematically integrated with Climate Variability/Climate Change Adaptation (CCA) projects and vice versa. The Centre shall formulate appropriate process and programme guidelines for integration of DRR in CCA projects and vice versa in respect of four natural disasters namely, floods, cyclones including saline intrusion, droughts and glacial lake outbursts for the guidance of the Member States.

(ii) *Technology Transfer* SDMC shall develop a Concept Paper on technology need assessment for integrating adaptation to climate variability and change into disaster risk reduction, especially those related to Early Warning Systems for drought and flood and submit the same to the National Governments and other relevant SAARC Regional Centres for their consideration.

(iii) *Finance and Investment* SAARC Disaster Management Centre shall study the potential application of Micro-credit, Micro-insurance and Crop Insurance for climate change adaptation in selected climate risk hotspots of the region.

(iv) *Education and Awareness* SAARC Disaster Management Centre shall develop tool kits on Climate Risks and Disasters for education and awareness of the people of the region.

(v) *Management of Impact and Risks Due To Climate Change* SDMC in collaboration with all relevant institutions shall develop Training Modules on Climate Risk Assessment relevant to the contexts of the South Asia region and conduct regional training programme for capacity building on climate risk assessment.

34.7 Conclusion

While the Hyogo Framework of Action has brought Disaster Risk Reduction to the central focus, Bali Action Plan has given impetus to Climate Change Adaptation. Both these sectors through share several common boundaries but not integrated fully especially in the context of high risk disaster prone areas. Such integration is a must for the sustainable development efforts in South Asia, which is emerging as risk 'hotspots' in the changing climate regime.

The SAARC as a platform provides a framework for regional cooperation in integration of Disaster Risk Reduction to Climate Change Adaptation and vice-versa. The SAARC Action Plan on Climate Change is a timely initiative in this direction. While regional cooperation does provide 'the extra mile', the efforts that promote such integration are always taken up at national/local levels. Climate 'proofing' to the development processes and Disaster Risk Reduction strategies does fill the gaps and brings in convergence between CCA and DRR. Besides promoting key applications such as coastal zone management, watershed development etc., SAARC Disaster Management Centre envisages bringing out the necessary tools and techniques to integrate DRR and CCA. Mandated to promote DRR and CCA integration, SAARC Disaster Management Centre has taken up concrete steps in this direction on the basis of having a framework for regional cooperation.

Acknowledgement The author would like to place on record the ideas and support from Mr. P.G. Dhar Chakrabarti, Director, SAARC Disaster Management Centre to develop this paper and enable the author to attend International Conference at Dhaka.

References

Allen MR, Ingram WJ (2002) Constraints on future changes in climate and hydrological cycle. Nature 419:224–232

Bruce JP, Lee H, Haites EF (1996) Climate change 1995: economic and social dimensions of climate change. Cambridge University Press, Cambridge

Burton I (2004) Vulnerability and adaptive response in the context of climate and climate change. Clim Change 36(1–2):185–196

Christoplos I, Mitchel J, Liljelund A (2001) Re-framing risk: the changing context of disaster mitigation and preparedness. Disasters 25(3):185–198

Climate Change and Disasters – Emerging Trends and Future Strategies (2008) SAARC Workshop, August 21–22, 2008, Kathmandu, Publication by SAARC Disaster Management Centre, New Delhi. http://saarc-sdmc.nic.in/pdf/publications/climate/cover.pdf

Gosain AK, Rao S, Basuray D (2006) Climate change impact assessment on hydrology of Indian river basins. Curr Sci 90(3):346–353

Government of India (2006) Guidelines for Hariyali (2003). Department of Land Resources, Ministry of Rural Development, Government of India, New Delhi. http://dolr.nic.in/HariyaliGuidelines.htm. Cited 18 Oct 2006

Ivey JL, Smithers J, De Loe RC, Kreutzwiser RD (2004) Community capacity for adaptation to climate induced water shortages: linking institutional complexity and local actors. Environ Manag 33(1):36–37

Kerr JM, Pangare G, Pangare VL (2002) Watershed development projects in India: an evaluation. Research Report 127, IFPRI, Washington, p 86

Lal M, Harasawa H, Murdiyarso D (2001a) Asia. In: McCarthy JJ, Canziani OF, Leary NA, Dokken DJ, White KS (eds) Climate change 2001: impacts, adaptation, and vulnerability: contribution of working group II to the third assessment report of the intergovernmental panel on climate change, Intergovernmental panel on climate change. Cambridge University Press, UK, pp 877–912

Lal M, Nozawa T, Emori S, Harasawa H, Takahashi K, Kimoto M, Abe-Ouchi A, Nkajima T, Takemura T, Numaguti A (2001b) Future climate change: implications for Indian summer monsoon and its variability. Curr Sci 81:1196–1207

Lim B, Spanger-Siegfried E (2005) Adaptation policy frameworks for climate change: developing strategies, policies and measures. Cambridge University Press, UK, p 258

McCarthy JJ, Canziani OF, Leary NA, Dokken DJ, White KS (2001) Climate change 2001: impacts, adaptation, and vulnerability: contribution of working group II to the third assessment report of the intergovernmental panel on climate change, intergovernmental panel on climate change. Cambridge University Press, UK, p 1008

Moench M, Dixit A (2004) Adaptive capacity and livelihood resilience: adaptive strategies for responding to floods and droughts in South Asia. Institute of social and environmental transition, Boulder and institute of social and environmental transition, Nepal, p 214

National Disaster Management Act (2005) Ministry of Law and Justice. Government of India, New Delhi, India, p 29

O'Brien K, Leichenko R, Kelkar U, Venema H, Aandahl G, Tompkins H, Javed A, Bhadwal S, Barg S, Nygaard L, West J (2004) Mapping vulnerability to multiple stressors: climate change and globalization in India. Glob Environ Chang 14:303–313

Pearce L (2003) Disaster management and community planning, and public participation: How to achieve sustainable hazard mitigation? Nat Hazards 28(2–3):211–228

Pittock AB, Jones RN (2000) Adaptation to what and why? Environ Monit Assess 61:9–35

Prabhakar S.V.R.K, Shaw R (2008) Climate change Considerations for drought risk mitigation: A perspective for India. Climatic Change. 88(2):113–130

Ranganath BK, Diwakar PG, Gowrisankar D, Jayaraman V (2006) Participatory watershed development using EO inputs – a working model for poverty alleviation and improved livelihood in rural India. 57th International Astronautical Congress, Sept 28–Oct 4, 2006, Valencia, Spain

Regional Studies on Cause and Consequences of Natural Disasters and Protection and Preservation of Environment (2008) Reproduction of SAARC Report, Reproduced by SAARC Disaster Management Centre, New Delhi, India, Nov 2008. http://saarc–sdmc.nic.in/publications.asp

Report on Implementation of The Hyogo Framework for Action (HFA): Asia, UN–ISDR/Gp/2007/Inf.5

Richards M (2003) Poverty reduction, equity and climate change: challenges for global governance. Nat Resour Perspect 83. http://www.odi.org.uk/nrp/index.html. Cited 18 Jan 2007

SAARC Action Plan on Climate Change, July 2008 Dhaka Declaration, SAARC Secretariat Kathmandu, Nepal

Smit B, Pilifosova O (2001) Adaptation to climate change in the context of sustainable development and equity. In: McCarthy JJ, Canziani OF, Leary NA, Dokken DJ, White KS (eds) Climate change 2001: impacts, adaptation, and vulnerability: contribution of working group II to the third assessment report of the intergovernmental panel on climate change, intergovernmental panel on climate change. Cambridge University Press, UK, pp 877–912

Smit B, Burton I, Klein RJT, Wandel J (2000) An anatomy of adaptation to climate change and variability. Clim Change 45(1):223–251

Smithers J, Smit B (1997) Human adaptation to climatic variability and change. Glob Environ Chang 7(2):129–146

Thomalla F, Downing T, Spanger-Siegried E, Han G, Rockstrom J (2006) Reducing hazard vulnerability: towards a common approach between disaster risk reduction and climate adaptation. Disasters 30(l):39

UN International Strategy for Disaster Reduction (UNISDR) (2007) Words into action: a guide for implementing the Hyogo Framework. Hyogo Framework for Action 2005–2015: building the resilience of Nations and Communities to Disasters. Geneva

UN International Strategy for Disaster Reduction (UNISDR) (2008a) Links between disaster risk reduction, development and climate change. Report prepared for the Commission on Climate Change and Development, Sweden.

UN International Strategy for Disaster Reduction (UNISDR) (2008b) Towards national resilience: Good practices of national platforms for disaster risk reduction, United Nations secretariat of the International Strategy for Disaster Reduction (UN/ISDR), Geneva, Switzerland, 72 p

Part IX
Need for Action

Chapter 35
Adapting to Climate Change: Research and Development Priorities

Rattan Lal

Humans are now carrying out a large scale geophysical experiment of a kind that could not have happened in the past nor be reproduced in the future.

Roger Revelle

Contents

Abstract Ever increasing demand for food and the basic necessities, driven by high demographic pressure and rapid economic development, necessitate strengthening the scientific knowledge of underpinning processes and creating the data bank for development of decision support system which policy makers can use to address urgent issues. The goal is to identify and implement strategies of natural resources management which adapt to the climate change with specific references to alterations in precipitation, temperature, risks of soil degradation, quality and quantity of water resources, and increase incidence of pests and pathogens. The adaptive management is crucial to addressing the complex issues of climate change,

R. Lal (✉)
The Ohio State University, 2021 Coffey Road, 422B Kottman Hall,
Columbus, OH 43210, USA
e-mail: lal.1@osu.edu

R. Lal et al. (eds.), *Climate Change and Food Security in South Asia*,
DOI 10.1007/978-90-481-9516-9_35, © Springer Science+Business Media B.V. 2011

environmental degradation, and food security. Other researchable topics include: (i) the fate of carbon transported by erosional processes, (ii) coupled ocean and climate studies, (iii) glacier dynamics in the Himalayas, (iv) monsoon science, and (v) drought management. There is a need for immediate action through a coordinated program implemented on a regional basis. Funding for long-term multi-disciplinary research must come from regional sources. Global environmental solutions require global funding for conducting solution-oriented research.

Keywords • Adaptation • Recommended management practices • Monsoon science • Drought management • Multi-disciplinary research

Abbreviations

SA	South Asia
LST	Land surface temperature
TIR	Thermal infrared
FAO	Food and Agricultural Organization of the United Nations
UNDP	UN Development Program
ADB	Asian Development Bank
ESCAP	Economic and Social Commission for Asia and Pacific
APAARI	Asia Pacific Association of Agriculture Resource Institutions
SAARC	South Asia Association for Regional Cooperation

35.1 Introduction

More than 3 billion people in Asia and 1.62 billion in South Asia (SA) depend on monsoon precipitation for water. Any change in the monsoon precipitation in SA may have strong impact on agriculture, food security and well being of the people. Drought, floods, and famines caused by extreme events have affected the well being of the population since the era of Indus Civilization that declined around 1900 BC. The problem has been equally severe during the Middle Ages which witnessed a drastic decline in the population of the sub-Himalayan region (refer Chapter 1).

The problem is made more severe during the twenty-first century than ever before. The SA is a home to ~25% of the world population, which is also increasing especially in this region where natural resources are already under great stress, soils are degraded and depleted, water is scarce in some parts and highly polluted in others, monsoons are highly variable in space and time and also unpredictable, and the climate is undergoing a rapid change at a pace to which ecosystems cannot adjust fast enough.

During 2008 and 2009, the SA region again exhibited signs of fragility to sustain the growing demand of food, feed, fiber and fuel. Food scarcity and increasing food prices, driven paradoxically by a rapid economic growth are affecting ~400 million people in the region. The problem of food insecurity is exacerbated by soil and environmental degradation driven by climate change and social, economic and political forces. This chapter briefly outlines research and development priorities on a regional basis to understand and apply the science to address urgent issues of food security and climate change in SA.

35.2 Adaptation to Climate Change

While the atmospheric concentration of CO_2 and other greenhouse gases must be stabilized and reduced through mitigation, the importance of adaptation to climate change cannot be over-emphasized especially in the SA region. Therefore, long-term, multi-disciplinary and inter-institutional studies are now needed throughout the SA region on identifying and implementing studies for adaptation to climate change. Successful adaptation necessitates enhanced understanding toward sustainable management of soil, water, crops/cropping systems and climatic/weather conditions (Fig. 35.1). The goal of identifying techniques for sustainable management of soil resources is to increase resilience of soils and ecosystems, while enhancing agronomic productivity and improving the use-efficiency of inputs.

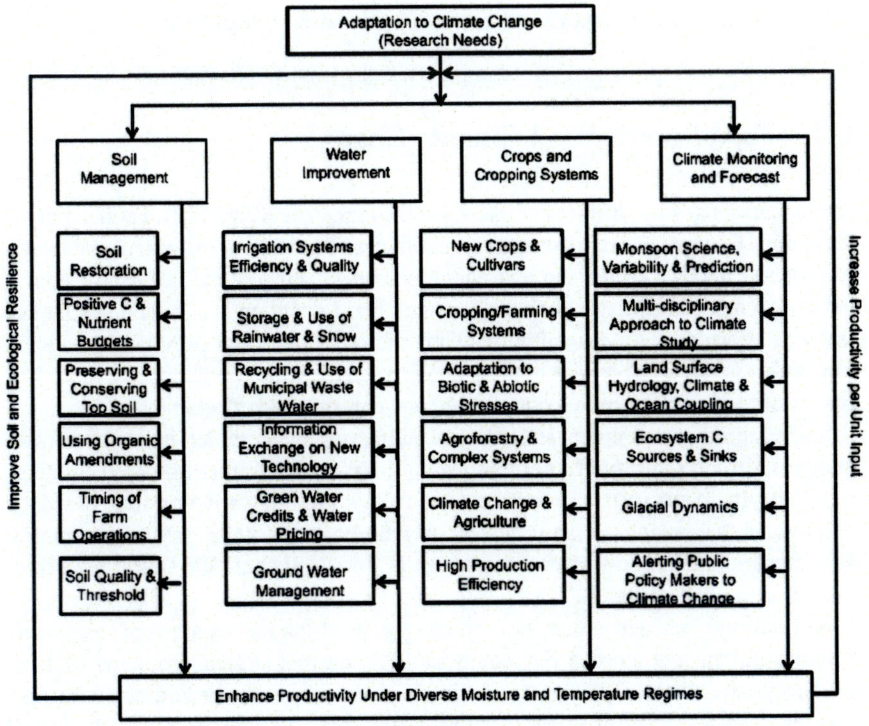

Fig. 35.1 Adaptation to climate change in South Aida necessitate focus on research and development

The most dramatic impact of climate change in the densely populated SA region is likely to be on: (i) soils, (ii) water resources, and (iii) vegetation. These may include increasing drought severity (especially in the northwestern regions of SA) because of rising temperatures, increasing evaporation, melting glaciers, decreasing soil organic C pool, and accentuating soil erosion and the attendant degradation and desertification of soil. It is therefore imperative that scientists, policy makers, and land managers work together to identify and implement climate change adaptation and response strategies. Strengthening and sharing the data on soil properties and water resources (hydrologic cycle and components) are needed to make planning decisions regarding the worst-case scenarios and identifying alternatives. Comprehensive water-balance studies, based on watershed monitoring, are needed for different watersheds and land use systems. Long-term observations on well equipped watersheds are needed for consumptive water use/evapotranspiration, soil moisture storage, runoff in relation to mean and extreme flows, and impact of change in land use especially urbanization.

It is important to develop the decision support systems that managers can use, especially those which allow users to comprehend the impact of climate change on the hydrologic cycle, soil quality, and agronomic/biomass production. There also exists need to redesign water systems and reservoirs, wastewater (grey water) and flood control systems to withstand the extreme events.

In this regard, the importance of vegetation management can never be over-emphasized. Afforestation and reforestation of denuded landscape, restoration of Sundarban and the deltaic/riparian forests is important to protecting the coastal ecosystems.

35.3 Food Security and Climate Change

Food security has four distinct but interrelated components (Fig. 35.2). Food supply is determined by strategies of increasing agronomic production and minimizing its losses. Food access involves increasing household income, creating ethnic and gender equity, and improving infrastructure including market. Food retention is a function of human health as affected by water quality (cholera) and other pests and pathogens (malaria). Food safety is essential because of the sanitation and hygienic conditions. All these four aspects of food security need a coordinated and concentrated effort in SA.

The adaptive management is crucial to addressing the complex issues of climate change and food security. Rather than using the static and rather outdated concepts (e.g., carrying capacity), it is important to follow the innovative options based on adaptive management of soil, water, vegetation and other natural resources. Adaptive and integrated water management (Pahl-Wostl et al. 2007) is essential for the water-scarce SA region.

Agricultural science must be effectively used as an engine of economic development in low-income SA countries. The current urgent problems of food insecurity, exacerbated by the risks of climate change, can only be addressed through a widespread adoption of science-based agriculture. Agriculture must be viewed in the context of rural livelihood involving both farm and non-farm activities

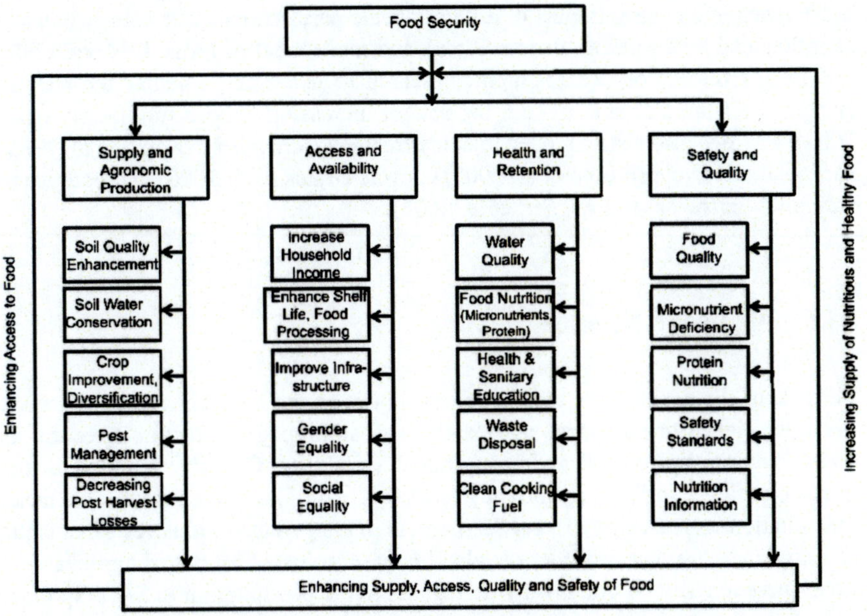

Fig. 35.2 Advancing food security in South Asia through research, development and education

(e.g., renewable energy, value addition, food processing) in alleviating poverty. Farmers must also be compensated by payments for ecosystem services as another income stream. Payments for ecosystem services include those for C sequestration in soils and trees, green water credits, and biodiversity enhancement.

35.4 Fate of Carbon Transported by Erosion and Sediment Deposition in the Ocean

The high topography of the Himalayas is causing higher rates of physical weathering, erosion and sedimentation. It is important to study the fate, residence time, and emission of greenhouse gases from transformation of C transported by erosional processes. In addition to taking core samples from the ocean floor to study historic patterns of sediment deposition, seasonal movement of sediment plumes can be studied by using ocean color sensors (Nezlin and DiGiacomo 2005; Sridhar and Ali 2008).

35.5 Coupled Ocean and Climate Studies

The Indian Ocean is a highly complex and dynamic system. It is also one of the most under sampled and least studied of the world's oceans (Hood et al. 2008). There is a strong need to study the control and fate of primary production, biogeochemical cycles, impact of anthropogenic perturbations, the energy balance as influenced by the Asian Brown Cloud, and monsoonal patterns. It is one of the last great frontiers of oceanographic research to determine whether the Bay of Bengal is a source or sink of CO_2 because of increasing temperature (Alory et al. 2007). There is also a strong need to study the biogeochemistry (Naqvi et al. 2006) such as the N cycle (Bange et al. 2005) C cycle (Bates et al. 2006) and ecosystem variability in the ocean (Wiggert et al. 2006).

35.6 Monsoon Science

Because of the dependence of a large population on the monsoons, it is important to study science of monsoons and natural and anthropogenic factors affecting its onset, pattern, and spatial and inter-annual variability. Capacity building in the monsoon science in SA is important. Wentz et al. (2007) observed that global mean precipitation increased by 7% per °C increase in temperature. However, what is the effective increase; because there is also increase in losses by runoff and evaporation. There are several unknown with regards to the precipitation in SA. Important among those are: (i) how will the precipitation change with the projected climate change? (Wentz et al. 2007; Previdi and Liepert 2008), (ii) How will the future change in climate affect the hydrological cycle at water shed, river basin, and the sub-continental (SA) scale? (iii) how will the land surface evaporation change with the projected climate change? (Brutsaert 2006), (iv) why does global warming cause more rain? (v) what is the impact of aerosols and black C on soot? and (vi) how is the radiation related to precipitation? There is also a strong need to create awareness of the seriousness and urgency of the problem among general public and the policymakers throughout the SA region. Enhancing climate science education can minimize the confusion and misunderstanding.

35.7 Glacier Dynamics in the Himalayas

Controversy about the fate of some 10,000 glaciers in the Himalayas (Bagla 2009) highlights the importance of long-term monitoring. The process has been hindered by political problems (Bagla 2006a, b), poor infrastructure, and weak institutional capacity. The glacial dynamics can be studied by using satellite imagery coupled with direct observations in the field (Shroder et al. 2008). Institutional and international/SA region cooperation is essential. Some key researchable issues in advancing the science and data bank of glacier dynamics in SA are (Shroder et al. 2008): (i) generating and sharing data, (ii) identifying benchmark glacier for long-term field-based study, (iii) standardizing methods used for monitoring and assessing glaciers, to facilitate cross-border comparison, (iv) developing basin-wide hydrologic/water scenarios across the region, and (v) studying status and trends of snow and ice resources in the region.

35.8 Drought Management

Soil moisture deficiency is an important factor affecting agronomic drought. Thus, enhancing soil moisture storage by reducing losses (evaporation, runoff) and improving the plant-water uptake (green water) is essential. Managing water is like hitting a moving target. Similar to soils, adaptive management of water resources is also crucial to addressing the complexities of the SA ecological systems. Being a regional phenomenon, land surface temperature (LST) is an important indication of the drought. The LST can be monitored by remote sensing data in the thermal-infrared (TIR)-band (8–14 μm) (Anderson and Kustas 2008). The LST-TIR data I a valuable index of biospheric stress resulting from soil moisture deficiencies. There is a strong inverse relationship between soil temperature and soil moisture reserve. Increase in canopy temperature because of the drought stress can be effectively detected by remote sensing techniques. The satellite data has numerous applications in: (i) monitoring drought, (ii) assessing consumptive water use, (iii) managing irrigation projects, (iv) predicting local and regional water demand, and (v) weather forecasting (Anderson et al. 2007). Increasing population in the SA region has severely strained the quantity and quality of water resources resulting in rapid drawdown in water resources in NW India (Kerr 2009; Rodell et al. 2009). The TIR-based data can be used to assess the ground water withdrawal and determinoptimal use for irrigation and other purposes (Bastiaannssen and Harshadepp 2005). The satellite data (Rodell et al. 2009) can be used to advise farmers and policy makers to shift away from water-intensive crops, and implement more efficient irrigation methods (e.g., furrow irrigation, sprinkler irrigation, drip irrigation). Coping with water scarcity in the rice-wheat system may involve switching to other crops and growing aerobic rice (Bouman et al. 2006; 2007). The remote sensing techniques can also be used to monitor ice mass in the Himalayan glaciers and assess land-based C sources and sinks (Canadell et al. 2007; Houghten and Goetz 2008).

35.9 Action Beyond the Brainstorming

The time for action is now. No-action or late-action can be dangerous because of the large population concentrated in the relatively small SA region. Coordinated action, through FAO and ESCAP, is needed for several rural/agricultural projects. Notable among those are afforestation, erosion control, restoration of eroded/degraded and desertified soils, construction of water reservoirs and good waterways, establishment of modern efficient irrigation systems, development of biofuel plantations, purification of water in rivers and streams, removing arsenic and other pollutants of water.

The SA region is in a dire need for implementation of community-based and bottom up sustainable development strategies. The trilemma of food insecurity, climate change, and economic development require a paradigm shift, in favor of multi-disciplinary and multi-scale approach. It involves uplift of the farm household within the context of national policies and adoption of improved and innovative technologies.

35.10 Funding for Multi-disciplinary Research

The SA climate is changing due to human activities. All inhabitants of the planet Earth are both the cause and the victim of this change, albeit to a varying degree. The adoptive multi-disciplinary and multi-institutional research needed for the region require commitment towards long-term allocation of resources. The humanity may have already moved the climate to an unstable point which will adversely impact the wellbeing of 1.6 billion inhabitants of SA. While the mitigation is needed, immediate action must be taken towards identification and implementation of adaptation strategies. Global environmental solutions require global funding for conducting solutions-oriented research.

The threats of food-insecurity, climate change, water scarcity and pollution, and loss of biodiversity do not respect national borders being debated and fought over by the SA countries. Research in these areas requires regional (and global) collaboration, and thus regional funding mechanism. Funding support from this must be managed by regional agencies (ESCAP, FAO, UNDP, ADB, etc.). The SA countries must create and contribute to a common fund to support the research activities outlined in this chapter. A Co-benefit of such a common activity would be creation of a peaceful and cooperative atmosphere among the neighboring nations. There is a precedent in the existence and functioning of South Asian Association for Regional Cooperation (SAARC), the Asia-Pacific Network for Global Change Research, and Association of Asia-Pacific Agriculture Resources Institution Network (APAARI). Such networks are needed to implement a wide-variety of risk-reducing livelihood strategies. Important among those are diversifying income sources, switching crops, adopting resilient production systems, developing climate forecasting systems. Resource-poor farmers of SA, remaining at the bottom of the social and economic ranks, deserve help and support from a coordinated effort implemented at regional scale.

35.11 Conclusion

Population of 1.62 billion and increasing is the driving force for several biophysical and socio-economic issues affecting the SA region. These issues include food-insecurity, natural resource degradation, deforestation and loss of natural ground cover, water pollution and eutrophication, drought, and poverty. Most of these issues are also being exacerbated by the anthropogenic climate change which will also create environmental refuges and cross border migration.

These regional issues require regional programs and funded through regional mechanisms. Adoption strategies, rather than static concepts of carrying capacity, designed to enhance soil/ecosystem/social resilience are needed for implementation at community level. Individual resource-poor farmers are not able to take action in view of the severe constraints. Adoption of recommended technologies can be promoted through payments for ecosystem services. While improving agriculture as an engine for economic development is given, improving household income through agri-business and non-farm activities is also important. Coordinated and long-term research is needed to study science of monsoons and factors affecting its variability, coupled processes of marine ecosystems and climatic factors, hydrologic cycle and impact of land use and climate change, glacial dynamics and water supply, soil quality and drought, and agronomic production in an era of warming climate and depleting water. This is a time for action.

References

Allen MR, Ingram WJ (2002) Constraints on future changes in climate and the hydrologic cycle. Nature 419:224–232

Alory G, Wijffels S, Myers G (2007) Observed temperature trends in the Indian Ocean over 1960 to 1999 and associated mechanisms. Geophys Lett 34:L02606. doi:10.1039/2006GL028044

Anderson M, Kustas W (2008) Thermal remote sensing of drought and evapotransportation. EOS 29(26):233–234

Anderson MC, Kustas WP, Norman JM (2007) Upscaling flux observations from local to continental scales using thermal remote sensing. Agron J 99:240–254

Bagla P (2006a) Pakistan gives geology conference the cold shoulder. Science 312:1117

Bagla P (2006b) Science for peace: across a political divide, researchers converge on Himalayan plan. Science 313:30–31

Bange HW, Naqui SWA, Codispoti LA (2005) The nitrogen cycle in the Arabian Sea. Prof Oceangr 65:145–148

Bastiaannssen WGM, Harshadepp NR (2005) Managing scarce water resources in Asia: the nature of the problem and can remote sensing help? Irrig Drain Sust 19:269–284

Bates NR, Pequignet AC, Sabine CL (2006) Ocean carbon cycling in the Indian Ocean I. Spatiotemporal variability of inorganic carbon and air-sea CO_2 gas exchange. Global Biogeochem Cy 20(3):GB3020. doi: 10.1029/2005GB002491

Bouman BAM, Xiaquang Y, Huagui W, Zhimin W, Junfang Z, Bin C (2006) Performance of aerobic rice varieties under irrigated conditions in North China. Field Crop Res 97:53–65

Bouman BAM, Lampayan RM, Toung TP (2007) Water management in irrigated rice: coping with water scarcity. IRRI, Los Banos, Philippines

Brutsaert W (2006) indications of increasing land surface evaporation during the second half of the 20th century. Geophys Res Lett 33:L20403. doi:10.1029/2006GL027532

Canadell JG, Le Quéré C, Raupach MR, Field CB et al. (2007) Contribution to acceleration CO_2 growth from economic activity, carbon intensity and efficiency of natural sinks. Proc Natl Acad Sci 104:18,866–18,870

Held IM, Soden BJ (2006) Robust responses of the hydrologic cycle to global warming. J Clim 19:5686–5699

Hood R, Naqui W, Goes J, McCreary J, Bates N, Karuppasamy PK, Mahowald N, Sybil NY, Myers G (2008) research opportunities and challenges in the Indian Ocean. EOS 89(13):125–136

Kerr RA (2009) Northern India's ground water is going, going, going. Science 325:798–799

Nezlin NP, DiGiacomo PM (2005) Satellite ocean color observations of storm water runoff plumes along San Pedro shelf during 1997–2004. Cont Shelf Res 25:1692–1711

Pahl-Wostl C, Kabat P, Möltgen J (eds) (2007) Adaptive and integrated water management: coping with complexity and uncertainty. Springer, Germany, p 440

Previdi M, Liepert BG (2008) Interdecadal variability of rainfall on a warming planet. EOS Trans AGU 89:193–195

Rodell M, Velicogna I, Famiglietti JS (2009) Satellite based estimated of ground water depletion in India. Nature 46:99–1002

Shroder JF, Bishop MP, Haritashya U (2008) Glacier and permafrost problems in high South Asia. EOS 89(22):27

Sridhar PN, Ali MM (2008) Seasonal occurrence of unique sediment plume in Bay of Bengal. EOS 89:22–23

Wentz FJ, Ricciardulli L, Hilburn K, Mears C (2007) How much more rain will global warming bring? Science 317:233–235

Wiggert JD, Murtugudde RG, Christian JR (2006) Annual ecosystem variability in the tropical Indian Ocean. Results of a coupled biophysical ocean general circulation model. Deep Sea Res II 53:644–676

Index